CUBIC ZIRCONIA AND SKULL MELTING

Yu.S. Kuz'minov, E.E. Lomonova and V.V. Osiko

Viva Books

New Delhi | Mumbai | Chennai | Kolkata | Bangalore | Hyderabad | Kochi | Guwahati

First Indian Edition 2010

VIVA BOOKS PRIVATE LIMITED

- 4737/23 Ansari Road, Daryaganj, New Delhi 110 002
  E-mail: vivadelhi@vivagroupindia.net, Tel. 42242200
- Plot No. 76, Service Industries, Shirvane, Sector 1, Nerul, Navi Mumbai 400 706
  E-mail: vivamumbai@vivagroupindia.net, Tel. 27721273, 27721274
- Jamals Fazal Chambers, 26 Greams Road, Chennai 600 006
  E-mail: vivachennai@vivagroupindia.net, Tel. 28290304, 28294241
- B-103, Jindal Towers, 21/1A/3 Darga Road, Kolkata 700 017
  E-mail: vivakolkata@vivagroupindia.net, Tel. 22836381, 22816713
- 7, Sovereign Park Apartments, 56-58, K. R. Road, Basavanagudi, Bangalore 560 004
  E-mail: vivabangalore@vivagroupindia.net, Tel. 26607409, 26607410
- 101-102 Mughal Marc Aptt., 3-4-637 to 641, Narayanguda, Hyderabad 500 029
  E-mail: vivahyderabad@vivagroupindia.net, Tel. 27564481, 27564482
- First Floor, Beevi Towers, SRM Road, Kaloor, Kochi 682 018, Kerala
  E-mail: vivakochi@vivagroupindia.net, Tel: 0484-2403055, 2403056
- 232, GNB Road, Beside UCO Bank, Silpukhuri, Guwahati 781 003
  E-mail: vivaguwahati@vivagroupindia.net, Tel: 0361-2666386

Published by arrangement with

Cambridge International Science Publishing
7, Meadow Walk, Great Abington
Cambridge CB1 6AZ, UK

ISBN 978-81-309-1234-9

Published by Vinod Vasishtha for Viva Books Private Limited, 4737/23, Ansari Road, New Delhi - 110 002. Printed and bound by Anand Sons, Delhi - 110 092.

CONTENTS

Preface*

Many branches of science, technology and industry require refractory oxide single crystals, glasses with the melting point higher than 2000°C, and fused ceramic materials. There is also a large number of important national economy and scientific–technical problems whose solution is delayed because of the absence of materials stable at high temperatures and in corrosive media. For example, at the moment, there are no electrical heating furnaces capable of working at temperatures higher than 2000°C in an oxidation atmosphere, because the conventional heaters made of silicon carbide or molybdenum disulphide cannot reach of these temperatures. At the same time, high-temperature heating systems, working in air or in the oxygen atmosphere, are urgently required for various technological processes: annealing of crystals, synthesis of luminofors, sintering of ceramic components, production of sitalls, etc. The development of materials with simultaneously high electrical conductivity and resistance to the effect of the oxidation atmosphere at temperatures of approximately 2000°C is a very important technological task. This also includes tasks of the development of highly homogeneous and corrosion-resisting refractory materials for the intensification of pyrometallurgical processes and increasing the quality of melted alloyed steels and alloys, solid electrolytes for chemical current sources, artificial jewellery stones, and many other areas.

The absence of these materials or their low quality are caused mainly by technological difficulties. The point is that the currently use methods of producing polycrystalline ceramic high-temperature materials have considerable shortcomings, reflected in the quality of components produced from them. For example, the main shortcoming of the method of sintering a powder charge at temperatures below the melting point is that the synthesis of compounds in the solid phase does not take place to the end and, consequently, the reaction product contains impurities of the phases of unreacted components. This greatly reduces heat resistance and other properties of ceramic materials. In the production of fused materials in arc furnaces the materials are contaminated with the products of combustion of the electrodes; heat fields are highly heterogeneous resulting in fractionation of the melt. Many substances as, for example, calcium oxides or compounds

**This preface was written by Nobel Prize Laureate A.M. Prokhorov for the book Fianites, published by Nauka, Moscow, in 2000.*

based on them, cannot be generally produced in a relatively pure condition by arc melting. Finally, to grow single crystals of high-temperature compounds from the melt, it is necessary to develop crucibles from materials capable of withstanding high temperatures and not reacting chemically with the melt. The range of these materials is extremely small, with the most suitable being metallic iridium which is in short supply. This material oxidises at elevated temperatures in air and, in addition to this, cannot be used at temperatures higher than 2500–2150°C. Other metals are characterised either by low (in comparison with iridium) melting point or chemically interact with the melt and contaminate it.

The difficulties can be overcome by the technology of fused ceramic materials and oxide high-temperature single crystals produced by direct high-frequency melting in a cold container. The method was developed at the P.N. Lebedev Institute of Physics of the Russian Academy of Sciences. Industrial equipment has been constructed, technologies of melting and directional solidification of a large number of oxide materials of simple and complicated compositions have been proposed, and the properties of new materials and components based on them investigated.

This book, written by V.V. Osiko and Yu.S. Kuz'minov, the well-known experts in the area of growth and investigation of crystals, generalises a large amount of experimental and theoretical material for the technology of melting in solidification of high-temperature dielectrics, produced both at the Scientific Centre of Laser Materials and Technologies of the General Physics Institute of the Russian Academy of sciences, and throughout the world. Undoubtedly, the book will be useful to chemical technologies, students of appropriate specialisations, and will help advances in the area of development of new materials and technologies.

Academician A.M. Prokhorov
Nobel Prize Laureate

Introduction

Single crystals, glasses and molten ceramic materials, based on nonmetallic high temperature compounds, are used widely in technology and industry. Components of optical devices, working elements of lasers and mazers, refractory components of metallurgical and glass furnaces, components of electrical engineering systems, electrodes and insulating elements of magnetohydrodynamic current generators – this list does not exhaust the entire range of applications of high-temperature nonmetallic materials.

In this book, the high-temperature materials refer not only to materials working at high temperatures (more than 1500°C), but also materials produced at high temperatures, irrespective of the application temperature. This type of material includes single crystals, glasses, polycrystalline ceramics based on oxides, fluorides, nitrides, and also borides.

The technology of high-temperature materials is an independent branch with its methods, equipment and, of course, unsolved problems. These problems include:

– the production of new single crystals, glasses and fused ceramics produced at temperatures higher than 2000°C. These temperatures require the development of completely new technological processes;

– the production of single crystals with a faultless homogeneous crystal structure. To satisfy this condition, it is necessary to develop technological processes with highly stable parameters over a long period of time;

– the production of single crystals, glasses and fused ceramics with high chemical purity. For example, the majority of laser crystals and glasses must not contain impurities of iron, copper, rare earth elements and other impurities with the concentration greater than $10^{-4}\%$, with the exception of intentionally added admixtures, and glasses for optical fibre communication should not contain some optically active impurities in the amount greater than $10^{-6}\%$. These conditions require special 'pure' technological processes in which the initial reagents are not contaminated;

– the production of materials in special oxidation–reduction gas media;

– the production of the required materials in large volumes with a minimum consumption of natural resources, *i.e.*, energy, water, etc. and also with minimum waste.

The combination of these problems is the main difficulty for technologies. For example, high temperatures can be easily produced in industrial processes by arc melting. However, in this melting method it is quite difficult to maintain the chemical purity of the material and create the required oxidation–reduction conditions, and also ensure a stable crystallisation process. On the other hand, melting using radiation furnaces results in the required high purity of the material and the required oxidation–reduction medium, but it is not possible to produce large quantities of materials. Considerable difficulties are also found in other technological processes of producing high-temperature materials. As an example, the process of growth of high-temperature single crystals from the melt by the Czochralski method will be considered [1]. A melt of a crystallising substance is produced in a metallic crucible, heated with a resistance furnace or HF current. The crucible plays not only the role of a container for the melt but also of a heating body from which the heat is transferred to the melt. This method is used in the production of the majority of synthetic crystals for science and technology. The possibilities of the method are restricted by crucibles which are produced using only a small number of materials: for the oxidation atmosphere (air, oxygen) these are the expensive materials of the platinum group; in the neutral and reduction atmosphere – metals tungsten and molybdenum or graphite. The crucible material should have a higher melting point than the melt placed in it, and should not interact with the melt and the surrounding atmosphere. Unfortunately, none of the materials mentioned previously is characterised by the entire set of the required properties and this creates considerable difficulties in the development of a technology of high-temperature crystals. Similar problems are also found in the technology of glasses and fused ceramic materials.

We believe that the technology described in this book may prove to be more efficient in solving the problem of production of high-temperature materials from the melt. It is based on the method of direct high-frequency melting in a cold container. The process consists of three main stages: 1) initiation of the process using a special procedure – starting melting, 2) interaction melting of the substance by high-frequency current, 3) maintaining the melt in a solid shell (in a skull) with the composition completely identical with the composition of the melt.

These main elements of the method were proposed earlier by different authors. For example, in 1905 it was proposed to melt metals in a solid shell produced from the same material [2]. This concept was so efficient that it became a basis for the extensive application of skull melting methods in pyrometallurgy. The energy sources in melting metals included mostly the plasma jet, the electric arc and also the electron beam [3]. The induction melting of a metal in a cold crucible was proposed at a considerably later time [4, 5]. However, the method was not used widely because of the special features of the method described below. It was also attempted to use high-frequency melting in a skull for the synthesis of semiconductors [6, 7]. In 1961, Monforte [8] used the method of high-frequency heating in a solid shell for melting ferrites. In Russia, work continued on the growth of iron garnet crystals [9].

Starting in the middle of the 60s, the work in Russia on the technology of high-temperature materials by the method of direct high-frequency melting, the construction of appropriate equipment and the production of crystalline materials was conducted at the Lebedev Institute of Physics of the Russian Academy of Sciences (FIAN) [10,11] where the first technological systems for melting high-temperature dielectrics were also constructed and the main relationships governing the process were described. This period indicates the start of a systematic investigation of the conditions of melting in solidification of different chemical compounds: fluorides, halides, simple oxides, complex oxide compounds. The synthesis conditions of single crystals, glasses, and fused ceramic materials, including polyphase compositions, were investigated. Ruby and leucosapphire crystals were grown for the first time by the Czochralski method at the FIAN. Large single crystals of stabilised cubic zirconia and hafnia were produced for the first time, in addition to single crystals of oxides of rare earth elements, scandium, yttrium, gadolinium and erbium. Many fused high-temperature compounds have better properties than identical materials produced previously by solid phase synthesis. They include the compounds such as mullite, beta-sodium-alumina, chromires of yttrium, lanthanum and rare earth elements, titanates, zirconates and hafnates of elements of the second and third group of the periodic table of elements, high-temperature glasses and other materials. The experience accumulated in these investigations shows that the new technology has a number of principal advantages in comparison with the conventional technologies. The main are:

– the temperature in the process may reach 3000°C without any principle restrictions;

– the absence of contact of the melt with chemically foreign substances guarantees the high ‘purity’ of the process. The synthesised materials contain impurities of secondary elements with the content at least not higher than the purity of the initial component. In many cases, the purity proved to be even slightly higher because high-temperature impurities, such as oxides of alkali elements, chlorides, nitrates, sulphates and many others are removed from the melt;

– the process can be realised in various gas atmospheres: in air, oxygen, neutral gases, and also in vacuum, so that it is possible to produce different oxidation–reduction conditions and carry out melting in oxidation, reduction or neutral atmospheres;

– the process is highly suitable for carrying out chemical reactions in the melt at high temperatures, for example, the synthesis of complex oxide compounds. Almost any type of chemical reaction can be carried out in a cold container in the melt, both in the single-phase and heterophase systems. Thus, the cold container is used successfully as a chemical reactor:

– the process of melting and crystallisation can be not only periodic but also continuous so that the process can be automated.

The previously mentioned advantages indicate that it may be possible to develop a new universal industrial process of producing high-temperature single crystals, glasses and fused ceramic materials.

Reports which appeared in recent years indicate that the new technology of melting high-temperature materials is used widely both in Russia and abroad. Intensive research is being carried out in the USA, North Korea, South Korea, Japan, and some other countries [12, 13].

In the book, special attention is given to the technology of zirconia crystals with a stabilised cubic structure. The method of directional solidification of this compound is used as a basis of the technologies for many other crystals. Zirconia crystals themselves are of considerable scientific and practical interest. Zirconia is characterised by several phase transformations, the last of which (monoclinic) results in complete fracture of the crystals. Therefore, the stabilisation of the cubic structure by yttrium, calcium, rare-earth elements and transition metals is an important scientific and technical task. The crystal lattice of zirconia has a relatively porous structure and permits the implantation of a large amount of secondary elements. Because of this, it is possible to produce a wide range of crystalline substances with interesting physical and mechanical properties, with jewelry stones being the most important products. Pure zirconia, stabilised by yttrium or calcium, can be used for imitating diamond crystals whose optical, mechanical (hardness) and other characteristics become closer and closer to the natural diamond. It may have diamond faceting, so that it is very difficult to distinguish them from real diamonds. The methods of identification of diamonds and of their imitators are also described in this book. Doping of zirconia crystals with rare earth elements and transition metals enables the production of an entire range of natural precious stones, including izumrud.

Large cubic zirconia crystals were synthesised and produced for the first time at the Lebedev Institute of Physics (FIAN) and referred to as **fianites** in honour of the Institute. In the world, they are referred to as cubic zirconia (CZ).

Investigation of the phase transformations in zirconia in stabilisation of the cubic structure by different elements, taken in different proportions, has shown three independent directions. The first direction are fianites in cases in which zirconia is fully stabilised and the crystals should have a stable cubic structure up to a melting point of approximately 2800°C. They are referred to as fully stabilised zirconia (FSZ).

Partially stabilised crystals are also interesting. In this case, we obtain crystals or, more accurately, intergrown polycrystals in which the different phases (cubic, tetragonal and monoclinic) are represented in different ratios, which depend on the nature of the stabilising element, the process kinetics and prior thermal history of the specimen. These materials are referred to as partially stabilised zirconia (PSZ). In the book, we examine in detail phase transformations, domain structure and mechanical properties of PSZ. A characteristic feature of these materials is their high mechanical strength and hardness so that they are regarded as promising materials for the fabrication of components of mechanisms in which these properties are of considerable importance.

The third direction are the crystals in which the amount of the stabilising element or of several elements is considerably greater than that required for the complete stabilisation of the cubic lattice (from 20 to 40 mol.% R_2O_3, where R – Y, Gd–Lu). The substances retain the cubic structure but because of the high density of the stabilising oxides with a large ionic radius, the optical properties of these substances differ from fianites. The investigated crystals are referred to as cubic oxides, C–OX). In the book, the physicochemical and mechanical properties of these substances are described and practical applications indicated. In addition, we examine the technology of oxide compounds produced for the first time in the single crystal state, and also of the well-known crystals, ceramics and glasses, synthesised by the cold container method. In most cases, the physicochemical properties of these compounds are considerably higher than those of the crystals produced by the traditional methods.

References

1. J. Czochralski, *Z. Phys. Chem.*, 92, 219, 1917.
2. W.V. Bolton, *Elektrochemie*, 11, 45, 1905.
3. A.A. Neustruev and G.L. Khodorovskii, Vacuum Skull Furnaces, Metallurgiya, Moscow, 1967.
4. A.A. Fogel, *Izv. AN SSSR, Metallurgiya Toplivo*, 24, 2, 1959.
5. G.H. Schipereit, *British Journal of Metals,* 13, 2, 140, 1961.
6. H.F. Sterling and R.W. Warren, *ibid*, 67, 404, 1963.
7. L.L. Kochergin, et al., *Byull. Izobret.*, 11, 52, 1961.
8. F.K.Y. Monforte, *Appl. Phys.*, 32, 959, 1961.
9. Yu.B. Petrov and E.I. Bezrukova, *Izv. AN SSSR, Neorg. Mater.*, 4, 1162, 1968.
10. V.B. Aleksandrov, et al., in: Current Topics in Materials Science, vol. 1, North Holland Publ., 1978.
11. V.I. Aleksandrov, et al., *Vest. AN SSSR*, 12, 29, 1973.
12. Yu.A. Polonskii, et al., *Ogneupory*, 7, 26, 1973.
13. J.F. Wenkus, Study, Design and Fabrication of a Cold Crucible System, Sci. Rep., Boston, No. 1, 1974.

1

Physical fundamentals of melting in a cold crucible

1.1. INDUCTION HEATING OF MATERIALS. DEVELOPMENT OF THEORY

The theory of induction heating, which has been studied in a large number of investigations [1–5], has been developed in detail and concerns mostly metals. However, the main conclusions of the theory are also fully applicable to the case of heating nonmetallic materials.

The method of induction heating is based on the interaction of an alternating electromagnetic field with a heated object without electrical contact. The variation of the electromagnetic flux with time t in the material, placed in the induction circuit, results in the formation of an electromotive force (EMF) which is determined by the Faraday electromagnetic induction law:

$$e = -k d\psi / dt \tag{1.1}$$

where e is the instantaneous value of the EMF, V; $\psi = \omega\Phi$ is the total flux linkage number of the circuit, consisting of ω turns and penetrated by magnetic flux Φ, Wb; k is the proportionality coefficient, in the Gauss unit system $k = 1/s$, in the SI system $k = 1$.The minus sign determines the direction of the induction flux in accordance with the Lentz rule. The induction EMF is equal to the work carried out in displacement of an unit charge along a closed circuit, carried out by the forces of a vortex electromagnetic field.

If the dependence of flux Φ on time is close to sinusoidal $\Phi_M = \sin 2\pi ft$, the actual value of the EMF in the material is.

$$E = 4.44 f\omega\Phi_m, \text{ V} \tag{1.2}$$

where Φ_m is the amplitude of the magnetic flux in the material, Wb, f is the frequency of oscillations of the field, Hz. Consequently, according to the Joule–Lentz rule, the thermal power, generated in the material, will be

$$p = E \cdot I \cos\varphi = 4.44 f\omega I \cdot \Phi_m \cos\varphi, \text{ W} \tag{1.3}$$

Here I is the current passing in the material, A.

Equation (1.3) shows that power P is proportional to frequency, with other conditions being equal. Increasing frequency increases the energy generated in the same volume, *i.e.* the power related to the unit volume increases.

In addition to frequency, the nature of absorption of the energy of the alternating electromagnetic field is determined by the electrical and magnetic properties of the material. The relationship of these properties of substances with the electromagnetic field is determined by the Maxwell equations. If the incidence of an electromagnetic wave onto a flat semi-bounded solid is examined, the transformation of the Maxwell equations and the transition to the coordinate system make it possible to obtain a relationship between the amplitude of the strength of the magnetic and electrical field:

$$H_m = H_{me} \exp(-x/\Delta), \tag{1.4}$$

$$E_m = H_{me}\left(\sqrt{2}/\gamma\Delta\right)\exp - x/\Delta, \tag{1.5}$$

where H_{me} is the amplitude of the strength of the magnetic field on the surface of the solid, x is the distance from the surface of the solid (depth), γ is the specific conductivity of the solid; Δ is the depth of penetration of current into the material.

The expression for the depth of penetration has the following form:

$$\Delta = 503(\rho/\mu f)^{1/2}, \text{ m} \tag{1.6}$$

where ρ is the specific electrical resistance of the material, Ohm·m, μ is the relative magnetic permittivity.

At $x = \Delta$ are: $J_m/J_{me} = E_m/E_{me} = H_m/H_{me} = 1/e = 0.368$, where J_m and $J_{me} = E_{me}/\Delta = H_{me}\sqrt{2}/\Delta$ are the density of alternating current at depth x and on the surface of the conductor, respectively. Thus, the value of Δ characterises a decrease in amplitudes E, H and J with the depth of penetration of the electromagnetic field. In addition to this, the active power in a layer with thickness x also depends on in the following manner [1]:

$$P_{0-x} = Sp_0\left[1-\exp\left(-2x/\Delta\right)\right] \tag{1.7}$$

where S is the surface area of the material through which irradiation passes; $p_0 = 0.25\left(J_{me}\right)^2 \rho\Delta$ is the specific power on the surface; at $x = \Delta$, expression (1.7) is transformed to the following form:

$$P_{0-x} = Sp_0\left(1-e^{-2}\right) = 0856Sp_0. \tag{1.8}$$

Consequently, in approximate calculations it may be assumed that the entire thermal energy is generated in the layer with thickness Δ.

For a solid infinite cylinder, situated in a homogeneous alternating electromagnetic field, the Maxwell equations are usually written in cylindrical coordinates. In this case, the distribution of H, E and P along the radius of the cylinder at constant frequency f is determined by the ratio $r_z\sqrt{2}/\Delta = d_z/\Delta\sqrt{2} = m$, where r_z, d_z is the radius and diameter of the cylinder, respectively. This relationship is referred to as the relative diameter of the cylinder and characterises the degree of transmittance of the cylinder for electromagnetic radiation. As the value of m decreases, the transmittance of material of the cylinder increases. In [1] it was shown that for a cylinder with height a_z, which is a part of the infinite cylinder (*i.e.* disregarding the edge effects), the equation of the power, generated in the cylinder, has the following form:

$$P_z = 0.5\pi\rho\left(m^2\right)A\left(H_{me}\right)^2 a_z,\ \mathrm{W} \tag{1.9}$$

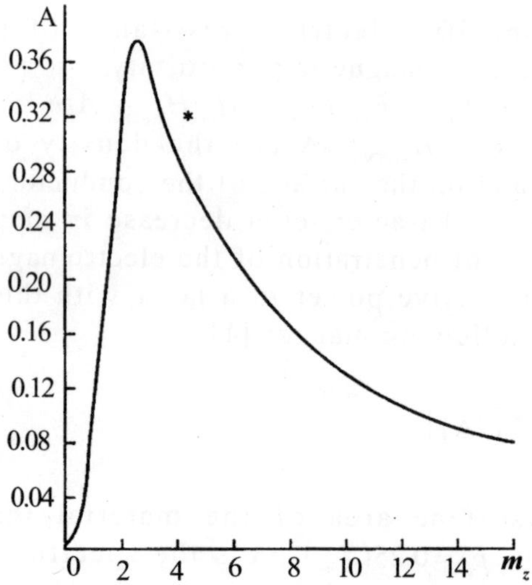

Figure 1.1. Graph of function A, calculated from equation (1.9).

where $A=(2/m)[\mathrm{ber}\ m \cdot \mathrm{ber}'m + \mathrm{bei}\ m \cdot \mathrm{bei}'m]/(\mathrm{ber}^2 m + \mathrm{bei}^2 m)$, where ber and bei are the Bessel cylindrical functions. The functional dependence of A is shown in Fig. 1.1.

We examine the distribution of current density and power in the cross-section of the cylinder for different frequencies of the electromagnetic field. At $m < 1$, the distribution of current density and the strength of the magnetic field depend only slightly on frequency and with a further decrease of frequency they change almost linearly, depending on the radius of the cylinder [1]. Consequently, the distribution of power in the cross-section of the cylinder also becomes independent of frequency. It is evident that the depth of penetration of current into the heated cylinder may be disregarded in these conditions. In this case, it is necessary to introduce the thickness of the active layer ξ and by analogy with the depth of penetration of current specify this depth in such a manner that 86.5% of the total amount of energy, supplied to the cylinder, is generated in the layer. Consequently, in the presence of a strong surface effect $\xi = \Delta$, and at low frequencies inequality $\xi<\Delta$ applies, and the depth of penetration of radiation Δ may exceed the radius or diameter of the cylinder. It was shown in [1] that at $m < 1$, the depth of the active layer is $\xi \approx 0.2\ d_z$, and the calculations carried out in [6] show that the approximate equality is already fulfilled at $m < 3.5$. In this case, with a decrease in

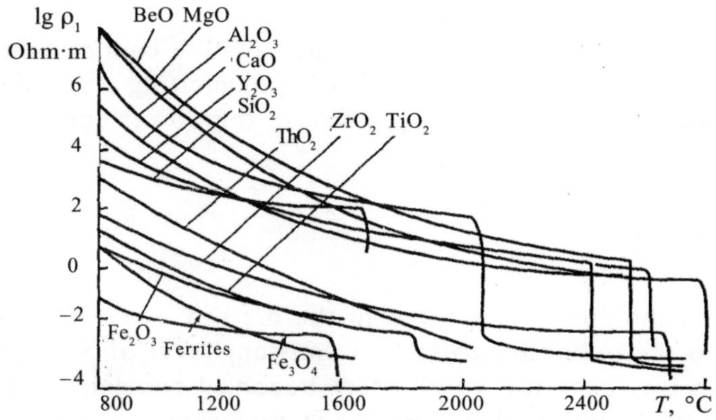

Figure 1.2. Temperature dependence of specific electrical resistivity of some oxides [9].

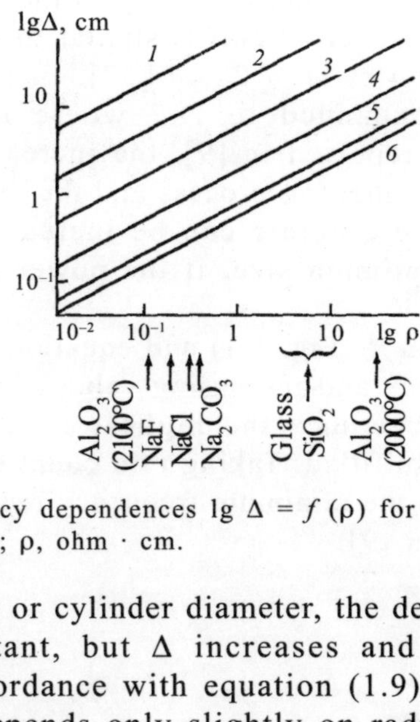

Figure 1.3. Frequency dependences lg $\Delta = f(\rho)$ for different substances [16]. 1,2) kHz; 3–6) MHz; ρ, ohm · cm.

current frequency or cylinder diameter, the depth of the active layer ξ remains constant, but Δ increases and generated power P_z decreases in accordance with equation (1.9). At $m > 6$, the depth of penetration depends only slightly on radius r, the intensity of surface absorption in these conditions rapidly increases and the dependences $H(r)$, $E(r)$, J(r) and P(r) have the same form as in the case of a semi-bounded medium with a flat surface. In this case, it may be approximately assumed that the entire energy of the electromagnetic field is generated in a hollow or cylinder with wall thickness Δ.

In [7, 8], a mathematical model was used for analysis of the electrical parameters of induction furnaces with a cold crucible. For the energy characteristics of the process of melting of oxides in a coldcrucible, it has been proposed to use the power coefficient of the induction coil:

$$\cos\varphi = P_m / (\eta_i S_i) \tag{1.10}$$

where P_m is the power generated in the melt; S_i is the total power supplied to the induction coil. At a constant level of S_i $\cos\varphi \sim P_m$ because in this case the value of η_i becomes constant and equal to 0.90–0.95 because of a very large difference in the specific electrical resistance values of copper, the material of the induction coil ρ_i and the heated material ρ [7]. It was established that the optimum conditions of generation of energy in the melt are fulfilled when the height of the induction coil is similar to the height of the melt in the crucible and at m = 2÷7 [4, 7]. In [8], the range of variation of m was expanded to 2–7 where heating is most efficient. However, as reported in [9], the increase of m is not a critical condition for the melting process and the temperature of the melt or diameter of the crucible can be increased 2–3 times in comparison with the optimum size, if the power of the generator allows this.

The graph of function A (Fig. 1.1) and equation (1.9) shows that with increasing frequency and increasing value of m, the efficiency of induction heating becomes the highest at $m \geq 2.5$ then the condition $d_z > 3.5\Delta$ is fulfilled. Taking into count equation (1.6), it is possible to determine the minimum frequency required for heating the body of the cylinder [9]:

$$f \geq 3 \cdot 10^6 \rho / (d_z)^2, \text{ Hz} \tag{1.11}$$

It is well-known [10–12] that at room temperature, the oxides are efficient dielectrics (Fig. 1.2). In order to ensure the generation of energy of the alternating field in the dielectrics, it is necessary to use very high frequencies. For example, for heating a cylinder with a diameter of d_z = 0.1 m from a material with $\rho = 10^5$ Ohm·m, in accordance with equation (1.11) it is necessary to use a frequency of $4.5 \cdot 10^7$ MHz, which corresponds to the infrared region of the spectrum. However, the frequencies used in practice of induction heating rarely exceed 10 MHz because of difficulties in the

Table 1.1. Electrical conductivity of several solid and molten nonmetallic materials [16]

Compound	State	Temperature, °C	Specific electrical resistivity, Ohm · cm
LiF	Molten	900	$1.2 \cdot 10^{-1}$
NaF	"	1020	$1.9 \cdot 10^{-1}$
KF	"	900	$2.6 \cdot 10^{-1}$
NaCl	"	800	$3.3 \cdot 10^{-1}$
MgO	Solid	2000	$5 \cdot 10^{2}$
SiO_2	"	1700	$9 \cdot 10^{3}$
ZrO_2	"	2000	1
Y_2O_3	Molten	2430	$5 \cdot 10^{-2}$
Y_2O_3	Solid	1500	$3 \cdot 10^{1}$
Al_2O_3	"	1875	$2.2 \cdot 10^{4}$
Al_2O_3	Molten	2200	$1 \cdot 10^{-1}$

generation and transfer of high power at these frequencies [9].

As already mentioned, in order to evaluate the possibility of melting oxide high-temperature materials, it is necessary to know the electrical properties in the solid state and in the melt. Table 1.1 shows the values of electrical conductivity of several oxides [13]. The dependence of lg Δ on lg ρ makes it possible to determine substances which can be melted by the induction method. The specific resistance of a melt of the substances should not exceed 10 Ohm·cm (Fig. 1.3), otherwise the optimum frequency is outside the technically rational ranges [13].

Since at room temperature the majority of oxide materials are dielectrics, heating of these substances to the melting point directly with a high frequency field is not possible. Therefore, it is necessary to carry out of the so-called starting melting, i.e. application of additional heating sources in order to obtain temperatures at which the material itself starts to absorb the energy of the high frequency field. There are a large number of methods of starting heating [8]. However, the main method used in this case is the method based on the application of an auxiliary electrically conducting (at room temperature) material, *i.e.* heating body. From the viewpoint of technology, it is most efficient to use small quantities of a metal of the appropriate oxide, included in the composition of the molten material. The metal is heated with a high frequency field and oxidised with atmospheric oxygen, and the generated heat case the surrounding material to the melting point. During heating, the charge are gradually melt, with the exception of the layer in the immediate vicinity of the watercooled walls of the container. The volume of the melt can be increased by adding new portions of the charge.

In order to determine scientifically substantiated methods of development of equipment for induction melting of oxide components, ensure the maximum productivity of this equipment with minimum energy losses, and also ensure the automatic control of the process of crystal growth in these systems, it is necessary to analyse thermal and electrodynamic processes taking place in the heated solid. In the literature concerned with induction melting of oxides, special attention is given to applied problems and the number of theoretical studies, concerned with examination of special features of the process, which greatly differs from the induction melting of metals, is extremely small. In particular, this is associated with a need to examine the electromagnetic field together with thermal processes in the melt. In this case, the thickness of the layer of the melt in which the energy of the high frequency (HF) field depends strongly on the main characteristics of the melt. In turn, the thermal and energy characteristics of the melt greatly change at high temperatures because of different types of processes taking place in the melt. The thermal and electrical characteristics of melt of oxides and oxide compounds and their temperature dependences in the temperature range 2000–3000 K are examined in only a small number of studies of guidance type. The absence of these data does not make it possible to calculate with sufficient accuracy the parameters of the actual system in heating dielectrics with a high frequency field. In addition, 'forced' assumptions, made in mathematical modelling of the system, cause that the resultant calculations are of qualitative nature and provide only approximate recommendations for the realisation of the actual process. However, these theoretical models make it possible to under stand more efficiently the relationships governing physical–chemical processes associated with direct high-frequency heating of oxide materials and use them for controlling the crystal growth process.

The process of melting and crystallisation in a cold crucible differs from the process of melting in a heated crucible. In commercial melting in a 'hot' crucible, layers of the charge adjacent to the walls of the crucible are the first to melt. The intensity of the heat flow to the interface depends on the rate of propagation of heat in the melt. In direct high-frequency heating in a cold crucible, the heat from the volume of the melt, adjacent to the interface, is transferred directly to the solid phase resulting in its melting. In turn, the melted areas themselves become sources of thermal energy. This schema of the process was used [14, 15] in examination of the process of starting melting – for determining the melting time and the speed of movement of the interface to the crucible walls, the energy losses in this stage of the process in relation to the frequency of the generator and the initial volume of the melt. The experimental results show that in cases in which is considerably smaller than the diameter of the region of starting melting, the melting time increases with frequency as a function of $f^{1/2}$, and the energy, required for melting decreases as a function of f^{-1}. The energy losses are almost completely independent of the initial diameter of the region of starting melting because the main part of energy is used in the final stage of melting when the melt reaches the walls of the crucible.

The authors of [7, 9, 19] investigated processes, associated with the energetic of the process of melting in a cold crucible, in particular, the conditions in which the energy of the HF field is efficiently absorbed by the oxide melt. This is important for economic aspects of the process of melting different materials by this method. The authors of the above studies showed that the maximum power, generated in the melt, is determined mainly by the power factor of the induction coil cosφ Figure 1.4a shows the dependence of cosφ on parameter $m = \sqrt{2r} / \Delta$, where r is the radius of the crucible with the melt. The presence of a distinctive maximum indicates that the heating conditions are optimum. Figure 1.4b shows the dependence of cosφ on the ratio of the height of the melt to the height of the induction coil. The dependence of cosφ on the position of the induction coil in relation to the level of the melt is identical. These experimental results show that it is possible to determine the optimum conditions of heating the melt: optimum frequency of the generator, position of the melt in relation to the induction coil and the ratio of the height of the melt and the

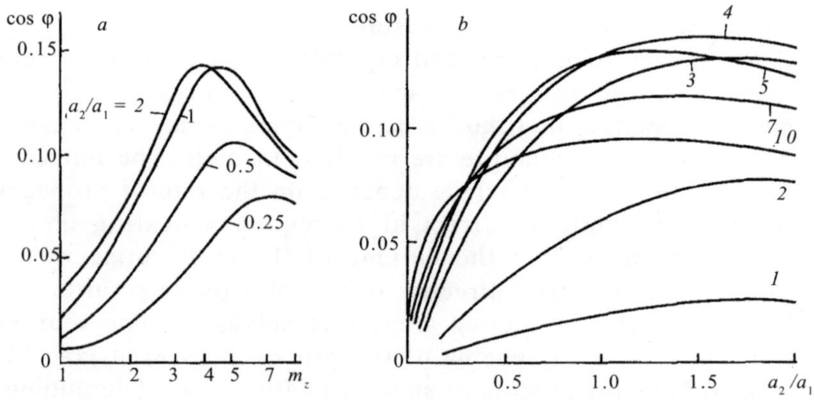

Figure 1.4. Dependence of cos φ on the radius of a crucible with a melt (in relative units) $m^2 = D(2)/\Delta\sqrt{2}$ (a) and on the ratio of the height of the melt to the height of the induction coil a_1/a_2 [9].

induction coil for specific materials.

In melting using HF heating of high-temperature nonmetallic materials, the spatial instability of the phase boundary is very important. This was mentioned for the first time in [17–20]. This problem is very important for controlling the process of crystallisation of the melt, especially in single crystal growth.

The formation of spatial instability is associated with specific features of direct HF heating and is determined by a rapid increase of the electrical conductivity of the material with increasing temperature and also with a large increase in the extent of heat transfer during melting. In induction melting of metals it was reported [4] that if the volume of molten metal has at least one variable parameter, for example, depth or diameter, which changes with the change in the level of supplied power, two types of heat balance appear: stable and unstable. The conditions of stability of the stationary thermal state are as follows:

1. The power balance $P_2 = P_1$, where P_2 is the power generated in the melt; P_1 is the power of heat losses [9, 15, 19–21];

2. $dp_1/da_2 > dp_2/da_2$, where a_2 is the height of the melt or, which is the same, $dp_1/dT_m > dp_2/dT_m, T_m$, where T_m is the temperature of the melt. The dependences $P_2 = f(a_2)$, $P_1 = f(a_2)$ are shown in Fig. 1.5. Point 3 and 4 characterise the conditions in which the balance of the powers (condition 1) is fulfilled. It should be mentioned that point 3 characterises the unstable state in which small deviations of the system from the equilibrium condition result

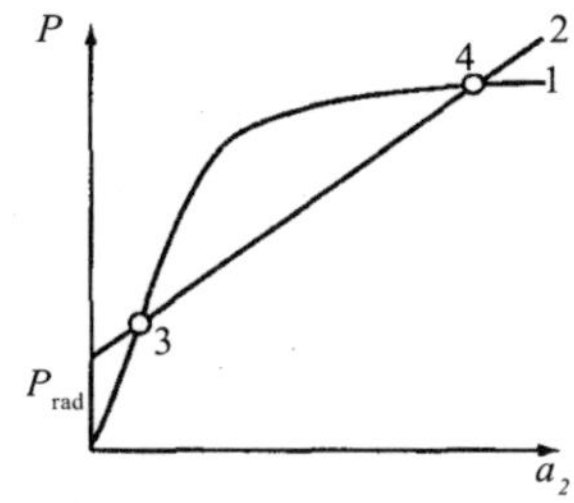

Figure 1.5. Dependences of generated power P_2 (1) and the power of thermal losses P_1 (2) on the depth of the melt $\mathbf{a}_2$ [9]; (points 3 and 4 characterize the conditions in which the power balance is established).

either in crystallisation of the melt as a result of inequality $P_1 > P_2$ or in transition of the regime to point 4 in which the regime of stable balance is fulfilled, because both conditions are fulfilled and any deviation of T_m or a_2 returns the system to the initial condition.

The existence of an instability of induction heating has been reported in the process of growth of crystals by internal melting [22, 23] which may be regarded as one of the modifications of direct HF melting. In this method, a pressed bar is melted inside the volume by means of direct heating in a high-frequency field. The resultant molten zone is situated inside the bar which is cooled on the external side as a result of radiant removal of heat. The molten zone moves along the axis of the bar and the resultant melt solidifies. In the above investigations, the authors presented a quantitative theory of the instability phenomenon formed in the process of induction heating of ceramic specimens. For this purpose, they solved a system of nonlinear differential equations determining the strength of the electromagnetic field and temperature in HF heating of dielectric materials whose. electrical conductivity depends exponentially on temperature: $\sigma = \sigma_0 \exp[-T_A / \mathrm{T}]$. An example of a solution by the iteration method is presented in Fig. 1.6, which shows the dependence of the reduced temperature of the specimen on dimensionless parameter A of the power supplied to the crucible.

The heat transfer from the specimen takes place as a result of radiation, which is especially intense at high temperatures, or by other transfer mechanisms. This dependence has the form of a double S-shaped curve whose upper part corresponds to the melting process. The variation of the shape of the curve in the upper part is due to a rapid increase of the electrical conductivity of the resultant melt. This is typical of oxide and many other dielectric

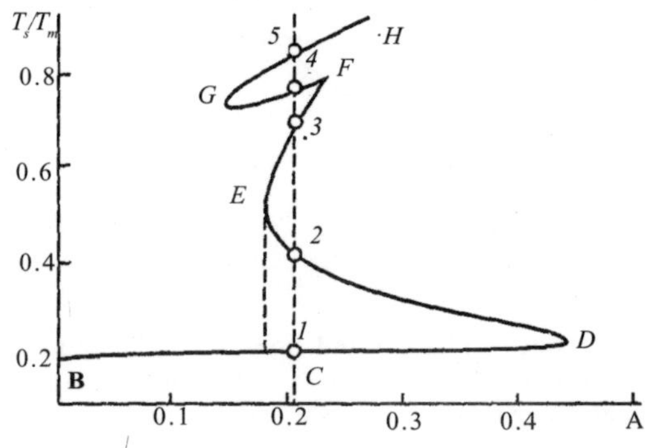

Figure 1.6. Surface temperature of the specimen T_s normalized to the melting point T_m, as a function of parameter A characterizing the power of the HF field supplied to the melt [25,37].

Table 1.2. Calculated parameters of systems for cylindrical specimens with a diameter of 3 cm [25]

Material	Frequency, MHz	Current, A	Power, kW	Material	Frequency, MHz	Current, A	Power, kW
BeO	45000	1	6	ZrO_2	5	16	7
MgO	450	2	13	HfO_2	5	22	14
Al_2O_3	90000	1	2	CeO_2	450	69	13
SiO_2	45000	1	1	ThO_2	45	240	16
TiO_2	230	1	1	UO_2	5	15	7

materials. At a power level corresponding to $A = 0.2$, there are five stationary states of which only state 2 is unstable, and the remaining states are metastable. The states 4 and 5 correspond to the minimum and maximum volumes of the liquid core of the heated bar.

To control the solidification process, it is necessary to investigate the behaviour of material in the temperature range close of the melting point. The authors of [22] investigated the stability of the processes of HF heating and cooling of the material in relation to a large number of parameters of the system and evaluated the possibilities of controlling the process and its optimum parameters for specific substances, presented in Table 1.2. This approach can be extended to systems with a cold crucible, if the removal of heat from the volume of the heated material is taken into account. This may lead to specific recommendations for controlling the crystallisation process. Using melting of a glass-forming material as an example, the parameters, characterising the real system, consisting of a generator, an induction coil and a melt, are presented below.

1.2. PARAMETERS OF DIRECT HF HEATING OF THE MELT

The method of mathematical modelling was used for analysis of the behaviour of the thermal and electrical parameters of a high-frequency furnace in relation to parameter $m = r_0\sqrt{2}/\Delta$. The authors established that the most critical parameter of induction melting in a cold crucible is the power factor of the induction coil $\cos\varphi = r_a / Zf$, where r_a and z_f is the active and total equivalent resistance of the induction coil, respectively [16]. The power factor of the induction coil characterises the active power generated in the melt at the maximum power of the induction coil *S*. Figure 1.4a shows the dependence cosφ = *f*(*m*) for different ratios of the dimensions of the induction coil and the volume of the melt. The extreme form of the curves indicates the existence of optimum conditions of generation of energy in the melt which are fulfilled at $r_0\sqrt{2}/\Delta = 5 \div 7$, where the height of the induction coil should be similar to the height of the melt. However, as shown in [9], in a real system the value of *m* should be higher than m_{opt} because of the need to ensure the unstable state of the melt.

In order to determine the main parameters of direct high-frequency induction melting of the material, *i.e.* the size of the crucible and the working frequency of the generator, it is necessary to know the electrical characteristics of the melt. It is well-known [26] that at room temperature the oxide glasses are typical dielectrics with a specific electrical resistance of 10^{15}–10^{17} ohm·cm, so that they cannot be heated directly by the high-frequency field at frequencies of 1–10 MHz. This frequency range is examined because the application of higher fre-quencies in practice is associated with problems in the generation and transfer of high powers at these frequencies. Starting materi-als, usually used for boiling of glass (SiO_2, $Na_2 CO_3$, $MgCO_3$, $CaCO_3$, *etc.*) are also dielectrics at room temperature. The specific electrical resistance of glasses of different grades (Fig. 1.7) decreases with increasing temperature and in the temperature range 1200–1400°C, used in most cases for boiling of glass, they are equal to 3–50 ohm/cm.

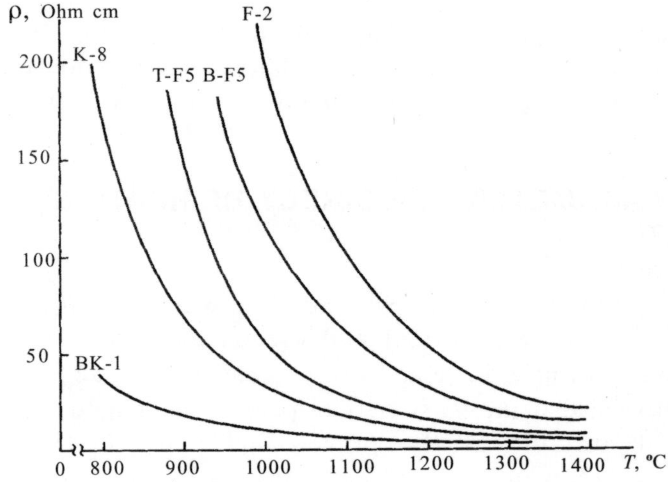

Figure 1.7. Temperature dependences of specific electrical resistivity of optical glasses [26].

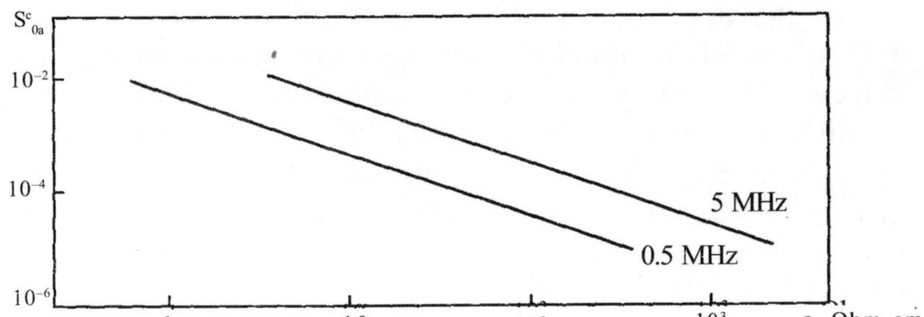

Figure 1.8. Power absorbed by the melt in relation to the electrical resistivity of the melt for frequencies of 0.5 and 5 MHz, crucible radius r_o = 2.5 cm [27].

Therefore, in direct induction melting of glass it is necessary to carry out preliminary heating with an additional heat source at temperatures at which the melt becomes sufficiently electrically conducting for direct absorption of the energy of the high-frequency field.

The required size of the crucible will be evaluated on an example of glass of the BK-1 type, with the composition given in Chapter 6. Figure 1.7 shows that at a temperature of 1400°C, the specific electrical conductivity of the melt is ~3 ohm·cm [24]. At a field frequency of 5 MHz, $\Delta \cong 3.9\text{cm}$; the minimum diameter of the crucible should be $d_{min} = 2\Delta = 7.8$ cm . The optimum diameter is

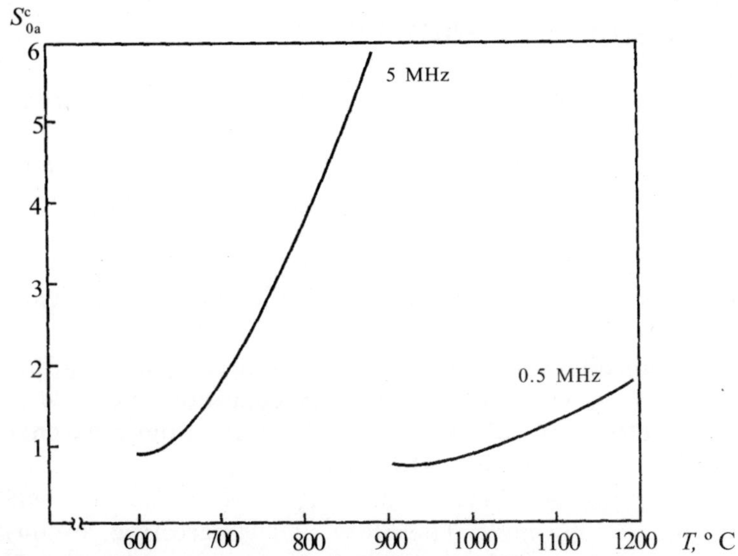

Figure 1.9. Power absorbed by the melt in relation to melt temperature. Chemical composition of the melt $20Na_2O{\cdot}80SiO_2$, r_0 = 2.5 cm [27]

determined by the relationship $r_0\sqrt{2}/\Delta = 5 \div 7$ and d_{opt}=(27.5–38.5) cm. Numerical calculations of the power, generated in the melt situated in a crucible with the radius $r(0)$ = 2.5 cm at two frequencies f_1 = 0.5 MHz and f_2 = 5 MHz (Fig. 1.8) were carried out in [25]. The graph indicates that at the same electrical conductivity of the melt, the power, generated in the melt increases by an order of magnitude with the frequency increasing from 0.5 to 5 MHz.

Using the temperature dependence of the electrical conductivity of the melt, it is possible to determine the dependence of the power, generated. in the melt, on the melt temperature. For a glass with the composition $20Na_2O{\cdot}80SiO_2$, this dependence is shown in Fig. 1.9 which shows that at a frequency of 0.5 MHz the glass must be treated to a temperature of 1000°C, and the preheat at a frequency of 5 MHz is 600°C.

Thus, an increase in frequency makes it possible to reduce greatly the temperature of starting heating of glass. However, it should be mentioned that at a frequency of 5 MHz, the power absorbed by the melt increases with increasing melt temperature more rapidly than at a frequency of 0.5 MHz and this complicates the regulation of the melt temperature.

1.3. THE MECHANISM OF MELTING OF DIELECTRICS BY THE HIGH-FREQUENCY FIELD

The heat generated during the passage of current through a material is calculated using the Joule–Lentz law. Q is the heat generated in a conductor with resistance R during time t in the course of passage of current I through the material and is given by the equation: $Q = aI_2Rt$, where a is a coefficient which depends on the selection of measurement units. When measuring the current in amperes, resistance in ohms and time in seconds, the value of Q is expressed in calories, $a = 0.239$, and in Joules $a = 1$.

Previous considerations show that the stages of starting melting and holding of the melt in the stationary condition, included in the process of melting of oxides, have been studied quite extensively from a theoretical viewpoint. However, in the cited investigations, almost no attention has been given to the mechanism of formation of the volume of the melt. Therefore, it is interesting to consider theoretical investigations of the induction melting of dielectrics carried out at the Institute of Chemical Physics of the Russian Academy of Sciences.

The process of melting in a cold crucible differs greatly from the process of melting in a heated crucible. In conventional melting in a 'hot' crucible, the melting process starts in layers of the charge adjacent to the wall of the crucible which is a heat source. The heat flow to the interface depends on the heat conductivity of the melt between the solid phase and the crucible wall. In induction heating, the heat generated in the conducting melt during the passage of electric current through the melt is transferred into the layers of the solid phase adjacent to the interface, heats them and causes them. to melt. In turn, the molten areas become new heat sources and melt subsequent layers. Thus, a melting wave propagates in the dielectric and the volume of the melt increases.

The authors of [26] assume that many processes, accompanied by heat generation, may take place in conditions identical with the combustion conditions. The combustion process is accompanied by the generation of heat during chemical transformations of matter, but the heat can generate not only as a result of the chemical reaction. The passage of electric current through the matter is also accompanied by the generation of heat (Joule heat). Another important property of the combustion wave is the strong nonlinear temperature dependence of the source of heat generation at the

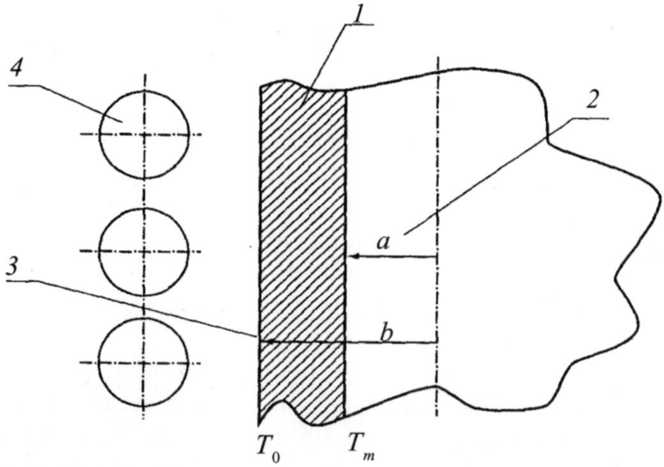

Figure 1.10. Schematic of an infinite two-phase cylinder in an infinite induction coil [18]. 1) solid phase, 2) melt, 3) crucible, 4) induction coil.

front of the wave. This is observed in induction melting of dielectrics as a result of the strong temperature dependence of electrical conductivity. Thus, the melting waves, propagating in the dielectric during heating of the melt due to Joule heat, are a sufficiently effective analogue of the combustion wave and can be described by mathematical facilities used widely in the theory of combustion and are regarded as an efficient means of examination of a wide range of thermal waves, similar to combustion waves [17, 20, 29].

Examination of the melting waves in induction heating was carried out on a model in which the region of the melt was in the form of an 'infinite' cylinder coaxial with the 'infinite' coil of the inductor, Fig. 1.10 [18]. The movement of the interface is accompanied only by a change of the radius of the region of the melt and its cylindrical shape remains unchanged. The propagation of the melting waves in the system requires a combined solution of a number of problems from the area of thermophysics and electrodynamics. However, the solution of the problem is simplified assuming that the conductivity of the solid phase is equal to zero and the conductivity of the melt remains constant and independent of temperature. In this condition, the mutually matched equations are separated and have independent solutions.

When solving an electrodynamic problem, an equation was obtained for the density of the thermal power at distance r from the axis of the cylinder:

$$Q_r = Q_0\left[J_1\left(\sqrt{2}ir/\delta\right)/J_1\left(\sqrt{2}ia/\delta\right)\right]^2, \quad r<a \tag{1.18}$$

where r is the cylindrical coordinate, a is the radius of the melt; δ is the depth of penetration of the field into the melt; J_1 is the Bessell function; Q_0 is the specific power of heat generation on the surface of the melt, which depends on the conductivity and the radius of the melt and also on the dimensions of high-frequency equipment (frequency, voltage, the size of the induction coil, and other factors).

Equation (1.18) shows that the heat in the melt generates in a nonuniform manner. The main part of the heat is generated in the surface layer with thickness $\delta = c_0/\sqrt{2\pi\gamma\omega}$, where C_0 is the velocity of light; γ is the conductivity of the melt; ω is circular frequency. In the limiting case of high frequencies, the source is a purely surface one, and at low frequencies the heat is generated in the volume of the melt. For a small thickness of the skin layer $(\delta/a \ll 1)$, the value of $Q(r)$ may be approximated by an exponential function $Q = Q_0 \exp\left[2(r-a)/\delta\right]$.

In a thermophysical problem of the propagation of a melting wave, the characteristic size is the width of the front of the wave ξ/v, where ξ is thermal diffusivity, v is the speed of the interface. In melting of oxides in cold crucibles, the radius of the volume, occupied by the melt, is large in comparison with the width of the melting wave, *i.e.* the inequality $\xi/v \ll a$ holds. Consequently, the curvature of the surface of the melt in the crucible may be ignored. In addition, this is accompanied by a disruption of the stationary nature of the process of melting: with increasing volume of the melt, the impedance of the melt increases and, correspondingly, the power of heat generation also increases and this in turn influences the time characteristics of the process. The last circumstance is indicated by the inequalities $\sim a/v \gg \xi/v^2$, where ξ/v^2 is the characteristic time during which the wave travels the distance equal to the width of the front of the melting wave. Consequently, at the given radius of the cylinder a, the instantaneous movement of the interphase boundary *da/dt*, is determined by power Q, generated in the melt. With the variation of the radius by the quantity $\sim\xi/v$, this power may be regarded as constant, *i.e.* the interface travels in the quasistationary regime. This assumption and also the ignorance of the curvature of the melt make it possible to calculate the mass speed of the melting wave:

$$u=(\lambda/c\delta)\left\{\left[1+(Q_0/\lambda)\delta^2/(T_m-T_0+L/c)\right]^{1/2}-1\right\}, \tag{1.19}$$

where λ is the heat conductivity coefficient of the melt, c is the specific heat capacity which is assumed to be identical for both phases; L is the specific heat of melting; T_0 is the temperature of the medium, T_m is the melting point. Mass speed u is linked with the speed of movement of the interface v by the relationship $u=\Omega v$, where Ω is the density of the dielectric. In a limiting case of a strong skin effect $(\delta \ll \xi/v)$:

$$u \approx (Q_0\delta/2c)\left[T_m-T_0+(L/c)\right], \tag{1.20}$$

The melting rate is independent of the heat conductivity of the melt. In this case, almost the entire generated heat is used for heating and melting the solid phase with a small superheating of the melt.

In the case of a weak skin effect $(\delta \ll \xi/v)$, equation (1.19) is transformed to the following form:

$$u \approx (1/c)\left\{\sqrt{Q}\,\lambda/\left[T_m-T_0+(L/c)\right]\right\}^{1/2}. \tag{1.21}$$

and for the speed of movement of the interface:

$$v \approx (1/\Omega c)\left[(Q\lambda)/(T_m-T_0+L/c)\right]^{1/2}. \tag{1.22}$$

In the structure, equation (1.22) is identical with the equation of the speed of combustion with a wide heat generation zone [27]. Similar features of the examined processes result from the fact that in induction heating in a heat wave moving together with the interface, an 'electrical' heat source, similar to the exothermic reaction in the combustion wave operates and, consequently, the melting process is similar to the propagation of the combustion wave with a wide zone of the reaction. The speed of the wave is determined by the power of the source at the front edge of the zone of heat generation (in a layer with a width of $<\xi/v$) and is independent of the size of the zone δ.

The authors of [15] estimated the speed of the thermal wave in melting of Al_2O_3: $v\sim 1$ mm/s, the width of the thermal wave $\xi/v\sim 10^{-1}$ cm, the depth of penetration of the field into the melt

$\delta \sim 1$ cm. In [15] it was reported that the regime (1.21) is realised for the high-frequency range at a mean specific power.

Using numerical values of the melting rate and the power of heat generation, the authors of [15] carried out numerical calculations in the relative units of the duration of movement of the melting wave to the walls of the crucible and unavoidable energy losses were also determined.

However, when deriving equations (1.19)–(1.22) it was assumed that the heat is transferred from the melt as a result of heat conductivity in the direction of propagation of the melting wave. Later, in [28–30], the authors detected and investigated a strong effect of capillary spreading (saturation) on the propagation of a combustion wave of mixtures with melting components. The spreading of a component with a lower melting point (for example, titanium) in the voids of the non-melting component (for example, soot) in the heating zone resulted in the intensification of the chemical interaction of the reagents and in an increase of the combustion rate. On the basis of an analogy between the combustion wave and the melting wave, the authors of [15] reported that if the solid-phase is a porous medium (for example, a powder), a more efficient mechanism of propagation of the melting wave in the form of the capillary spreading of the melt through the voids of the solid phase may operate. The theoretical examination of the behaviour of the melting wave in a porous dielectric at a given power of heat generation in the melt heated by the high-frequency field has been carried out in detail in [31].

The capillary spreading of the melt is determined by the property of wetting of the natural solid phase by the liquid. In this case, the spreading of the melt through the voids corresponds to the transfer of heat sources.

The presence of non-conducting particles does not prevent the heating of the melt by induction currents because in any plane there is an infinite number of the methods of producing a closed contour between any two points of the liquid. It was also shown in [31] that the two-phase front is characterised by the solidification of the liquid phase and not by the melting of the solid substance. The leading edge of the spreading melt enters continuously into contact with new portions of the solid phase at a constant temperature T_0. The density of the heat flow into such a solid phase particle may be very high, and at the moment of contact the density may tend to infinity. Therefore, at any temperature of the melt at the leading edge of the melting wave, the arrival of heat to the surface of the solid particles is not capable of compensating the losses for the heating of these particles and solidification of the melt takes place. With heating, the intensity of interfacial heat exchange. decreases and at a certain distance from the front edge, the generation of heat fully compensates the heat absorbed by the solid phase. The direction of the phase transition changes. Thus, the melting front propagates.

Since the energy of the high-frequency field is absorbed only by the melt, partial crystallisation results in weakening of the heat source. The stable propagation of the heat wave by means of the capillary spreading process is possible only at the generated power exceeding the threshold value, determined by the equation [33]:

$$Q_t = 4\lambda(1-\varepsilon)(T_m - T_0)^2 / d^2\varepsilon^2[T_m - T_0 + L/c], \tag{1.23}$$

where λ is the coefficient of heat conductivity of the melt; ε is the initial porosity (prior to melting) of the solid substance; d is the diameter of the voids or particles of the charge. Thus, if the density of power Q, generated in the melt, is higher than Q_t, the melting regime is stationary. If $Q < Q_t$, the spreading regime of the melt is disrupted as a result of the disruption of the interfacial heat balance, associated with a decrease of porosity resulting in weakening of the heat source below the threshold value $m > (m/2)$ [31]. Consequently, the melt in the remaining voids solidifies and a layer of a continuous solid substance forms between the charge and the melt. This layer melts slowly as a result of heat conductivity. After melting of the layer, the melt enters into contact with the charge, spreads and solidifies, forming a new solidified area which then melts as a result of the heat conductivity of the melt. In this case, the mean rate of melting is determined by the slow conductivity stage and may be calculated from equation (1.19) [15]. The melting regime, characterised by the alternation of the stages of conductive melting and rapid spreading, is referred to as quasiperiodic. Equation (1.22) may be written in the following form:

$$v \simeq \sqrt{\xi} / \tau, \tag{1.24}$$

where $\xi = \lambda\Omega c$ is the thermal diffusivity coefficient; $\tau = (T_m - T_0 + L/c)(\Omega c/Q)$ is the characteristic time required for melting of the substance and the given power of the heater. The same authors [31] obtained an equation for the melting rate in the stationary regime which differs from the quasiperiodic equation (1.24) by the substitution of the coefficient of thermal diffusivity by the effective transfer coefficient:

$$K_{eff} = \alpha \cdot d/\eta, \tag{1.25}$$

where α is the surface tension coefficient; d is the diameter of voids or particles of the charge; η is the viscosity of the melt. The dimension of K_{eff} is that of thermal diffusivity.

As an example, attention will be given to the values of the coefficients of thermal diffusivity ξ, the effective transfer coefficient (1.25) and the threshold density of power Q_t for aluminium oxide Al_2O_3 with the diameter of the voids of d = 1 μm, porosity ε = 0.5 μm of the initial powder. The aluminium oxide was selected because the required information for this oxide is available in the liteature. The liquid aluminium oxide has the following properties [11,32]: T_m = 2054°C, c = 138.16 J/mol deg; L = 113.04 kJ/mol; $\lambda = 2.05(1 - 5.3 \cdot 10^{-4}\, \Delta T)$, W/m·deg, where ΔT is the superheating of the melt above T_m; $\Omega = 3.03 \cdot 10^3 - 1.08\Delta T$, kg/m^3; $\alpha = 0.66 - 0.195 \cdot 10^{-3}\, \Delta T$, J/m^2; $\eta = 0.06$ Pa·s at T = T_m; $\eta = 0.024$ Pa·s at 300°C. The values obtained in the calculations at T_m are as follows: $\xi = 5 \cdot 10^{-7}\ \mathrm{m^2/s}$, $\alpha \cdot d / \eta = 1.1 \cdot 10^{-5}\ \mathrm{m^2/s}$, and the values resulting from calculations at ΔT=300°C, $\xi = 4.7 \cdot 10^{-7}\ \mathrm{m^2/s}$, $\alpha \cdot d / \eta = 2.5 \cdot 10^{-5}\ \mathrm{m^2/s}$.

The calculations show that the value of the effective transfer coefficient is tens of times higher than heat conductivity and this difference increases with increasing melt temperature. Thus, the melting of the solid particles by the melt spreading through the pores is considerably more efficient than heating of the solid phase as a result of heat conductivity.

The evaluation of the density of the threshold power gives: Q_t = 2.4·10^{13} kW/m^3 = 7.9·10^9 kW/kg. It should be mentioned that these equations are of the approximate nature and make it possible to calculate only the order of magnitude. An important circumstance is the formation of threshold values of power for the given melting conditions. However, the previously presented estimate of the value of Q_t for the aluminium oxide cast doubts on the possibility of establishing the stationary conditions in melting of the porous absence in actual conditions.

Thus, the large number of various assumptions made in the

mathematical modelling of the melting waves and the limited literature data on the electrical and thermal properties of the melts of hightemperature oxide and also other refractory materials indicate that the theoretical investigations make it possible to obtain only the qualitative pattern of the process of melting of oxides. Nevertheless, these estimates are very important, in particular in the absence of experimental investigations, for the understanding of the processes taking place during induction melting of dielectrics. This understanding will support a successful solution of a number of problems of induction melting and growth of single crystals in a cold crucible [23] (see also Chapters 4 and 5).

References

1. Slukhotskii A.E., Ryskin S.E., Induction coils for induction heating of engineering components, Moscow and Leningrad, Energiya, 1974.
2. Vainberg A.M., Induction melting furnaces, Moscow and Leningrad, Energiya, 1967.
3. Gude Zh., Industrial electronics, Moscow and Leningrad, Energiya, 1960.
4. Babat G.I., Induction heating of metals, Moscow and Leningrad, Energiya, 1965.
5. Zagaevskii T., et al., Industrial electronics, Moscow, Energiya, 1976.
6. Razorenov G.A. Izv. LETI im. B.I. Ul'yanova (Lenin),1940, 37-50.
7. Petrov Yu.B., et al., *Elektrotekhnika*, 1982, No. 8, 16-19.
8. Petrov Yu.B., Kanaev I.A., Induction furnaces for melting of oxides, Leningrad, Politekhnika, 1991.
9. Petrov Yu.B., Induction melting of oxides, Leningrad, Energoatomizdat, 1983.
10. Krzhizhanovskii R.E., Shtern Z.Yu., Thermophysical properties of nonmetallic materials, Leningrad, Energiya, 1973.
11. Samsonov G.V. (editor), Physico-chemical properties of oxides, Metallurgiya, Moscow, 1978.
12. Bol'shakova N.V., Borisanova K.S., Burtsev V.I., Materials for electrothermal equipment, a handbook, Moscow, Energoatomizdat, 1987.
13. Borik M.A., Lomonova E.E., Osiko V.V., Prokhorov A.M., in: Problems of crystallography, Moscow, Nauka, 1987, 165–189.
14. Merzhanov A.G., et al., *Dokl. Akad. Nauk SSSR*, **221**, No. 5, 1114–1117.
15. Merzhanov A.G., et al., *Dokl. Akad. Nauk SSSR*, 1980 253, No. 2, 330–334.
16. Nezhentsev V.V., et al., *Izv. LETI*, 1981, No. 299, 14–20.
17. Aleksandrov V.I., et al., *PTE*, 1970, No. 5, 222–225.
18. Aleksandrov V.I., et al., *Vestn. AN SSSR*, 1973, No. 12, 29-39.
19. Aleksandrov V.I., et al., Current Topics in Material Sciences, Amsterdam, 1978, 421–480.
20. Aleksandrov V.I., et al., *Usp. Khimii*, 1978, **47**, No. 3, 385– 427.
21. Slukhotskii E.A. (editor), Using high-frequency currents in electrothermal applications, Leningrad, Mashinostroenie, 1973.
22. Hartzell R.A., Sekerka R.F., *J. Crystal Growth*, 1982, **57**, No. 1, 27–42.

23. Sekerka R.F., Hartzell R.A., Farr B. J., *J. Crystal Growth*, 1980. Vol. 50. N 4. P. 783–800.
24. Demkina L.I. (editor), Physico-chemical fundamentals of production of optical glass, Leningrad, Khimiya, 1976.
25. Scott B., Rawson H., *Glass Technology*, 1973, **14**, No. 5. 115–124.
26. Merzhanov A.G., Rumanov E.N., Combustion without fuel, Moscow, Znanie, 1978.
27. Aldushin A.A., et al., *Fizika Goreniya i Vzryva*, 1972, **8**, No. 2, 202–212.
28. Shkiro V.M., Brovinskaya I.P., *Fizika Goreniya i Vzryva,* 1976, **12**, No. 6, 945–948.
29. Aldushin A.A., et al., *Fizika Goreniya i Vzryva,* 1976, **12**, No. 6, 819–827.
30. Nekrasov E.A., et al., *Fizika Goreniya i Vzryva,* 1978, **14**, No. 5, 26–32.
31. Rumanov E.N., Melting wave of porous matter, Preprint of the Institute of Chemical Physics, Russian Academy of Sciences, Chernogolovka, 1982.
32. Murakh M.A., Mitin B.S., Liquid refractory oxides, Moscow, Metallurgiya, 1972.
33. Shteiberg A.S., Raduchev V.A., *Dokl. Akad. Nauk SSSR*, 1991, **317**, No. 2. 370–373.

2

Formation and treatment of the melt in the cold crucible

2.1. SPECIAL FEATURES OF HEATING MATTER IN THE COLD CRUCIBLE

Attention will now be given to several special features of the process of direct high-frequency heating of nonmetallic materials in a cold crucible, determined by the electrical properties of this type of material. Figure 2.1 shows the typical temperature dependences of the electrical resistance of a nonmetallic material (Fig. 2.1a) which is dielectric in the solid state, and of a metal (Fig. 2.1b).The figure indicates the principal difference between these two types of materials: in the first case, the electrical resistance of the melt is several orders of magnitude smaller than that of the solid phase, and in the second case, the melt is characterised by a resistance which is considerably higher than that of the solid metal. Figure 2.1c, d shows for the two cases schematically the nature of absorption of energy of the high frequency field by a two-layer cylinder, whose outer shell is represented by the solid phase, and the internal part is occupied by the melt of the same composition as that of the shell. This model accurately corresponds to the actual experimental conditions. Because in the case of metal the electrical resistance of the solid phase is lower than that of the melt, then in accordance with equations (1.1) and (1.2), presented in Chapter 1, the metal absorbs a larger part of the energy of the high-frequency field and screens the melt (Fig. 2.1d).

A different situation is observed in the case of a material with a 'drooping' temperature dependence of electrical resistance.In this

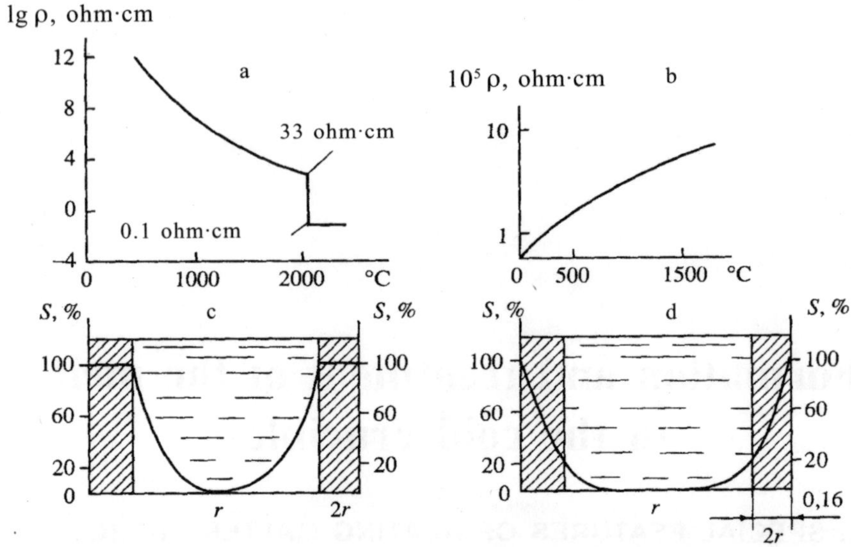

Figure 2.1. Temperature dependence of resistivity (a) and radial distribution of the energy of the high-frequency field (s) absorbed by a two-layer cylinder consisting of solid (outer cross-hatched layer) and liquid (inner part) phases with the same chemical composition [5]. a) aluminium oxide; b) steel; c) aluminium oxide cylinder; frequency of the electromagnetic field 5 MHz; d) a steel cylinder, frequency of the electromagnetic field 2500 Hz.

case, the solid shell is in principle 'pervious' to the electromagnetic field, does not prevent its penetration into the melt and is not heated by the field (Fig. 2.1c). Thus, direct induction heating in a cold crucible appears to be specially designed for materials characterised by the decreasing temperature dependence of electrical resistance and is not suitable for metals. In direct high frequency melting, the entire amount of the metal is melted and no solid shell forms between the melt and the walls of the cold crucible. This is the most significant advantage of this method in comparison with other methods: *i.e.* retention of the chemical purity of the initial material.The specific distribution of the power, generated in the melt, results in a situation in which the melt heated to a high temperature and the cooled crucible are always separated by a layer of the solid phase, not heated by the field, with the entire temperature gradient formed in this layer. It will be shown later that this factor results in a decrease of the energy losses in direct high frequency melting of nonmetallic materials in a cold crucible.

2.2 STARTING HEATING

The process of melting in a cold crucible consists of three stages. The first stage is the production of a 'seed' melt which, as already mentioned, may be produced only by using an additional heat source. The following methods are used for this purpose: 1. melting by supplying radiant energy, 2. using flames (gas or plasma torches), 3. electric arc, 4. melting by the introduction of a secondary electrically conducting material (heating body) into the material to be melted. The most suitable method for melting oxide and some other types of material is the method of starting melting based on the application of small amounts (usually several grams) of the metal included in the composition of the molten material. For example, it is aluminium in the case of aluminium oxide, yttrium in the case of yttrium oxide, and so on. Small pieces of metal, introduced into the charge, are heated by the energy of the high-frequency field and melt the surrounding charge. With melting, the metal is oxidised by atmospheric oxygen and the resultant oxide is included in the composition of the melt, without introducing secondary impurities into it. The physicochemical properties of the metals, used for starting heating, are presented in Table 2.1, [1].

Zirconium and yttrium are characterised by the highest heat of combustion. Zirconium can be easily ignited in air only in the refined condition (in the form of shavings or powder), compacted metallic zirconium is ignited in air only at a very high temperature [2].

After reaching the critical volume of the starting melt under the effect on the high-frequency field, the entire charge is gradually melted, with the exception of the layer which is in contact with the cooled walls of the crucible. The second, transition stage of the process is the addition into the crucible of new portions to the required volume of the melt. Complete melting of the charge is accompanied by the establishment of an equilibrium condition which may be maintained for a long period of time. This is the third stage of the melting process.When the generator is switched off, rapid mass crystallisation or formation of a glassy phase takes place, *i.e.* polycrystalline ingots or glasses are produced. In slow directional solidification of the melt by means of a gradual increase of the power of the generator or slow lowering of the crucible with the melt in relation to the induction coil it is possible to produce large single crystal blocks. Single crystals may also be produced by

Table 2.1. Physico-chemical properties of metals used for starting melting [5]

Metal	Atomic mass, %	Melting point, °C	Heat of melting	T_0 – start of oxidation	Heat of formation of oxides, J/kg at 2000°C
Y	88.9	1585	113.2	In air, burns at 470°C	7430
Zr	91.22	2125	230.5	Burns, refined at 500°C	7880
Gd	157.3	1799	71.2	Ignites at 400°C	4400
Yb	173.04	1097	46.1	As above	4021

drawing of seed crystals. The processes of crystallisation and glass formation will be examined in greater detail in sections concerned with specific materials.

Thus, the process of high-frequency melting of refractory dielectrics' consists of heating the material from room temperature to the melting point and of holding of the material in the molten condition by supplying the energy of the high-frequency field.

Special attention will now be paid to the first stage of the process, the so-called process of starting melting (Fig. 2.2a). Initially, it is necessary to determine the amount of initial solid material which must be melted in order to ensure that the resultant melt is independently heated by the high-frequency field and its volume increases he condition of its 'viability' is determined by the inequality:

$$V \cdot P > K\left(T_m - T_0\right)\pi D^2, \tag{2.1}$$

where V is the volume of the starting melt; P is the power, absorbed by the unit volume of the melt, K is the coefficient of the transfer from the melt to the heat carrier (cooling water), T_m is the temperature of the melt, T_0 is the temperature of the heat carrier, D is the diameter of the droplet. In order to simplify the calculations, the heat of phase transition is not taken into account because, ideally, the melting process can be carried out at any suitable slow rate. When the droplet diameter D is considerably smaller than the internal diameter of the crucible, it is assumed that specific

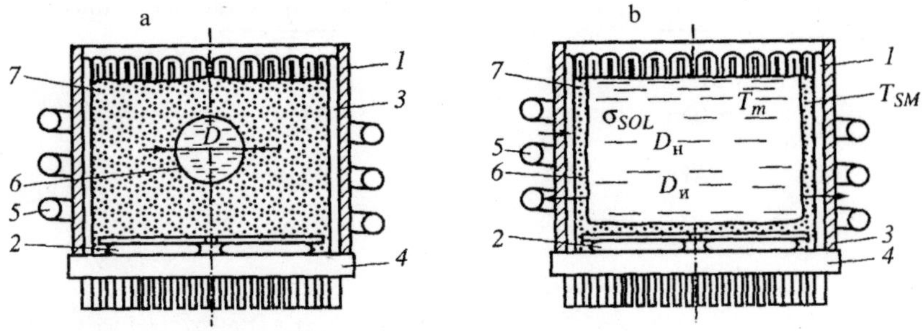

Figure 2.2. Scheme illustrating melting (a) and equilibrium state (b) of the melt–solid phase system (b) [6].

absorbed power P is constant and equation (2.1) is simplified:

$$D > 6K / P(T_M - T_0). \tag{2.2}$$

The inequality shows that the critical size of the droplets of the startingmelt depends, at the given strength of the electromagnetic field and electrical conductivity of the melt which determine the value of P, on the heat-insulating properties of the solid phase.Thus, when condition (2.2) is fulfilled, a 'viable' droplet melt forms and, absorbing the energy of the high-frequency field, the droplet will grow as a result of melting of the surrounding solid-phase.

The authors of [3] propose to evaluate the amount of metal for the starting melt on the basis of different considerations. The starting zone of the melt is represented in the form of a disk with a diameter d_{ST}, height h_{ST} and volume V_{ST}, positioned in the plane of the induction coil. According to [4], in heating in crucibles with an internal diameter of $d_K < 0.2$ m, the size of the zone should be: $d_{ST} = (0.7–1.0)\ d_K$, $h_{ST} = (0.5–0.1)\ d_K$, $V_{ST} = (0.4–0.08)d_K^3$. In crucibles with $d_K > 0.2 <$ volume V_{ST} should be smaller than the recommended volume if the starting zone is produced in the form of a torus with the external diameter and the height corresponding to the previously presented relationships and with the radial width not smaller than the depth of the active layer ξ (see Chapter 1). Induction heating is also possible in the case of smaller volumes of the starting melt, but this requires additional power of the generator which, however, can be used in the process of further melting and sustaining the melt in the stationary state [5].

According to [2], the amount of metal G, required for the reaction of oxidation with a calorific effect, sufficient for producing

Table 2.2. Specific ignition values of metals forming refractory oxides [1]

Metal	Z, kg/m^2	Metal	Z, kg/m^2
Be	420	Nb	1170
Al	480	La	1210
Sc	515	Ce	1460
Ti	550	Sm	1490
Cr	1220	Hf	2270
Y	985	Th	2570
Zr	1150	U	3375

the volume of the melt V_{ST}, can be determined from the equation

$$G = Z \cdot V_{ST}, \qquad (2.3)$$

where Z is the characteristic of starting heating, referred to as specific ignition, kg/m^3. This characteristic represents the mass of combusted metal, required for heating and melting 1.0 M_3 of the oxide of this metal. Table 2.2 presents the values of the specific ignition for a number of metal forming refractory oxides. The oxides of Mg, Ca, Sr and Ba cannot be melted by combustion of their metals because the temperatures of ignition of their vapours are lower than the melting point of their oxides [6]. For starting heating of these oxides it is essential to use other methods of preheating.

2.3. MAINTAINING THE MELT IN THE STATIONARY STATE

To answer the question whether the process of melting will continue to complete melting of the entire solid phase or whether it will be interrupted in a certain stage, special attention will be given to the phase equilibrium condition, *i.e.* spatial stability of the interface between the melt and the solid phase (Fig. 2.2b). The phase equilibrium indicates that the temperature at any point of the phase boundary is equal to T_{SM} and does not change with time. The amount of heat, transferred by the melt at the boundary of the solid phase under the condition of complete mixing of the melt, is determined by the expression: $Q_1 = \alpha_1 (T_M - T_{SM}) \cdot F \cdot \tau$, where α is the heat transfer coefficient, T_{SM} is time, and the remaining symbols are the same as previously. The amount of heat transferred to the heat carrier through the layer of the solid phase with thickness σ_S and with heat conductivity γ is: $Q_2 = \gamma/\sigma_S (T_{SM} - T_0) \cdot F{\cdot}\tau$. For the ideal stationary regime $T_M = T_{SM}$, $Q_1 = Q_2$. Consequently, we obtain $\alpha_1 (T_M - T_{SM}) = (\gamma/\sigma_S)(T_{SM} - T_0)$, or:

$$(\lambda/\sigma_S)\alpha_1 = (T_M - T_{SM}) = (T_{SM} - T_0). \tag{2.4}$$

Here $T_M - T_{SM} = \Delta T$ is the quantity, characterising the superheating of the melt. At $\Delta T \to 0$, $\sigma_S \to \infty$; This means that to sustain phase equilibrium, the melt should be superheated. It is assumed that $\gamma\alpha_1$ and T_0 independent of σ_S and, consequently, we obtain:

$$\sigma_S \Delta T = \text{const.} \tag{2.5}$$

Equations (2.4) and (2.5) show that when the temperature of the melt is increased or reduced, the system reacts by decreasing or increasing the thickness of the layer of the solid phase, respectively. On the other hand, complete melting of the solid phase requires infinitely large superheating. Any processes taking place in the solid shell, leading to a change of its heat conductivity (for example, sintering), disrupt the phase equilibrium and result in a change of either the temperature of the melt or the thickness of the solid layer (skull). The phase equilibrium condition is also affected by the intensity of mixing the melt which determines the value of the heat transfer coefficient. The requirement for superheating, determining the very existence of the melt in the investigated system, is the main feature by which the cold crucible differs from a heated crucible in which the substance can be heated to any temperature close to the melting point, and in some cases it is even possible to supercool the melt. Unavoidable superheating of the melt in the cold crucible result in important consequences in crystallisation of the melt, in particular, in single crystal growth (Chapter 4).

Attention will also be given to the reaction of the 'cold crucible–solid shell–melt' system to a reduction in absorbed power ΔP (caused, for example, by an uncontrolled reduction in the power of the high frequency generator). A decrease in the power results immediately in a decrease in melt temperature T_M which disrupts the phase equilibrium: the thickness of the solid shell starts to increase, the ratio of the diameter of the load *(D_N)* to the diameter

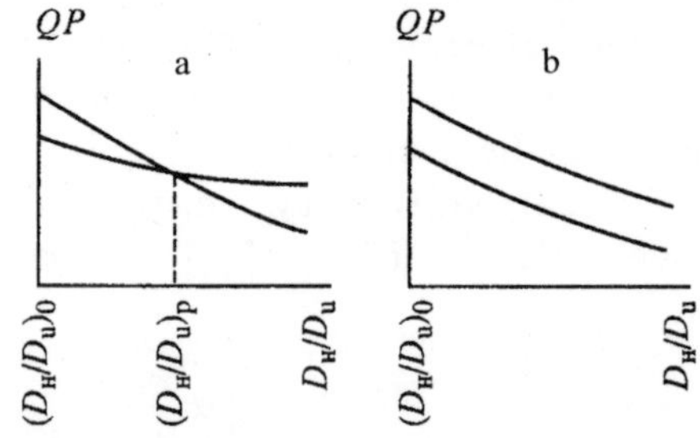

Figure 2.3. Reaction of the melt–solid phase system in a cold crucible to a decrease of the generator power [5]. a) stable system, b) unstable system. D_H is the diameter of the crucible with the melt; D_u is the diameter of the induction coil.

of the induction coil *(D_i)* decreases. Two situations may occur in this case: 1.The establishment of a new equilibrium state with the value of σ_S higher than previously; 2. uncontrollable cooling of the melt. The point is that an increase in σ_S is also accompanied by a decrease in the diameter of the region of the melt representing the load of the induction coil of the high-frequency generator; the efficiency of con-nection with the induction coil decreases. On the other hand, with an increase in the thickness of the solid shell, thermal insulation increases and the heat flow from the melt to the cold crucible decreases. The rates of decrease in the power absorbed by the melt and in the heat losses differ. Figure 2.3a, b shows schematically two possible variants of the reaction of the system to a decrease in the generator power. The result of these competing processes is controlled by the heat conductivity of the solid phase: as the heat conductivity in-creases, the stability of the system decreases. The formation of the spatial instability of the phase boundary in high-frequency melting of refractory nonmetallic materials is a new factor not observed in the high-frequency heating of metals. The latter factor has a significant effect on the technical circumstances and technological regimes ofmelting and crystallisation of high-temperature materials.

After successful starting melting, the volume of the melt gradually increases as a result of melting of the adjacent solid phase represented by a charge or a mixture of a charge with crystal fragments from previous melts. This continues until the entire charge is melted, with the exception of the layer of the charge that is in contact with the walls of the crucible which together with the layer of the solidified melt forms the so called skull preventing direct contact of the melt with the crucible fittings cooled by water.

In addition to the skull, a layer of the melted charge in the form of a cupola forms above the melt. This layer greatly screens thermal radiation from the surface of the melt. To add new portions of the charge to the melt, a small hole is made in the cupola to ensure that the charge falls directly into the melt. After loading, the hole is covered with the charge. When the required volume of the melt is produced, the melt is held for some time at a constant level of the power supply to the melt to ensure that the entire system is transferred to the stationary condition.

2.4 DISTRIBUTION OF TEMPERATURE IN THE MELT VOLUME

The distribution of temperature in the cold crucible was examined on an example of glass-forming melts using a thermocouple. There are certain special features of the measurement of temperature in semi- transparent media. This problem has been examined in detail in [10] for the case of measurement of the temperature of glass mass in a glass-boiling furnace. If a thermocouple is immersed into the glass mass, the protective sheath of the thermocouple absorbs the radiation from the surrounding hotter sources. These sources may be represented by both the refractory lining of the furnace and by the regions of the glass mass with higher temperature. This absorption may result in errors in the measured temperature. To evaluate this error, investigations were carried out into the process of heat ex-change between a thermocouple and a glass mass, including radiant heat exchange and heat conductivity. The following assumption were made in the calculations: 1) the junction of the thermocouple has the form of an isothermal sphere with radius r_0; 2) convective heat transfer between the glass mass and the thermocouple is insignificant; 3) at a wavelength greater than 2.7 μm it is assumed that the glass mass is an absolutely black body and in the case of shorter wave-lengths it is a 'grey' body. Under these conditions, the following equation was obtained for the relative error in the determination of the melt temperature [8]:

$$\left(T_0 - T_f\right)/T_f = \left(r_0 \cdot \varepsilon n^2 \cdot \sigma \cdot T_f\right)/K_f \left\{1/2[(T_s/T_f)^4 + 1] - (T_0/T_f)^4\right\}, \qquad (2.6)$$

Here T_0 is the temperature of the thermocouple, T_f is the actual temperature of the melt, T_S is the temperature of the heating source, K_f is the heat conductivity coefficient of the melt, ε is the

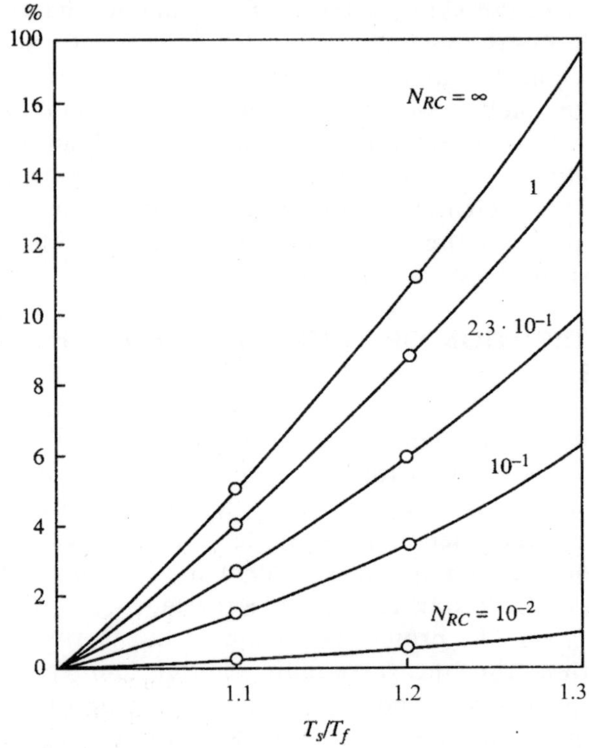

Figure 2.4. Relative errors in the measurement of melt temperature in relation to parameter N_{RS} and the T_s/T_f ratio of the temperature of the source to glass temperature [8].

emissivity of the material of the thermocouple in contact with the melt, n is the refractive index of the melt (the glass mass in a specific case), sis the Stefan–Boltzmann constant: the multiplier in the right-hand part of the equation $N_{RC} = (r_0 \cdot \varepsilon \cdot n^2 \sigma \cdot T_f^3)/K_f$ is a dimensionless coefficient characterising the ratio of the heat transfer by radiation to the transfer by heat conductivity. The graphical solution of equation (2.6) is shown in Fig. 2.4 which shows that the error of measurement of temperature depends on two parameters: N_{RC} and the T_s/T_f ratio. In the case of low values of N_{RC} the heat conductivity is controlling and the error of measurements decreases. The ratio T_s/T_f is determined by the temperature gradients in the melt and the error of measurement also decreases with a decrease in temperature gradients. Here, the error of measurement is given for the following parameters of the melt [8,9]: T_0 = 273 K and T_f = 1673 K, T_s/T_f = 1.3, ε = 0.6;

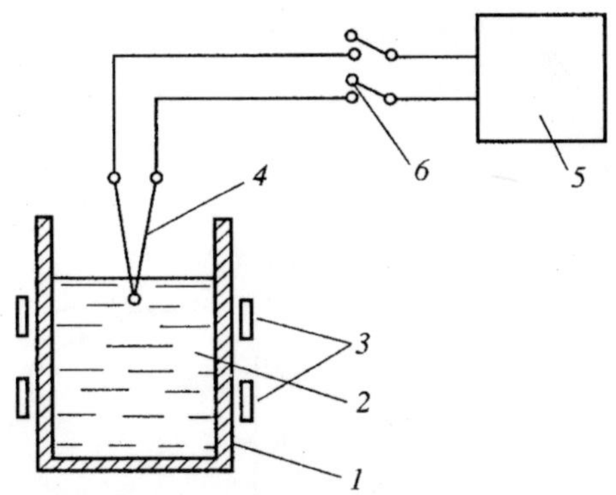

Figure 2.5. Measurement of melt temperature [16]. 1) water-cooled rucible; 2) melt; 3) induction coil; 4) thermocouple; 5) potentiometer; 6) thermocouple switch.

n = 1.5, K_f = 1.7 W/m K, and the error of measurement at r = 1 cm and $N_{RC} \approx 2.5 \cdot 10^{-2}$ is ≈ 2%.

Thus, the application of a small thermocouple without protection greatly decreases the error of measurements. For this reason, the experiment described here were carried out using an unprotected thermocouple Pt/Rh-30–Pt/Rh-6 with r = 0.025 cm. It should be mentioned that the values of T_S and T_f, selected for the evaluation of the error of measurement, correspond to the unfavourable measurement conditions.

The main factors complicating the measurement of temperatures in the melt in the case of direct HF heating are: firstly, the presence of high-frequency interference resulting in induction heating of the thermocouple and, secondly, the fact that the melt is insulated from the body of the crucible by a dielectric layer (skull) and is under a voltage whose value may reach several kilovolts. The measurement of temperatures in the given conditions greatly complicates the measuring system as a result of the introduction of high-frequency filters and insulation of the recording device. Therefore, the melt temperature was measured immediately after switching the equipment off using the simplest circuits (Fig. 2.5). With the circuit open, the thermocouple was immersed in the melt with operating equipment.Simultaneously, when the high-frequency generator was switched off, the circuit of the thermocouple was closed and the potentiometer recorded the variation of the electromotive force of the thermocouple during the measurement period. The true temperature of the melt was determined by extrapolating the cooling curve of the thermocouple

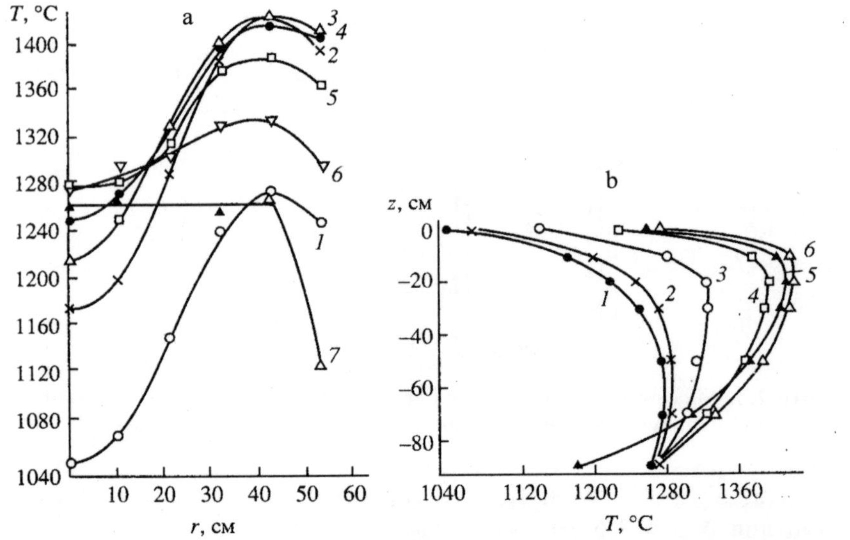

Figure 2.6. Variation of glass melt temperature along radius *t* [9]. a) at the depth of the melt of 0 (1), 10 (2), 20 (3), 30 (4), 50 (5), 70 (6), 90 (7) mm; b) at depth *z* and the distance from the axis of the crucible of 0 (1), 10 (2), 20 (3), 30 (4), 40 (5), 50 (6) mm (*r*, *z* along the axes, mm). The numbers are the numbers of the curves.

to the moment of switching off the equipment. In order to create reproducible conditions, after taking measurements at any single point and subsequent activation of equipment, the melt must be returned again to the initial condition. In order to maintain the melt in the stationary conditions, measurements were taken using a system for the stabilisation of the power of the high-frequency generator with respect to anodic current which transferred the melt to the initial condition. Multiple measurements of the temperature of the melt at any single point showed that the results are reproducible if the time between two measurements was at least 5 min. The accuracy of temperature measurements was determined by the error made in the recording thermal electromotive force on the diagram paper of the device and ±5°C. The temperature of the melt is also affected by the variation of the flow rate of water for cooling the crucible. However, special measurements of the melt temperature show that with the variation of the flow rate of water by a factor of 1.5, the temperature of the melt remains constant in the range ±5°C.

Figure 2.6 shows the results of the measurements of temperature along radius *r* and in the depth of the melt z. The radial distribution of temperature to a depth of 70 mm is in qualitative agreement with the distribution of temperature on the surface of the aluminium oxide and with the theoretical calculation of the temperature distribution in a glass melt [10]. The radial distribution of temperature has a maximum whose position is constant up to a specific depth. The largest decrease of the melt temperature at $r>40$ mm is determined by the effect of the water-cooled crucible walls and is characteristic of the melt in the cold crucible. It should be mentioned that the observed temperature maximum increases with increasing depth of the melt and at the depth greater than the critical value it completely disappears as a result of the rapid removal of heat through the bottom of the crucible and also due to a reduction of the intensity of heating in the lower part of the crucible.

A characteristic feature of the temperature distribution in the depth of the melt is the presence of temperature gradients in the surface layer of the melt at a depth of 10–15 mm, determined by the large heat losses from the open surface of the melt. The temperature distribution in the melt is shown in Fig. 2.7 a–d. Curves a–c correspond to different positions of the induction coil in relation to the surface of the melt at a constant volume of the melt and constant generated power. Analysis of the temperature distribution in the melt makes it possible to separated two regions of the melt: 1. The surface layer with a depth of 10–15 mm, 2. The remaining volume of the melt. With loading of the container, the axial gradient is greatly increased and the radial gradients of temperature in zone 1. The temperature distribution in region 2 becomes more uniform, but the range of the maximum temperature is displaced into the bulk of the melt and the size of this region decreases.

The observed changes may be explained using a model which takes into account the special features of direct high-frequency melting of materials in which the energy of the electromagnetic field is absorbed only by the layer of the melt whose thickness is determined by the working frequency of the generator and by the electrical conductivity of the melt. The remaining part of the melt is heated as a result of the heat transfer from this hot zone of the melt. As a model, induction heating may be regarded as heating from a heat source in the form of a hollow cylinder, immersed in the melt. The height of the cylinder is determined by the height of the induction coil, and the diameter of the cylinder by the thickness of the active zone of the melt. The lowering of the crucible is accompanied by the displacement of the heat source into the depth

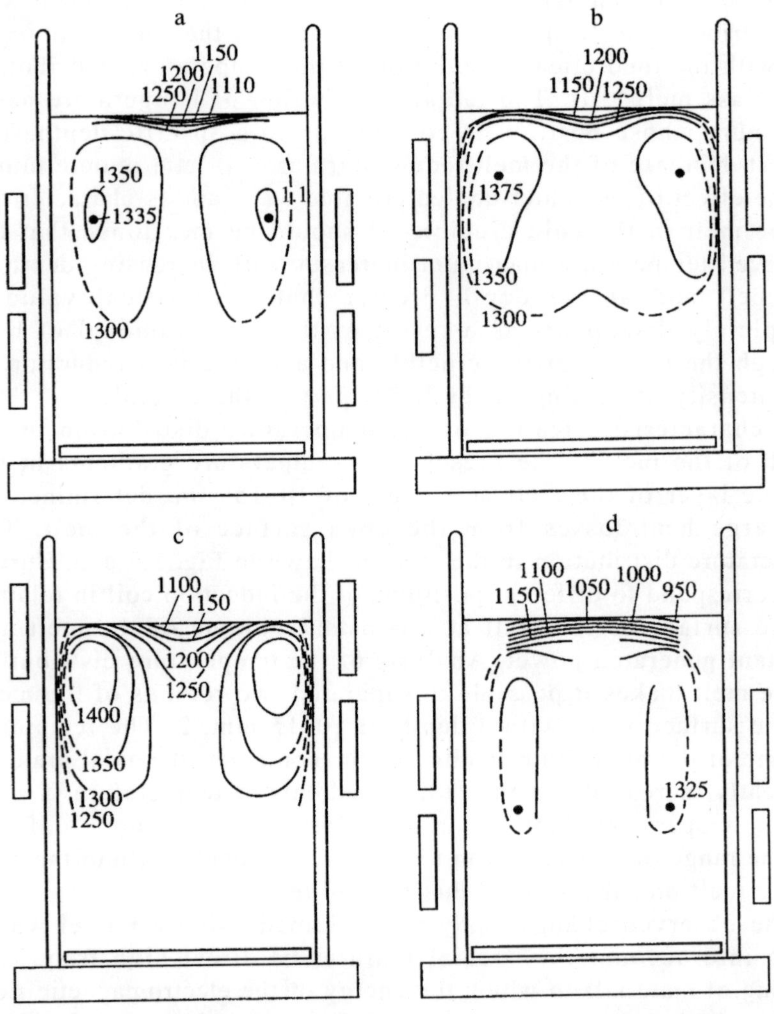

Figure 2.7. Distribution of temperature (°C) in a glass melt in relation to the position of the induction coil in relation to the melt surface and the power absorbed by the melt Q_i [9]. a) $l = -35$ mm, $Q = 12.1$ kW; b) $l = -15$ mm, $Q = 12.1$ kW; c) $l = +20$ mm, $Q = 12.1$ kW; d) $l = 35$ mm, $Q = 9.9$ kW.

of the melt, the surface layers cool down, the viscosity of the melt increases and the intensity of convective heat transfer in the surface layers inсreases. The latter circumstance results in a rapid increase of the axial gradient in region 1. A decrease in the radial temperature gradient in this region is also caused by a decrease in the strength of the effect of the heat source with movement of the

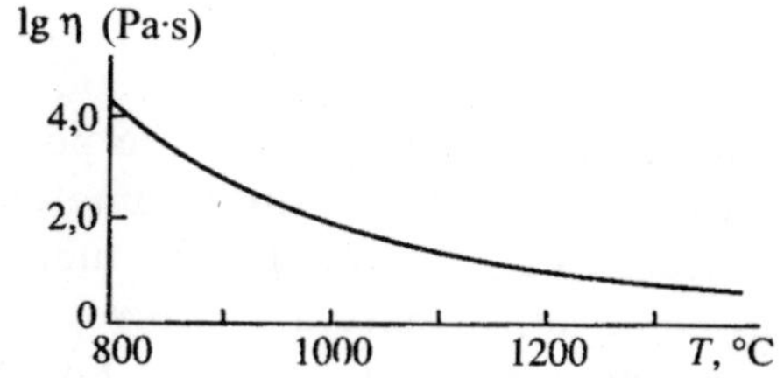

Figure 2.8. Temperature dependence of the viscosity of the melt of sodium-borosilicate glass [9].

source away from the surface. Figure 2.7 a,d also shows the temperature distribution in the melt with the variation of the power absorbed by the melt. A decrease in the power, absorbed by the melt, greatly changes the form of the temperature field in the melt: the melt temperature decreases, the region of maximum temperature is displaced into the depth of the melt, the axial gradient in region 1 becomes deeper, and the sign of the radial temperature gradient in the surface layer with a depth of 20 mm changes. To explain these phenomena, it is necessary to examine the temperature dependence of the viscosity of the melt, shown in Fig. 2.8. Since the heat source is relatively far away from the surface of the melt, a decrease in the intensity of the heat source increases the viscosity of the melt, especially in region 1. For example, when the melt temperature is reduced from 1300 to 1000°C, the viscosity of the melt increases by more than an order of magnitude. Therefore, the intensity of convective heat transfer from the depth of the melt decreases and the controlling factor in region 1 is now the removal of heat by the water-cooled walls of the crucible. This results in the inversion of the sign of the radial temperature gradient. In region 2, the temperature remains sufficiently high for the efficient mixing of the melt and equalisation of the temperature gradient.

Thus, the viscosity of the melt is an important factor determining the temperature distribution in the melt which must be taken into account in melting materials in a cold crucible.

2.5. EFFECT OF POROSITY OF THE INITIAL MATERIAL ON MELTING CONDITIONS

In the production of large, optically perfect crystals, a controlling role is played by the conditions in the stage of formation of the volume of the melt, for example, the shape of the solidification front, the structure and geometry of the heat-insulating layer (skull) at the bottom and the wall of the cold crucible. In turn, the physical–chemical characteristics of the skull depend on the initial technological parameters of melting, with the porosity of the initial charge being one of the most important parameters. On an example of zirconia, the authors of [11] showed by means of experiments that, in the actual conditions, there are two regimes of the melting of dielectrics — stationary and quasiperiodic. For the same dielectric with the same initial technological parameters, the type of the melting conditions and the direction of propagation of the melt are determined by the porosity of the initial charge. For this purpose, the parameters of the generator determined in the melting process were: the power, generated in the melt, heat losses in the cold crucible, and also the mass and configuration of the melt.

The investigations were carried out in Kristal-401 Fianit equipment, with an induction coil with a diameter of 250 mm and a copper cold crucible, diameter 190 mm. Two types of charge, differing in the shape and the grain size, which filled the specific volume of the crucible, were used in the investigations. The porosity of the charge was calculated from the equation:

$$\varepsilon = (V - V_0)/V,$$

Here V is the total volume of the charge, V_0 is the volume occupied by the particles of the charge whose porosity was $\varepsilon_1 = 0.78$ and $\varepsilon_2 = 0.86$. Although melts were produced in the same, initial conditions: the operating regime of the generator, the position of the crucible and the starting metal in relation to the induction coil, the size and mass of the tablets of the starting metal represented by the pressed shavings of zirconia. During melting, the operating parameters of the generator changed due to a change in the load characteristics [7]. It was established that the increase of anodic current to the maximum value corresponds to the increase of the power, generated in the melt and, correspondingly, to an increase of the mass of the melt. In the same experiments, the indicator of the regime of the melting process was represented by the anodic current of the generator lamp characterised by the high sensitivity to changes of the load. Figure 2.9 a,b,c shows by points: the start of increase of the constant component of anodic current I_{ao}, determined by the formation of the melt (point 0), the start of rapid linear increase (A, D), establishment of the maximum value (B and E), and also the maximum frequency of the generator f whose

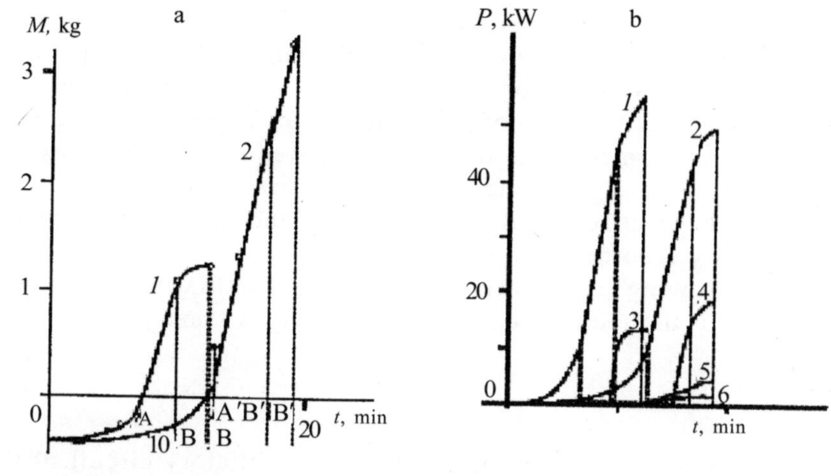

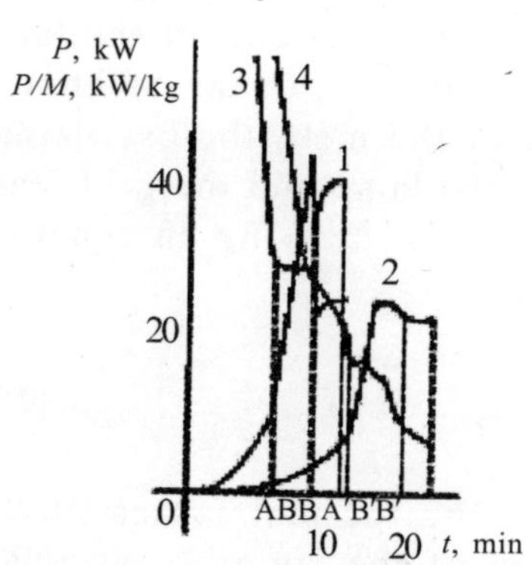

Figure 2.9. Time dependences of: a) mass of the melt in melting of a charge with $\varepsilon_1 = 0.78$ (1) and $\varepsilon_2 = 0.86$ (2); b) power generated in the melt and heat losses in a cold crucible in melting of the charge with ε_1 ($1 - P_{p1}$, $3 - P_{w1}$) and ε_e ($2 - P_{p2}$, $4 - P_{w2}$, $5 - P_{b2}$, $6 - P_{rad2}$); c) power used for melting the charge with ε_1 ($1 - P_{m1}$) and ε_2 ($2 - P_{m2}$) and power density ($3 - P_{m1}/M_1$, $4 - P_{m2}/M_2$).

increase is also associated with an increase in the melt volume (C and F).

In order to determine the mass of the melt M melting was carried out to one of the characteristic moments of the procedure of loading and propagation of the melt in the crucible (Fig. 2.10 *a–g*); when these moments were reached, equipment was switched off, the solidified melt was separated from the charge and weighed. The amount of zirconia, formed from the starting metal, was deducted from the mass of the melt. Subsequently, the crucible was again filled with the investigated material and melting continued up to the next characteristic state, and so on. A similar procedure was used for producing a series of melt of the investigated material with different porosity. All working parameters of the generator and heat loses in the crucible were recorded during these experiments. The heat losses in the crucible were determined by separate measurements, using differential thermocouples, of the temperature of the cooling water prior to and after passage through the wall (P_w) and the bottom (P_b) of the crucible at a constant flow rate. A radiation pyrometer was used for measurements of the losses of radiation power P_{rad} through the open surface of the melt. The readings of the differential thermocouples and the pyrometer were continuously recorded in a potentiometer.

The following procedure was used for calculating the power generated in the melt. According to [12], for a lamp generator:

$$P(\sim) = P_o - P_a \tag{2.7}$$

where P(~) is the power, generated in the oscillatory circuit of the generator, P_a is the power used for heating the anode of the lamp, $P_o = (E_{ao} \cdot I_{ao})$ is the power supplied to the anode circuit of the lamp (E_{ao} is anodic voltage, I_{ao} is the constant component of anodic current). In turn, $P(\sim)$ is used for heating the melt, the oscillatory circuit of the induction coil, the grid of the lamp and the grid leak. Consequently, equation (2.7) may be used for deriving an equation for the power generated in the melt:

$$P_p = P_o - P_a - P_k - P_{ind} - P_c - P_g, \tag{2.8}$$

Here P_k, P_{ind} and P_c are the power of the thermal losses in the circuit, the induction coil, and the grid of the lamp; $P_g = I^2_{co} R_g$, where R_g is the power of the thermal losses in the grid leak, where I_{co} is the constant component of grid current, R_{gr} is the electrical resistance of the grid leak. The numerical value of R_g is available, the values of E_a, I_{ao}, I_{co} were continuously recorded in a potentiometer; P_a, P_k and P_{ind} were determined in the same manner as the power of the heat losses in the crucible, because the anode of the lamp, the circuit and the induction coil were cooled with water. In addition to this, the frequency meter was used for recording the frequency f of the generator. The error of measurements of the absolute values of P_w, P_b and P_p did not exceed 5%, and that of P_{ind} was not higher than 10%. In this case, the convective losses of the crucible may be ignored, and the losses as a result of the radiation of the layer of the charge above the melt (cupola) were calculated.

Figure 2.9a shows that the form of the curves 1 and 2 is identical, *i.e.* a small increase of the melting rate to the moments A and A', followed by a more rapid linear increase of the mass of the melt M; the curve 1 tends to saturation, and the curve 2 continues to increase at a lower rate; the mass of the melt M at the moments A', B', B' is higher than in the corresponding moments A', B', B'. Consequently, Fig. 2.9b,c shows that power P_{p1} (curve 1), generated in the first melt is higher than the power generated in the second melt P_{p2} (curve 2); it may also be seen that the power P_{w1} (curve 3) of heat removal from the crucible wall is smaller than the total sum of the powers of heat removal from the wall P_{w2} (curve 4), the bottom of the crucible P_{b2} (curve 5) and as a result of radiation P_{rad2} (curve 3). Consequently, the power used directly for melting $P_m = P_p - P_w - P_b - P_{\mathrm{rad}}$ is considerably higher in melting a charge with a lower porosity (Fig 2.9c).

In order to explain the experimental results, the authors used the theoretical model of induction melting of the dielectrics [13,14]. The rate of frontal melting depends on the heat flow to the inter-face. Because the heat is generated in the volume of the melt, the controlling parameter is power density Q, scattered in the melt. In [13] it was proposed that heating of the solid phase takes place as a result of heat conductivity. However, as shown in [14], the heating of the solid particles is far more efficient by the melt spreading through the pores which wets the solid phase. The stable propagation of the melting front by means of capillary spreading is possible only if the supplied power is higher than a specific threshold value. However, if the supplied power is lower than the threshold value, the regime of capillary spreading is unstable. The melt, penetrated into the pores, solidifies, forming a region of the continuous solid phase separating the porous charge from the melt. The rate of melting of this region is lower as a result of heat conductivity. When the melt enters again into contact with the charge rapid spreading takes place and a new solidifies region forms, and so on. Thus, a quasiperiodic melting process takes place. In the stationary regime above the threshold, the melt–charge boundary moves in a uniform manner. According to [14], the threshold power density is determined by the equation:

$$Q_1 = 4\cdot\lambda(1-\varepsilon)(T_m - T_o)^2 / r_n^2\varepsilon^2(T_m - T_o + L/C), \qquad (2.9)$$

Here λ, T_m, T_0 and ε is the heat conductivity, melting point, initial temperature and initial porosity (prior to melting) of the solid substance, respectively; r_p is the diameter of the pores, L is the heat of melting of the substance, C is the heat capacity of the melt (for both types of charge, the values of T_p, T_0, L and C are identical). Because the charge is polydisperse, the determination of the radius of the pores r is difficult, but the diameter of the pores may be related to porosity ε. Thus, in the present case, Q_1 is higher for the charge with ε_1 in comparison with that for the charge with ε_2. This is in agreement with the time dependences of the power density P_m/M (Fig. 2.9c, curves 3, 4) used directly for the melting of the charge, and with the schema. of propagation of the melt shown in Fig. 2.10b–g, which correspond to the characteristic moments A–F.

As indicated by Fig. 2.7c and 2.8, the ratio $P_p(l)$/M(l) in the time period O–B (charge with ε_1) initially rapidly decreases (to A) and subsequently decreases with a constantly decreasing rate, because the increase of P_m (l) starts to catch up with the increase of M(l). This takes place as a result of the high speed of propagation of the melt preferentially in the direction to the crucible wall resulting in a rapid change of the coefficient of filling of the window of the

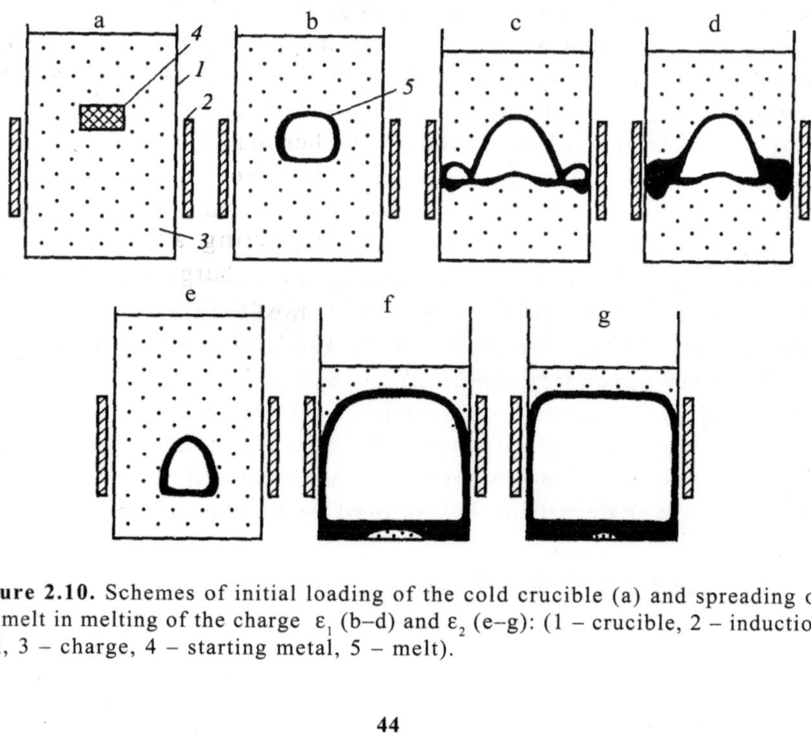

Figure 2.10. Schemes of initial loading of the cold crucible (a) and spreading of the melt in melting of the charge ε_1 (b–d) and ε_2 (e–g): (1 – crucible, 2 – induction coil, 3 – charge, 4 – starting metal, 5 – melt).

crucible by the melt K_f; a small increase of the coefficient results in a large increase of P_p [3]. This explains the fact that $P_{p1}>P_{p2}$ and $P_{m1}>P_{m2}$ in all characteristic moments, with the exception of A and D, where the filling coefficients are close in values (Fig. 2.8 b,e) and consequently $P_{p1} \approx P_{p2}$ and $P_{m1} \approx P_{m2}$ (since P_{w1}, P_{w2}, P_{b2}, P_{b2} are still absent).

At the moment of time B, the melt reaches the crucible wall and this results in a rapid increase of P_w and a decrease of P_{m1}/M_1. Consequently, in the section B–C, an increase of the mass of the melt M_1 rapidly slows down. Evidently, at the moment of time B there is a transition from the stationary melting regime to quasiperiodic regime in which the heating of the solid phase takes place as a result of heat conductivity and the melting rate decreases. This indicates that the value of the threshold power density for the charge ε_1 is in the range 22–30 kW/kg. However, it must be remembered that this definition leads to a relatively a rough estimate, because the power density was calculated without taking the skin effect into account.

Additional experiments showed that the melting of the charge takes place mainly on the horizontal ring in the vicinity of the crucible wall. The layer of the solid phase, separating the melt from the charge, locally melts through, the melt again comes into contact with the charge and decreases to a new level. The melting process takes place up to reaching the bottom of the crucible and, subsequently, the central part of the charge melts in the quasiperiodic regime. Melting by this mechanism leads to an increase in the power generated by the generator during a long period of time in the vicinity of the crucible wall. This often results in damage of the skull layer and in irreversible changes in the conditions of the crystal growth process.

In melting of the charge with porosity ε_2, the superheating of the melt is less marked in comparison with the first case because the power density P_{m2}/M_2 (Fig. 2.9c, curve 4) and decreases uniformly. The melting process propagates mainly in the direction to the bottom of the crucible and the value of K_f decreases. After passage through. point C, the melt drops below the edge of the induction coil and the extent of absorption of the energy of the high-frequency field in the volume of the melt de reases. However, the inequality $M_2 > M_1$ holds in all characteristic points. Melting in the stationary regime takes place up to complete melting of the main bulk of the load. Up to the moment of passage through point D, the vertical direction of propagation of the melt is a consequence of a low threshold power and, consequently, the stable regime of capillary spreading is maintained in the vicinity of the centre of the crucible.

Thus, high-frequency heating in the cold crucible is characterized by two types of melting regime of the dielectrics – stationary and quasiperiodic. The realisation of a specific regime is determined by the ratio of the power density to the threshold density. The threshold power density depends strongly on the porosity of the initial charge and for a charge with $\varepsilon_1 = 0.78$ it is in the range 22–30 kW/kg.

References

1. Aleksandrov V.I., et al., *Usp. Khimii,* 1978, **47**, No. 3. 385-427.
2. Borik M.A., et al., In: Problems of crystallography, Moscow, Nauka, 1987.
3. Byndin V.M., et al., *Ogneupory*, 1983, No. 2, 41–45.
4. Petrov Yu.B., Kanev I.A., Induction furnaces for melting of oxides, Leningrad, Politekhnika, 1991.
5. Petrov Yu.B., Induction melting of oxides, Leningrad, Energoatomizdat, 1983.
6. Khaneev N.P., Dissertation, General Physical Institute, Russian Academy of Sciences, Moscow, 1994.
7. Cooper A.R., *J. Noncrystalline Solids*, 1977, **26**, No. 1–3, 27–37.
8. Eryon N.D., Glicksman L.R., *J. Heat Transfer*, 1972, **94**, No. 2, 224–230.
9. Aleksandrov V.I., et al., *Neorganicheskie Materialy*, 1983, **19**, No. 3, 443–447.
10. Aleksandrov V.I., et al., *Current Topics in Materials Science*, 1978, **1**, 421–480.
11. Aleksandrov V.I., et al., *Izv. AN SSSR, Neorgan. Mater.*, 1991, **27**, 983–987.
12. Donskoi A.V., et al., High-frequency electrothermal equipment with valve generators, Leningrad, Energiya, 1974.
13. Merzhanov A.G., et al., *Dokl. Akad. Nauk SSSR*, 1980, **253**, No. 2. 330–334.
14. Rumanov E.N., Melting wave of porous matter, Preprint of the Institute of Chemical Physics, RAS, Chernogolovka, 1982.

3

Main sections of equipment for direct radiofrequency melting in the cold crucible

3.1. RADIOFREQUENCY GENERATORS. COLD CRUCIBLE

At present, a series of radiofrequency (RF) equipment for induction melting of dielectric materials in the cold crucible of the Kristall-400 type has been developed and is manufactured by industry [1–4]. In these systems, equipment Kristall-401-Fianite, Kristall-403 and Kristall-405 differ in the power of the generator and the volume of the crucible and are designed for the melting of oxides and growing single crystals by directional solidification. All these systems have been constructed on the basis of the block principle using unified sections.

Equipment Kristall-401-Fianite is in fact the first industrial equipment for the growth of single crystals of oxides by directional solidification. The main technical characteristics of equipment, specially developed for the cold crucible method, are presented in Table 3.1, [5]. Equipment is powered from a three-phase AC mains with a voltage of 380 V, frequency 50 Hz. Equipment consists of the following units: a rectifier, a generator, a furnace, and a control unit. The rectifier unit, supplying power to the anode circuit of the generator, contains a thyristor regulator, a power transformer and a rectifier. The generator unit contains the anode (primary) circuit. The furnace includes a condenser bank of the loading (secondary) circuit with an electric drive for auto-setting of frequency, an induction coil and a crucible, placed inside a water-cooled chamber,

Table 3.1. Simple and complex compounds melted and synthesized in a cold crucible

Compound (product)	T_m, °C, T_{synth}	Comment	Compound (product)	T_m, °C, T_{synth}	Comment
LiF	842		Tb_4O_7	2330	
NaF	995		Dy_2O_3	2485	
CaF_2	1403		Ho_2O_3	2395	
NaCl	800		Er_2O_3	2400	a)
KCL	770		Tm_2O_3	2390	
MgO	2880		Yb_2O_3	2420	
CaO	2585		Lu_2O_3	2470	
SrO	2430		TiO_2	1855	a)
BaO	1923		ZrO_2	2690	
Al_2O_3	2046	a)	HfO_2	2790	
Sc_2O_3	2405	a)	V_2O_5	670	
Y_2O_3	2410	a)	Nb_2O_5	1510	
La_2O_3	2230		Ta_2O_5	1877	
CeO_2	-		Bi_2O_3	817	
Pr_6O_{11}	2040		Cr_2O_3	2300	
ND_2O_3	2315		MnOx	1700	
Sm_2O_3	2320		Na_2CO_3	851	
Gd_2O_3	2395	a)			

a) Single crystals were produced; b) Doped with 5% CaO, Cr_2O_3; c) Synthesis atmosphere: air and nitrogen, specific conductivity (Ω cm) 40 and 10^4 respectively; d) As above; e) Optically transparent glass was produced

and a mechanism for moving the crucible. The generator and furnace units are screened in order to reduce emitted radio interference. The panel of the control block is used for controlling the entire equipment and also includes systems for stabilisation and automatic reduction of anodic voltage.

Since the parameters characterising the condition of the melt vary in a wide range, from the complete absence of the load (in starting heating) to the maximum load, the electrical parameters of the induction system also vary in a wide range. Therefore, the generator must have a wide range of regulation of the oscillatory power and output voltage, with auto-setting of frequency and a wide range of variation of the feedback coefficient.

Kristall-401 equipment contains a two-circuit of the generator based on a GU-23A generator lamp with auto-transformer connection of the loading circuit to the anodic circuit and a capacitance feedback. Figure 3.1 shows the electrical diagram of the generator which is typical of all industrial systems designed for the induction melting of oxides in the cold crucible.

The anodic circuit consists of capacitance C_1 and circuits with inductance coefficients L_1, L_2, L_3. The voltage in the loading circuit is lower than in the anode of the generator lamp and is determined by the ratio of the inductance coefficients $L_2/(L_1 + L_2)$ which are variable parts of the inductance of the anodic circuit. The values of L_1 and L_2 depend on the spatial position of the controlling short-circuiting coil L_3. Varying the position of the coil, it is possible to change the value of the voltage transformation coefficient in the range 1.1–2.0. The excitation voltage in the grid of the generator lamp is generated using a feedback with a capacitance divider C_3, C_4. The bias voltage on the grid of the lamp is generated using a grid leak R_g, connected to the circuit through which the constant component of the grid current passes. The voltage in the grid is regulated by changing inductance L_3 by means of the short-circuited coil L_4 as a result of a change in the feedback coefficient. To set the working frequency of the generator, inductor L_5 and variable

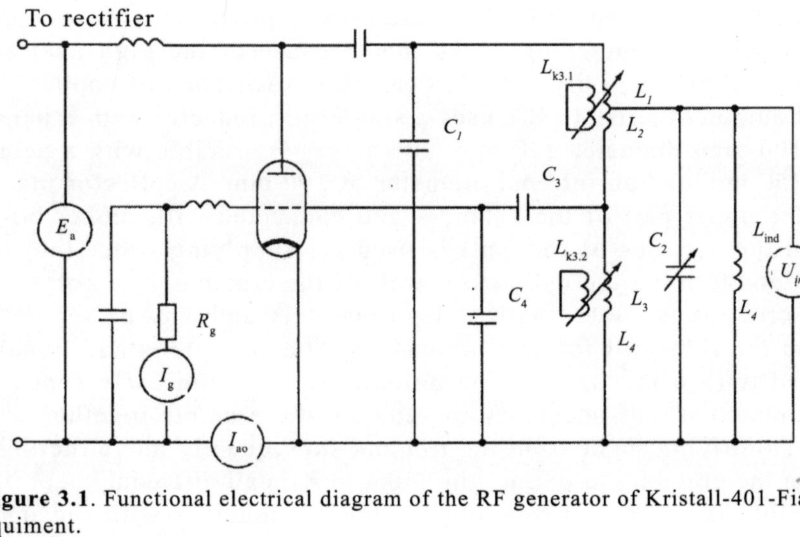

Figure 3.1. Functional electrical diagram of the RF generator of Kristall-401-Fianite equiment.

capacitance C_2 are included in the loading circuit.

The control panel contains regulators of anodic voltage, frequency setting, power (as a result of changes in the inter-circuit connection coefficient), the feedback coefficient and the speed of movement of the crucible. The operating regime of the generator is controlled by ammeters and voltmeters connected to the circuit (Fig. 3.1). The following parameters are controlled in the control panel: anodic voltage E_a, voltage in the inductor U_{ind}, the constant component of anodic current I_{ao}, the constant component of the grid current I_g. The circuit also includes a digital frequency meter for controlling the frequency of the generator.

The growth chamber is designed for protecting personnel against the effect of the high-frequency electromagnetic field and thermal radiation. The chamber is cylindrical with double walls made of stainless steel and is cooled with water. Easy access to the chamber is ensured by a door with a latch and a screw clamp, which is also cooled with water. The upper part of the chamber contains a hole for supplying the initial material, and also two holes for moving rods to which the crucible is secured. The mechanism for lowering the crucible is on the top part of the chamber.

The crucibles used for melting high-temperature materials consist of individual sections electrically insulated from each other and positioned parallel to the magnetic flux of the induction coil (Fig. 3.2). This design of the crucible ensures the highest permittivity for the RF field. The crucible and the inductor (consisting of one or many turns), cooled with water, are usually produced from a copper pipe with a diameter of 12–14 mm because of the high heat and electrical conductivity and also chemical resistance of copper.

Equipment Kristall-401 uses a single-turn inductor with a height of 100 mm, diameter 250 mm, and a copper crucible with a height of 350 mm and an internal diameter of 190 mm. A collector placed in the upper part of the chamber and connected with rubber hoses with the sections of the wall is used for supplying water into the sections forming the cylindrical wall of the crucible. The bottom of the crucible is cooled with water separately and may be separated from the side wall for easy unloading. The crucible is additionally fitted with a mechanism for unloading. If in the early types of equipment it was necessary to separate the crucible together with the solidifying ingot from the rod and subsequently move the ingot from the crucible, to extract the ingot in Kristall-401 equipment, the crucible is lifted upwards and the bottom together with the ingot

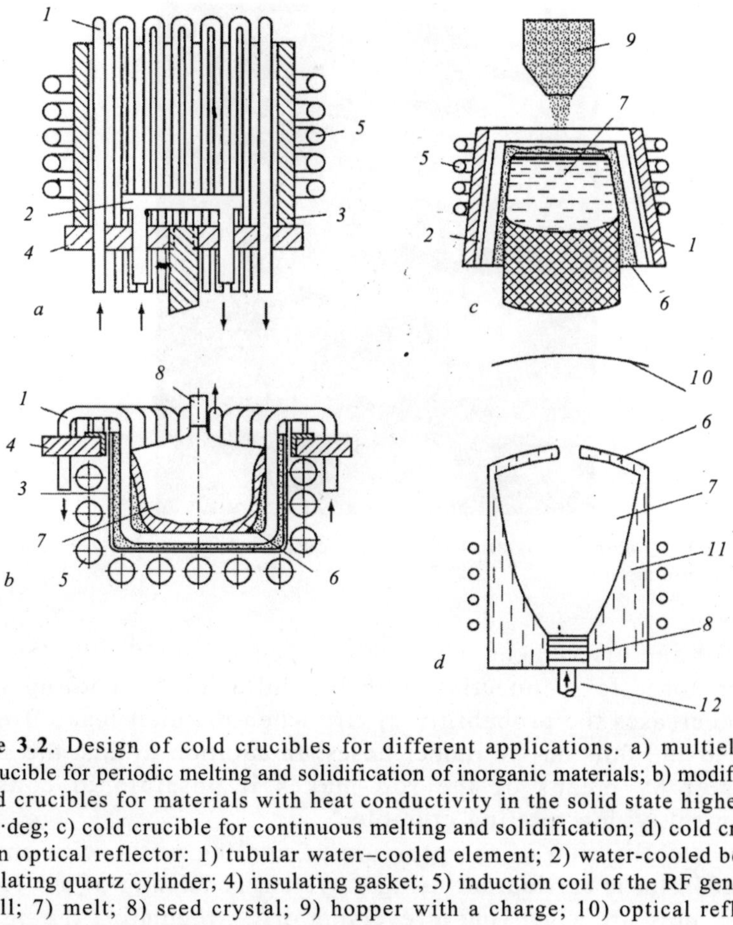

Figure 3.2. Design of cold crucibles for different applications. a) multielement cold crucible for periodic melting and solidification of inorganic materials; b) modification of cold crucibles for materials with heat conductivity in the solid state higher than 6 W/m·deg; c) cold crucible for continuous melting and solidification; d) cold crucible with an optical reflector: 1) tubular water–cooled element; 2) water-cooled bottom; 3) insulating quartz cylinder; 4) insulating gasket; 5) induction coil of the RF generator; 6) skull; 7) melt; 8) seed crystal; 9) hopper with a charge; 10) optical reflector; 11) unmelted charge; 12) cooling water [5,7].

remain stationary. Equipment contains a system for automatic loading the charge from a hopper. The external appearance of equipment is in Fig. 3.3.

Improved furnaces of the Fianite-1 type are used for the production of crystals by directional solidification of the melt (in particular, fianites), and also fused ceramic materials and glasses. The chamber is produced in the form of a right-angled box with opening doors ensuring free access to the crucible which is secured to a rod through a threaded joint. Water is supplied to the sections of the crucible and removed through the bar. This improvement makes it easier to replace the crucibles and load the molten

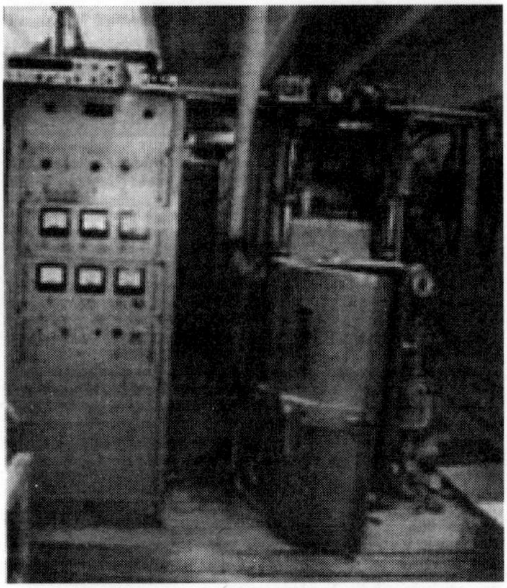

Figure 3.3. Industrial Kristall-401-Fianite equipment for growing crystals by the cold crucible method.

material and also makes it possible to reduce greatly the length of rubber hoses for connection with the collector for cooling water. This decreases the probability of fire when the melt leaks from the crucible and falls on the rubber hoses. In addition to this, the supply of water by means of the rod makes it possible to construct equipment with a rotating crucible.

The stabilisation of the pressure of cooling water with an error of ±0.05 atm is carried out using a system consisting of a electrical contact pressure gauge and a reversing drive, regulating the position of a valve on a water line.

Industrial variants of the equipment for RF melting of dielectrics include also equipment of the type Donets-4 and Kristall-402-Fianite. Donets-4 equipment is designed for growing crystals from a melt by both the method of directional solidification and by the Czochralski method. This equipment may also be used to produce fused materials and glasses. The power source in equipment is represented by a generator on non-controlled diodes. The rectified voltage is controlled using a thyristor regulator included in the primary winding of the transformer. A special programme is used for automatic setting of the given regime of variation of generator power. The upper part of the working chamber contains a withdrawal mechanism, the lower part the mechanism for lowering

Table 3.2 Technical parameters of equipment for direct HF melting of dielectrics [5]

Parameter	Parameters of equipment		
	Kristall-401	Kristall-403	Kristal-405
Rated power, kW	100	200	280
Oscillatory power, kW	60	160	160
Current frequency, MHz	5.28	1.76	1.76
Crucible diameter (inner), mm	200	400	500
Crucible height, mm	300	700	1000
Working temperature of the melt, K	3000	2900	2900
Productivity, kg/h: fianites, melted oxides	0.15	3.0	3.0
		25	
Water consumption, m^3/h	6	10	8
Working stroke speed, mm/min	0.02–20.0	0.02–10.0	0.05–0.5
Rated speed, mm/min	100	200	200
Weight of equipment, t	3.1	5.5	8.4
Area occupied by equipment, m^2	13	26	21

and lifting the crucible, the speed of displacement of both bars is regulated smoothly in the range 1–100 mm/h. The drives of the mechanisms for lifting and lowering the crucible are fitted on a special stand. The chamber can be sealed and crystals can be grown in any atmosphere and also in vacuum. This is achieved using a system for pumping and controlling the gas pressure in the chamber. There is also a system for stabilising the flow rate of cooling water with a pressure of up to 2 atm. The upper and lower rods are also cooled. The water for cooling the crucible is supplied to using rubber hoses connected to the collector. The technical parameters of equipment for direct RF melting are presented in Table 3.2.

The design of the crucible should be discussed quite extensively. The design of the cooled crucible has a number of specific requirements. Primarily, because the energy source is outside the crucible, the crucible must be 'pervious' to electromagnetic waves generated by the generator. This means that the crucible should be made of a dielectric material or contain slits preventing the closure of the eddy currents. The crucibles are produced from materials with high heat and electrical conductivity. The crucibles shown in Fig. 3.2 are more suitable for melting and solidification of refractory nonmetallic materials. One of these crucibles, Fig. 3.2a, makes it possible to melt and solidify a large number of different materials. Some of these materials are summarised in Table 3.2. For solidification of materials whose conductivity in the solid state is higher than 6 W/m deg, it is recommended to use a different crucible (Fig. 3.2b) in which the electromagnetic energy is supplied not only from the side surface of the melt but also from the bottom of the crucible. The crucible shown in Fig. 3.2c is used for continuous melting and solidification of materials. The direct experimental determination of the efficiency of these systems shows that in the crucible with the melt up to 50–60% of electrical power supplied to the equipment is generated.

The technical–economic parameters of these systems are determined by both their technological schema and the nature of processed material. For equipment with periodic melting regime the productivity is up to 3 kg/h with a specific consumption of electric energy of 5–10 kW h/kg. In continuous melting, the productivity reaches 10–20 kg/h, and the specific consumption of electric energy is (1.5–7.0) kW h/kg.

The experience accumulated in the service of Kristall-403 equipment was utilised in its modernisation. A new variant of equipment is known as Kristall-405. The regulators of anodic voltage, frequency setting, power (the coefficient of inter-circuit connection), the feedback coefficient and the rate of lowering the crucible are installed on the control panel of equipment. The operating regime of the generator is controlled on the basis of readings of devices included in the circuit and displayed on the control panel. The following parameters are controlled: 1. Anodic voltage E_a, voltage in the inductor U_{ind}, the constant component of cathodic current I_{ao}, the constant component of grid current of the generator lamp I_{ao}. In addition to this, a digital frequency meter is used and the readings are utilised in the control of generator frequency f. The block diagram of Kristall-405 equipment is shown in Fig. 3.4. The distribution of the units is more compact and they occupy a smaller area.

The volume of the cold crucible is approximately three times larger, the speed of displacement is lower and the stability of

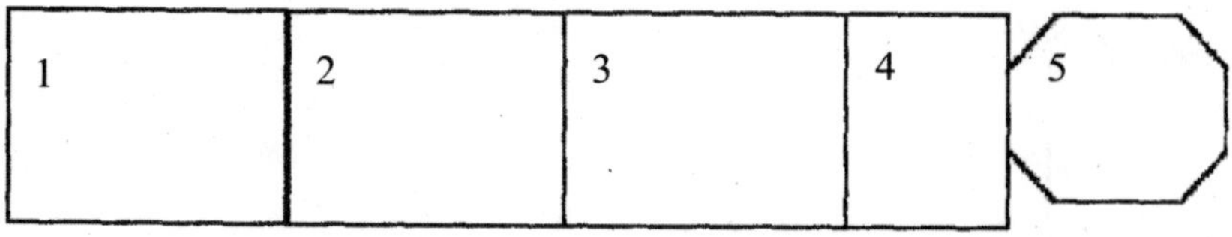

Figure 3.4. Flow chart of Kristall-405 equipment: 1) rectifier; 2) filter unit; 3)

displacement is improved. Equipment includes a system for automatic control of the lamp generator. The system makes it possible to carry out automatic operation of equipment in accordance with the given programme from the activation of the incandescent lamp to the unloading of the crystal block from the cold crucible. Table 3.1 gives the technical characteristics of the main types of industrial equipment.

3.2. METHODS AND DEVICES FOR CONTROLLING AND REGULATING THE TEMPERATURE OF THE MELT

As already mentioned, in the melting of dielectric materials the energy of the RF field is generated mainly in the melt and, consequently, the parameters of operation of the generator are determined by the composition and volume of the melt. Some of these parameters may be controlled using measuring devices included in the circuit of the generator (see section 2.1). Previously, it was reported in [6] that the constant component of the anodic current and grid current of the generator lamp are parameters sensitive to the change of the volume of the melt in the stage of formation of the melt. However, the authors of [2,4] reported that I_{ao} may be used as an indicator of the condition of the melt only if the degree of superheating of the melt is not large and the depth of the melt may change. In the case of large superheating of the melt and a constant volume of the melt, the temperature of the melt is determined by the generated power. In this case, the condition is indicated by the voltage in the inductor U_{ind}. However, in the above studies, the authors did not describe the criteria for application of a specific parameter as the controlling parameter. For example, the degree of superheating at which I_{ao} or U_{ind} can be used as an indicator was not determined. The superheating of the melt may be evaluated (with a large error) on the basis of the electrical parameters of the generator. The authors of [16] controlled the process of growth of single crystals on the basis of

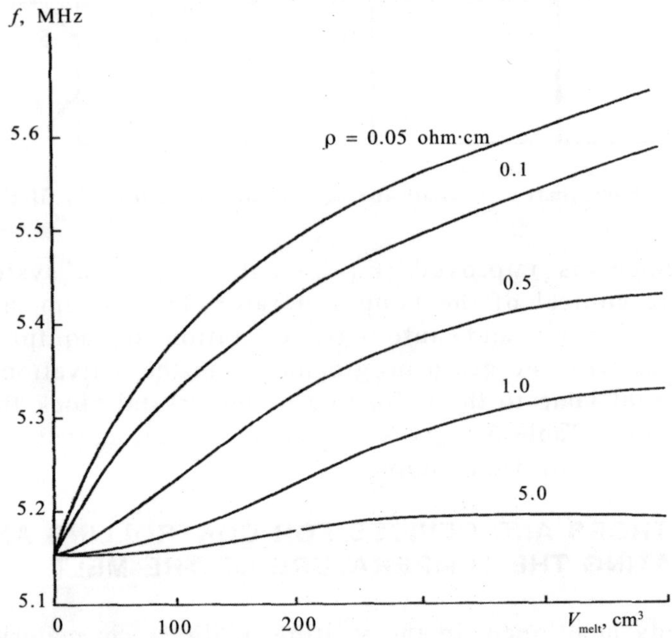

Figure 3.5. Dependence of the frequency of the HF generator on the volume and electrical resistance of the melt in a cold crucible [7]

the results of measurements of the power generated in the anodic circuit of the generator lamp ($E_a I_{ao}$). Because of the complicated nature and nonlinearity of the energy processes taking place in the generator and in the load, the power required by the generator lamp does not characterise unambiguously the stability of heating the melt in the cold crucible. In the melting and solidification stages, all electrical and thermal parameters of the generator–load the system are characterised by large changes.

In service of the initial prototypes of equipment for melting oxides it was noted that the frequency characteristic of the generator depends on the volume of the melt. In subsequent stages, this was confirmed in [7] in which the frequency dependences of the secondary circuit of the RF generator on the volume of the melt in the crucible for different specific electrical resistance of the melt were determined. Figure 3.5 shows the frequency characteristics of the RF generator in relation to the electrical resistance and the volume of the melt. The graph indicates that the increase of the frequency is proportional to the increase of the volume of the melt in the crucible. However, in the course of melting it is not possible to determine the volume of the melt on the basis of the frequency of the generator f because of the following reasons: 1. The temperature dependence of the specific electrical resistance of the melt remains unknown; 2. In the melting process, the operating regime of the generator is regulated by means of the transformation coefficient, feedback and the value of the variable capacitance in the secondary circuit. Each of these parameters affects the frequency of generation. Nevertheless, as reported in [2], the generator frequency is the most sensitive function of the variation of the volume of the melt. An increase of frequency f indicates an increase of the volume of the melt and a decrease of frequency indicates that the solidification process prevails over the melting process. Thus, the generation frequency f may be a sensitive indicator of the state of the melt and of the start of solidification.

In induction heating of the dielectrics in the cold crucible the energy of the electromagnetic field is generated in the melt. The state of the melt is determined greatly by the power absorbed in the melt P_p. The control and examination of the processes of melting and solidification can be carried out efficiently by measurement of the value of P_p. This can be carried out using an additional contactless method based on the measurement of the power factor of the inductor $\cos\varphi$, which has no restrictions with respect to temperature. This dimensionless coefficient changes in accordance with the variation of P_p at a constant level of the total power supplied to the inductor. Therefore, the proposed method of control may be used only for the stationary or quasistationary condition of the melt [7]. In this case, in the determination of the appropriate value of P_p using the available values of $\cos\varphi$, it is necessary to consider the parameters of the generator–load system which are often not available. In addition to this, the measurement of $\cos\varphi$ at frequencies higher than 1.0 MHz and at low absolute values of the measured quantity are associated with large difficulties. Industry does not produce standard devices for measuring $\cos\varphi$ and consequently, it was necessary to develop a laboratory prototype of a device which was described in [8] where the authors reported a large error in the measurement of this parameter. The dependence of $\cos\varphi$ on the parameters of the system is presented in Fig. 1.4, and the notations used in the figure are given in Chapter 1.

The power P_p can be determined in a simple manner and with sufficient accuracy by means of experiments with the measu rement of the power taken by the cooling water from the crucible and the power losses as a result of radiation from the open surface of the melt [6, 9]. However, these investigations can be carried out only for the stationary condition of the melt. Therefore, it is not justified to assume that the value of P_p is the same as the losses of power removed by cooling water and radiation, as proposed in [10].

Thus, the previously mentioned electrical parameters of the generator (the control of these parameters is taken into account in the design of industrial equipment for the production of single crystals and fused materials by direct RF heating in the cold crucible) does not make it possible to characterise unambiguously the condition of the molten material and are only an indicator of the state. The other previously examined methods of controlling the state of the melt have a number of restrictions and shortcomings preventing experimental examination and control of not only the non-stationary (in melting and crystallisation) but also the stationary state of the melt. In addition to this, these methods do not make it possible to determine the dependence of the condition of the melt on the operating regime of the generator.

On the current level of technology, the regulation of the operating regime of the generator is carried out manually in the process of formation of the volume of the melt. In the stage of the process of directional solidification, it is necessary to maintain a constant value of E_a by means of the system for stabilising the anodic voltage. In this case, the stationary nature of the working parameters of the generator in the process of melting and the quality of grown crystals are controlled by the operating experience of the operator of equipment, i.e. control of the process is unfortunately subjective.

Modernised equipment of the type Kristall-405 uses a system for the automatic control of the lamp generator on the basis of a microprocessor controller which enables automatic operation of equipment in accordance with a programme, starting from the point of heating of the generator lamp to the working displacement of the crucible [2]. Automation is based on the application of the criteria of the optimality of functioning of the subsystem consisting of a generator with a load which is a multi-parameter system. However, even this approach does not take into account completely the entire variety of the processes taking place in the load. These processes may have a significant effect on the operating regime of the generator.

As an example, we shall examine the time dependences P_p, P_w,

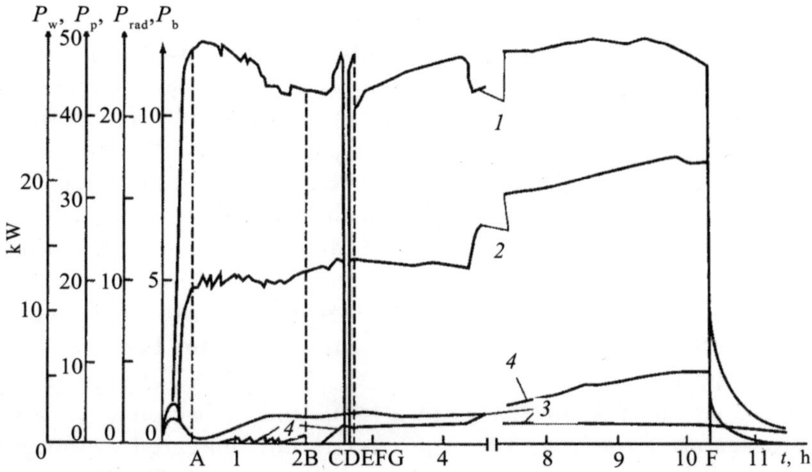

Figure 3.6. Time dependences of the power generated in the melt P_p (1) and the power of heat losses: through the wall P_w (2), the bottom of the cold crucible P_b (3) and by radiation P_{rad} (4) [11].

P_b, P_{rad} shown in Fig. 3.6 (curves 1, 2, 3, 4, respectively). These relationships were obtained in the process of melting and growth of single crystals of zirconia stabilised by yttrium oxide. The values of P_{rad} were measured in the experiments using a radiation pyrometer [11].

When the generator is switched on, P_w and P_b rapidly reach the level (curves 2 and 3) corresponding to the power of natural losses of the cold crucible, in the absence of the melt in the crucible. The energy is used basically for heating the crucible and starting metal.

After the formation of the melt, P_p rapidly increases, and P_w and P_n decrease as a result of the greater (in comparison with copper) flux linkage of the RF field with the melt, due to a large difference in the values of the specific electrical resistance. This decrease takes place until the melting front approaches the bottom and the wall of the crucible and, subsequently, they are heated as a result of heat removal from the melt. The power of natural losses of the crucible after filling with the melt in accordance with [11] is 1–2% of P_p. For the crucible used in [11], this power was approximately 1% of P_p.

After forming the starting melt, new portions of the material for increasing the volume of the melt were loaded in the time period A–B. It is interesting to note that when a hole is made in the cupola of the charge above the melt, the value of P_{st} rapidly decreases (curve 2). This is caused by melting of the melt as a result of radiation through the resultant hole and by the energy losses for melting of the newly supplied material. Subsequently, P_{st} smoothly increases to the next charge, and so on. The value by which P_w decreases depends on the position and the size of the hole, the time during which the hole remains open, and also on the volume of the melt. For example, in the examination of curve 2 in the time period A–B, there is a tendency for a decrease of the extent of oscillations of P_w, associated with making the holes for loading. This takes place as a result of an increase in the volume of the melt, with other parameters of the generator being constant. During the time when the hole in the layer of the charge remains open, P_{rad} (curve 4) increases in a jump and, subsequently, rapidly decreases after loading. These jumps are not indicated in the graph because they are very short. The oscillations of P_{rad} in the time period A–B are caused by gradual heating and melting of the layer of the charge, forming a cupola above the melt, and also additions of the new layers of the charge. Making the hole in the cupola and the supply of new portions of the charge in the melt affect the value of P_p (curve 1) because in direct RF heating of the melt, the energy of the electromagnetic field of the inductor is almost completely transformed to heat and, consequently, the state of the melt affects the working parameters of the generator.

The bottom of the crucible is quite far away from the surface of the melt and, consequently, making the hole in the layer of the charge and additions of new portions of the material does not have any significant effect on the value of P_b (curve 3). The section of the curve A–B is characterised by a decrease of power P_p. Subsequently, as a result of uncontrollable circumstances, the power emitted by the generator is greatly increased. This results in local melting of the skull and at the moment of time *C* a certain amount of the melt is discharged through the gaps between the sections of the crucible. This is followed by emergency shutdown of equipment and, consequently, P_w starts to rapidly decrease. After cooling the melt and formation of a skull, equipment is again switched on and the previous value of P_p prior to shutdown (the moment of time D) is restored. In this case, P_w reached the value close to the value detected prior to discharge of the melt.

At the moment of time E power P_p was also reduced and at F the crucible with the melt was lowered at a rate of 15 mm/h. The graph indicates that during lowering of the crucible, power P_b

decreases because the bottom of the crucible travels away from the heating zone. In this stage of the process, the values of P_p, P_w, P_{rad} gradually increase and their total stable decrease at the end of the process indicates that the entire amount of the melt almost completely solidified and the generator was switched off. The largest decrease of the power removed from the crucible takes place on its walls. The curves 2–4, shown in Fig. 3.6, indicate the cooling of the crystal ingot.

Thus, the dependences make it possible to evaluate the interaction between the component elements of the generator–crucible–melt–solid–phase system and also the reaction of the system to external effects. The analysis carried out previously can be used for a detailed determination of the characteristics of the melt for different operating regime of the generator.

The application of the method has shown its high sensitivity and information content which make it possible to control the process of melting of crystallisation of the material and exert efficient effects on these processes. For example, after making a hole in the cupola and adding the charge, it is possible to evaluate the degree of cooling of the melt and, if necessary, increase the power supply to the melt. In addition to this, the value of this power and also the power removed from the crucible may be used to estimate the volume of the melt and determine the moment of the start of crystal growth.

Detailed consideration and analysis of the heat removed from the crucible and of the power generated in the melt would enable researchers to investigate different processes taking place in the melt and the solid phase, independently of the melting point of the material.

References

1. Petrov Yu.B., et al., *Vysokochistye veshchestva,* 1989, No. 3, 136–140.
2. Petrov Yu.B., Kanaev I.A., Induction furnaces for melting of oxides, Leningrad, Politekhnika, 1991.
3. Byndin V.M., et al., Induction heating in production of special purity materials, Leningrad, Mashinostroenie, 1980.
4. Byndin V.M., et al., *Ogneupory*, 1983, No. 2, 41–45.
5. Borik M.A., et al., in: Problems of crystallography, Moscow, Nauka, 1987.
6. Aleksandrov V.I., et al., *Neorganicheskie Materialy*, 1983, **19**, No. 3, 443–447.
7. Aleksandrov V.I., et al., Synthesis of laser materials from the melt by direct

HF heating in a cold container, Report of FIAN, Moscow, 1968.

8. Lomonova E.E., Production and examination of single crystals of solid solutions based on ZrO_2 and HFO_2, Dissertation, Moscow, 1980.
9. Petrov Yu.B., et al., *Elektrotekhnika*, 1981, No. 5, 55–59.
10. Petrov Yu.B., et al., Induction melting of oxides, Leningrad, Energoatomizdat, 1983.
11. Borik B.A, Development of technology and examination of oxide glasses for electronics, Dissertation, Moscow, 1985.
12. Aleksandrov V.I., et al., *PTE*, 1991, No. 3, 231–234.

4

Synthesis and crystallisation of refractory materials in the cold crucible

4.1. THE COLD CRUCIBLE – EFFICIENT CHEMICAL REACTOR

The rapid development of new branches of technology, for example, electronics, requires the application of precision ceramics. Consequently, it was necessary to produce high purity oxide substances and materials with specific physical-chemical properties: amorphous, finely dispersed, crystalline, fused with a specific phase composition, granulated, etc.

The most important element of the technology of high-temperature materials is chemical synthesis because the majority of materials are complicated chemical compounds: complex oxides, oxygen-containing salt-forming substances, complex fluorides, mixed or simple compounds into which additions were added. In order to produce single crystals, glass or polycrystalline ceramics, the initial simple substances must be synthesised to produce the required compounds. In most cases, synthesis is carried out by the method of solid-phase reactions (the sintering method). For this purpose, the initial substances in the form of fine-dispersion powders are thoroughly mixed and subsequently heated to temperatures below the melting point of the interaction product. Usually, the temperature is (0.5–0.7) T_m. The mechanisms of chemical reactions in the solid phase are complicated and vary greatly. However, regardless of the components, an unavoidable stage of the process is the diffusion of one of the components (or atoms of all components) through the layer of the solid reaction products. The coefficients of diffusion in the solids are not high and usually equal 10^{-6} to 10^{-12} cm^2/s. Thus, the diffusion process is a 'bottleneck' of the solid-phase reaction, restricting the rate of the reaction.

The consequence of the lower rate of solid-phase reactions is the long duration of the process of synthesis and the problems in completing the process. Consequently, the reaction products usually contain impurities of unreacted components, i.e. the synthesis products are not homogeneous [1]. These shortcomings are not found in the synthesis of compounds in the melt. In the interaction of components in the melt, diffusion is no longer a bottleneck because, firstly, the rate of diffusion in the liquid is considerably higher than in solids and, secondly, the liquid can rapidly move under the effect of a force or by means of convection and, consequently, the rate of the reaction increases. The liquid products of the reaction can be solidified and, consequently, their chemical purity can be increased and structural perfection improves. Regardless of the obvious advantages, the synthesis of high-temperature compounds in the melt is used only on a limited scale. There is only a small amount of investigations in which attention was given to melt synthesis in a solar furnace [2], in a plasma furnace [3] or in arc furnaces [4]. The cold crucible is a universal high-temperature chemical reactor which does not contaminate the reaction products. In the cold crucible, the melt of the initial substances and the reaction products are heated with high-frequency currents. Using this method, it was possible to synthesise a large number of complex oxide compounds which together with the equations of the reactions are presented in Tables 4.1 and 4.2. The majority of the reactions given in the tables related to the reactions of the compound and take place in a homogeneous system. Some of the reactions are combined: in addition to the interaction of liquid components, a heterogeneous reaction with the participation of the gas phase also takes place. Suitable examples of these reactions are synthesised solid solutions of zirconium and hafnium dioxides, stabilised with the oxides of rare-earth and transition metals with a variable valency. Examples of these are in Table 4.1. The nature of the reaction and the composition of reaction products change in relation to the composition of the gas atmosphere. Another characteristic example of the combined reaction is the synthesis of solid solutions of chromites of yttrium, lanthanum and rare-earth

Table 4.1. Compounds synthesized in a cold crucible

Compound	T_m, °C	T_{synth}, °C
$Na_2O \cdot 11Al_2O_3$	1923	1970
$MgO+Al_2O_3 \rightarrow MgAl_2O_4$	2130	2300
$CaO + Al_2O_3 \rightarrow CaAl_2O_4$	1600	2300
$BaO + Al_2O_3 \rightarrow BaAl_2O_4$	2000	2300
$3Y_2O_3 + 5Al_2O_3 \rightarrow 2Y_3Al_5O_{12}$*	1930	2100
$La_2O_3 + 5Al_2O_3 \rightarrow 2LaAlO_3$	2100	2300
$3Al_2O_3 + 2SiO_2 \rightarrow 3Al_2O_3 \cdot 2SiO_2$	1910	2100
$2Al_2O_3 + SiO_2 \rightarrow 2Al_2O_3 \cdot SiO_2$	1850	2100
$SrO + TiO_2 \rightarrow SrTiO_3$	2038	2200
$R_2O_3 + 2TiO_2 \rightarrow R_2Ti_2O_7$*	-	2000–2300
(R = Sc, Y, La, Nd, Sm, Eu, Gd, Dy, Ho, Er, Tm, Yb) $CaO + ZrO_2 \rightarrow CaZrO_3$	2345	2600
$Sc_2O_3 + 2ZrO_2 \rightarrow Sc_2Zr_2O_7$*	2700	2900
$SrO + ZrO_2 \rightarrow SrZrO_3$	2470	2600
$La_2O_3 + 2ZrO_2 \rightarrow La_2Zr_2O_7$	-	2400
$SrO + HfO_2 \rightarrow SrHfO_3$	-	3000
$Y_2O_3 + Cr_2O_3 \rightarrow 2YCrO_3$**	2290	2500
$MgO + Cr_2O_3 \rightarrow MgCr_2O_4$	2150	2300
$CaO + HfO_2 \rightarrow CaHfO_3$	2700	2900
$La_2O_3 + Cr_2O_3 \rightarrow 2LaCrO_3$**	2430	2600
$CaO + MoO_3 \rightarrow CaMoO_4$	1449	1700
$SrO + MoO_3 \rightarrow SrMoO_4$	1468	1800
$CaO + WO_3 \rightarrow CaWO_4$	1580	1800
$SrO + WO_3 \rightarrow SrWO_4$	1566	1800
ZrO_2 + (8÷50 mol.%) MgO, CaO, SrO, Sc_2O_3, Y_2O_3, La_2O_3, TR_2O_3*	-	2700–2900
HfO_2 + (8÷50 mol.%) MgO, CaO, SrO, Sc_2O_3, Y_2O_3, La_2O_3, TR_2O_3	-	2800–3000
R_2O_3–Al_2O_3–SiO_2 (R = Sc, Y La, Nd, Sm, Eu, Gd, Tb, Er, Yb)***	-	1800–2400

* Single crystals were produced
** Admixture 5 wt.% $CaCr_2O_4$
*** Optically transparent glasses

Table 4.2. Physical and mechanical properties of refractory ceramics synthesized by solidification in a cold crucible

Properties	$3Al_2O_3 \cdot 2SiO_2$	Sc_2O_3	Y_2O_3	Gd_2O_3	Er_2O_3
Purity, %	99.7	99.0	99.9	99.9	99.9
T_m(°C)	1910	2405	2410	2395	2400
Calculated density, g/cm³	3.15	3.86	5.03	8.33	8.66
Measured density, g/cm³	2.5-2.6	3.28	4.25	6.8	7.5
Porosity, %	17-18	15-19	14.8	18.4	15.5
Limit of mechanical stability	500	1400	190	300–350	300–350
Pressure, kg/cm²					
20 °C	150		120	-	-
1000 °C	-		100	100–200	100–200
1300 °C			125	(20-1400 °C)	(20-1400 °C)
Elasticity moduli $E \cdot 10^6$, kg/cm	0.37	-	1.5	-	-
Temperature of deformation under load of 2 kg/cm²	1760-1810	1950	1960	1700–1800	1700–1800
Deformation in vacuum, mm/mm·h	-	-	T = 1650°C	-	-
			σ = 14.79 kg/cm₂		
			$9.8 \cdot 10^{-4}$		
Thermal expansion coefficient, Gr-$1 \cdot 10^{-6}$ (20-800 °C)	4.4 (850°)	-	7.3	8.4	7.3
Resistivity, Ω · cm					
vacuum 1000°C			$2.5 \cdot 10^6$		
1500°C			$3.0 \cdot 10^3$	10^4-10^5	
vacuum 1000°C			$1.0 \cdot 10^6$		
1500 °C			$3.0 \cdot 10^3$		0.75
Heat conductivity, kcal/m·h·gr (1200 °C)					
Thermal stability					
1300 °C water cycles			12-14		
Rates of evaporation (vacuum), g/cm² · s					
1700 °C	$0.5 \cdot 10^{-7}$				
1800 °C	$1.0 \cdot 10^{-7}$				
1900 °C			$1.6 \cdot 10^{-8}$	$6.0 \cdot 10^{-5}$	$1.2 \cdot 10^{-5}$
2000 °C			$1.5 \cdot 10^{-7}$	$2.0 \cdot 10^{-5}$	$1.9 \cdot 10^{-5}$

elements with additions of chromites of magnesium, calcium, strontium and barium. The electrical conductivity of these substances greatly changes in relation to the composition of the atmosphere.

Examination of the kinetics of reactions of the compounds in the melts shows that the chemical interaction of the initial components takes place at a rate equal to the rate of dissolution of solid particles of the substance in the melt, i.e. with the melting rate. In the typical cases, the melting rate is 0.1–1.0 kg/min, depending on

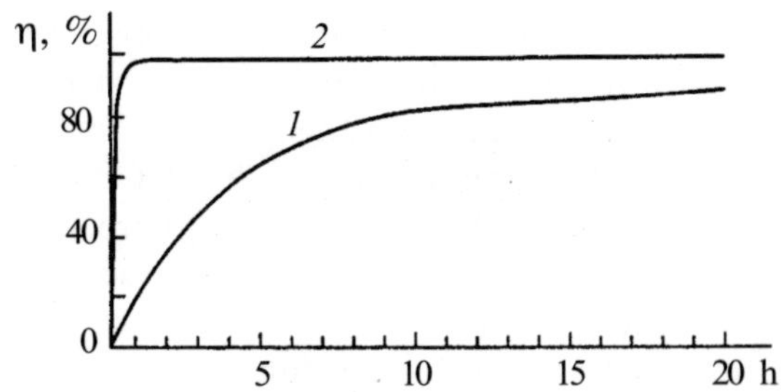

Figure 4.1. Kinetic curves (yield of the reaction product η in relation to time): 1) reactions in a mixture of solid phases $CaCO_3 + Fe_2O_3 \rightarrow CaFe_2O_4$; 2) reaction in themelt $CaO + ZrO_3 \rightarrow CaZrO_3$ [5].

the degree of superheating of the melt, viscosity and the heat of phase transition in melting initial substances. Thus, homogeneous chemical equilibrium in the melt is established during the time of several seconds to several minutes. Examination of the products of interaction by means of x-ray diffraction, microscopic and electron microscopy and spectral analysis methods shows clearly that for the strictly stoichiometric composition of the initial mixture of the components the content of the initial substances in the final product does not exceed 0.01–0.1 vol%. Figure 4.1 compares the kinetics of synthesis in the melt (2) with the appropriate dependence in solid-phase synthesis (1); comparison is in favour of synthesis in the melt.

The situation in the case of heterophase reactions is more complicated. In this case, the molten substance reacts with the gas atmosphere. Two limiting cases should be examined.

1. The acting component of the gas atmosphere does not dissolve in the melt. In this case, the contact surface is restricted by the open surface of the melt. The reaction rate is limited by the processes of diffusion of gas molecules through the phase boundary and by diffusion of the reaction products in the volume of the melt, if there is no mixing of the melt or if mixing can be ignored. Therefore, the viscosity of the melt plays a significant role. The dependence of the time of establishment of equilibrium on the viscosity of the melt can be clearly examined in comparing the rate of burnout of the starting metal and the products of reduction of the oxides of lanthanum, mullite ($3Al_2O_3{\cdot}2SiO_2$) and a glass with the composition $La_2O_3{\cdot}Al_2O_3{\cdot}6SiO_2$ from the melt. The viscosity of the melt rapidly increases in transition from the lanthanum oxide to mullite and, in particular, to the glass. In full correspondence with this, the rates of burnout of metallic aluminium and the products of

reduction, dispersing the volume of the melt, greatly differ. From the La_2O_3 melt, aluminium burns out within 5–10 min, from the mullite melt in 15–20 min, and from the glass melt, aluminium does not burnout in 10 h at a temperature of 2000°C.

2. The acting component of the gas atmosphere is soluble in the melt; in this case, the rate of the heterophase reaction is not restricted by the rate of diffusion of the reaction products. A suitable example of heterophase reactions with the dissolution of the gas component is the oxidation of metallic aluminium and the products of reduction in the melts of Al_2O_3, $YAlO_3$, $Y_3Al_5O_{12}$, $MgAl_2O_4$ and some other compounds. The experiments show that the melts of these substances absorb large quantities of gaseous oxygen [6]. Thus, the melt of aluminium oxide in contact with air dissolves more than 40 vol% of oxygen. In this case, heterophase equilibrium is established within 5 min in the volume of a 0.5 l melt. In rapid solidification of the melts, the dissolved oxygen is rapidly released, the melt 'boils up', and the solidified block is penetrated with a large number of large pores. It is important to note that as a result of combining 'the pure' method of heating and 'pure' method of melting in the skull, this method has no restrictions as regards the elemental composition of the synthesised compounds. Thus, in the melt synthesis of compounds in the cold crucible, it is possible to obtain high rates and 100% synthesis of the given chemicai compound. This may be accompanied by additional purification of the product to remove impurities in comparison with the starting material.

The application of additional purification may be indicated on an example of the synthesis of an aluminium–magnesium spinel and solid solutions with the structure of the spinel in the $MgO–Al_2O_3$ system, being promising materials for the production of high-density ceramics. The initial materials were reagents of the oxides of the synthesised compounds included in the composition having 'analytical' and 'special purity' grades. The synthesised product solidified in the cold crucible, produced from copper pipes. The temperature of the melt was maintained in the range 2200–2250°C. The x-ray examination of the phase composition of the synthesised material showed only the presence of spinel. The impurity composition was determined on the basis of the content of the impurities of alkali and transition metals which were the main characteristics of the material and determined its suitability for practical application. The results of analysis presented in Table 4.2

show that the content of the impurities in the synthesised material is on the level of the content of these impurities in the initial reagents and has a tendency for a decrease in the content of iron, chromium and alkali metals.

Special attention should be paid the impurity of copper in the melt because of the induction coil and the crucible, used for melting, are produced from copper tubes. An increase of the copper content to $(10^{-4}–10^{-2})\%$ is typical of the examined process and shows that the skull, formed between the melt and the cold crucible, reliably protects the melt against contact with the crucible. According to the authors of [7], the main source of the increase of the copper content is the chemical interaction of the outgoing gases, containing 'inherited' impurities of the compounds of sulphur, chloride and other elements, with the surface of the crucible above the melt, and the adsorption of compounds of copper by the powders supplied into the crucible for melting. However, the impurities of copper in the melt, when the initial material 'inherits' nitrogen compounds, may be excluded using a crucible produced from aluminium or stainless steel. The application by coatings of chemically resistant metals is also an efficient method.

The qualitatively new characteristic of the materials, synthesised in induction heating in the cold crucible, is the production of these materials in the form of a crystalline block with a circular or other section, granules, dispersed powder, with the given phase composition, depending on application. Figure 4.2 shows the diagrams of different systems for the production of products of synthesis in the cold crucible. A crystalline oxide material in the form of a block should be used, for example, for loading into the crucible in the growth of single crystals from the melt by the Czochralski method. The blocks of aluminium oxide were used in the growth of optical leucosapphire by the Musatov method. The application of crystalline blanks makes it possible to simplify the preparatory stages of the technology of growth of single crystals and reduce the content of impurities. An important direction of the application of fused polycrystalline special purity materials are film-forming materials, compounds for spraying different optical coatings.

The induction melting of oxides in the cold crucible is a promising method for the production of polycrystalline refractory materials and cast refractory component [8]. This technology is used widely for remelting of powder oxides into a crystalline material. The

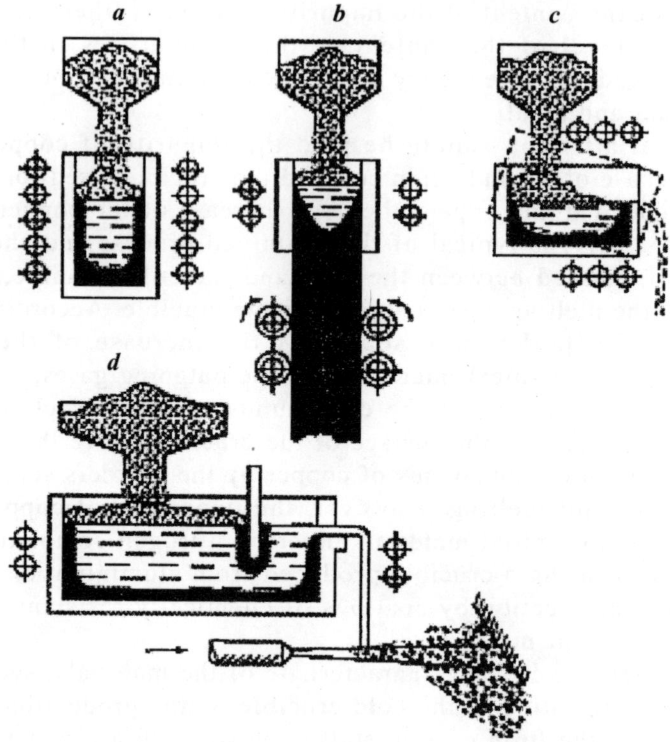

Figure 4.2. Design of cold crucibles for various applications [7]. a) production of a block of molten oxide in a cold crucible; b) continuous melting 'into a block'; c) periodic discharge of the melt from a horizontal cold crucible; d) continuous discharge of the melt from a crucible with a constant level and blowing of the jet of the melt stream for producing finely dispersed molten powder.

efficiency of remelting is determined by the high thermodynamic stability of the grains, produced in disintegration and milling of the crystalline materials, in comparison with the initial or sintered powder. The ceramics produced from the fused grains, is characterised by small shrinkage and deformation in sintering, constant properties in service, and sufficiently high heat resistance.

The application of the cold crucible for the remelting of oxides is justified by the high purity of final products which is maintained on the level of the purity of initial materials because the 'pure' induction method of heating and melting in the skull is used.

The moltem oxides are produced by the 'melting into a block' and 'melting for discharge' methods. The melting into a block method is realised by the crystallisation of the melt in the cold crucible with the close bottom or by continuous melting of a block in a continuous cold crucible with a seed crystal in the lower part of the crucible. In continuous melting of the block, the productivity of the process is higher and the specific energy losses are lower. Equipment for continuous melting is designed in the form of a load-carrying frame whose internal part contains a moving frame with a seed crystal, travelling in the vertical direction, Fig. 4.2 [9]. In the process, it is necessary to match the heating conditions, the intensity of supply of the charge and the speed of travel of the seed crystal. The matching criterion is the constant position of the level of the melt in relation to the plane of the induction coil and, consequently, the parameters of equipment are regulated independently. In the design shown in Fig. 4.2, the height of the crucible is approximately equal to (1.5–2.0) D_i, where D_i is the diameter of the induction coil; the bottom of the crucible is lowered below the induction coil to at least $D_i/2$. At this distance, the strength of the electromagnetic field is small and, consequently, the skull does not melt. Melting could result in discharge of the melt. The rate of flow of the water in the sections is in the range 1.0–1.5 m/s and is determined taking into account the need to prevent boiling in the jet cooling the wall of the section. The minimum flow speed of the water in the section is determined by the equation, V_{min}, m/s:

$$V_{min} \geq 5.5 \cdot 10^{-10} p^{-2/3} q_{max}^2, \qquad (4.1)$$

here p is the pressure of water in the cooling system, N/m^2; q_{max} is the maximum specific heat flow from the melt to water, W/m^2, which is determined by calculations on the level of the melt surface.

In order to melt the centre of the ingot, the relative diameter of the crucible m should be in the range:

$$1.0 \leq m \geq 7.0, \qquad (4.2)$$

where m is determined from the equation:

$$m = D_T / \Delta_P \sqrt{2}, \qquad (4.3)$$

Here D is the crucible diameter, m; Δ_p is the depth of penetration of current into the melt, m;

$$\Delta_P = 503\sqrt{\rho / f}, \qquad (4.4)$$

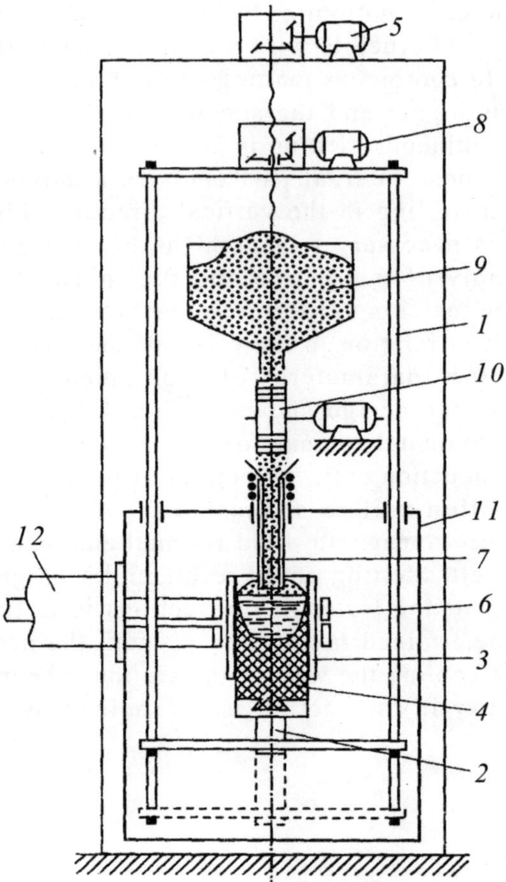

Figure 4.3. Diagram of equipment for induction melting in a cold crucible to produce blocks of molten oxides [9]. 1–7) notations in the text; 8) displacement mechanism of the seed crystals; 10) charge dosing device; 11) working chamber; 12) high-voltage cable; for the rest of the notations see the text.

where ρ is the specific electrical resistance of the melt, ohm m; f is the current frequency of the generator, Hz. If $m < 1$, the efficiency of the induction coil decreases and there is a risk of 'freezing' the melt. At $m > 7$ it is necessary to superheat the melt and this may have a detrimental effect on the thickness of the skull, with an increase of the probability of electrical breakdown between the crucible sections.

Table 4.1 shows the parameters of equipment for melting a number of oxides in cold crucibles of different diameters. To

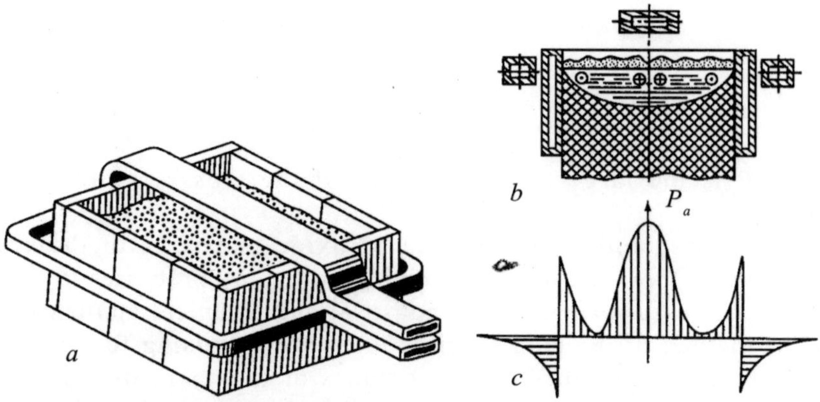

Figure 4.4. Inductor with a transverse busbar for producing a large block of molten oxide. a) crucible with an inductor; b) directions of circulation of currents in the melt; c) curve of specific power generated in the melt.

characterise the energy losses of the process, the Table gives the total specific losses of energy ΔW, determined by experiments, and the specific energy of melting ΔW_T, calculated from the equation:

$$\Delta W_{\mathrm{T}} = (H_m - H_{293} + \Delta H)/3600M, \qquad (4.5)$$

Here H_m is the enthalpy of the oxide at the melting point, kJ/mol; ΔH is the specific heat of melting, kJ/mol; M is the molecular mass of the oxide, kJ/k·mol. The ratio $\Delta W_T/\Delta W$ corresponds to the total efficiency of equipment which depends on the nature of the oxide and the melting conditions. The Table shows that the total efficiency is minimum in the melting of oxides of rare-earth elements and is maximum for corundum. The total efficiency in melting in arc furnaces is higher in comparison with induction heating but the factors determining the quality of material make the latter method more attractive.

To increase the transverse dimensions of the ingot and melt the centre, it is recommended to use an induction coil with a transverse bus shown in Fig. 4.4. The presence of the induction lead (transverse bus) above the melt results in superheating of the melt in the centre of the crucible. This results in melting of the centre of the ingot and intensifies heat transfer from the melt to the charge, thus increasing the rate of the process. For the induction coil with the transverse bus, the ratio (4.2) is retained, and expression (4.3) transforms to the following form:

$$m = B / 2\Delta_p \sqrt{2}, \qquad (4.6)$$

where B is the large transverse size of the cross-section of the crucible, m.

Thus, using induction melting of oxide materials in the cold crucible, it is possible to realise high productivity processes in order to produce fused materials for industrial and technical applications.

The method of the cold crucible is highly promising for methods (Fig. 4.4c, d) using different techniques of treatment of the jet of the melt which make it possible to produce granules, disperse powders, fibres from high purity oxide materials and also quench the metastable phases at a high cooling rate of the melt. The application of individual oxides and compounds in the form of related materials makes it possible to improve the efficiency of the processes of growth of crystals and casting of glasses, and the purity of the products increases.

The processes of production of high purity crystalline materials may be realised at present on the experimental or industrial scale. Information on equipment for these purposes is presented in Chapter 2.

4.2. PRODUCTION OF MOLTEN MULLITE

In this and subsequent sections, specific examples of the melt synthesis of materials in the cold crucible will be discussed. Because of the development of new metallurgical procedures, for example, vacuum treatment of steel, there has been special interest in high-alumina materials of the Al_2O_3–SiO_2 system, in particular, mullite, usually produced by electric melting [10]. The mullite, in accordance with the equilibrium diagram, forms phases with a variable composition [11]. In the synthesis of mullite by the Verneuil method, investigators produced single crystals with the composition $2Al_2O_3 \cdot SiO_3$, confirming the congruent nature of melting [12]. The mullite with the composition $3Al_2O_3 \cdot 2SiO_2$ melts without dissociation, but this compound could not be produced by melt synthesis in air [13, 14]. Equipment for melting mullite in air is shown in Fig. 4.5 [15]. The process is characterised by successive crystallisation of the block (*1*) in local melting of the powder charge (*2*) on its upper surface inside the crucible (*3*) situated inside the

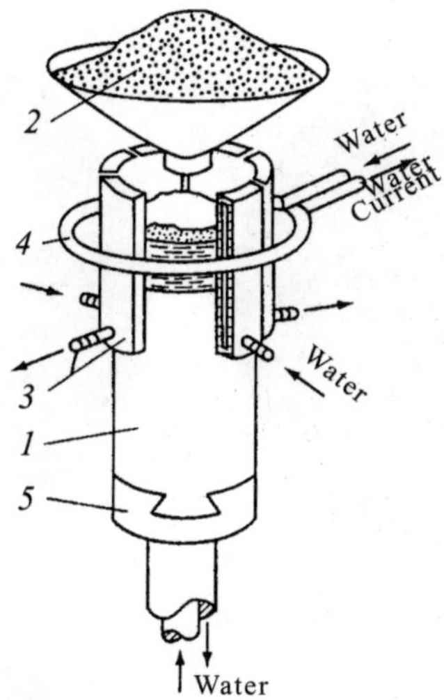

Figure 4.5. Schematic of the equipment for producing mullite by induction melting in a cold crucible [15].

high-frequency induction coil (*4*). When using a continuous crucible and a moving baseplate (*5*), it is possible to produce blocks whose length is determined by the design features of equipment.

In consecutive growth of the crystalline block it is possible to ensure the conditions in which the given ratio of the oxides and the degree of purity of the initial material is maintained. These conditions include a small volume of the melt (20–100 ml), short duration of holding material in the liquid phase (30–100 s), uniform heating of the melt and small superheating above the melting point (not more than by 50–100°C). The rate of cooling and crystallisation of the melt is 3–5°C/s and depends on the volume of the melt and the distance from the cold wall of the crucible. The speed of movement of the solidification front found is not greater than 5 mm/min and is similar to the quenching conditions. The results of microscopic analysis show that the main phase consists of mullite and only the samples with a high content of aluminium oxide contain corundum. The characteristic specimens of fused mullite are shown in Fig. 4.6. The specimens contained grains of efficiently solidified mullite, separated by thin glass layers. The amount of the glass phase is small and is determined by the ratio of the oxides of aluminium and silicon in the initial charge. The chemical composition of the synthesised mullite is presented in Table 4.3 which shows that the method of melting in the cold crucible makes it possible to

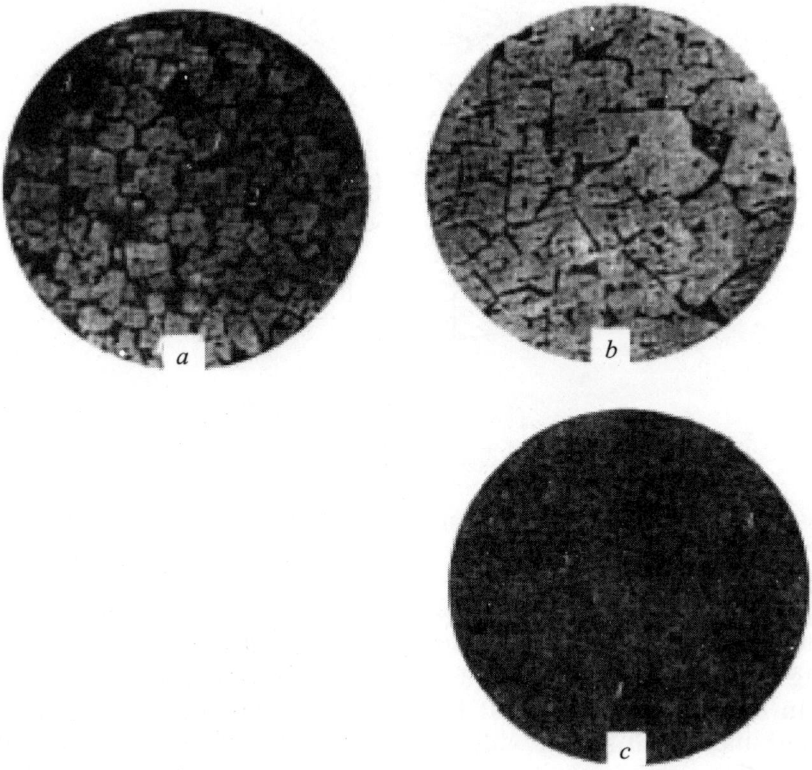

Figure 4.6. Photographs of sections of molten materials of the Al_2O_3–SiO_2 with the Al_2O_3 content of 60.5% (a), 64.9% (b), 70.4% (c): 1) mullite; 2) glass; 3) pores, ×110 [15].

produce compounds covering as regards composition a wide range of the solid solutions from mullite $3Al_2O_3 \cdot 2SiO_2$ to pragite $2Al_2O_3 \cdot SiO_2$. The comparatively high content of the glass phase indicates the nonequilibrium conditions of solidification of the melt and crystallisation of mullite. The parameters of the elementary cells of the crystal lattice of the solid solutions of mullites are presented in Table 4.4. The x-ray diffraction patterns of different zones of the fused mullite block are shown in Fig. 4.7.

It is also important to mention another important aspect of the method of successive crystallisation in the single ingot for the examination of the phase composition of the products of synthesis of complex oxide systems in the subliquidus region which is difficult to examine. As a result of melting charges of different composition

Table 4.3. Chemical composition of molten mullite

Specimen No.	Al_2O_3 content in sample, %	Amount of insoluble residue, %	Al_2O_3 content in soluble residue, %	Composition of mullite κ*, according to the results of analysis	
				chemical	X-ray**
1	not determined	77.3	73.0	1.59	1.49(2)
2	60.5	77.1	73.8	1.66	1.69(3)
3	63.8	81.9	74.6	1.73	1.71(3)
4	67.9	86.4	75.4	1.82	1.75(2)
5	64.9	83.9	75.6	1.83	1.82(3)
6	69.4	88.8	75.3	1.80	1.88(4)
7	70.4	88.7	76.4	1.92	1.92(2)
8	73.6	91.6	77.8	2.06	2.09(3)***
9	75.9	93.3	78.6	not determined	2.12(4)***
10	not determined	94.7	not determined		2.04(4)***
11	As above	95.2	As above		not determined

*κ – Al_2O_3 : SiO_2 ratio in mullite, i.e. its basicity
** Values in brackets are random errors corresponding to the fiducial probability of 0.9
*** Samples contained corundum: sample No. 8 < 1%, No. 9 and 10 ≥ 5%

Table 4.4. Parameters of the crystal lattice of molten mullite

Mullite composition,	Lattice parameters, Å					
	a	Δ *a*	*b*	Δ *b*	*c*	Δ *c*
1.49	7.5705	0.0028	7.7038	0.0015	2.8899	0.0006
1.69	7.5648	0.0036	7.7014	0.0018	2.8913	0.0002
1.71	7.5645	0.0032	7.6975	0.0015	2.8897	0.0012
1.75	7.5596	0.0028	7.6975	0.0013	2.8927	0.0016
1.82	7.5541	0.0018	7.7032	0.0012	2.8927	0.0016
1.88	7.5540	0.0008	7.7016	0.0010	2.8906	0.0004
1.92	7.5634	0.0078	7.6958	0.0006	2.8886	0.0020
2.09	7.6038	0.0013	7.6927	0.0006	2.8929	0.0006
2.12	7.5997	0.0012	7.6926	0.0012	2.8925	0.0003
2.04	7.5987	0.0035	7.6957	0.0014	2.8942	0.0005

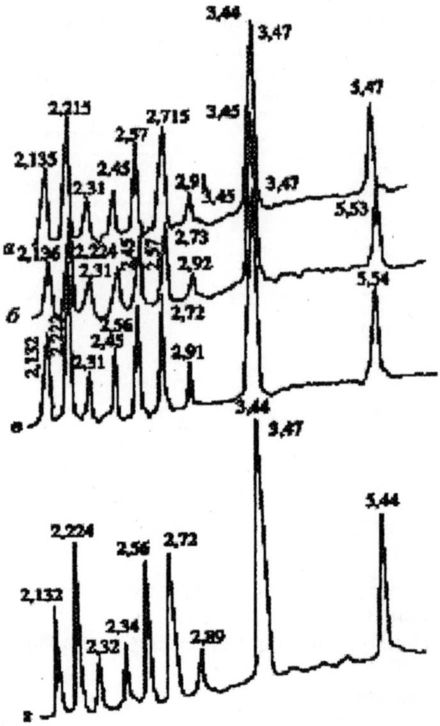

Figure 4.7. X-ray diffraction patterns of different zones of a molten mullite block [15]. a) upper, b) central, d) lower, d) peripheral zones.

in the same ingot, it is possible to ensure all theoretically possible ranges of concentration, including systems with volatile components, as shown on the example of the Al_2O_3–SiO_2 system. The proposed methods can be used efficiently for constructing equilibrium diagrams of the physical–chemical systems used partially in the synthesis of high-temperature glass (Chapter 6).

4.3. POLYCRYSTALLINE ELECTRICALLY CONDUCTING CHROMITES OF YTTRIUM AND LANTHANUM

An outstanding examp le of the efficient application of the cold crucible as a chemical reactor is a study carried out by a group of scientists at the Institute of General Physics and the Institute of High Temperatures of the Russian Academy of Sciences in the search for the optimum composition of electrically conducting lanthanum chromite for the electrodes of powerful MHG generators and heaters of high-temperature furnaces. The cold crucible method makes it possible to synthesised quite rapidly tens or hundreds of chemical compounds and investigate their physical–chemical properties and, consequently, save resources in comparison with the application of traditional solid-phase synthesis.

The currently known refractory oxides of 17 elements of the periodic table of elements form 680 binary compounds which have been investigated sufficiently. Further increase in the complexity of the system as a result of increasing the number of components is evidently irrational because already in the four-component systems the melting point decreases in accordance with the equation

$$T_{max} \approx \mathrm{A} - \mathrm{B} \cdot n \text{ for } n \leq 5, \tag{4.7}$$

where A = 5100, B = 700.

The materials, produced by the new technology, are characterised by significant advantages in comparison with the materials used until recently for the fabrication of electrodes of MHG generators. Firstly, these materials are characterised by high chemical purity and, secondly, they are thermodynamically equilibrium at high temperatures because the processes of melt synthesis take place to a high degree at temperatures considerably higher than the working temperature in the channel of the MHG generator. Since the material was synthesised in air, it is in equilibrium with the oxidation atmosphere and retains the stability of its properties in the plasma flow characterised by the oxidation properties.

The most promising materials for the development of new electrodes of MHG generators are polyphase composite materials based on high-temperature oxides. Using these materials, it is possible to improve greatly the physical and physical–chemical properties of the electrodes. This method is promising and the following results were obtained using this method.

The new components for 'construction' of electrodes were represented by the following materials:

1. The single crystals of stabilised zirconia.
2. Lanthanum chromite with an additives of the oxides of alkali-earth metals in the concentrations from 2 to 15 mol%
3. Yttrium chromite, pure or with an addition of alkali-earth elements (2–15 mol%).

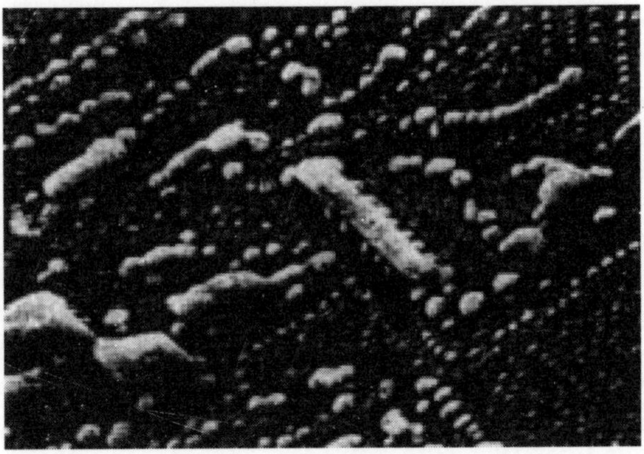

Figure 4.8. Electron micrograph of a $LaCrO_3$ ingot with an addition of 7 mol% of calcium chromite (for explanation see the text) [5].

4. Metallic chromium in the form of a powder.

The production and properties of single crystals of stabilised zirconia will be described in chapter 7. The synthesis conditions and a number of physical–chemical properties of the lanthanum chromites have been presented in [6]. Each of these substances has specific physical–chemical properties. For example, zirconia is characterised by the highest melting point (2750°C) and relatively high electrical conductivity at temperatures above 1200°C. The single crystals of cubic zirconia are characterised by high hardness (7.5–8.5 on the Moss scale) and strength which they retain at temperatures above 2000°C. The melted chromites of lanthanum and yttrium are characterised by high electrical conductivity already at room temperature and the ceramic materials produced from them have high thermal stability. In addition to this, the yttrium oxide which may be present in yttrium chromite after synthesis, does not interact with water and water vapours at high temperatures. Metallic chromium is also characterised by high electronic conductivity. The positive properties of this material include the relatively low capacity for oxidation as regards metals. The combination of these properties of the materials made it possible, using the cold crucible method, to develop electrodes for MHD generators and high-temperature furnaces. Figure 4.8 shows the electron micrograph of a cleavage of an ingot of lanthanum chromite doped with of 7 mol% of calcium chromite. The photograph shows clearly the crystalline grains of the solid solution of lanthanum, calcium chromite and the droplet of the eutectic

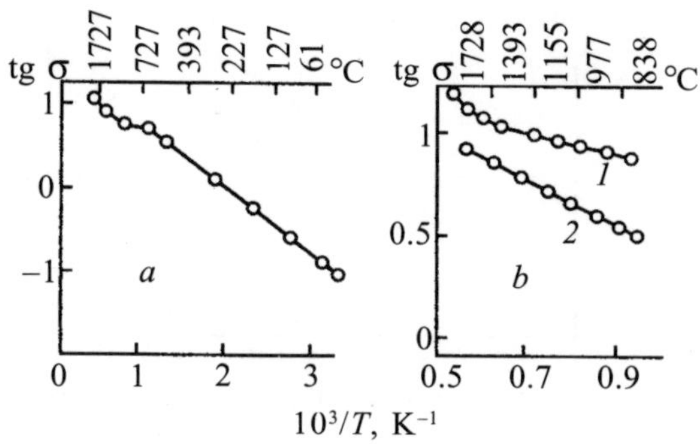

Figure 4.9. Temperature dependence of the electrical conductivity of lanthanum chromite with the addition to 7 mol% of calcium chromite [5].

composition, enriched with calcium oxide displaced to the boundaries between the crystals. The ingot of the chromites was refined and the produced powder was used for production of ceramic materials by ceramic technology methods. Figure 4.9 shows the temperature dependences of the specific electrical conductivity of these specimens. This material has high electrical conductivity already at room temperature. The special feature of the ceramics produced from lanthanum chromite is the strong dependence of electrical conductivity on the oxidation–reduction properties of the atmosphere. The high electrical conductivity in the temperature range from room temperature to 1000°C is obtained only after heating the specimen in the oxidation atmosphere. Heating in a redusing or neutral atmosphere, or in vacuum, greatly increases the specific resistance of ceramics in the low temperature range, as indicated by Fig. 4.9. Repeated heating in the oxidation atmosphere completely restores the initial electrical conductivity. X-ray diffraction examination of molten lanthanum chromite indicates that the metal subjected to reducing annealing contains a single phase $LaCrO_3$. After heating in the oxidation atmosphere, new lines, indicating the formation of the impurity of the new phase, appear. Thus, it may be concluded that electronic electrical conductivity below 1000°C is associated with the formation in the material of the impurities in which the valency of chromium is higher than 3. The fused chromites are efficiently combined with other metallic and nonmetallic phases, forming a new series of polyphase materials [14].

One of the main special features of the heaters produced from lanthanum chromite, modified with oxides of rare-earth elements,

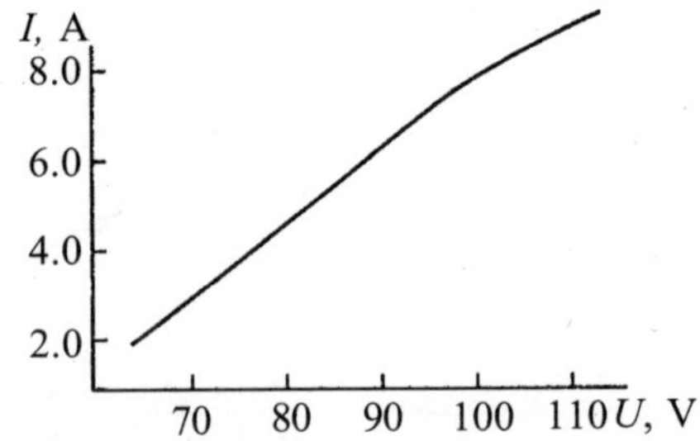

Figure 4.10. Volt–ampere characteristic of a lanthanum chromite heater [17].

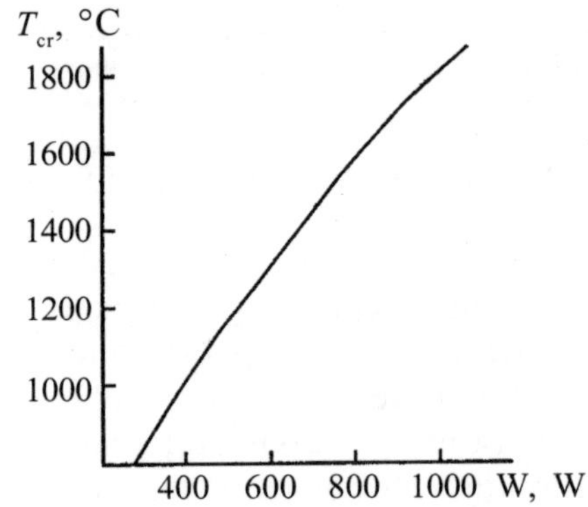

Figure 4.11. Dependence of the brightness temperature of the active part of a heter on the required power [17].

is the relatively low specific electrical resistance at room temperature (≈10 ohm·cm) and, consequently of these, heaters can be used starting from 20°C. The high resistance of chromite heaters results in stable operation not only in the continuous but also quasi-periodic regime at high heating and cooling rates [17]. The high thermal resistance of the electrodes is the result of application of porous ceramics (approximately 25%) with the granular structure. This type of ceramics is characterised by high strength, low elasticity modulus and, consequently, is used for the production of heaters. Figures 4.10 and 4.11 show respectively the volt–ampere characteristic and the dependence of brightness temperature on the required power of the heaters produced from lanthanum chromite illustrating the high physical parameters of the electrodes produced from lanthanum chromite.

The furnaces with heaters produced from lanthanum chromite require efficient high-temperature thermal insulation. Disruption of thermal insulation results in overheating and failure of the heaters.

Table 4.5. Main parameters of a lanthanum chromite heater

Parameter	Numerical value
Total length, mm	from 350 to 700
Length of active part, mm	to 250
Diameter, mm	from 16 to 35
Scatter of resistivity, %	± 20
Ratio of resistance of leadouts and active part	1:5
Total porosity, %	~25
Coefficient of linear thermal expansion (20-1000°C)$\cdot 10^{-6}$ /deg	8.0
Permissible working temperature on internal surface of heater, °C	1850
Permissible working temperature in furnace, °C	1750
Change of resistivity in the first 600 h of operation, %	≤ 15

The heat-insulating materials in the furnaces are represented by layers of dense corundum, low-dense alumina, or caolin cotton based material. The main parameters of heaters produced from lanthanum chromite are presented in Table 4.5.

References

1. Foex M., *Rev. Haute temp. Refract.*, 1966, **3**, 309.
2. Cohn W., Braught G., *J. Opt. Soc. Amer.*, 1954, **44**, 45.
3. Foex M., Third Colloque National du Centre National de la Recherche Scientifique Sur la Chimie des Hautes Temperature, Paris, 1963, 117.
4. Ol'shanskii L.Ya., Pryanishnikov V.P., Chemical electro-chemistry, Moscow and Leningrad, Goskhimizdat, 1962.
5. Aleksandrov V.I., et al., *Usp. Khim. Nauk*, 1978, **10**, No. 3,385–427.
6. Aleksandrov V.I., et al., *Izv. Akad Nauk SSSR, Neorgan. Mater.*, 1973, **9**, 236.
7. Petrov Yu.B., et al., *Vysokochistye Veshchestva,* 1989, No. 3, 136.
8. Ganyuchenko V.M., et al., *Steklo i Keramika*, 1975, No. 9.
9. Petrov Yu.B., et al., *Ogneupory*, 1980, No. 10, 41–46.
10. Cichy P., *Interceram.*, 1974, No. 4, 1975, No. 1.
11. Toropov N.A., et al., Equilibrium diagrams of silicate systems, Moscow and Leningrad, 1965.
12. Barta R., Barta Ch., *Zh. Prikl. Khimii*, 1956, **29**, No. 3.

13. Petrov Yu.B., Ratnikov D.G., Cold crucibles, Moscow, 1972.
14. Aleksandrov V.I., et al., *Vestn. AN SSSR*, 1973, No. 2, 29.
15. Sokolov A.N., et al., *Ogneupory*, 1979, No. 9, 35–39.
16. Aleksandrov V.I.,et al., Synthesis and crystal growth of refractory materials by RF melting in a cold container, Curr. Top. Mater. Sci., Amsterdam, 421–480, 1978.
17. Balkovich V.L., et al., *Elektron. Prom-st'*, 1981, No. 7/8, 103–104.

5

Growth of single crystals

5.1. GROWING CORUNDUM, RUBY AND OTHER CRYSTALS, USING A SEED (THE CZOCHRALSKI METHOD)

To grow single crystals by drawing using a seed crystal, by the Czochralski method, it is necessary to:

1. Produce a strictly determined distribution of temperature in the melt and at the solidification front, in particular in the central part of the melt surface the temperature should be lower than at any other point of the melt;

2. Ensure high stability of temperature at the solidification front. The application of the Czochralski method for the growth of crystals from a cold crucible is of considerable scientific and applied interest. However, in this case, there are difficulties of the principal nature, caused by specific features of direct melting using high-frequency current in a cold crucible, which include: 1. The need for superheating the melt to ensure its stable maintenance in the crucible; 2. The regulation of melt temperature to compensate phase instability in direct high-frequency heating; 3. The formation of radial temperature gradients on the surface and in the volume of the melt; 4. The existence of turbulent convection flows in the melt due to steep temperature gradients in the volume of the melt and the presence of internal heat generation sources.

The previously presented isotherms (Fig. 2.6 and 2.7) provide information on the nature of temperature distribution in the melt and show that the change of the position of the induction coil in relation to the surface of the melt and the variation of the power supplied to the melt have strong effects on the temperature gradients in the melt. The factors affecting the formation of temperature gradients in the melt also include the following technological parameters: the dimensions of the crucible, the working frequency of the generator, the presence of thermal screens or additional heat sources, the configuration of the induction coil, and other.

The efficiency of a specific method of the formation of thermal fields in the melt is determined mainly by the physical characteristics of the melt and the growing crystal. Therefore, there is no universal recipe for obtaining the required temperature distribution in the melt and for every specific case it is necessary to select appropriate technological conditions.

The formation of the required radial temperature gradient and the stabilization of convective flows in the melt are achieved by increasing the generator frequency and the diameter of the melt surface. Figure 6.3 shows the temperature distribution on the surface of a Al_2O_3 melt in a crucible with a diameter of 80 mm at a frequency of the field of 5.28 MHz. Such radial temperature distribution corresponds to the condition necessary to grow crystals by the Czochralski method. However, even in this case, the fluctuations of the convective flows with time remained significant. The majority of produced melts are characterised by low viscosity and only some of them, which include glass-forming components in their composition, such as SiO_2, are viscous. The absence of the convective flows on the surface of the melt is associated with high melt viscosity.

Using a number of examples, special attention will be given to the application of different technological procedures in the growth of single crystals from a cold crucible using a seed crystal. For example, the authors of [1,2] described the growth of dislocation-free silicon crystals from a cold crucible. The optimum thermal conditions were achieved by selecting the size of the crucible, the frequency of the generator, and the position of the inductor in relation to the surface of the melt. It should be mentioned that as a result of the electromagnetic interaction of the silicon melt with the high frequency field, the melt is displaced from the walls of the crucible and the surface of the melt becomes sferical. According to the authors of [3], the growth of silicon by this method is highly promising because it is possible to obtain a high degree of purity in comparison with the crystals produced by the Czochralski method from a conventional crucible and a higher uniformity in comparison with the crystals produced by zone melting.

One of the simplest methods of the formation of temperature

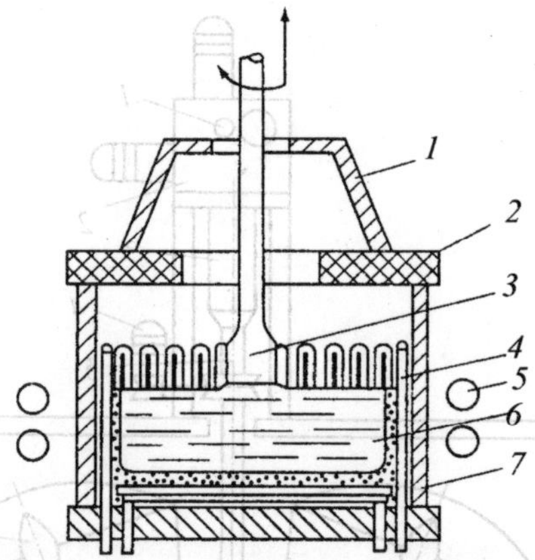

Figure 5.1. Schematic of a cold crucible for growing $SrTiO_3$ crystals [4,5]/ 1,2) heat screens; 3) growing crystal; 4) cold crucible; 5) inductor; 6) melt; 7) quartz pipe.

gradients on the surface of the melt is the application of thermal screens which were used successfully in [4,5] in the growth of crystals of strontium titanate. The diagram of the thermal section for the growth of crystals is shown in Fig. 5.1. The cold crucible is accurately centered in relation to the induction coil and the thermal screens are positioned in such a manner that the thermal convection flows on the surface of the melt converge in the centre, forming a melt spot with a slightly reduced temperature. The melt temperature is regulated by the variation of the power supplied to the melt by the high-frequency generator. The stability of generator power is ensured by stabilising the anodic current of the generator lamp.

A further development of the method of growth of crystals from the cold crucible by the Czochralski method is the combination of direct high-frequency melting with radiation heating of the surface of the material [6,7]. The diagram of equipment for this process is shown in Fig. 5.2. Powerful optical concentrators with arc xenon lamps as light sources are used for starting heating. Equipment provides for vertical displacement and rotation of the crucible. The presence of the light sources makes it possible to control the configuration of the thermal fields in the melt by varying the distance of the focal spot from the axis of rotation of the crucible and by regulating of the power of the light flux. In order to reduce the temperature gradients in the growing crystals, and also for subsequent annealing, a platinum electrical resistance furnace and

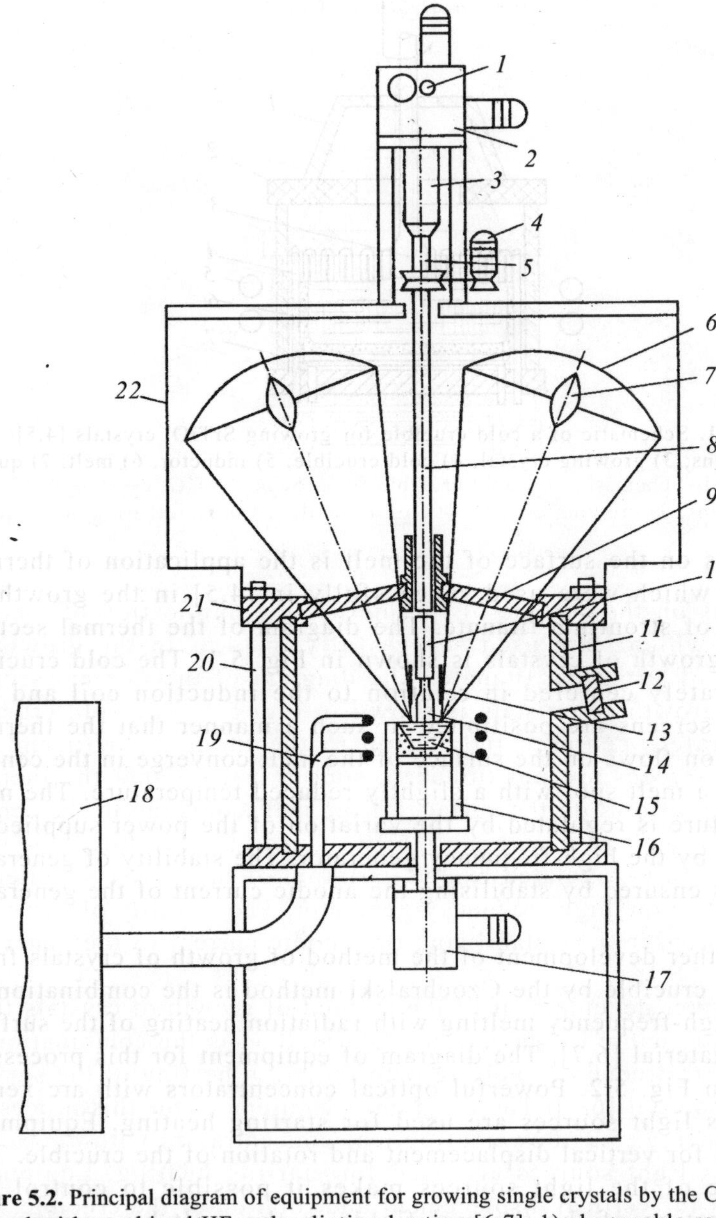

Figure 5.2. Principal diagram of equipment for growing single crystals by the Czochralski method with combined HF and radiation heating [6,7]. 1) changeable reductor; 2) drive of the upper rod; 3) coupling; 4) drive for rotating the upper rod; 5) magnetic coupling; 6) elliptical reflector; 7) source of light radiation; 8) upper rod; 9) quartz glass; 10) seed holder; 11) chamber; 12) quartz window; 13) annealing furnace; 14) crystal; 15) melt; 16) cold crucible; 17) drive for the lower rod; 18) HF oscillator; 19) inductor; 20) pin; 21) chamber lid; 22) jacket.

heat screens made of aluminium oxide are placed above the melt.

In the investigated system, the presence of radiation heat sources at the effective ratio of the powers of the high-frequency fields and light radiation makes it possible to produce thermal fields in the melt for stable growth of the crystals on a seed. The experimental results show that convection becomes the dominant mechanism of heat transfer in the melt. Changing the direction of movement of the melt flows, it is possible to control the configuration of the thermal fields and the shape of the solidification front. However, it should be mentioned that the control of the heat flows in order to obtain the required structure of the thermal fields remains the most complicated process and further work is required to solve this problem.

One of the first crystals grown successfully by drawing on a seed crystal from a cold crucible were leucosapphire and ruby crystals [8]. The process of growth of these crystals from the cold crucible is controlled by: 1. Relatively low electrical conductivity of the Al_2O_3 melt; 2. The low viscosity of the melt; 3. The high solubility of gaseous oxygen in the melt, and 4. High heat conductivity of solid corundum. The first two features lead to the formation of a fluctuating convection pattern, associated with the variation of temperature at the solidification front. The third specific feature leads to the release of oxygen at the liquid–solid interface of the oxygen-saturated melt. Hence, even the smallest fluctuation of temperature, small oxygen bubbles are trapped by the growing crystals and form light-scattering inclusions.

The high heat conductivity of corundum, on the one hand, is favourable for supports stable growth of the crystal because it ensures rapid removal of heat and prevents tangential growth. On the other hand, rapid removal of heat from the melt through the skull`to the walls of the cold crucible resulting in the spatial instability of the interface between the melt and the skull and requires extensive superheating of the melt. Therefore, to ensure stable long-term growth of the sapphire and ruby crystals, it is necessary to use crucibles and induction coils shown in Fig. 3.2b. This configuration of the crucible results in uniform heating of the melt not only from the side but also from the bottom. Figure 5.3 shows the photograph of a sapphire crystal growing from the cold crucible. The diameter of the crystal was regulated by changing the position of moving screens, situated above the melt. The rate of withdrawal was 10–30 mm/h, the speed of rotation was varied in

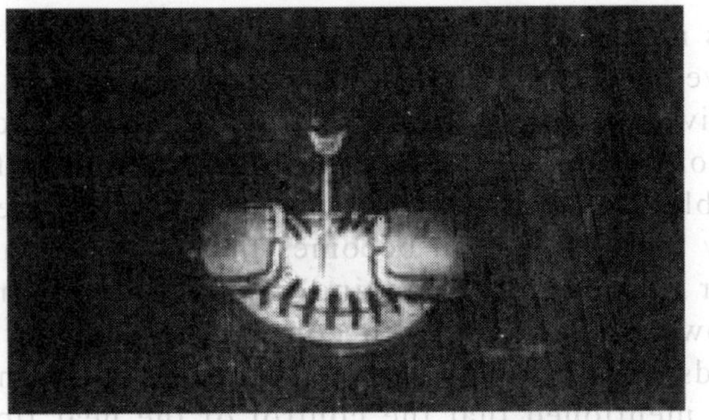

Figure 5.3. Photograph of a growing sapphire crystal [8].

the range 20–140 rpm. In cases in which the convection pattern greatly fluctuated, it is necessary to employ the planetary rotation of the crystals. In this type of rotation, the crystals intersect a large number of areas with different temperatures which are averaged-out at the solidification front. In this case, there is no capture of the oxygen bubbles. In these crystals, the solidification front is almost completely flat.

This method was used to produce single crystals of leucosapphire and ruby with different orientations, up to 160 mm long, with a diameter of up to 35 mm. The grown crystals were characterised by special features determined by the features of preparation.

The density of dislocations in the basal plane was 10^2–10^3 cm^{-2}. The flat solidification front resulted in the uniform radial distribution of chromium atoms in the ruby crystals. However, the distribution of chromium along the length of the crystals was non-uniform, the chromium concentration decreased during drawing. This was associated with the evaporation of chromium oxide from the overheated melt in superheating. The concentration of chromium along the length of the crystal was equalised using the programmed 'saturation' of the melt with chromium oxide powder from a special hopper.

5.2. DIRECTIONAL SOLIDIFICATION OF THE MELT IN A COLD CRUCIBLE

In most cases, single crystals are produced from the cold crucible by the directional solidification of the melt. This preference is associated with the relatively simple procedure of the method because it does not require an additional heating source and complicated design features of the thermal section.

As already mentioned, the method is based on the production, by means of direct high frequency heating, of a relatively large volume of the melt in a cold crucible with subsequent solidification at a specific rate. Crystallisation can be initiated most efficiently by lowering the crucible with the melt in relation to the induction coil. With exit of the lower part of the melt from the induction coil, the efficiency of coupling of this part with the high-frequency field decreases and, consequently, less power is supplied to the melt. The crystallisation process takes place as a result of gradual cooling of the bottom part of the melt. Less frequently, the melt is cooled by a programmed decrease of the generator power. However, the latter method is less reliable because it may result in the formation of a spatial instability and uncontrolled solidification of the melt. Consequently, the solidification method is only suitable for melts characterised by high stability in relation to the variation of the temperature of the melt which is determined by its physical–chemical properties [9].

In comparison with other methods of preparation of refractory single crystals (Verneuil method, zone melting), the method of directional solidification in the cold crucible has a number of advantages. The large volume of the melt facilitates convective mixing of the matter and impurities during growth. Crystals can be grown in any atmosphere. In addition to this, the open surface of the melt makes it possible to regulate the composition of the melt by adding components and also supports the removal of highly volatile impurities and gases from the melt.

However, the method has also shortcomings. The most important shortcoming is that it is not possible to examine visually and control the crystallisation process by measuring the temperature of the melt and the crystal directly during growth. The possibility of regulating the melt temperature in a relatively wide range by regulation of the power supplied to the melt is relatively limited because it is determined by the conditions of equilibrium and stability of the phase boundary [8,10]. In addition to this, it is not possible to anneal and cool to room temperature the crystal immediately after growth without using additional heating devices.

The feasibility of using the method for the growth of single crystals of refractory oxides is determined by the physical–chemical properties of the materials in the melt and the solid phase. In particular, this relates to the materials whose high melting point and corrosive effect do not make it possible to select a suitable crucible material.

The possibility of application of the examined method also depends on the electrical and thermophysical properties of the material. They determine the very possibility of the formation and stable maintenance of the melt, as already mentioned in Chapters 1 and 2. For example, to ensure a stable process, a low conductivity of the solid phase of the material is desirable. High rates of evaporation of the components of the melt may become a reason for a decrease in the volume of the melt and this may also disrupt the stability of the system and of the crystallisation process. The volatility of the individual components also disrupts the stoichiometry of the grown crystals. In this case, it is necessary to correct the composition of the melt or use an atmosphere preventing evaporation.

In most cases, a cold crucible consists of several sections, separated by small gaps. When the skull is melted through the crucible melt may be poured out thus disrupting the process of growth of the crystals. Pouring out of the melt depends on the surface tension of the melt, the width of the gap between the sections and the forces of molecular bonding of the melt with the material of the crucible wall. Consequently, it is necessary to use melts with a relatively high viscosity.

The directional solidification of the melt in the cold crucible greatly differs from the growth of crystals in a conventionally heated crucible, for example, by the Bridgman method. Firstly, this is associated with the existence of a skull layer on the water-cooled sections of the crucible with which the melt is in contact; secondly, it is linked with the specific distribution of temperature inside the cold crucible which is determined by the absorption of the HF power directly in the melt and the directional heat removal from the solidification front to the cold walls of the crucible.

These special features will now be discussed in greater detail. As a result of directional crystallisation of the melt in a cold crucible without the application of special seed crystals, one obtains ingots consisting of a large number of crystals. The design of the cold crucible does not restrict the amount of nuclei formed. However, during growth, the number of these crystals decreases as a result of selection on the basis of the growth rate. Competition of this type results in the survival of crystals in which the direction of 'rapid' growth is close to the normal to the solidification front. In this case, the crystals in the ingot are elongated and distributed parallel to each other, forming a columnar structure. Attempts have been made to use seed crystals. However, difficulties in controlling the melting of the charge and the absence of any means of checking that the 'seed' has not melted, make this process unreliable. Thus, the number and, correspondingly, size of the crystals depend on the composition of the melt and growth conditions.

The growth of large and high-quality crystals is based on solving problems of controlling temperature and its gradients in the melt and the solid phase. Therefore, the distribution of temperature fields in the cold crucible is extremely important. However, as already mentioned, high temperatures (>2000°C) and the high-frequency field prevent direct measurement of the melt temperature by contact methods. Only the radiation method can be used in the present case, i.e. in the growth process only the temperature of the melt surface can be measured. The qualitative pattern of the radial distribution of temperature in a viscous glass melt was indicated in [11]. In [12], temperature fields in an oxide melt upon direct high-frequency heating were experimentally studied, and the effect of several technological factors on the temperature distribution was also investigated. This melt is characterised by high viscosity and relatively low-temperature (≈1500°C) and the results of the experiments provide a characteristic pattern of the temperature field in the melt in which the energy of the high-frequency field is directly converted into thermal energy.

Figure 2.7 shows the distribution of temperature in a melt for different positions of the induction coil in relation to the surface of the melt and the variation of the absorbed power. It may be seen that by changing these parameters it is possible to form efficiently the temperature fields in the volume and on the surface of the melt. The nature of the temperature distribution, shown in Fig. 2.7, is in agreement with the results published in [13] in which the conditions of crystallisation are linked with the distribution of the components of the ZrO_2–Y_3O_3 solid solution and defects affecting the optical homogeneity of these single crystals. The pattern of the temperature distribution in the melt (Fig. 2.7) explains qualitatively the mechanism of formation of these defects and is in agreement with the experiments carried out on a modelling melt [12].

Very high melt temperatures in the growth of the crystals by the method of the cold crucible result in large thermal losses from the surface of the melt. Intensive radiation heat removal from the surface of the melt results in the formation of a layer of the sintered charge whose presence has a positive effect decreasing the heat losses and increasing the stability of the solidification process [14, 15]. However, when correcting the composition of the melt in the growth process, the sintered layer prevents the pouring of the charge on the surface of the melt. Mechanical failure of this layer reduces the stability of the equilibrium of the melt–crystal system and has a negative effect on crystal growth. A screen-reflector has been proposed for reducing this undesirable effect, Fig. 3.2, which prevents the formation of a 'roof'' above the melt surface and which reduces the heat losses due to radiation of the melt.

The method of directional solidification has been used for growing a large number of single crystals of simple and complicated oxides and also solid solutions based on them (see Table 5.1). The most successful application of this method is the synthesis of a new class of single crystals – fianites (cubic zirconia). The technology and properties of this material are described in detail in Chapter 7. The first reports on the production of the single crystals of zirconia with a size of several millimetres using the cold crucible methods were published in [16,17]. At the same time, Russian investigators produced large single crystals of different compositions with a number of applications in practice. This technology is used extensively throughout the world and has stimulated the application of the method of direct high-frequency melting for the growth of other single crystals.

In addition to the crystals based on ZrO_2 and HfO_2, this method has been used successfully for the production of large crystals of oxides of transition elements [18–20] and for growing a large number of complex oxides [8, 10, 20–23]. Investigations of the oxides of the transition elements show that, at high temperatures, many of these elements are characterised by the electrical and thermophysical properties that permit efficient heating by high-frequency current, melting, maintenance in the molten condition and solidification of the melt in the cold crucible. The cold crucible method has been used for growing, in the air atmosphere, crystals of CoO, Fe_2O_3 and TiO_2 with a volume of up to 3 cm^3 [19]. The colour of produced TiO_2 crystals indicates a disruption of stoichiometry with respect to oxygen which, however, is completely restored by annealing at temperatures of 1000–1300°C in an oxygen atmosphere.

Table 5.1. Single crystals grown from a cold crucible

Material	T_m,°C	Methods and growth parameters*	Crystal dimensions
Al_2O_3	2046	Czochralski method, crucible with heated bottom D_c = 80 mm, V_g = 10÷30 mm/h V_r = 20÷140 rpm Directional solidification	Diameter d = 10÷35 mm length l = 160 mm Several millimetres
TiO_2	1855	Directional solidification W = 60 kW, f = 5.28 MHz D_T = 50÷80 mm, air	Volume 1-3 cm³
Fe_3O_4	1560	Directional solidification W = 60 kW, f = 5.28 MHz D_T = 50÷80 mm, air W = 20 kW, f = 3 MHz D_T = 60÷85 mm, CO/CO_2 (flow-through)	Volume 1-3 cm³ Columnar crystals l = 2 mm
$M_{1-x}O$ (M = Mn, Co, Fe, Ni)	~2000	Directional solidification W = 20 kW, f = 3 MHz, D_T = 60÷85 mm, CO/CO_2 (flow-through), V_p = 7.5÷15 mm/h	Columnar crystals l = 1 cm
CoO	1805	Directional solidification W = 60 kW, f = 5.28 MHz, D_T = 50÷80 mm, air	Volume 1-3 cm³
CeO_2	2700	Directional solidification W = 60 kW, f = 5.28 MHz, D_T = 50÷80 mm, air	Volume 1-3 cm³
Pr_2O_3	2250	As above	Plate-shaped crystals
Nd_2O_3	2300	As above	l = 1.5 mm
Sc_2O_3	2405	Directional solidification	Several millimetres
Y_2O_3	2410	Czochralski method	Volume 1-3 cm³
Cd_2O_3	2395	As above	As above
Er_2O_3	2400	As above	"
$ZrO_2(HfO_2)$-(8-50%) MgO, CaO, SrO, Sc_2O_3,	-	Directional solidification W = 20 ÷ 160 kW, f = 0.44; 1.76; 5.28 MHz, air	Mass < 2 kg

Material	T_m,°C	Methods and growth parameters*	Crystal dimensions
Y_2O_3, Ln_2O_3		$V_p = 1\div40$ mm/h	
Fe_3O_4- Fe_2TiO_4	-	Directional solidification $W = 20$ kW, $f = 3$ MHz $D_T = 60$; 85 mm, CO/CO_2	Columnar crystals $l = 1$ cm
$CaTiO_3$	1970	Directional solidification	Columnar crystals
$CaWO_4$	1560	$W = 60$ kW, $f = 5.28$ MHz	5 × 5 × 10 mm
$BaWO_4$	1490	$D_T = 80$; 120 mm, air, $V_p = 1\div10$ mm/h	
$SrTiO_3$	1910	As above Czochralski method $D_T = 130$; mm, air, $V_p = 5\div10$ mm/h $V_{BP} = 40\div60$ rpm	As above $d = 15$ mm $l = 30$ mm
$Ln_2Ti_2O_7$ (Ln = Sc, Y, La, Nd, Sm, Eu, Dy, Tb, Gd, Ho, Er, Yb)	1600- 1900	Directional solidification $W = 60$ kW, $f = 5.28$ MHz $D_T = 80$; 120 mm, air, $V_p = 1\div10$ mm/h	2-12 mm
		Czochralski method	$d = 3$ mm $l = 8$ mm
Ln_2NiO_4 (Ln = La, Pr, Nd)	-	Directional solidification $W = 20$ kW, $f = 3$ MHz $D_T = 60$; 85 mm, CO/CO_2	4 × 4 × 6 mm
$LaCrO_3$	2430	Directional solidification	Needle crystals
$YCrO_3$	2290		
$ErCrO_3$	2300	$W = 60$ kW, $f = 5.28$ MHz $D_T = 80$; 120 mm, air,	1 × 1 × 10 cm
$Y_3Al_5O_{12}$	1930	Directional solidification	Several millimetres
$Sc_2Zr_2O_7$	2700	As above	As above
$Bi_{12}GeO_{20}$	930	Czochralski method $W = 5$ kW, $f = 1\div10$ MHz, $V_p = 1\div150$ mm/h $V_{BP} = 0\div100$ rpm	$d = 20$ mm $l = 70\div80$ mm
$MnZnFe_2O_4$	-	Czochralski method with additional light heating	$d = 10\div17$ mm $l = 100$ mm

Material	T_m,°C	Methods and growth parameters*	Crystal dimensions
$Nd_3Ga_5O_{12}$	-	Czochralski method W_1 = 100 kW, f_1 = 1÷7 MHz, W_2 = 75 kW, f_2 = 7÷10 kHz, D_T = 80 mm, V_p = 4 mm/h V_{BP} = 0÷100 rpm	d = 35 mm l = 92 mm
Si	1417	Czochralski method f = 470; 250÷400 kHz, D_T = 31; 75 mm, argon, V_p = 12÷168 mm/h	d = 15 mm l = 25 mm

*Growth parameters: power of HF generator W, frequency f, crucible diameter D_T, atmosphere, growth rate V_p, speed of rotation of the seed V_r

Up to the present time, high-quality crystals of magnetite Fe_3O_4 have been efficiently grown by the Bridgman in platinum crucibles. However, this method is expensive, because after growth, the crucible must be sectioned in order to release the crystal and, in addition to this, iron diffuses into platinum and disrupts this metal. Consequently, the method of direct high-frequency melting has a number of obvious advantages in this case. From the viewpoint of the retention of the stoichiometry of magnetite with respect to oxygen, it is important to create an oxygen-enriched atmosphere in the solidification chamber and control its composition. A CO/CO_2 buffer mixture has been used for this purpose. The process of crystal growth consists of two stages. The first stage consists of the production of the required volume of the melt in air, and the second stage is the evacuation of the solidification chamber and passage of the gas mixture through the chamber. The melt is held in this atmosphere for 90 min, in order to establish equilibrium and, subsequently, the crucible is lowered. In order to retain the equilibrium of the melt with the solid and gas-forming phases, the rate of growth is reduced to 7.5 mm/h, this also increases the size of the crystals.

In the cold crucible, it is also possible to grow successfully crystals of oxides of transition metals $Mn_{1-x}O$, $Co_{1-x}O$, $Fe_{1-x}O$ and $Ni_{1-x}O$ [20], characterised by interesting electrical, optical and magnetic properties. Equipment for the growth of single crystals in the cold crucible was the same as in the previous case. As a result of examination of the effect of the atmosphere used for the growth process on the phase composition of the single crystals it was established that the presence of the inclusions of M_3O_4 is associated with a surplus of oxygen in the melt. In a reduction atmosphere, the oxide is reduced to metal. The investigations [20] also showed the significant role of the subsolidus annealing of the crystals which improves the homogeneity of the crystals and results in the complete disappearance of the M_3O_4 phase.

After selecting a suitable buffer oxygen atmosphere, single crystals of titano-magnetite $(Fe_3O_4)_{1-x}(Fe_2TiO_4)_x$ were grown in the chamber. The size of the crystals reached 1 cm^3 [21,24]. The atmosphere was controlled throughout the process, including melting. Examination of the phase composition of the crystals grown in different atmospheres showed that this method is suitable for producing single crystals of ferrites of complicated composition. In addition to the previously mentioned groups of materials, the same equipment and procedure was used for the growth of crystals of Ln_2NiO_4, where Ln = La, Pr, Nd [23].

The above examples of the growth of single crystals by directional solidification in the cold crucible show that the method is highly promising. As an example, Table 5.1 shows the far from complete list of the single crystals, produced by the cold crucible method.

In many cases, the application of single crystals depends greatly (and in some cases almost completely) on their geometrical dimensions, the shape, the homogeneity of the optical, mechanical and other properties in the volume of the crystal. The simplest method of increasing the size of the crystals grown by solidification from the melt and of increasing their homogeneity is to increase the volume of the solidifying melt. However, this method has its limitations determined by the possibilities of processing equipment and economic efficiency. In addition to this, the increase of the volume of the melt on its own does not solve the problem of producing single crystals of the required shape and dimensions and the reproducibility of the growth process.

Another method of increasing the size and improving the quality of single crystals is the formation of the optimum conditions for crystal growth. However, the solution of these problems is made more complicated by the shortcomings typical of the cold crucible method, with the main disadvantages being the fact that it is not possible to examine visually the process of melting and solidification, and the absence of reliable methods for obtaining complete information on the state of the melt directly during the solidification process. In order to control the process, it is necessary to determine the dependence of the condition of the melt on the operating conditions of the high-frequency generator.

Both examined directions of the development of directional crystallisation in the cold crucible supplement each other in terms of increasing the size and improving the quality of crystals. However, only the second direction results in a principal improvement of the method and offers extensive possibilities for producing large crystals in the automatic regime.

A special feature of the process of directional solidification in the cold crucible is that, with lowering of the crucible with the melt in relation to the induction coil, the block of the crystals grows. For crystal technology, it is interesting to investigate the technological factors affecting the number, size and shape of the crystals forming the block, *i.e.*, the block structure changes.

Initially, attention will be given to the effect of the shape of the solidification front on the structure of the block. In the cold crucible method, this factor is very important because the melt in the entire surface of the phase interface is in contact with the polycrystalline phase of the same composition. This surface contains ready solidification centres on which crystals begin to grow already in the stages of formation of the volume of the melt and its homogenising.

It is assumed that the crystals grow in the direction normal to the solidification front, and the boundaries between the crystals are also perpendicular to the solidification front. In the initial variant of the model, the phenomenon of displacement of some crystals by others is not taken into account. Under the given assumptions, the schema demonstrating the growth of a block of crystals in the cold crucible for different shapes of the solidification front is shown in Fig. 5.4*a*, *b*, *c*. The concave and convex solidification fronts (SF) are indicated by arcs of circles, and the flat front is indicated by a straight line. The solid lines indicate the initial state of the solidification front and the position in which the front is found a certain period of time after the start of crystal growth. The dotted lines indicate the intermediate position of the solidification front during the crystal growth, with the radius of the curvature of the front remaining constant.

For a concave solidification front (Fig. 5.4*a*), the crystals nucleated on the side surface of the phase boundary grow into the volume of the melt during lowering of the crucible. The number of growing crystals consequently increases and the cross sections of the crystals decrease. The block forms from the crystals grown both from the bottom of the crucible and on its side. In this case, all crystals grow in the final analysis to the upper surface of the melt.

If the solidification front is convex (Fig. 5.4*b*), the number of crystals growing from the bottom of the crucible decreases during growth an d the transverse dimensions of the crystals increase. The crystals, growing at the periphery of the front gradually reach the side surface of the phase boundary and stop to grow. Thus, not all the crystals that started their growth from the bottom of the crucible with the melt reach the upper surface of the block. In the final analysis, the block consists of two regions, *i.e.* the skull a region of the crystals, grown on the side surface prior to the start of lowering of the crucible, and the central region, consisting of crystals grown from the bottom during lowering of the crucible. In both cases, with the concave and convex solidification fronts, the growth of crystals is distorted. In the case of the concave solidification front, it is distorted in the direction of the surface of the melt, and in the case of the convex front, in the direction of the side wall of the crucible. In the presence of a flat solidification front (Fig. 5.4*c*) there is no bending of the axis of growth of the crystals and the number and size of the crystals remain constant. In the final analysis, the block consists of a cluster of parallel columnar crystals (Fig. 5.5).

In practice, the shape of the solidification front is more complicated as a result of the generation of heat directly in the melt in the vicinity of the crucible wall. The number and shape of the crystals formed in the entire solidification front may differ, but in every separate region the crystals grow in accordance with the previously examined cases. The shape of the solidification front may change as a result of the variation of thermal conditions during the growth process. For example, the front may be initially concave and gradually transform into flat and then into convex. Consequently, in addition to the change of the number and transverse dimensions of the crystals, there may also be extensive bending of the axis of growth of the crystal as a result of the effect of the sum of heat flows. In practice, it is always necessary to try to produce a flat or slightly convex solidification front.

Thus, the shape of the solidification front has a strong effect on the number and, consequently, the size of crystals in the grown block. With the concave solidification front, the lowering of the

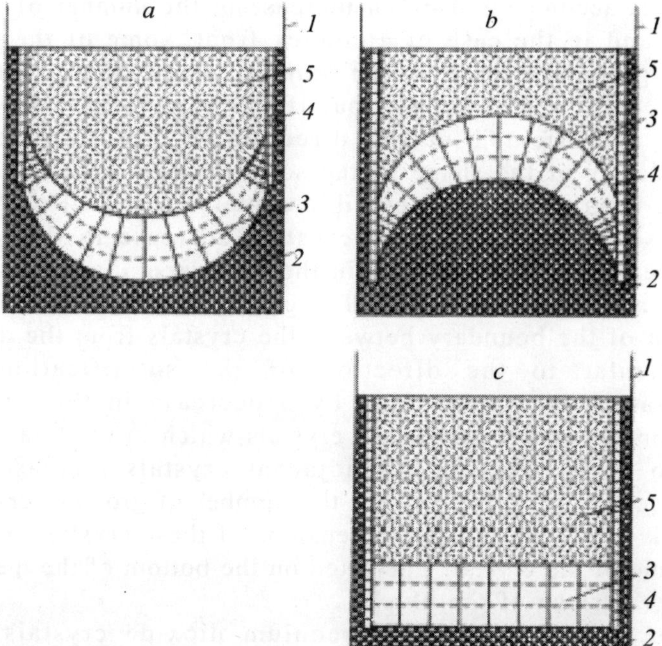

Figure 5.4. Growth of a crystal block in a cold crucible [18,25]. a) with the concave solidification front, b) with the convex solidification front, c) with a flat solidification front, 1) crucible, 2) skull, 3) crystals growing from the crucible bottom with the melt, 4) crystals growing from the side surface of the crucible with the melt, 5) melt.

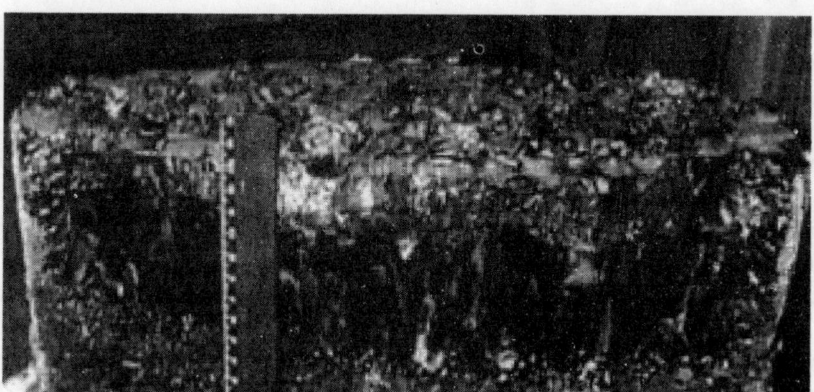

Figure 5.5. Block of crystals of cubic zirconia produced in a cold crucible.

crucible is accompanied by an increase in the number of growing crystals and in the case of a convex front, some of the crystals cease to grow as a result of the limited volume of the melt.

Previously, it was shown that if the boundaries between the crystals are oriented in the direction of heat transfer, then independently of the shape of the solidification front the crystals continue to grow until the crucible contains a melt. However, the transverse dimensions of the crystals decrease in this case. A second reason for the decrease in the transverse dimensions of the crystals may be the process of degeneration, associated with a deviation of the boundary between the crystals from the direction perpendicular to the direction of the solidification front. Degeneration is accompanied by a decrease in the transverse dimensions of only a part of the crystals which in the final analysis cease to grow, whereas the adjacent crystals increase in the transverse size (Fig. 5.6). Thus, the number of growing crystals in the block decreases and the dimensions of these crystals increases. Only some of the crystals nucleated on the bottom of the melt reach the upper surface of the block.

In the growth of cubic zirconium dioxide crystals it was established that one of the factor affecting the degeneration process is the degree of superheating of the melt in the stationary condition. The degree of superheating of the melt is determined by the density of the power generated in the melt in the stationary condition:

$$Q_{stat} = P_{m.stat} / M_{stat} \tag{5.1}$$

where $P_{m.stat}$ is the power generated in the melt in the stationary state; M_{stat} is the mass of the melt in the stationary condition. Parameter Q_{stat} determines the level of superheating of the melt in the crystallisation process. In order to determine the effect of Q_{stat} on the degeneration process, a number of experiments was carried out with the growth of ZrO_2 crystals stabilized with Y_2O_3 from melts at different values of Q_{stat}. These experiments were carried out using the measuring-calculation complex described in Chapter 2, controlling the state of the generator–load system. Consequently, it was possible to calculate the value of P_m directly during the melting process and determine the value of $P_{m.stat}$ and also control the change of the melt volume. In order to ensure the reproducible crystallisation conditions, the dimensions of the crucible and the induction coil, the level of the load and the rate of lowering of the

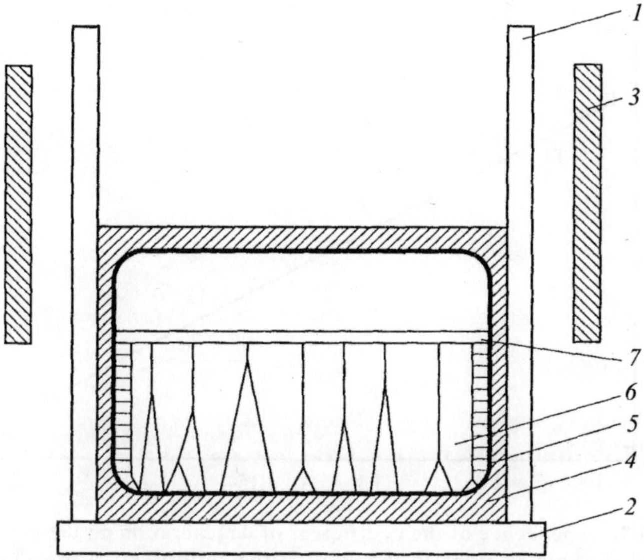

Figure 5.6. Cold crucible with a grown block of a cubic zirconia crystal [26]. 1) wall and 2) bottom of the cold crucible, 3) inductor, 4) charge, 50 solidified melt, 6) crystals growing from the bottom of the crucible with the melt, 7) the single crystal grown from the upper surface of the melt.

cruciblet, and also the purity of the charge were maintained constant in all experiments. In this case, the technological parameters were selected in such a manner that the solidifying melts where similar in the mass and the absorbed power of the melt P_m, and the bottom of the crucible was flat in order to eliminate the effect on the shape of the solidification front on the experimental results. In the crystallisation process, anodic voltage was stabilized. The values of P_m and $P_{m.stat}$ in all experiments were varied in almost the same manner and at the end of the process the shape of the solidification front was distorted only slightly.

After completion of the growth process, the height of the block of the crystals was measured and the number of crystals starting to grow from the bottom of the crucible with the melt that reach the upper surface of the block was determined. The solidified melt was weighed and the value of Q_{stat} calculated. In [25], to investigate the process of degeneration of part of the growing crystals, the number of these crystals was counted at different distances from the start of growth. In order to characterise the process of degeneration of crystals throughout the entire block, a

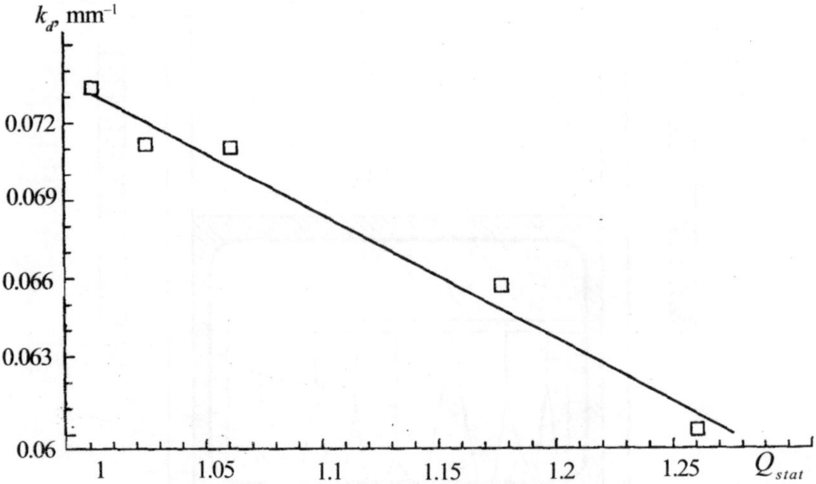

Figure 5.7. Dependence of the coefficient of degeneration on the density of power generated in the melt prior to the start of the solidification process [26].

generalised parameter, i.e. the degeneration coefficient, was introduced:

$$k_d = (n_1 - n_2)/n_2 h, \quad (5.2)$$

Here n_1 and n_2 are respectively the number of crystals which started to grow from the bottom of the crucible and the number of crystals which reached the upper surface of the block; h is the height of the block of the crystals. This coefficient characterises the mean (with respect to height and radius) number of the degenerated crystals normalised to the unit height of the block per a single crystal that reached the upper boundary of the block. Parameter k_d also characterises the mean angle of growth of the crystals, under the condition that degeneration takes place uniformly along height h. The larger k_d increases, the larger the angle of growth of crystals.

Figure 5.7 shows the experimental dependence of k_d on Q_{stat} in relative units. The minimum value of Q_{stat} differs from the maximum value by ~25%. It is evident that the application of the linear dependence as an approximating function is in agreement with the experimental data. Figure 5.7 also shows that the value of k_d increases with decreasing Q_{stat}. The latter circumstance indicates that for a lower degree of superheating of the melt the rate of degeneration is higher and the crystals are characterised by a larger growth angle.

A satisfactory explanation of the process of degeneration of cubic zirconia crystals with a flat solidification front may be provided on the basis of Tiller's theory [27]. According to this theory, grooves form at the boundary between the growing crystals. These grooves are formed by the surfaces of the boundaries of the crystals and, consequently, the free energy per unit volume in them is higher than in the sections of the crystal outside these boundaries. Consequently, the temperature of the liquid–solid phase transition inside the grooves is lower and the surface on which the crystals join has a lower temperature in comparison with the remaining interfacial surface and, consequently, a cavity forms between the crystals. The dimensions of the cavity depend on the temperature gradients in the melt and the solid phase and on the direction of growth of the contacting crystals.

According to Tiller's theory, the slope of the edges of the grooves depends on the orientation of crystals. The steepest slope is characteristic of a crystal oriented in the <111> direction in relation to the growth axis. Figure 5.8 shows two intergrown crystals, a bicrystal, with the misorientation of these crystals being the result of rotation around the axis of growth. The crystals were in the same growth conditions, but one of them had the interfacial surface in the plane <*hkl*> and the other one in the (111) plane. The distance δx of the base of the groove of the <*hkl*> bicrystal in relation to the base of the <111> groove is the consequence of the wedge-shaped form of the <*hkl*> groove. The growth of the

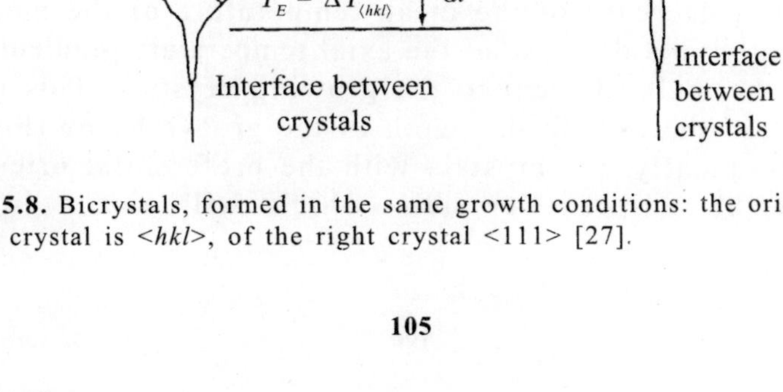

Figure 5.8. Bicrystals, formed in the same growth conditions: the orientation of the left crystal is <*hkl*>, of the right crystal <111> [27].

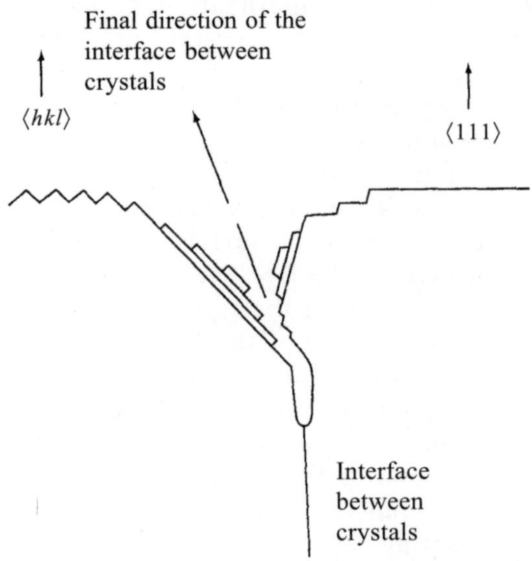

Figure 5.9. Bicrystal consisting of two crystals with the orientations <*hkl*> and <111>. The scheme indicates how the crystal <111> displaces the <*hkl*> crystal [27].

layers of the walls of the groove of this bicrystal occurs faster than the growth of the <111> layers of the crystal. This may take place if the mean temperature of the layers of the <*hkl*>-crystal is lower in comparison with that of the <111> crystal, because the rate of implantation of the atoms into the crystal lattice is a function of supercooling of the melt.

However, if the bicrystal forms from <*hkl*> crystal and <111> crystals, then the groove at the boundary of the crystals is asymmetric (Fig. 5.9). The normal growth of the layers on the walls of the groove results in the sagging of the <111> crystal above the <*hkl*> crystal. Thus, the <111> crystal will 'step' on the <*hkl*> crystal and in the final analysis will displace this crystal.

Taking into account this mechanism, an increase of k_d with a decrease of Q_{stat} may be explained as follows. A decrease of Q_{stat} results in a decrease of the mean temperature of the melt and, consequently, in a decrease of the axial temperature gradient in the layer of the melt adjacent to the growing crystals. This in turn results in an increase in the depth of the groove by δx (Fig. 5.9) and, consequently, the crystals with the preferential orientation displace the adjacent crystals which will result in their degeneration.

Detailed examination of the grown crystals shows that the grooves actually form at the boundary between the cubic zirconia crystals. They form either during rapid solidification of the remainder of the melt when the solidification front is exposed, or during the crystallisation of the melt when the single crystals, growing in the upper part of the melt in the direction from top to bottom, come into contact with the crystals growing from the bottom (Fig. 5.6). In this case, the front of growth of the crystals, growing from the bottom, 'imprints' on the lower surface of this single crystal, covering the entire surface of the melt. Figure 5.10 shows the photograph of a fragment of such a single crystal indicating the imprint of the groove. Figure 5.11 shows the cross-section of the upper crystal by the plane normal to the boundary between the crystals (the arrow indicates the point of joining of the crystals).

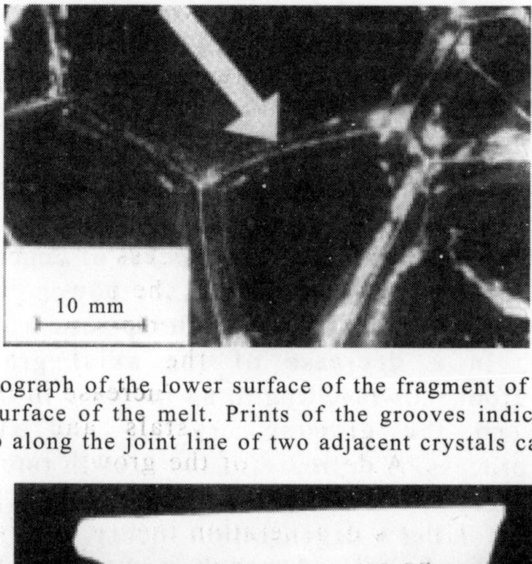

Figure 5.10. Photograph of the lower surface of the fragment of a single crystal grown from the surface of the melt. Prints of the grooves indicating elongated cavities with a rib along the joint line of two adjacent crystals can be seen [26].

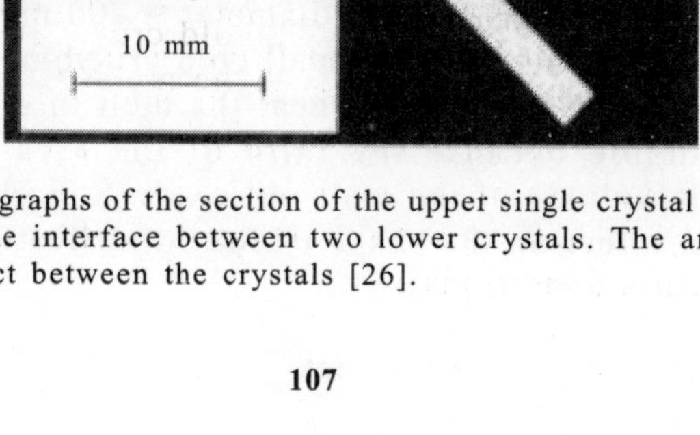

Figure 5.11. Photographs of the section of the upper single crystal in the vertical plane normal to the interface between two lower crystals. The arrow indicates the point of contact between the crystals [26].

Thus, the experiments show that in the growth of crystals in the cold crucible by the method of directional solidification with a flat solidification front, the process of degeneration takes place by the mechanism proposed by Tiller [27]. On the basis of this theory, it is also possible to explain the increase of the transverse dimensions of the crystals with a decrease of the rate of growth [27] or a decrease of these dimensions with an increase in the content of impurities in the melt. If the content of the impurities in the melt is insufficient for achieving concentrational supercooling resulting in the formation of a cellular structure, the interfaces between the crystals play the role of a sink of impurity atoms and supercooling in the groove is released. The deposition of the impurities, in particular Al_2O_3, on the side surfaces of the cubic zirconia crystals was reported in [28]. The (111) planes trap the impurities and keep them at the base of the groove consequently delaying their growth and resulting in complete interruption of growth at a specific impurity concentration. Consequently, the groove becomes deeper because of the poisoning of the (111) planes by the impurities thus preventing their sagging above the <*hkl*>-crystal. In this case, displacement is less intensive or is completely interrupted.

The effect of the power supplied to the melt on the process of degeneration of the crystals was investigated in [25]. The results show that in the examination of the process of degeneration on the basis of Tiller's theory, a decrease of the power generated in the melt results in a decrease of the mean temperature of the melt and, consequently, in a decrease of the axial gradients at the solidification front thus resulting in an increase in the depth of the groove between the growing crystals and supporting the degeneration process. A decrease of the growth rate also increases the size of the crystals causing their degeneration [27, 29]. From the viewpoint of Tiller's degeneration theory, the ratio of the rate of displacement to the rate of growth in the axial direction in the case of low growth rates is higher and the efficiency of displacement should be higher.

The size of the crystals is also affected by the volume of the cold crucible. In large crucibles (diameter > 200 mm), it is possible to grow larger crystals [30]. In small cold crucibles (diameter ~90–100 mm) it is necessary to superheat the melt in order to maintain it in the crucible because the ratio of the area of the cooling surface to the volume of the melt increases. Superheating has also a detrimental effect on the shape of the solidification front and on the degeneration conditions.

In addition to the technological conditions, the degeneration process is affected by the presence of secondary impurities and also by the type and concentration of the stabilizing oxide. It is well known that in the case of the directional solidification of multicomponent melts with the distribution coefficients of the components differing from 1 the component which decreases the crystallisation temperature is displaced into the melt and accumulated ahead of the solidification front (at $K < 1$). As already mentioned, the grooves at the boundaries between the crystals operate as sinks for certain impurities, decrease the rate of lateral growth of the crystals, increase the depth of the grooves and reduce (or completely interrupt) the degeneration process. The effect of several secondary impurities on the size and quality of single crystals will be considered in Chapter 7.

5.3. MODIFICATION OF THE PROCESS OF CRYSTALLISATION IN THE COLD CRUCIBLE

To grow crystals of neodymium–gallium garnet $Nd_3Ga_5O_{12}$, Mateika *et al* [31] have improved the method of the cold crucible introducing into the melt a bottom heater and an additional crucible immersed in the melt. A schematic of the modified cold crucible is shown in Fig. 5.12. A cylindrical crucible was produced from 40 water-cooled copper tubes. The bottom of the crucible consists of tubes spaced at 0.3 mm from each other and coated with metallic rhodium. A ring of aluminium oxide is used for maintaining the cylindrical shape of the crucible when the volume of the charge increases during heating. Springs are used for protecting the quartz sheet against failure during compensation of vertical pressure. The external appearance of the crucible with the water-cooled bottom, with the height and internal diameter of 120 mm, is shown in Fig. 5.13. The cylindrical induction coil, surrounding the crucible, was connected to a high-frequency generator operating in the frequency range (1–7) MHz. The maximum power of the generator was 100 kW. The second induction coil of the helical shape, situated below the bottom of the crucible, was connected to a 75 kW generator, operating at a frequency of 10 kHz. Inside the crucible there was an iridium ring placed on pins and resting on a quartz sheet. The iridium crucible, which will be referred to as the growth crucible, was suspended on a support in the upper part of the cold crucible. Its internal diameter is 80 mm and the crucible has a conical bottom

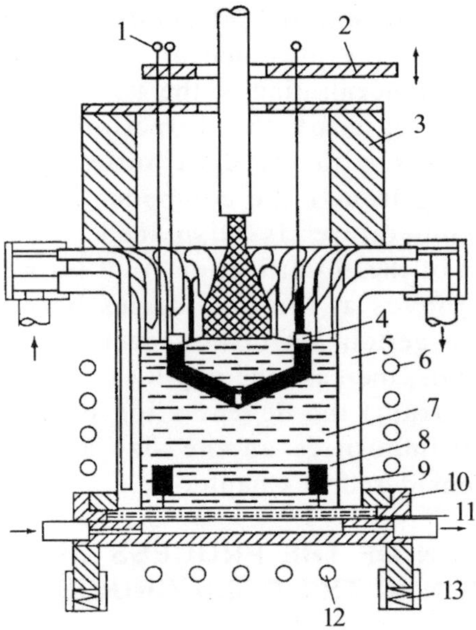

Figure 5.12. Schematic of a double crucible [31]. 1) detector of the melt surface; 2) crucible support; 3) thermal insulation; 4) growth crucible; 5) cold crucible; 6) HF heater (1–7 MHz); 7) melt; 8) skull; 9) bottom heater; 10) quartz ring; 11) quartz sheet; 12) HF heater (7–10 kHz); 13) shock absorbers.

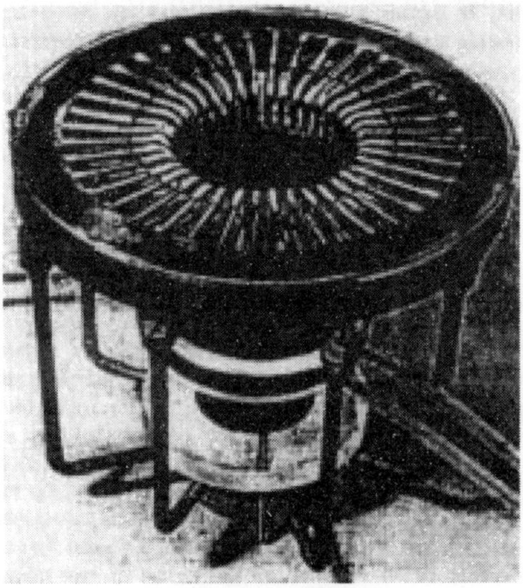

Figure 5.13. External view of a double cold crucible [31].

with a hole in the centre. The walls of the crucible contain several diametrical slits. The support of the crucible moves in the vertical direction by means of an electric motor. An iridium feeler measures the lowering of the surface of the melt in the crucible during crystal growth by contacting the voltage source with the growth crucible. A ceramic tube in the upper part of the crucible decreases the extent of radiation losses. The iridium ring on the quartz sheet is heated with a field with a frequency of 8 kHz by an induction coil below the bottom of the crucible. As soon as a melt appears in the region of the ring, the field of the cylindrical induction coil with a frequency of 2 MHz is activated and the volume of the melt increases. When the free surface of the melt is produced, the melting stage is completed. The growth crucible is lowered into the melt and filled through the slits and the hole in its bottom section. The temperature of the melt during seeding of the crystal is regulated by controlling the power of the two generators. The growth process takes place by the well-known Czochralski method. The diameter of the growing crystal is controlled by regulating the power of the high-frequency generator, and the power of the generator at a low frequency is maintained constant. In the growth of the neodymium–gallium garnet, a gas mixture, consisting of 98% of nitrogen and 2% of oxygen, is continuously pumped through the chamber. During crystal growth, the detector continuously controls the level of the melt. The growth crucible is lowered using the mechanism for maintaining the level of the melt constant. When the required length is reached, the crystal is separated from the melt and withdrawn from the growth crucible. The generator power decreases to 0 during 10 hours.

The melts, heated directly by high-frequency currents, are characterised by a tendency for spontaneous solidification. This takes place if the volume of the melt decreases below the critical level required for coupling the high-frequency field with the melt. In order to prevent these effects, it is necessary to use the iridium ring (Fig. 5.12) situated at the bottom of the cold crucible. It is also used as a heating element at the start of the melting process. In subsequent stages, the hotter ring is used as an additional heating element in order to stabilize the required volume of the melt. This operation makes it possible to control the crystal diameter by changing the power supplied by the high-frequency generator. Thus, the melt remains stable throughout the entire process of crystal growth. Figure 5.14*a* shows the flow pattern on the surface of the melt of the neodymium–gallium garnet prior to lowering the growth crucible into the melt. Examination shows small irregular bubbles moving in different directions. This indicates the presence of turbulent convection of the melt and the radial distribution of temperature with a minimum in the centre of the surface. The turbulent convection of the melt in the immediate vicinity of the growing crystal is an unexpected phenomenon observed when using the growth crucible. The melt inside this crucible is heated mainly as a result of the melt surrounding it. Only a small volume of the melt is poured out through the slits in the wall of the internal crucible. The application of the internal crucible creates suitable conditions for the growth of the crystal by the traditional Czochralski method and a regular pattern should form on the surface of the melt. Figure 5.14*b* shows the patterns of the flow lines in the growth crucible. The patterns, identical with those in Fig. 5.14*a*, were detected on the surface between the two crucibles, and inside the growth crucible the flows of a specific shape travelled into the centre of the melt. This clearly indicates the radial temperature gradient with a minimum in the centre. The hole in the bottom of the growth crucible is used as an output hole for the melt in order to ensure the stability of circulation of the melt from one crucible into another (Fig. 5.15). This disposition of the crucibles stabilises the circulation of the melt flows from the walls of the hotter crucible to the centre of the surface.

A significant special feature of this modified technology is the possibility of regulating the growth process. The height of the melt in the growth crucible is maintained constant as a result of the presence of a large volume of the melt in an external crucible. This is combined with the efficient restriction of the volume of the melt in the zone of the growing crystal. The proposed method also

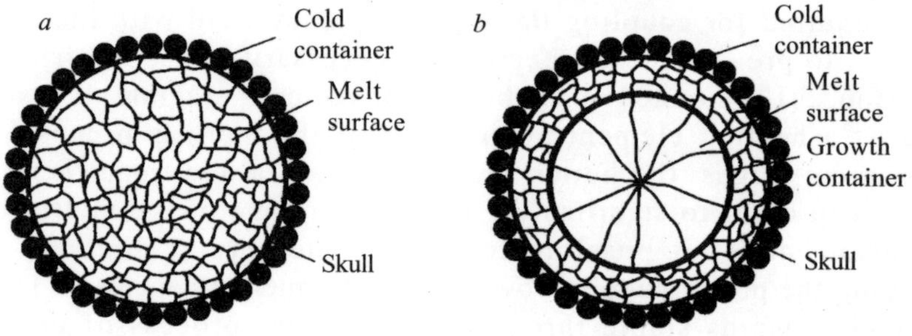

Figure 5.14. Schematic of the flow lines on the surface of a neodymium–gallium garnet melt [31]. a) without the growth crucible; b) with the growth crucible.

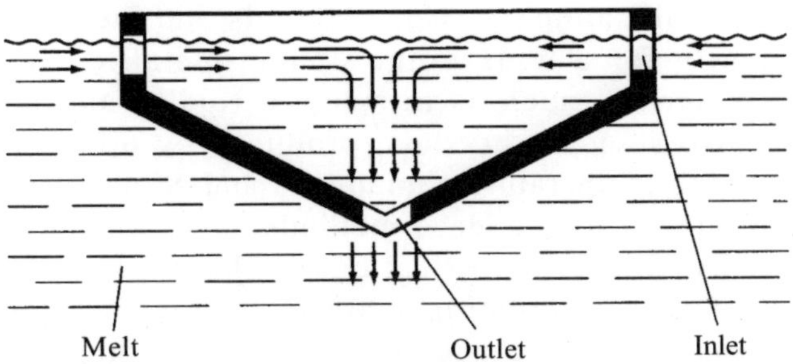

Figure 5.15. Forced convection of the melt inside the growth crucible [31].

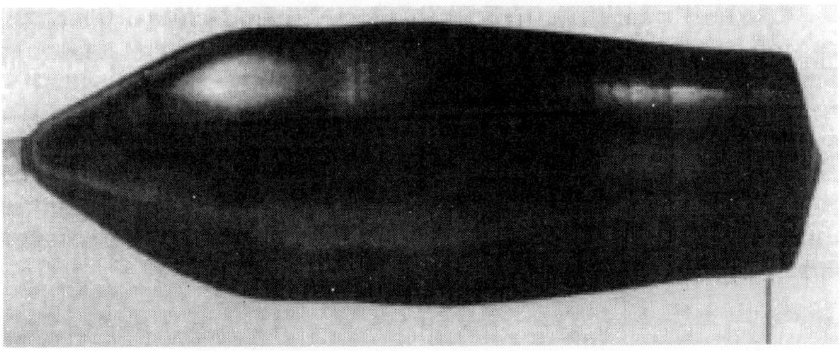

Figure 5.16. Neodymium–gallium garnet grown in a double crucible, scale 1 cm [31].

makes it possible to save noble metals. The iridum ring and the growth crucible are produced from sheet metal. Consequently, this technology has several advantages in comparison with the conventional method of growing large single crystals.

A single crystal of the neodymium–gallium garnet 92 mm long with a diameter of 35 mm was grown at a withdrawal rate of 4 mm/h, the speed of rotation was 10 rpm. The growth surface of the end part of the crystal was conical. The drawn crystal is shown in Fig. 5.16.

From the viewpoint of crystal technology, it is important to note that the proposed method is very complicated because the synchronisation of simultaneous operation of two high-frequency generators is a complicated technical task.

The presented review of the methods for growing single crystals from the cold crucible makes it possible to propose the most promising directions of further development of the method.

The method offers extensive possibilities of producing high-temperature non-metallic single crystals. The authors investigated mainly the crystallisation of oxide materials, but the method is also used efficiently for a wide range of non-metallic, in particular refractory, materials, such as nitrides, carbides, and borides and others, and also standard crystals, produced by other methods, as well as new high-temperature compounds and solid solutions which cannot be produced by standard methods.

Undoubtedly, further investigations in this area will lead to the production of new interesting materials which will be used in different areas of science and technology. The most convincing example is the development of a new family of crystals, i.e. fianites (cubic zirconia).

It should be mentioned that the method of production of single crystals should be selected for each specific material taking into account the physical–chemical, electrophysical and thermal characteristics of the melt and the crystal. In this case, in addition to the already investigated methods (directional crystallisation, the Czochralski method), it is evident that special interest is attracted by the combination of the method of the cold crucible with growth methods such as the Stepanov method, growth from the solution in the melt, and other methods.

References

1. Ciszek T.F., *Cryst Growth*, 1984, Vol. 70, No. 1/2, 324–329.
2. Wenckus J.F., Menashi W.P., Final technical report RADC-TR-82-171, Rome, Harsom. Air Develop. Cent., 1982.
3. Kern E.L., Vaggy L.S., V[th] Amer. Conf. Crystal Growth, Abstracts. San Diego, 1981, 75–76.
4. Aleksandrov V.I., Vishnyakova M.A., et al., *Izv. AN SSSR, Neorgan, Mater.*, 1983, Vo.l. 19, No. 1, 104–107.
5. Aleksandrov V.I., Barygov S.Kh., *ibid*, No. 2, 265–268.
6. Balbashov A.M., Zavartsev Yu.D., *Tr. Leningrad Elektrotekhn. In-ta*, 1080. No. 455, 97–106.
7. Egorov S.K., Zavartzev Yu.D., et al., Europ. Meeting Cryst. Growth 82, Mater. Electron. Abstr. Phaque, 1982, 205–206.
8. Aleksandrov V.I., Osiko V.V., et al., *Usp. Khimii*, 1978, Vo. 47, No. 3, 385–427.
9. Borik M.A., et al., Problems of crystallography, Nauka, Moscow, 1987, 165–189.
10. Aleksandrov V.I., Osiko V.V., Prokhorov A.M., Tatarintsev V.M., *Curr. Top. Mater. Sci.*, 1978, Vol. 1, 421[a]480.
11. Scott B., Rawson H., *Glass Technol.*, 1973, Vol. 14, No. 5, P. 115–124.
12. Aleksandrov V.I., Blinov A.L., et al., *Izv. AN SSSR, Neorgan. Mater.*, 1983, Vol. 19, No. 3, 443–447.
13. Aleksandrov V.I., et al., *ibid*, 1980, Vol. 16, No. 1, 99–104.

14. Marchall R.C., Larkin J.J., Armington A.F., Abstracts of Fourth Amer. Conf. on Crystal Growth, Gaitherburg, 1978, 66–67.
15. Petrov Yu.B., Induction melting of oxides, Energoatomizdat, Leningrad, 1983.
16. Michel D., Perez-Y-Jorba M., Collonques R., C. r. Acad. sci., 1968,Vol. 266, No. 23, 1602–1604.
17. Roulin Y., Vitter G, Deportes C., *Rev. Intern. Haut. Temp. et Refract.* 1969. Vol. 6, N 3, P. 153-158.
18. Harmon H.R., Aragon R., *Mater. Res. Bull.* 1978, Vol. 13, No. 11, P. 1097–1104.
19. Melekh B.T., Andreeva A.A., et al., *Izv. AN SSSR, Neorgan. Mater.*, 1983, Vol. 18, No. 1, 98–101.
20. Harrison H.R., Aragon R., Sandberg C.J., *Mater. Res. Bull.*, 1980, Vol. 15, No. 5, 571–580.
21. Harrison H.R., Aragon R., IVth Amer. Conf. Crystal Growth, Abstracts. Gaithersburg, 1978, 64–65.
22. Ershova L.M., Ignate'v B.V., et al., *Izv. AN SSSR, Neorgan. Mater.*, 1972, Vol. 13, No. 11, 2042–2945.
23. Buttrey D.J., Harrison H.R., Honig J.M., Schartman R.R., *J. Solid State Chem.*, 1984, Vol. 54, No. 3, 407–413.
24. Aragon R., Harrison H.R., McAllister R.H., Sandberg C.J., *J. Cryst. Growth*, 1983, Vol. 61, No. 2, 221–228.
25. Khaneev R.P. Development of a method for controlling and investigating processes of direct high-frequency melting and directional crystallization, Dissertation, Moscow, 1994.
26. Osiko V.V., Penyaz D.L., Khaneev N.P., *J. Cryst. Growth*, 1993, Vol. 128, 1193–1196.
27. Tiller W.A., *J. Metals*, 1957, Vol. 9, 847–855.
28. Lomonova E.E., Production and investigation of single crystals of solid solutions based on ZrO_2 and HfO_2, Dissertation, Moscow, 1980.
29. Aleksandrov V.I., et al., A method for producing single crystals based on solid solutions of refractor oxides, Author's Cert. 1048857, USSR, Publ. 15.06.1983.
30. Aleksandrov V.I., et al., A method for producing single crystals based on stabilised zirconia, Author's Cert. No. 1365739, USSR, Publ. 08.09.1987.
31. Mateika D., Laurien R., Liehr M.J., *Cryst. Growth,* 1983, Vol. 65, No. 1/3, 237–242.

6

Production of refractory and special purity glasses

6.1. ELEMENTS OF GLASS PRODUCTION TECHNOLOGY

Glass is one of the most important artificial materials, used in various areas of science and technology. The range of application of glasses is continuously widening. The formation and development of branches such as quantum electronics, radioelectronics, missile and nuclear technology, semiconductor of physics, fibre optics, etc, has greatly widened the area of application of glasses. This has resulted in the development and production of new types of glass. They include laser glasses, activated with rare-earth ions; electric vacuum glass; semiconductor halide glasses; glasses transmitting or absorbing x-radiation, gamma radiation and also neutrons; glass-crystalline materials whose strength is considerably higher than the strength of commercial glass and can withstand steep temperature gradients; special purity glasses for fibre optics communication lines, and many other types.

The production of glasses of complicated composition, 'unsuitable' for the commercial technology of glass boiling, and also the increase in the requirements on optical homogeneity and chemical purity of glass present continuously new tasks to technologists such as, for example, the production of glasses with very high melting points and special-purity glasses.

High-temperature glasses, whose synthesis temperature is higher than 1600°C, are characterised by high temperatures of the start of the process of deformation and by heat resistance and, consequently, they are irreplaceable for a large number of practical applications in high-temperature technology. Special attention is paid by investigators, in particular, to high-temperature systems of glasses: R_2O_3–Al_2O_3–SiO_2 (where R is Y, La, rare-earth elements). The experimental results show [1–5] that they are characterised by high refractive properties, high mechanical strength and resistance to the effect of alkali.

The glasses produced by industry at present are synthesised at temperatures below 1600°C. This limitation is associated with the possibilities of the currently available methods of synthesis in gas and electric furnaces. The development of methods of synthesis at higher temperatures will make it possible to produce optical materials with unique properties.

The glasses, satisfying stringent requirements on chemical composition, in particular, ultrapure glasses, are essential for fibre optics communication lines and laser technology. For example, in glasses for fibre optics communication lines, the content of impurities of transition metals should not exceed 10^{-5}–10^{-6} wt%. The problem of synthesis of glasses with a low content of OH^- hydroxyl groups has not as yet been solved completely in the technology of production of laser glass. This problem is especially acute in the technology of highly concentrated neodymium and erbium glasses. The content of OH^- groups in these glasses, of the order of 10^{19} cm^{-3}, results in the quenching of luminescence of active ions. At present, the methods of purification of the initial materials for the glasses used in fibre optics communication lines makes it possible to produce materials satisfy the requirements on the content of the impurities. However, the commercial boiling methods increase the content of impurities by more than an order of magnitude. In the stage of synthesis of laser phosphate glasses which are boiled in platinum crucibles because of the corrosive effect of the melt, platinum penetrates into the glass. Platinum inclusions in glass greatly reduced the beam strength of the active elements produced from the glass of this type. It should be mentioned that the hydroxyl groups penetrate into laser glass in the synthesis stage.

To develop highly concentrated neodymium glass, it was necessary to develop a method for synthesis of laser glasses with the composition close to the stoichiometric compositions of crystals, i.e. the prototypes of glasses. In glasses of this composition, the neodymium content may reach $\approx 4\cdot10^{21}$ cm^{-3} whilst maintaining a higher quantum yield of luminescence of neodymium. These glasses can be used to produce highly efficient small solid-state lasers.

The melts of these glasses are characterised by high crystallisation capacity. This greatly complicates the synthesis of these glasses. They include glasses with compositions similar to that of crystals M Nd P_4O_{12} (here M is Li, Na, K, Rb, Cs). A change in the composition of a glass in order to reduce the crystallisation capacity is undesirable because of a decrease in the spectral-luminescence characteristics.

In this chapter, attention is given to the special features of the synthesis and boiling of high-temperature glasses in a cold crucible [6].

Crucibles

Evidently, the first publication on the synthesis of glass by the method of direct RF melting was [7], published in 1968, describing the synthesis of glass in quartz crucibles cooled with water. The design diagram of the system is shown in Fig. 6.1. Induction heating of glass in cooled quartz or ceramic crucibles was also used for vitrification of radioactive waste [8].

The further development of the method of synthesis of glass was associated with the problem of synthesis of high-purity multicomponent glasses with small optical losses for fibre optics communication lines. In [9, 10], glasses were boiled in air-cooled quartz crucibles resulting in the formation of up to 600 g of glass. The method makes it possible to increase the yield of glass to several kilograms. A significant disadvantage of the application of quartz crucibles is the fact that it is necessary to regulate efficiently the temperature of the melt because at high temperatures and insufficient cooling the walls of the crucible may fail resulting in the penetration of water into the melt and in ejection of the latter. In addition to this, the application of quartz crucibles restricts the boiling temperature of glass and complicates the extraction of glass mass. Significant disadvantages also include the fact that quartz crucibles can be used only once.

The investigations show that the most promising method for the synthesis of glass is the application of metallic water-cooled crucibles described in [2, 11, 12]. The design diagrams of several crucibles are shown in Chapter 3.

Metallic crucibles, used for the boiling of glass, in contrast to crucibles for the growth of crystals and the synthesis of polycrystalline materials, have a number of design features represented by the following [13]:

–a quartz vessel is placed in a copper crucible to prevent the contamination of the melt with copper. Aluminium crucibles are also used. The aluminium oxide does not colour the glass and is usually included in the main composition of the glass.

–assembled crucibles contain a large number of sections for preventing electric breakdowns in gaps.

–the crucibles are produced from split sections in order to be able to extract rapidly the glass block for annealing. An example of such a crucible produced from aluminium pipes is shown in Fig. 6.2.

Starting melting

As mentioned previously, the simplest and most efficient method of starting melting oxide materials is a method based on the introduction, into the charge, of small quantities of a metal which is included in the composition of the melted material. However, this method can not be used for glass boiling because of the uncontrollable contamination of the melt with metallic aluminium and products of production of aluminium, dispersed in the volume of the

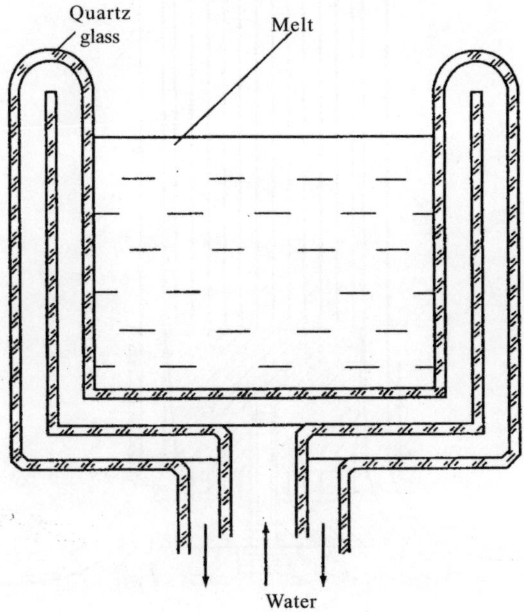

Figure 6.1. A water-cooled quartz crucible for boiling glass [7].

melt. Because of the high viscosity of glass forming melts, metallic aluminium is not oxidised during long periods of time [11]. Other metals show similar behaviour. Therefore, other methods of starting melting are used for boiling of glass. In [7], the starting volume of the melt was produced by melting a portion of a charge in a platinum crucible in a resistance furnace. Subsequently, the melt was poured into a crucible with the charge. In [9, 10, 12], starting melting was realised by introducing a secondary electrically conducting material, the heating body, into the charge. The heating body was represented as either by graphite or low-resistance silicon [8, 16], placed in a quartz capsule.

To prevent the evaporation of low-melting components of glass (salts of alkali metals), the authors of [16] have proposed to realise the process using the following procedure. The first stage consists of melting low melting components and, subsequently, refractory components are introduced into the melt and the melt temperature is gradually increased. Shortcomings of this method of loading the crucible include the high probability of production of glass with

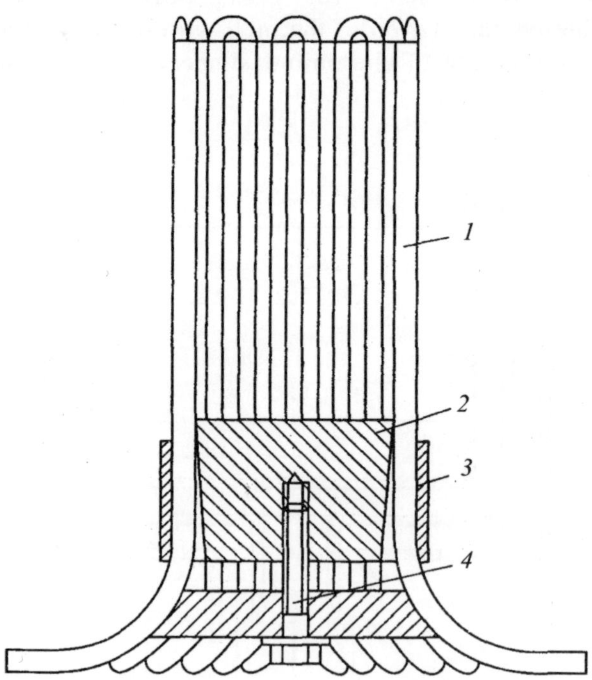

Figure 6.2. Aluminium water-cooled crucible [15]. 1) tubular sections, 2) spacing cone, 3) split supporting holder, 4) tightening bolt.

optical defects because the chemical composition of the skull (see Chapter 2) differs from that of the glass mass.

Homogenising of the melt

Homogenising of the melt is an important process essential for producing glasses with high optical quality. In [10], homogenising was carried out using either an air-cooled quartz stirrer or by the passage of a gas (oxygen or helium) through the melt. The melt of high-temperature glasses of the system R_2O_3–Al_2O_3–SiO_2 (where R is Se, Y, La) was stirred with a water-cooled stirrer. However, the application of the stirrer may have a negative effect on the quality of glass because it results in the formation of steep temperature gradients and increases the viscosity of the melt because of reducing its temperature [15]. The stirrer is therefore used at the beginning of the homogenising process, and in the final stage of homogenising takes place as a result of holding the melt at a high temperature. The presence in the melt of steep temperature gradients results in intensive convective mixing of the melt. However, experience shows that the efficiency of this stirring method is not sufficient for complete homogenising of the melt [17, 18].

Extraction of a glass block from a crucible and annealing

In the synthesis of glass in a glass boiling furnace, the glass block is annealed in the same crucible as the one used for boiling. This method can be used in boiling of glass using induction heating,. The point is that in cooling of the melt its electrical conductivity decreases and at a certain temperature the melt ceases to absorb the energy of the high temperature field. This is accompanied by rapid vitrification of the melt and, consequently, the glass cracks. Therefore, after cooling the block below some temperature which depends on the type of glass, the block must be transferred into a special annealing furnace [11, 15].

In melting in metallic water-cooled crucibles, the separation of the glass block from the crucible is relatively easy because a porous layer of the charge forms at the periphery of the block which is then easily separated from the crucible walls. When using quartz crucibles, a crucible together with the glass is placed in the annealing furnace. To prevent the formation of cracks in the glass block, as a result of the difference in the coefficients of thermal expansion of the glass and quartz, it is usually recommended to use thin-wall crucibles [19]. In annealing, the quartz crucible cracks and separates and the glass blocks remains monolithic. If the glass block is used for withdrawal of the fibre, the quartz crucible with glass is transferred into the furnace for annealing and at a specific temperature glass rods with a diameter of up to 10 mm are drawn. On the outside, the rods are covered with a quartz layer [9, 10, 17].

In [6], the synthesis of glasses was carried out in Donets-4 equipment. The general operating principles of these systems and the main characteristics were presented in Chapter 3.

An important element of equipment is the water-cooled cold crucible whose design has a number of special features, resulting from the specific features of direct RF melting of glass. Glass mass is boiled using multisection tubular crucibles with a diameter of 130 mm and 90 mm, produced from aluminium or copper pipes. To extract a glass block from the crucible for subsequent annealing, two sections were replaced with U-shaped water-cooled elements which were easily removed from the crucible together with the glass block.

Water-cooled stirrers of different design, produced from aluminium tubes, are used for glass homogenising. The stirrer is secured to the upper rod. The speed of rotation of the stirrer is regulated smoothly from 0 to 100 rpm.

Distribution of temperature in glass-forming melts

The distribution of temperature in the melt is an important technological parameter having a strong effect on all glass formation processes, clarification and homogenising of the glass mass [20]: the temperature distribution in the melt determines the heterogeneity of the viscosity of the melt which in turn determines the direction and intensity of flows in the volume of the melt [21].

The distribution of temperature in the melt was already partially discussed in Chapter 2, where the method of determination of the melt temperature was described. In this chapter, problems typical of the production of high-quality glass are examined.

A large number of investigations, for example [22–24], have been concerned with the calculation of temperature distribution in glass melts in industrial glass boiling furnaces of different types. On the other hand, the temperature fields in melts have been examined in detail in the growth of crystals from the melt because this parameter is most important since it determines the conditions of growth of high-quality crystals. The investigations of temperature distribution and the hydrodynamics of the melt during the growth of crystals by the Czochralski method were carried out using the method of mathematical modelling [24–25], and experimental modelling was carried out on transparent liquids [28, 29].

The main special feature of the method of direct RF melting of materials in the cold crucible is the direct heating of material by the energy of the high frequency field. In this case, the material is in the water-cooled metallic crucible. Because of these circumstances, the thermal conditions in the melt greatly differ from the temperature distribution in the melt in the commercial methods of melting of materials in 'hot' crucible. The radial distribution of temperature on the surface of an Al_2O_3 melt is shown in Fig. 6.3 [10]. The given pattern of the radial distribution of temperature was obtained on the basis of the theoretical examination of direct RF heating assuming that the specific electrical conductivity is constant throughout the entire volume of the melt. The graph shows that the distribution of temperature along the radius of the crucible has a maximum whose position is determined by the depth of penetration of current into the melt. The temperature in the vicinity of the crucible walls decreases as a result of high thermal losses through the crucible walls. This pattern was obtained in

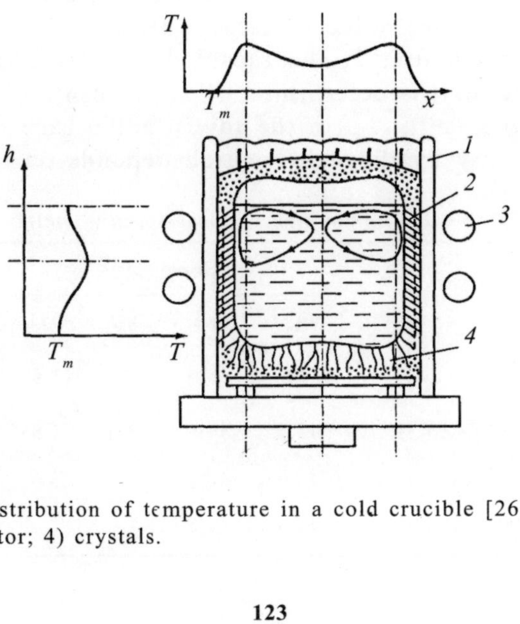

Figure 6.3. Distribution of temperature in a cold crucible [26]. 1) crucible; 2) skull; 3) inductor; 4) crystals.

examination of direct RF heating in a quartz crucible cooled with air.

Effect of electrical conductivity and viscosity of the melt on temperature distribution

Table 6.1 shows the composition of glasses used for the determination of temperature distribution in the melt. Temperature dependences of the viscosity and electrical conductivity for the compositions of the Na_2O–SiO_2 system are available [30], and they are also available for industrial glasses of complicated composition which are of practical interest for fibre optics communication lines. Figure 6.4 shows the temperature dependences of the viscosity and electrical conductivity of the investigated melts.

The typical dependences of the temperature distribution in the direction of radius $T = f_r$ and the depth of the melt $T = f_z$ in a melt of sodium–boron–silicate glass (NaBSi) and also in melts of other glasses, are presented in Fig. 2.7. The identical dependences, obtained for melts of other glasses, have many common features with those shown in Fig. 2.7. All the curves of the radial distribution of temperature to a certain depth (in the examined case to z = 70 mm) show a maximum whose position remains constant for the investigated melts. The position of this maximum is determined by two factors: the nature of distribution of the active power absorbed by the melt and the special features of removal of heat through the watercooled crucible walls.

As shown in Chapter 1, the radial distribution of the power absorbed by the melt is determined by relationship (1.3) and for a crucible with the radius r_0 (in the investigated case r_0 = 6 cm) it is determined only by the value which depends on the electrical

Table 6.1. Composition of glass-forming melts

Glass No.	Glass	Composition, wt%				
		Na_2O	SiO_2	PbO	Al_2O_3	B_2O_3
1	OM-1	20	58.5	–	1.5	20
2	Zh-31	11.7	39.5	32.6	1.8	14.4
3	Na_2O–SiO_2	45.6	54.4	–	–	–
4	Na_2O–SiO_2	55.4	44.6	–	–	–

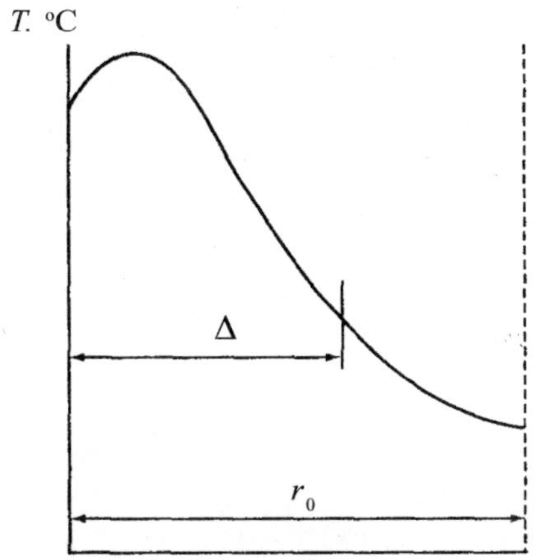

Figure 6.4. Radial distribution of temperature in a glass melt in direct HF heating [36]

conductivity of the melt. The removal of heat through the side walls of the cold crucible is determined by the thermophysical characteristics of the glass mass and the skull layer adjacent to the crucible wall.

The stability of the position of the temperature maximum of the glass mass of a specific composition in different conditions of the melting process indicates that the intensity of heat removal through the crucible walls is constant. The author of [6] investigated the temperature distribution in a melt in relation to two factors – the properties of the melt and the thermal conditions of the technological process. Figure 6.5 shows the radial temperature distribution in the melts of different glasses at $z = 0$ which indicates that the position of the maximum on the curve $T = f_r$ changes from $r = 35$ mm (melt Zh-31) to $r = 50$ mm (glasses 3 and 4). The dependence $T = f_r$ in the volume of the melt is the same.

According to the author of [6], the variation of the position of the maximum is determined mainly by different electrical conductivities of the investigated melts. This conclusion is confirmed by the similar position of the maximum for the glasses 3 and 4 characterised by almost the same electrical conductivity. Evidently, the change of the thermophysical characteristics of the melt and the skull layer for glasses of different composition is not very large and, consequently, the intensity of heat removal through the side surfaces of the crucible is not controlling for the radial distribution of temperature in the melt. It should be mentioned that the

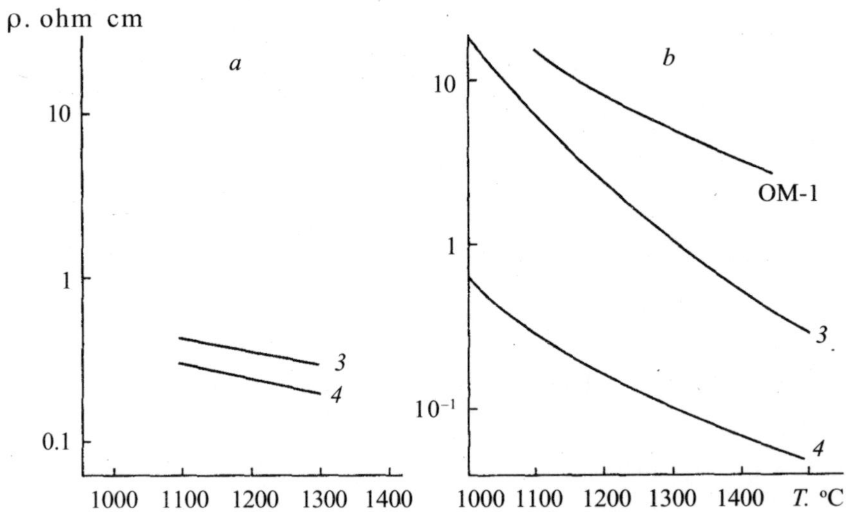

Figure 6.5. Temperature dependence of electrical conductivity and viscosity of glass melts [30]. Numbers of the curves correspond to glass compositions in Table 6.1.

temperature dependence of the electrical conductivity of glass-forming melts, in contrast to melts of simple oxides, has distinctive maxima and, consequently, by changing slightly the temperature of the melt it is possible to change the position of the maximum on the dependence $T = f_r$ However, it will be shown later that the author of [6] did not manage to record the displacement of the maximum of the radial distribution of temperature with the variation of the power absorbed by the melt. Possibly, the change of the position of the maximum was not greater than the error of determination of its position. Figure 2.7 shows that the magnitude of the detected maximum on the $T = f_r$ curves decreases with increasing depth z, and at any specific value of z it completely disappears. This is determined by the rapid removal of heat through the water-cooled crucible bottom and also by a decrease in the efficiency of heating the layers of the melt situated below the induction coil. In the surface layer of the melt with a thickness of 10–15 mm, the distribution of temperature in the depth of the melt (Fig. 2.7) is characterised by the presence of temperature gradients reaching 150°C/cm caused by the removal of heat through the open surface of the melt.

The temperature distribution in the melt is strongly affected by the melt viscosity. Figure 6.7 shows the distribution of isotherms in

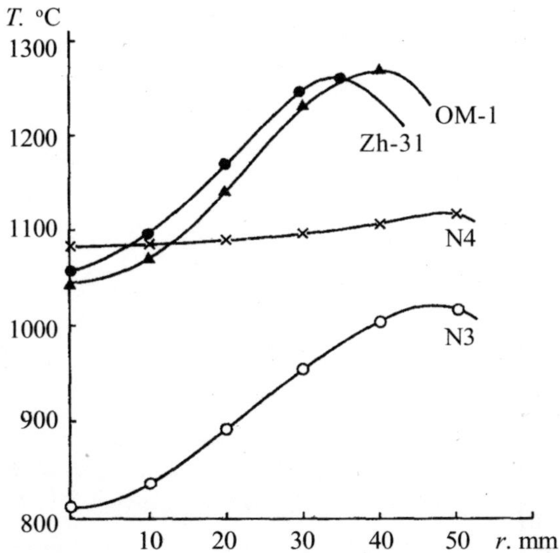

Figure 6.6. Radial distribution of temperature in glass melts of different composition on the melt surface [6]. Glass compositions are presented in Table 6.1.

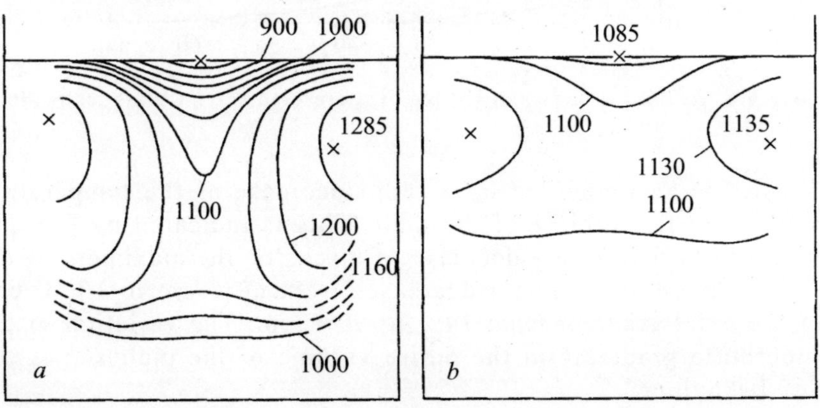

Figure 6.7. Temperature distribution in glass melts with different viscosity [6]: a) 45.6 Na_2O·54.4 SiO_2 (No. 3); b) 55.4 Na_2O·44.6 SiO_2.

the melts of glasses 3 and 4 which, having similar values of electrical conductivity, greatly differ in viscosity. For example, in the temperature range 1085–1140°C, the viscosity of the glass melt 4 changes from 0.4 to 0.2 Pa s, and the viscosity of glass 3 in the temperature range 1000–1285°C changes from 12 to 1 Pa s. Comparison of Figs. 6.7 (a) and (b) indicates that a decrease in the

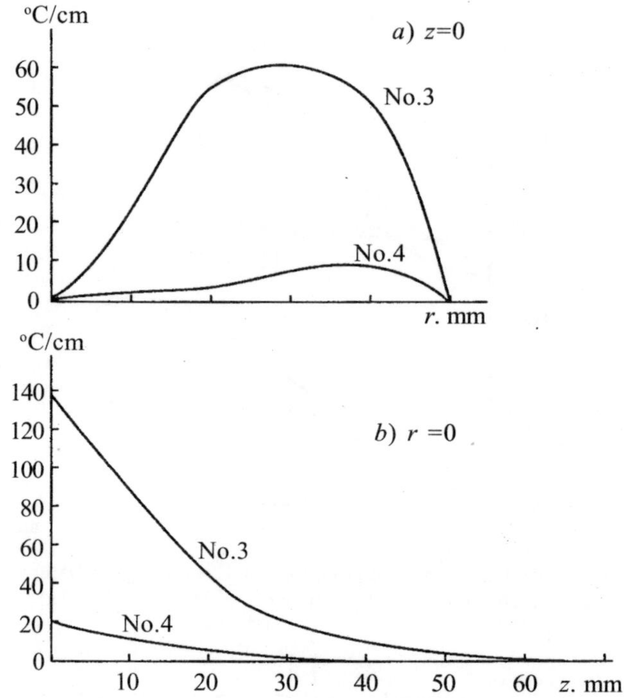

Figure 6.8. Radial (a) and axial (b) temperature gradients in glass melts No. 3 and 4 [6].

viscosity is accompanied by a rapid decrease of the temperature gradients in the volume of the melt. This is indicated by Fig. 6.8 which shows that with a decrease of viscosity, the maximum of the radial temperature gradient decreases from 60°C/cm to 10 °C/cm, and the axial gradient from 140° to 20°C/cm. The variation of the temperature gradients in the entire volume of the melt shows the same behaviour.

Evidently, the most probable reason for this large decrease of the temperature gradients in the melt with a decrease in melt viscosity is an increase in the intensity of convective mixing of the melt resulting in the equalisation of the temperature gradients.

Heat shields are used extensively in crystal growth for the shaping of the temperature gradients on the melt surface. In order to explain the effect of screening the surface of the melt on the distribution of temperature in the melt volume, experiments were carried out with the melt of glass 1. A heat shield made of fire clay was placed on the edge of the cold crucible. All other parameters:

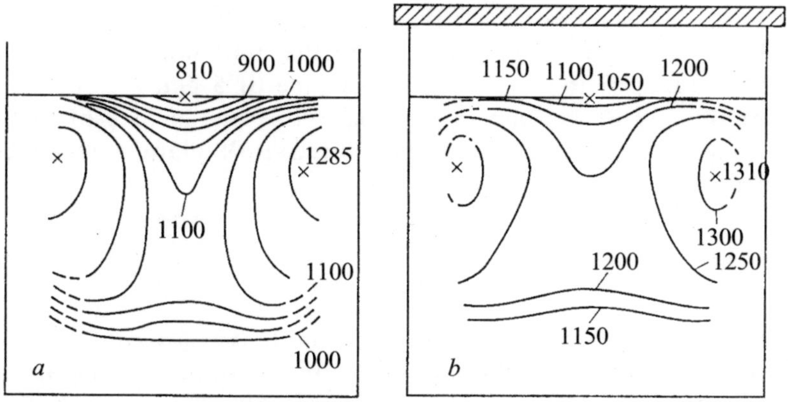

Figure 6.9. Temperature distribution in a glass melt with composition 45.6 Na_2O·54.4 SiO_2 with an open surface (a) and with a heat shield above the melt (b) [6]

the volume of the melt, the position of the induction coil in relation to the surface of the melt, the power, required by the RF generator, remained the same for all experiments. The measurement results are presented in Fig. 6.9. As indicated by the graph, the introduction of a heat shield has a significant effect on the temperature distribution in the volume of the melt:

–the maximum temperature on the surface of the melt increases by approximately 150°C;

–the temperature in the centre of the crucible on the surface of the melt greatly increases (≈250°C);

–the maximum temperature in the volume of the melt slightly increases, by ≈25°C;

–the region of the maximum temperature is slightly displaced into the bulk of the melt.

The measurements of the power, removed from the melt through the water-cooled crucible walls and the crucible bottom shows that the introduction of a shield increases this power from 6.8 to 8.1 kW. For the given crucible and the temperature of the glass melt of 1000–1200 °C the heat losses as a result of radiation from the free surface of the melt were ≈15% of the power absorbed by the melt.

The power of the heat flow of radiation (q) from the surface of the melt is calculated using the equation:

$$q = \{Q_{scr} - Q_O\} / \pi r^2, \tag{6.1}$$

where Q_{scr} is the power generated in the melt when using a heat shield; Q_0 is the power, generated in the melt without the screen. For the radius of the investigated surface of 5 cm, q = 17 W/cm^2.

The density of the power of radiation of the surface of the melt may also be evaluated using the Stefan–Boltzmann law which is strictly valid for the absolutely black body:

$$Q = \sigma \varepsilon T^4, \tag{6.2}$$

Here σ is the Stefan–Boltzmann constant, $\sigma = 5.67032 \cdot 10^{-8}$ W/m$^2 \cdot$ K^4 (in 1982), ε is the emissivity of the melt, T is the absolute melt temperature. For ε = 0.6 and T = 1500 K q = 17.7 W/cm^2, *i.e.* the calculated value of the density of the heat flow from the melt surface is in good agreement with the experimental value.

The effect of the position of the induction coil in relation to the melt and also of the value of the active power, generated in the melt, on the temperature distribution in the melt was investigated in a series of experiments with sodium–boron–silicate glass (NaBSi) at a constant volume of the melt. The main parameters, characterising the process conditions and also the temperature of the melt at certain points, are presented in Table 6.2.

To analyse the distribution of temperature in the melt, two regions should be selected: 1. The subsurface layer with a thickness of 10–15 mm, 2. The remaining volume of the melt. When the induction coil was lowered, the axial temperature gradients in the first region

Table 6.2. Main parameters of the process of melting and temperature distribution in the melt of sodium borosilicate glass [6]

Parameter	Parameter value			
Position of induction coil in relation to melt, z, mm	–35	–15	+20	–35
Power absorbed by melt, kW	12.1	12.1	12.1	9.9
Maximum surface temperature at point r mm, °C	1200; 40	1190; 40	1275; 40	975; 0
Temperature in crucible centre, r = 0, mm, °C	1060	1090	1040	975
Maximum temperature of melt at points r, z mm, °C	1355; 40; 50	1380; 40; 30	1425; 40; 20	1329; 37; 90

Comment: r is the distance from crucible centre, z is the depth of the melt

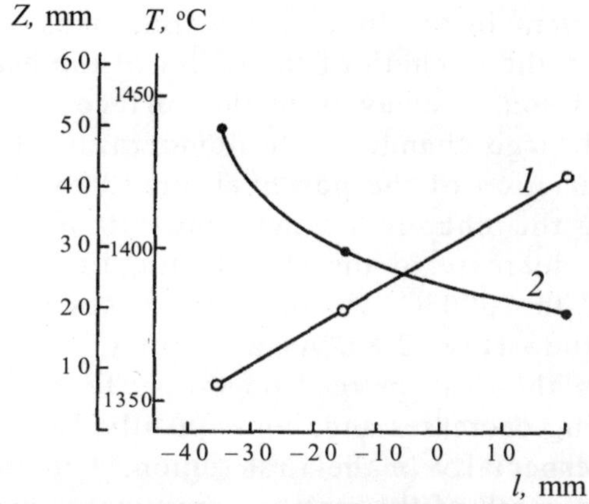

Figure 6.10. Dependence of the maximum temperature in the melt of sodium–borosilicate glass (1) and depth of the melt with maximum temperature (2) on the distance between the inductor and the surface of the melt: (+ – upper turn of the inductor above the melt surface) [6].

rapidly increased whereas the radial temperature gradients decreased. The distribution of temperature in the second region became more uniform and, in this case, the region of maximum temperature is displaced into the bulk of the melt and its size decreases (Fig. 6.10).

As already mentioned, the value Δ is linked with the specific electrical conductivity of the melt by the relationship:

$$\Delta = 5.03 \cdot 10^3 (\rho / \mu f)^{1/2} \tag{6.3}$$

Thus, the detected change in the position of the maximum of the radial distribution of temperature of the melt of glasses of different compositions (Fig. 6.5), within the framework of the investigated model which considers a cylindrical heat source instead of the induction coil, is determined by the variation of the diameter of the heat source d_{hs}.

A decrease in the level of the induction coil in relation to the surface of the melt corresponds to the displacement of the heat source into the bulk of the melt. In this case, the regional maximum temperature is also displaced into the bulk of the melt, the surface layers cool down and this results in an increase in the viscosity of the melt and convective heat transfer in the surface layers of the melt becomes more complicated. This results in a rapid increase of the axial gradient in the first region. A decrease in the radial temperature gradient in the investigated region is also associated with a decrease in the strength of the effect of the heat source with movement of the source away from the surface.

The observed large change in the temperature distribution the melt with the variation of the power absorbed by the melt is also associated with the change of the viscosity of the melt. For example, with a decrease of the melt temperature of the (NaBSi) glass from 1300 to 1000°C, viscosity increases by more than an order of magnitude (Fig. 2.8). As a result of a relatively large distance between the heat source and the surface of the melt, the intensity of heating decreases and, consequently, the viscosity of the melt increases, especially in the first region. The convective heat transfer from the depth of the melt is complicated and the removal of heat through the watercooled walls of the crucible becomes controlling in the first region. This results in a change in the sign of the radial temperature gradient.

Thus, in cases in which the induction coil is situated well below the level of the melt, the variation of the power absorbed by the melt makes it possible to control the radial temperature gradient on the surface of the melt. This is very important in growing single crystals by the Czochralski method.

6.2. OPTICAL HOMOGENEITY OF GLASS

Optical glass may contain the following defects: crystallites, air bubbles and ripples [20]. In this section, special attention is given to the mechanism of formation of ripples in the glass produced in a cold crucible. The ripples have the form of the filament-like thin layers of glass with different refractive properties in comparison with the surrounding mass of the glass because of differences in the chemical composition. In boiling of glass in ceramic potters kilns and bath furnaces, there are two sources of formation of ripples: the boiling vessel and the free surface of the glass melt. At a high temperature, the glass melt interacts with the ceramic material of the vessel with the formation at the walls of a boundary layer enriched with interaction products. This layer is partially trapped by the moving melt, mixes with it, forming the ripples of 'ceramic' origin.

In the molten condition, glass becomes volatile. Some components of the glass evaporate very rapidly, others remain almost completely in the melt. Consequently, a layer of a melt with changed composition, with the refractive index differing from the index of the main mass of the glass, forms on the surface. In the mixing, the surface layer is displaced into the volume of the melt and, consequently, ripples appear.

By analogy with the above, it is possible to define the sources of formation of ripples in the synthesis of glass by the methods of direct RF melting in the cold crucible. In the melt, the walls of the cold crucible are covered with a skull layer, consisting of a centre charge and viscous glass. As a result of the heterogeneity of the chemical composition of this layer, it may act as a source of ripples. However, it is important to note the principal difference in the formation of ripples in the methods used in commercial technology and in cold crucible technology. The point is that in boiling glass in a cold crucible, the formation of the boundary layer with the chemical composition different in comparison with that of the main mass may be avoided using, for example, a highly homogeneous charge, produced by the method of co-precipitation from solutions, or the boiling process is realised in such a manner that a relatively thick layer of viscous glass of the same composition as that of the main mass of the melt forms in the vicinity of the crucible walls in the homogenising stage.

The formation of ripples as a result of surface disproportionation takes place by the same mechanism during boiling of glass in ceramic and cold crucibles. (NaBSi) glass will be examined as an example: glass was boiled in a crucible with a diameter of 90 mm for 3 hours at a temperature 1400°C. The total losses of glass were 4.76 wt%. They consisted of the losses of 1.91 wt% B_2O_3 and 2.85 wt% Na_2O. Thus, in boiling, a layer depleted in boron and sodium oxides forms on the surface. Mixing with the remaining bulk of the glass, this layer results in the formation of ripples.

The optical quality of produced glasses was investigated in shadow equipment with a bright spot (Twayman equipment) [31]. In examination of shadow patterns it is necessary to note two aspects of the problems of the optical quality of glass. The first is the information on the distribution of convection flows in the glass mass, the second is the intrinsic optical quality of the produced glasses. The actual pattern of the ripples on shadow photographs characterises convection flows in the glass mass. However, when cooling a glass block in a cold crucible, the shape of the ripples is distorted because of the formation of the block (formation of a shrinkage cavity), and because of the distribution of flows during

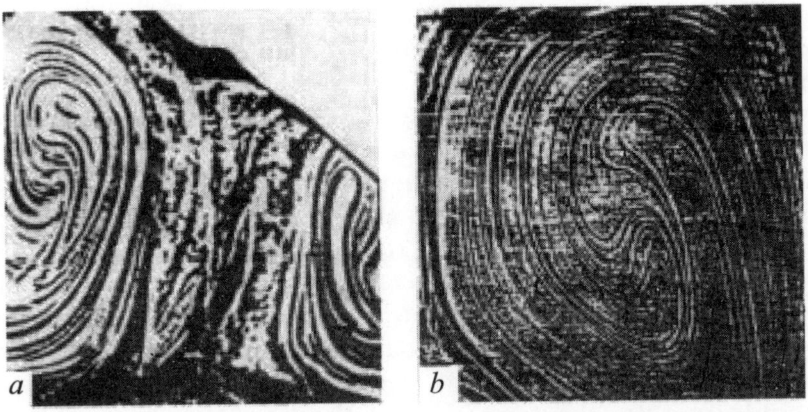

Figure 6.11. Schlieren photographs of glass containing lead [6]. a) without mixing the melt; b) the same glass with the melt mixed with a water-cooled stirrer. Speed of rotation 30 rpm, mixing time 3 h

nonuniform cooling, determined by the redistribution of thermal fields in the melt volume. The peripheral parts of the glass block are the first to cool down, i.e. the regions adjacent to the walls and bottom of the cold crucible. The pattern of the distribution of ripples in these regions reflects the direction of convection flows in the glass mass. The largest distortion is detected in the regions of glass solidifying last, i.e. the regions in the centre of the glass block. Since the cooling rate of the block is relatively high (the internal regions of glass are cooled to a temperature of 600–700 °C within 3–5 min), it is assumed that the shadow patterns provide relatively complete formation also on the convective flows in the glass mass. Figure 6.11 shows the shadow photograph of a glass containing lead, synthesised in a cold crucible for 4.5 hours without forced mixing of the melt. The photograph shows that the glass specimen is characterised by high optical heterogeneity and the nature of the ripples in the volume of the glass differs. Randomly distributed optical heterogeneities were found in the centre and at the periphery of the glass block, whereas in the remaining volume of the melt, the ripples are oriented according to the convection flows in the melt. This nature of distribution of the ripples is in good agreement with the temperature distribution in the melt (Fig. 2.6). For example, glass regions, distributed at of the periphery of the block are at a relatively low-temperature and, consequently, are characterised by high viscosity. Therefore, the convection stirring of the melt in these regions is difficult and this affects the optical

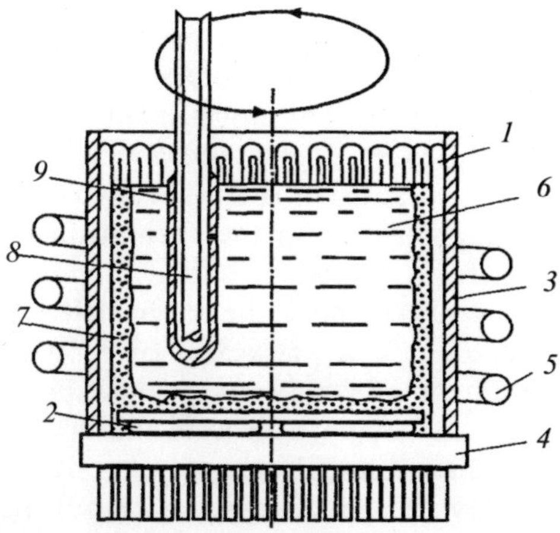

Figure 6.12. A crucible for producing high-temperature glasses. 1) a tubular water-cooled element; 2) water-cooled bottom; 3) insulating quartz cylinder; 4) securing ring made of an electrically insulating material; 5) inductor of HF generator; 6) melt; 7) charge; 8) water-cooled stirrer; 9) skull.

quality of glass. The central regions of the glass melt are characterised by small temperature gradients resulting in only slight convection mixing in this part of the melt. The remaining part of the melt has a higher temperature and is characterised by a steep temperature gradient. Therefore, in this region the ripples are oriented distinctively in the direction of convection flows. The formation of loop-shaped 'whirls' in this part of the block is evidently determined by the previously examined effects taking place during cooling of the glass block.

Thus, in boiling glass in the cold crucible, the melt contains 'dead' zones with insufficiently intensive homogenising of the melt. Therefore, the melt should be homogenised by means of forced stirring, using water-cooled stirrers. Figure 6.12 shows the method of production of high-temperature glass from a cold crucible using a water-cooled stirrer for homogenising the melt. The application of forced stirring results in the formation of high-intensity flows in the melt which determine the orientation of ripples. In this case, the volume of the 'dead' zones rapidly decreases, and only the 'column' in the centre of the crucible remains. The column is characterised by the formation of an ordered structure of ripples.

The application of forced stirring of the melt (within the limits of the given boiling and stirring conditions) is reduced to the ordering of ripples throughout the volume of glass. However, stirring of the melt does not make it possible to produce glass completely free from ripples. The application of stirrers of the previously mentioned type also does not lead to any significant change in the nature of distribution of ripples in glass specimens. In the case of NaBSi glass forced mixing results in the local concentration of ripples in small volumes of glass and in the formation of large sections of glass almost free from ripples.

It is also necessary to note the large difference in the shadow photograph of specimens of glasses of type Zh-31 and OM-1 (Table 6.1). In the specimens of Zh-31 glass, the ripples are characterised by higher contrast in comparison with the OM -1 glass, as a result of differences in the chemical composition of the glass. Since PbO is characterised by a high partial contribution to the refractive index, even small local changes in the PbO content of the glass result in a large change of the refractive index. The optical properties of the NaBSi glass depend only slightly on its components and, consequently, local changes in the concentration of some component result in a less marked change of the refractive index.

The optical homogeneity of the glass is also affected by the grain size composition of the initial charge. A melt of a charge containing SiO_2 grains with a diameter of 1–2 mm contains lack of boiling defects, situated in the vicinity of the skull which are evidently the reason for the formation of distinctive randomly distributed ripples (Fig. 6.13).

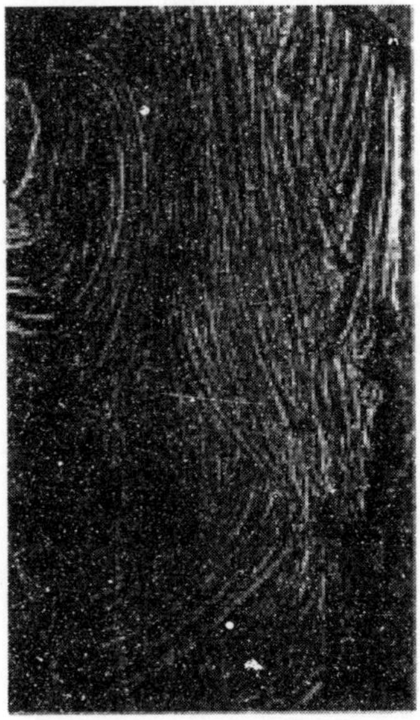

Figure 6.13. Schlieren photograph of sodium–borosilicate glass, produced from a charge containing large (1–2 mm) SiO_2 grains [6].

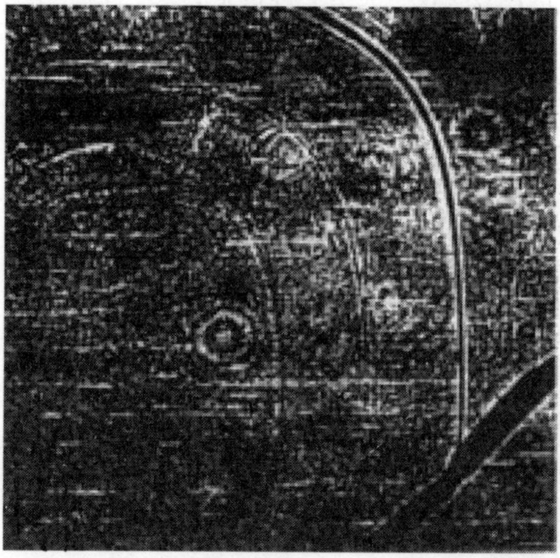

Fig. 6.14. Schlieren photograph of sodium–borosilicate glass produced using a water-cooled lid–cooler [6].

In order to improve the optical quality of glasses, the author of [6] proposed a method for removing ripples formed as a result of the selective evaporation of glass components from the melt surface. A melt of OM-1 glass was held at a temperature of 1400 °C to complete disappearance of particles of the charge and bubbles detected usually on the surface of the melt. The duration of melting of the charge and clarification of the melt should be minimum because in this stage the volume of 'surface' ripples increases. In the examined case, this time was 1.5 hours. Subsequently, the surface of the melt was cooled using a cover-refrigerator in the form of a coil consisting of water-cooled aluminium pipes. The power of the RF generator was reduced and the melt was homogenised below the cooled upper layer of the glass for 2.5 hours. As a result of this treatment, the quality of glass greatly improved and there were only single ripples in the bottom part of the melt (Fig. 6.14). Further improvement of the optical quality of glass is manifested in a decrease of the contribution of the skull bottom part of the melt to the formation of ripples. This may be achieved by 'freezing' the bottom part of the melt as a result of lowering slightly the crucible in relation to the induction coil.

Chemically homogeneous impurities

It is well-known that the main source of 'contamination' in the melting of initial powder components of glass and subsequent

Table 6.3. Physico–mechanical and electrical properties of high-temperature glass [2]

Property	Glass composition, wt.%	Glass composition, wt.% SiO_2: Al_2O_3:Sc_2O_3 = 60:30:10
Density, g/cm³	2.20–2.21	2.578
Refraction index	1.4588	1.542
Young modulus, kg/mm²	7450–7140	12740
Shear modulus, kg/mm²	3200	5100
Poisson coefficient	0.16	0.249
Microhardness, kg/mm²	795	780
Linear expansion coefficient, deg⁻¹	$5 \div 7 \cdot 10^{-7}$	$27 \cdot 10^{-7}$
Temperature of start of softening, °C	1250	1200–1250
Specific resistivity, ohm·cm		
at 20 °C	$1 \cdot 10^{19}$–$1 \cdot 10^{18}$	$3 \cdot 10^{13}$
at 600 °C	$6 \cdot 10^{7}$–$6 \cdot 10^{6}$	$1.7 \cdot 10^{5}$
at 1000 °C	$1 \cdot 10^{6}$–$2.5 \cdot 10^{5}$	$1.1 \cdot 10^{4}$
Dielectric permittivity at 20°C	3.2	6.8
Tangent of the angle of dielectric losses in the frequency range 30 kHz – 10 MHz	$2 \cdot 10^{-4}$	$5 \cdot 10^{-4}$

homogenising of the melt is the crucible material. Contamination of the melt with the crucible material may take place by both the direct interaction of the crucible with the melt and by the transfer of material of the crucible into the melt through the gas phase. The dominant contamination mechanism is determined by the crucible material and the specific conditions of the boiling process (temperature, working atmosphere).

Platinum is a classic crucible material used widely in optical glass boiling. Platinum is resistant to the effect of the majority of

Table 6.3 (continued)

Glass composition, wt.%		
SiO_2:Al_2O_3:Y_2O_3=50:20:30	SiO_2:Al_2O_3:La_2O_3=54:21:25	SiO_2:Al_2O_3:Er_2O_3=60:30:10
3.05	3.016	2.833
1.592	1.573	1.555
11040	9650	9080
4510	3940	3660
0.225	0.225	0.240
1050	735	930
$70 \cdot 10^{-7}$	$28 \cdot 10^{-7}$	$10.3 \cdot 10^{-7}$
1000	1100–1200	1150–1200
$2 \cdot 10^{13}$	$4 \cdot 10^{13}$	$1 \cdot 10^{13}$
-	$1 \cdot 10^{5}$	$1.7 \cdot 10^{7}$
-	$5 \cdot 10^{3}$	$5 \cdot 10^{3}$
7.0	7.1	6.9
$2 \cdot 10^{-4}$	$1 \cdot 10^{-4}$	$2 \cdot 10^{-4}$

glass melt and, consequently, homogeneous glasses may be produced. However, in glasses synthesised in platinum crucibles there are often clusters of inclusions of metallic platinum.

As stressed a number of times, one of the advantages of the method of direct RF melting in the cold crucible is the high purity of the method determined by the absence of direct contact of the melt with the crucible material. Therefore, it was natural to investigate the variation of the chemical composition of glass in comparison with the composition of the initial reagents after boiling in the cold crucible. For this purpose, experiments were carried out with the boiling of OM-1 and Zh-31 glasses using initial reagents of high purity. The content of the main colouring elements in the initial reagents did not exceed several units, 10^{-6} wt%, and only the iron content was an order of magnitude higher. The boiling of glasses, carried out in a crucible produced from copper pipes, showed that when using the initial charge containing nitrate salts, the glass becomes blue, with this colour typical of the presence of Cu^{2+} in the glass. The absorption spectrum of the OM-1 glass is characterised by the presence of a wide absorption band with a maximum in the range 750 nm. When the initial materials are represented by oxides resulting in colouring of the glass, the absorption band does not form in this region of the spectrum. The comparison of these two factors was used for proposing a mechanism of contamination of glass with copper. When heating a charge, the initial stages of synthesis of glass are characterised by thermal dissociation of nitrate salts with the precipitation of nitrogen oxides. These components in the gas form react with the copper oxide film formed on the surface of copper pipes resulting in the formation of copper nitrate, with the latter falling into the melt when loading the charge into the crucible. Thus, copper is transferred from the crucible walls into the melt. Therefore, glass should be boiled in crucibles made of aluminium pipes. Aluminium is not a colouring addition in glass and is included in the main composition of many synthesised glasses. The aluminium oxide layer on the surface of pipes is resistant to the effect of nitrogen oxides. In [6] it was established that synthesised glasses are characterised by an increase in the content of iron, copper and vanadium in comparison with the initial charge. One of the possible channels of contamination of glass with iron and copper is the preliminary heat treatment of the charge. The need to carry out his operation was caused by special features of direct RF melting in the cold crucible because the initial charge contains H_3BO_3 which decomposes during starting melting, with the generation of water vapours. The vapours condense on the cold pipes of the crucible and close tubular sections, resulting in electrical breakdown and failure of the crucible walls. It is therefore necessary to introduce an additional operation: prior to loading into the cold crucible, the charge is baked in a resistance furnace at a temperature of ≈400 °C in order to remove water vapours.

The application of a quartz crucible for the baking process greatly decreases the content of undesirable impurities in glass. The quality of produced glass was inspected on the basis of measurements of absorption losses by the colourimetric method at a wavelength of 1.06 μm [32]. However, it should be noted that the absorption losses were slightly higher in comparison with multicomponent glasses. This was due to a relatively high content of the impurities in the given reagents. The surrounding atmosphere in which glass synthesis is carried out may be a source of contamination of glass in heating with RF currents and, consequently, the synthesis of high purity glasses must be carried out in specially built facilities with the efficient removal of dust from the air.

6.3. REFRACTORY GLASSES OF THE R_2O_3– Al_2O_3– SiO_2 SYSTEM (R = Sc, Y, La, Nd, Er)

The glasses based on alumosilicate systems of yttrium and lanthanum are characterised by valuable properties for application. The combination of the high refractive index with high mechanical characteristics and deformation temperature and also the relatively low values of the thermal expansion coefficient and the very high resistance to the effect of alkali make these glasses promising optical materials capable of operation at high temperatures in chemically corrosive media.

The most promising method of production of these types of glass is the method of direct RF melting in the cold crucible. The method has no principal restrictions as regards the synthesis temperature and can also be used for producing glasses in volumes required for practical application without extensive contamination of glass with the crucible material.

The authors of [2] investigated glasses of the R_2O_3–Al_2O_3–SiO_2 system (here R is Y, Sc, La, Er) synthesized by the method of direct RF melting in the cold crucible. Glass synthesis was carried out at temperatures of approximately 2000°C for several hours. This resulted in the formation of glass blocks weighing up to 1.5 kg with almost no inclusions and components of the charge which have not managed to react. The compositions of the glasses corresponding to the section of the ternary phase diagram of the $Al_2O_3{\cdot}3SiO_2$–$R_2O_3{\cdot}3SiO_2$ system. The results show that at a content of up to 5 mol% ($R_2O_3{\cdot}3SiO_2$) glass remains transparent up to temperatures of 1100–1200 °C. With an increase of the content of $R_2O_3{\cdot}3SiO_2$, there is intensive crystallisation of glass at temperatures of 1050–1150°C. Crystallisation of a glass with a low content of oxides of rare-earth elements results in the formation of mullite ($3Al_2O_3{\cdot}2SiO_2$). An increase of the content of the rare-earth oxide after the formation of mullite leads to the formation of silicates of rare-earth elements, for example $Er_2(Si_2O_7)$ or

Table 6.4. Values of the coefficient of thermal expansion and temperature of the start of deformation and glass transition of La_2O_3–Al_2O_3–SiO_2 system

Composition of glass, mol%			$\alpha_{20-800} \cdot 10^7$, deg^{-1}	$T_{s.d.}$,°C	T_g, °C
SiO_2	Al_2O_3	La_2O_3			
50.1	32.8	17.0	64.0	855	820
55.0	28.4	16.6	62.5	860	825
59.1	34.7	6.3	43.2	895	850
71.6	26.4	2.0	33.7	920	900
72.0	21.5	6.3	38.9	910	865
68.3	25.6	6.1	39.1	905	860

alumosilicates with a complicated composition (Chapter 4).

Recently, high-temperature glasses have attracted attention because of possible use as laser materials. For example, the authors of [5] synthesised glasses with the composition 44.7 SiO_2–35.8 Al_2O_3–19.5[$(1-x)$ La_2O_3 + $x$$Nd_3O_3$] and investigated their physical–chemical and spectral–luminescence properties. Glass from a charge prepared by the co-precipitation method was synthesised. The charge was baked in corundum crucibles in a gas furnace at a temperature of 1550°C for 1.5 hours. The physical and mechanical properties of the synthesised glasses are presented in Table 6.3 shows that the values of the refractive index, density, Young modulus, shear modulus and the Poisson coefficient for high - temperature glasses of this compositions are considerably higher than for quartz glasses. Figure 6.14 shows the dependence of the density, refractive index and microhardness on the lanthanum oxide content of glass. Table 6.4 shows the values of the thermal expansion coefficients in the temperature range 20–800°C. These values are important for practical application of the glasses.

Glass was synthesised in equipment Donets-4, which was described in Chapter 2. Crucibles were produced from copper or aluminium and their diameter was 130 and 90 mm.

The process of synthesis of glasses in the cold crucible may be divided into the following main stages: preparation of the charge and placing the charge in the crucible; starting heating and melting; clarification (holding) of the melt; cooling of the melt and annealing of the glass blocks. The initial reagents were baked in a silit-rod furnace at a temperature of 1000–1100°C for 8–10 hours, weighed in corresponding proportions and thoroughly mixed in a china mortar. The amount of the prepared charge for crucibles with diameters of 130 and 90 mm was 4.5–7 kg and 1.5–2.5 kg, respectively.

The charge was placed in the crucible so that 80–90% of the volume of the crucible was filled. To produce a starting melt, a graphite rod with a diameter of 50 mm and 70 mm long was placed in the centre of the crucible. The rod was used as a heating source in order to produce the starting volume of the melt. The graphite rod was heated and melted the charge which was in direct contact with it. During melting of the charge a cavity formed around the rod which was then filled with the next portions of the charge up to obtaining the critical volume of the melt sufficient for the effective absorption of the energy of the RF field directly by the melt. This was controlled on the basis of the variation of the intensity of anodic current I_a. In the conventional operating conditions of the RF generator, the intensity of the anodic current in the idling regime is 2.5 A, and the increase of current to 3.5–4 A indicates the start of direct RF melting of the charge. The critical value of current $I_{a\,cr}$ depends on the composition of the synthesized glass, mainly on the SiO_2 content. For example, for glasses with 70–75 mol% of silicon oxide, the critical current is $\approx$ 4 A, and for glasses with a lower content of SiO_2 it is $\approx$ 3.5 A. It may be assumed that the melts of the glass with a high content of silicon oxide have a lower specific electrical conductivity than the glasses with a lower content of SiO_2 and, consequently, the starting volume of the melt in the first case should be larger (or it should be heated to a higher temperature). This also results in an increase of $I_{a\,cr}$.

The moment of the start of direct RF heating of the melt is determined visually when the temperature of the melt in the vicinity of the crucible walls becomes higher than in the central part of the crucible. When the critical volume of the melt has been reached, the graphite heater is removed and small portions (100–200 g) of the charge are poured into the crucible, up to obtaining the required volume of the melt. The duration of starting melting is usually 15–20 min and the melting process of the charge lasts approximately 1 hour.

Starting melting and placing of the charge are characterised by complicated physical–chemical processes: melting of the individual components of the charge, the formation of the liquid phase whose composition corresponds to the eutectic with the lowest melting point on the equilibrium diagram, the occurrence of solid-phase reactions and reactions in the melt leading to the formation of silicates, dissolution of SiO_2 grains in the resultant mixture of the silicates.

The next stage of the process is the clarification of the melt usually represented by the removal of gas bubbles from the melt [20]. In the stage, the chemical composition of the melt is equalised. After loading the charge, the melt contains a large number of bubbles and insoluble particles of the charge. In order to accelerate the dissolution of charge particles and remove the bubbles, it is necessary to increase the power of the RF generator up to the temperature of the melt (1600–1700)°C. As already mentioned, a characteristic feature of direct RF melting of glasses is the presence of a steep temperature gradient in the melt.

The condition of the melt in the cold crucible is characterised by the maximum temperature which is usually measured with an optical pyrometer. The nonuniform distribution of temperature in the melt has a strong effect on the processes of clarification of the glass mass and dissolution of solid charge particles. The initial process is the removal of bubbles from the zone with the maximum temperature situated at a distance of 10–20 mm from the crucible walls. After holding for 20 min this zone of the melt no longer contains any bubbles. The remaining bubbles are concentrated in the vicinity of the crucible centre.

An important consequence of the steep temperature gradient in the melt is the high-intensity convection of the melt resulting in the clarification of the latter. Figure 6.13 shows the direction of the convection flows in the vicinity of the melt surface decorated with the trajectories of movement of the bubbles in the melt. The layer of the melt, adjacent to the crucible walls, contains charge particles that are in direct contact with the walls of the cold crucible and with the layer of viscous glass whose thickness depends on the operating regime of the generator and is usually 3–5 mm. Because of small variations of the generator power, the skull melts periodically with the generation of bubbles which either 'freeze' in the layer of viscous glass or penetrate into its main volume. In order to eliminate this potential source of bubbles, after holding the melt for 1–1.5 hours, the generator power is slightly reduced. This results in an increase in the thickness of the layer of the viscous glass melt and, consequently, in a decrease of the probability of penetration of bubbles from the skull into the melt. The presence of the skull layer increases the probability of crystallisation of the glass because the charge particles are potential nuclei situated in direct contact with the melt and, under specific conditions, may initiate the crystallisation of the whole volume of the glass. Thus, if a glass has a high crystallisation capacity, it is very difficult to produce this glass in the cold crucible.

The conditions of production in the cold crucible of glasses with higher crystallisation capacity depend on a large number of factors: the melt volume, the thickness of the layer of viscous glass adjacent to the crucible walls, the viscosity and electrical conductivity of the melt, and also the cooling rate of the melt and the rate of crystallisation of the melt. In practice, it is very difficult to take all these factors into account. However, it may be concluded quite confidently that if the rate of crystallisation in the 'dangerous' temperature range is sufficiently high to ensure that during cooling of the melt the crystal can 'grow through' the entire volume of the glass, then a solidified ingot is produced.

The melt is homogenised using a rod copper water-cooled stirrer, with the speed of revolution being 30–40 rpm. After clarification and homogenising of the melt, the stirrer is withdrawn from the melt and this is followed by a slower decrease of the generator power and the temperature of the melt resulting in a decrease of the electrical conductivity of the melt and, consequently, in an increase of the value of Δ (for the explanation of the value Δ see Chapter 1). When Δ becomes comparable with the crucible radius $\Delta \approx r_0$, the energy absorbed by the melt rapidly decreases and the melt cools down spontaneously as a result of heat transfer through the walls and bottom of the crucible.

A glass with a specific composition has a characteristic temperature dependence of electrical conductivity which determines the lower temperature up to which the cooling rate of the glass may be regulated. For the majority of investigated high-temperature glasses, this temperature is in the range 1400–1500°C. When the temperature is reached, RF equipment is switched off and the glasses cooled in the cold crucible. Cooling of the glass is nonuniform: the peripheral section of the block cools at a considerably higher rate than the central part of the block. The duration of withdrawal of the glass from the crucible is determined by two mutually excluding circumstances: in order to prevent cracking, the glass block must be withdrawn from the crucible as soon as possible; however, this operation may result in heating of

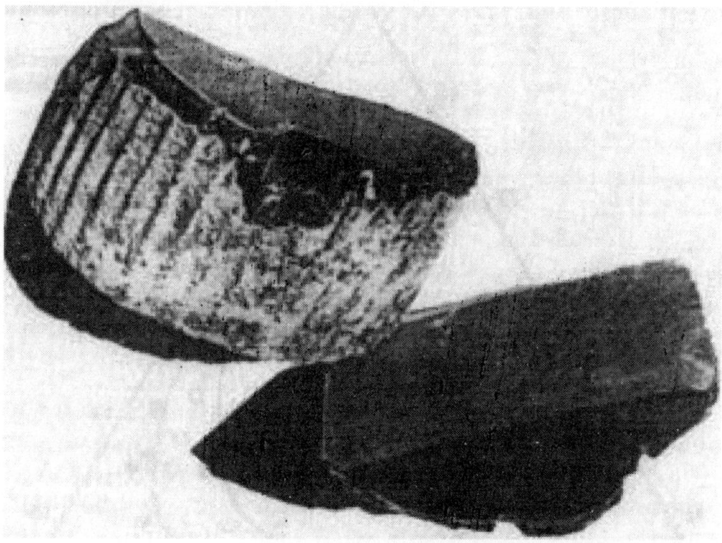

Figure 6.15. A block of alumosilicate glass, active with neodymium produced in a cold crucible [6].

the peripheral part of the ingot as a result of the internal hot region. This may result in softening of the peripheral part of the block and escape of glass mass. Experiments show that in extraction of the glass from the crucible at the surface temperature (700–800°C the block does not crack and is not deformed. The glass, withdrawn from the crucible, is annealed in a furnace at a temperature of 800°C for 4–6 hours and is then cooled by inertia in the furnace to room temperature.

The distribution of ripples in the produced glasses was identical with that in NaBSi glass, but the ripples were characterised by a lower contrast, probably as a result of less extensive evaporation of the glass components during synthesis. This technology was used to produce glass blocks weighing 5–7 kg free from lack of boiling defects and bubbles. Figure 6.15 shows a typical casting of neodymium–alumosilicate glass.

Examination of the range of glass formation in the La_2O_3–Al_2O_3–SiO_2 and Y_2O_3–Al_2O_3–SiO_2 systems

The method of direct RF melting in the cold crucible was used in [6] for the determination of the ranges of glass formation in the La_2O_3–Al_2O_3–SiO_2 and Y_2O_3–Al_2O_3–SiO_2 systems. It should be mentioned that the method of synthesis in the cold crucible proved to be highly suitable for this purpose because it makes it possible to examine the range of glass formation in no more than two experiments.

The following procedure was used for the preparation of experimental specimens. The two–component SiO_2 + Al_2O_3 mixture of a specific composition was loaded into the cold crucible. For starting melting, 3–5 g of aluminium in the form of powder was placed in the initial charge. The powder melted the surrounding charge. After formation of the critical volume of the melt, the melt was heated directly by the energy of the RF field. The melt was held at a temperature of 1800°C up to complete burning-out of the metal floating on the surface of the melt. Usually, this time is ≈ 1 hour. Subsequently, the prepared SiO_2 + $La_2O_3(Y_2O_3)$ mixture of the required composition was added in small portions to the melt (30–50 g). After adding each portion of the charge, the melt was held for 10 min. Subsequently, several droplets of the melt weighing 1–3 g were taken using a quartz stick. The droplets were subsequently quenched on a water-cooled metallic plate. Using this method, it was possible to produce specimens of glasses with compositions in different ranges of the investigated phase diagram.

The proposed method can be used to produce specimens for investigations of the three-component system. It is important to mention one significant principle which must be taken into account in the preparation of specimens for the examination of glass formation ranges. The initial composition of the phase diagram should have a higher liquidus temperature than every following portion. Otherwise, a crust forms on the surface of the metal preventing the averaging of the chemical composition of the melt and the production of specimens with a homogeneous composition. The range of glass formation was determined by the methods of x-ray and microscopic analysis of the specimens and also by visual examination. Figures 6.16 and 6.17 show the areas of glass formation in the investigated systems. In these systems at an SiO_2 content in the range 65–75 mol%, an addition of approximately 2 mol% of lanthanum or yttrium oxides results in the formation of transparent glasses. At an $SiO_2 \leq 60$ mol%, the boundary of the range of glass formation is limited by the compositions characterised by surface crystallisation, and at a higher silicon content, the range of glass formation borders with the compositions of glasses characterised by luminescence. The detected widening of the ranges of glass formation in investigated systems is caused evidently by a

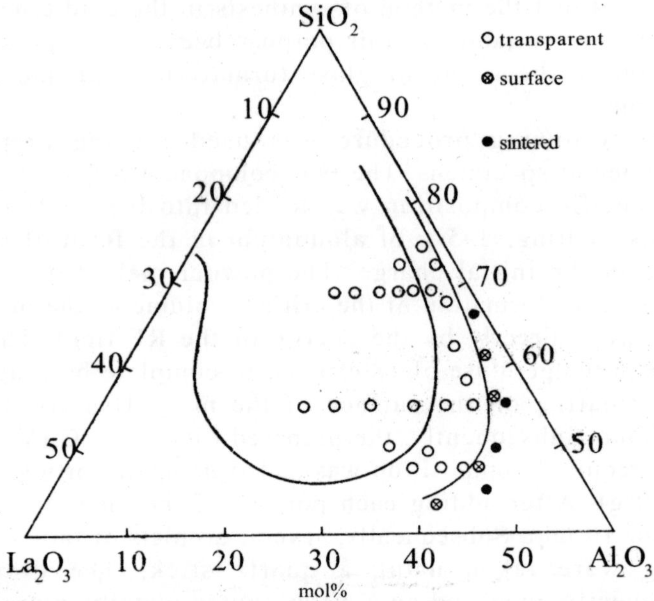

Figure 6.16. Region of glass formation in the La_2O_3–Al_2O_3–SiO_2 system. Thick line – synthesis temperature 2000°C [6]; thinner line – synthesis temperature higher than 1600°C [35].

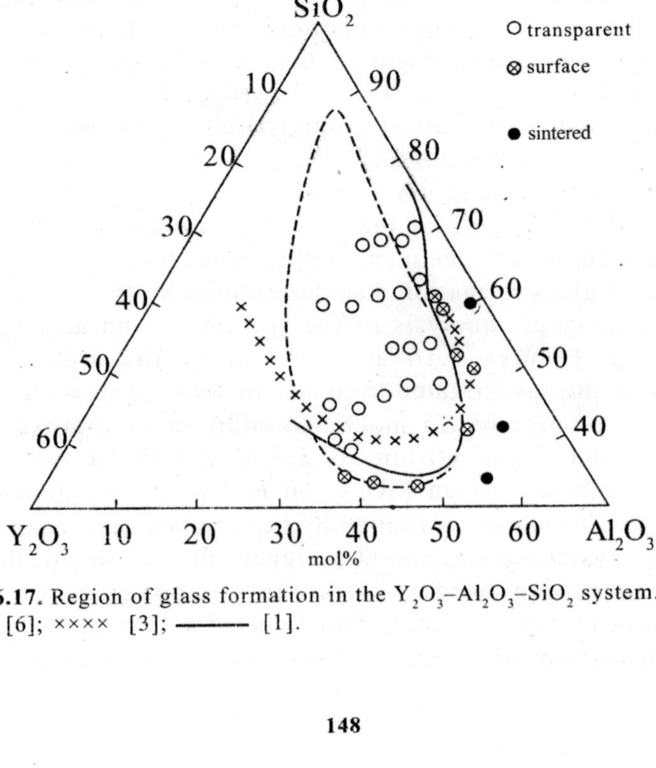

Figure 6.17. Region of glass formation in the Y_2O_3–Al_2O_3–SiO_2 system. Data from: - - - - - [6]; ×××× [3]; ——— [1].

Table 6.5. Boiling conditions for glass of different compositions

Glass composition	Anode voltage, kV	Anode current, I_a, A	Grid current, I_c, A	Melt temperature, °C	Diameter of cold crucible, mm
$50SiO_2 \cdot 25Al_2O_3 \cdot 25Tb_2O_3$	7.0	5.5	1.15		130
$49.5SiO_2 \cdot 24.7Al_2O_3 \cdot 25.8Nd_2O_3$	7.4	5.2	0.95	1730	130
$49.5SiO_2 \cdot 24.7Al_2O_3 \cdot 25.8Y_2O_3$	7.8	5.2	1.1	1820	130
$47.3SiO_2 \cdot 27.5Al_2O_3 \cdot 22.7Yb_2O_3 \cdot 2.5Ho_2O_3$	7.8	5.5	0.85	1750	130
$54SiO_2 \cdot 27Al_2O_3 \cdot 19Y_2O_3$	8.0	5.0	0.8	1770	130
$54SiO_2 \cdot 27Al_2O_3 \cdot 19La_2O_3$	8.4	5.0	1.15	1770	130
$10Sm_2O_3 \cdot 30Al_2O_3 \cdot 60SiO_2$	9.0	4.0	1.2	1950	90
$10La_2O_3 \cdot 30Al_2O_3 \cdot 60SiO_2$	9.7	5.2	1.1	2100	90
$10Nd_2O_3 \cdot 30Al_2O_3 \cdot 60SiO_2$	9.5	4.7	1.0	2050	90
$15Y_2O_3 \cdot 20Al_2O_3 \cdot 65SiO_2$	8.5	4.3	1.1	2000	90
$15Dy_2O_3 \cdot 20Al_2O_3 \cdot 65SiO_2$	8.9	5.1	0.9	2300	90

high synthesis temperature and a high cooling rate of the analysed samples. The compositions and conditions of boiling of the synthesised glasses are presented in Table 6.5.

References

1. Sakharov B.A., Sedykh G.S., Ranevskaya E.P., *Trudy Giredmeta*, 1974, Vol. 52, 83–87.
2. Aleksandrov V.I., et al., *Fizika i Khimiya Stekla*, 1977, Vo. 3, No. 3, 177–180; 1980, Vol. 6, No. 2, 170–173.
3. Makishima A., Tamura Y., Sakuino T., *J. Am.Ceram.Soc.*, 1978, Vol. 61, No. 5–6, 247–249.
4. Makishima A., Shimohira T., *J. Non.-Crystaline Sol.*, 1980 Vol. 39/40, part 11, 661–666.
5. Malashkevich G.E., et al., *Zh. Prikladnoi Spektroskopii,* 1982, Vol. 37, No. 2, 261–265.
6. Borik M.A., Development of technology and examination of oxide glasses for electronics, Dissertation, General Physics Institute, Moscow, 1985.
7. Synthesis of laser materials from the melt by direct HF heating ina cold container, Institute of Physics, Russian Academy of Sciences, Moscow, 1985.
8. Moulin J., Reboux I., *Verres et Refractaires*, 1972, Vol. 26, No. 4–5, 123–127.
9. Scott B., Rawson H., *Glass Technology*, 1973, Vol. 14, No. 5, 115– 124.
10. Scott B., Rawson H., *Opto-electronics*, 1973 Vol. 5, No. 4, 285–288.
11. Aleksandrov V.I., et al., *Uspekhi Khimii*, 1978, `Vol. 47, No. 3, 385–427.
12. Nezhentsev V.V., et al., *Izv. LETI, Nauchnye Trudy*, 1980, No. 273, 51–56.
13. Petrov Yu.B., Cold crucibles, Metallurgyia, Moscow, 1972.

14. Aleksandrov V.I., et al., *Current Topics in Materials Science*, 1978, Vol. 1, 421–479.
15. Petrov Yu.B., Induction melting of oxides, Energoatomizdat, Leningrad, 1983.
16. Nezhentsev V., et al., A method of boiling refractory glass and an induction furnace for this purpose, Author' Cert. No. 872 465, USSR, Byul.. Izobr. 1981, No. 38, 123.
17. Gossink R.G., 11 International Congres on Glass, Prague, 1977, 114–157.
18. Gossink R.G., Topical Meeting on Optical Fiber Transmission, Williamsburg, 1975, TaB 4, 1–5.
19. Parent I. P., LeSergent C., Lerner P., Galay S., Brehan C., Second European Conference on Optical Fibre Communication, Paris, 1976, 33–35.
20. Demnika L.I. (editor), Physico-chemical fundamentals of production of optical glass, Khimiya, Leningrad, 1976.
21. Rawson H., *Physics in technology*, 1974, Vol. 5, No. 2, 91–114.
22. Mase H., Ocda K., *J. Non-Crystalline Solids*, 1980, Vol. 38/39, part 2, 807–817.
23. Rawson H., J. Non-Crystalline Solids, 1977, Vol. 26, 1–25.
24. Plumat E., *J. Non-Crystalline Solids*, 1977, Vol. 26, 179–201.
25. Kobayashi N., *J. Crystal Growth*, 1978, Vol. 43, 337–363.
26. Kobayashi N., Arizumi T., *J. Crystal Growth*, 1980, Vol. 49, No. 3, 419–425.
27. Kobayashi N., *J. Crystal Growth*, 1981, Vol. 52, No. 2, 339–344.
28. Shiroki K., *J. Crystal Growth*, 1977, Vol. 40, No. 1, 129–138.
29. Brandle C.D., *J. Crystal Growth*, 1977, Vol. 42, No. 1, 400–404.
30. Mazurin O.V., et al., Properties of glasses and glass-forming melts, Handbook, Vol 1, Nauka, Leningrad, 1973.
31. Doladugina V.S., Equipment and methods fo rexamining the homogeneity of quartz, in: Growth of crystals, Institute of Crystallography, Vol. 3, Moscow, 1961, 481.
32. Bryushkova T.I., et al., *Kvantovaya elektronika*, 1976, Vol. 3, No. 11, 2500–2503.
33. Bogdanova G.S., Kozel'skaya E.S., Glass-forming systems and new glasses, Khimiya, Moscow, 1971, 49–50.

7

A family of crystals – fianites

Single crystals – large crystals of simple and complicated substances, play a significant role in science, technology and industry. They are used in the development of lasers and optical devices, electronic circuits and sections of computers, and different analytical devices. In the group of single crystals important for practice, an important position is occupied by the crystals based on oxide and complex oxide compounds with high melting points and exceptionally high chemical resistance. These are crystals of quartz, corundum, yttrium–aluminium and gadolinium–gallium garnets, lithium niobate or tantalate, and other compounds. Many of crystals of this type are present in the nature in the form of appropriate minerals and follow their traditional names: ruby, garnet, alexandrite, apatite, and so on. However, others have no natural analogues and are known only as artificially produced crystals. In the group of synthesized crystals, new crystals, fianites, are of considerable importance. In recent years, the production of fianites in Russia and throughout the world has reached the level of the production of the 'mass-produced' synthetic crystals.

Fianites is the name for the entire family of the crystals. Their chemical base is represented by dioxides of rare metals, zirconium (Zr) and hafnium (Hf), and the stabilising components are represented by oxides of other elements: yttrium (Y), calcium (Ca) and rare-earth elements characterised by high melting points. The properties of fianites are truly unique. The exceptionally high melting point (depending on composition, the melting point varies in the range 2700–2850°C) is combined with high hardness and chemical resistance. Fianites are optically isotropic because of the highly symmetric cubic structure. They are characterised by higher transparency in a wide wavelength range, highly refractive index values (2.0–2.2) and high dispersion. Because of their unique properties, fianites are used in technology, medicine, jewellery production since they efficiently simulate diamonds.

7.1. POLYMORPHISM AND THE STRUCTURE OF PHASES OF PURE ZIRCONIA

Pure zirconia has three polymorphic modifications: monoclinic (*m*–$P2_1/C$), tetragonal (*t* – $P4_2/nmc$) and cubic (*c*–Fm_3m). The region of thermal stability of the monoclinic phase extends up to 1140°C, the tetragonal phase from 1160 to 2370°C, and the cubic phase from 2370 to the melting point of 2680°C.

The cubic phase ZrO_2 is a non-distorted structure of the fluorite type (Fig. 7.1). The spatial group is Fm_3m; the coordination number of oxygen is 4, the coordination number of zirconium in relation to oxygen is 8 [1, 2].

The structural type of fluorite is one of the most widely encountered types of inorganic crystalline materials [3]. The main reason for such large occurrence of the fluorite motive MX_2, is the high stability of its cation sublattice, resulting from the distribution of all metallic atoms by the type of the densest cubic packing-in the nodes of the face-centred F-lattice. In particular, this cation structural motive changes constantly in all derivatives of the structure of CaF_2, whereas the oxygen matrix often undergoes extensive transformations. The cation sublattice usually has no vacancies, i.e., it is completely filled.

The second special feature of this structural type is the constancy of the coordination of the anions each of which is always surrounded by four metallic atoms, forming together a tetrahedron distorted to various degrees. In Fig. 7.2*a* the fluorite structure is depicted in Pauling polyhedrons showing clearly the primitive cubic packing of the anions distributed in the form of layers in the nodes of the quadratic cell. Half of all anion cubes is populated in accordance with the three-dimensional checkered law. The cations are characterised by the eightfold coordination. Each node of the cation cell has four filled and four empty polyhedrons.

If some atom X of the anion packing is eliminated from the investigated motive, the four polyhedrons of the cube, previously situated in the close vicinity of the selected node, loose one apex

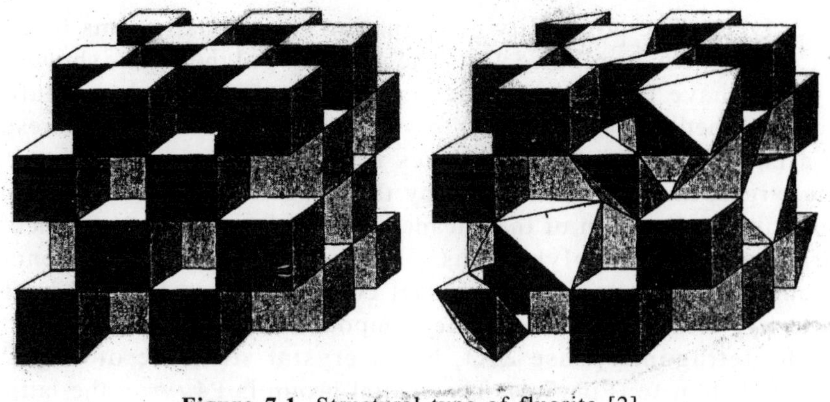

Figure 7.1. Structural type of fluorite [2].

a *b*

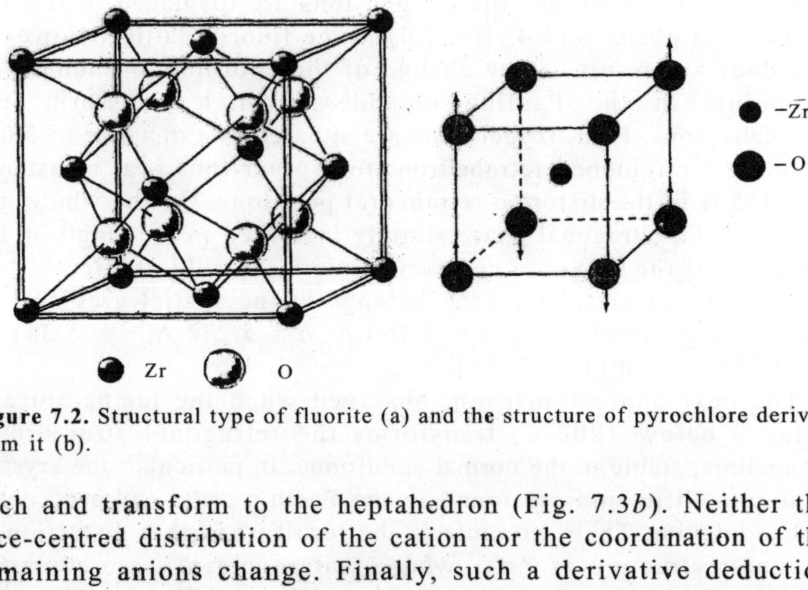

Figure 7.2. Structural type of fluorite (a) and the structure of pyrochlore derived from it (b).

each and transform to the heptahedron (Fig. 7.3*b*). Neither the face-centred distribution of the cation nor the coordination of the remaining anions change. Finally, such a derivative deduction structure is favourable fo incorporation of the metal ions with different chemical nature and geometrical parameters. As a result of the transformation of the vacancies in the X-sublattice, the initial M-cube may be transformed into new polyhedrons (Fig. 7.3*a*–*e*). Investigations of specific structures have confirmed the existence of cubes (*a*), hexahedrons (*b*) and also of polyhedrons of the type (*d*) and (*e*), with the octahedrons found most frequently in the two latter types. The possibility of formation of trigonal prisms (*c*) has not as yet been explained [7]. The Pauling polyhedrons, shown in Fig. 7.3, have the limiting (idealised) form. In the real structures, these polyhedrons are distorted to various degrees which is the result of a large number of compromises in the produced crystal lattice. The largest movement is shown by the oxygen atoms which usually move in the direction of the vacancies, resulting in a more isometric form of the M-polyhedrons. In particular, the 'flattened' octahedrons (*d*) acquire their usual configuration, observed in the majority of the known inorganic compounds.

The tetragonal phase ZrO_2 has a crystal structure of slightly distorted fluorite (Fig. 7.4). The spatial group is $P4_2/nmc$, the lattice parameters are $a = b = 5.085$A, $c = 5.166$; $a/c = 1.016$ [4–6].

When the temperature is reduced to 2370°C, the cubic phase becomes unstable and transforms to tetragonal modification as a result of the distortion of the fluorite structure. Due to small movement of the atoms, the oxygen ions are displaced in relation to the ideal position (1/4, 1/4, 1/4) in the fluorite lattice. However, this does not result in any change of the coordination number of zirconium in the Pauling octahdera, which transform into dodecahedrons. Four oxygen ions are situated at a distance of 2.065 A in the straightened tetrahedron, four other ions – at a distance of 4.455 A in the distorted tetrahedral positions. On the whole, the lattice of the tetragonal phase slightly increased in the length in the direction of the *c* axis in comparison with the cubic lattice.

The monoclinic phase ZrO_2 belongs to the spatial group $P2_1/c$, the lattice parameters are $a = 5.169$ A, $b = 5.232$ A, $c = 5.341$ A, $\beta = 99°15'$, $Z = 4$ [1–3, 7–10].

The next phase transition, observed when the temperature is reduced below 1200°C, transforms the tetragonal structure to monoclinic, stable in the normal conditions. In particular, the crystals of this modification are found in the Earth's crust and are known as baddeleyite. Their structure is the result of further distortion of the cubic structure of ZrO_2, whose rearrangement is so extensive that a completely new structural type forms. The coordination number of zirconium decreases to 7 (Fig. 7.5*a*) but not as a result of the previously mentioned 'truncation' of the M-cubes, and is due to the unique packing of the oxygen atoms whose number corresponds to the initial stoichiometry MO_2. In this packing, half of all the square networks remains unchanged (only slightly deformed), and the anion motive of the fluorite structure may be

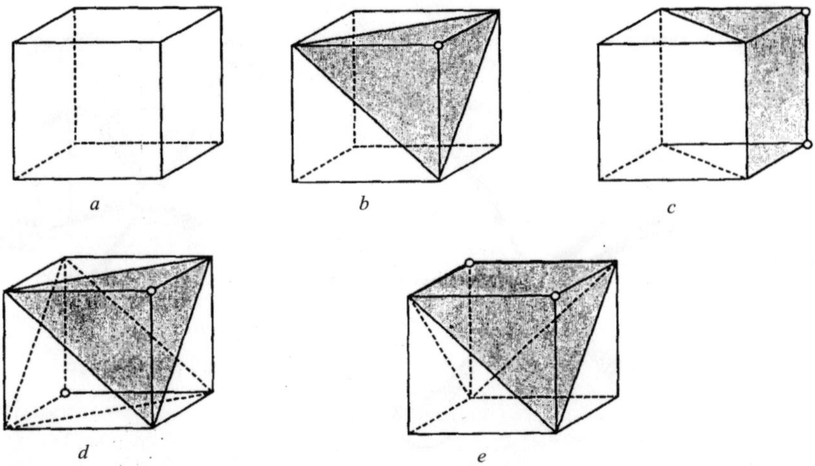

Figure 7.3. Pauling polyhedra formed from the cube as a result of 'truncation' of its tips (anion vacancies) [3]. a) cube; b) heptahedron; c) trigonal prism; d) octahedron; e) hexahedron with a single quadratic face.

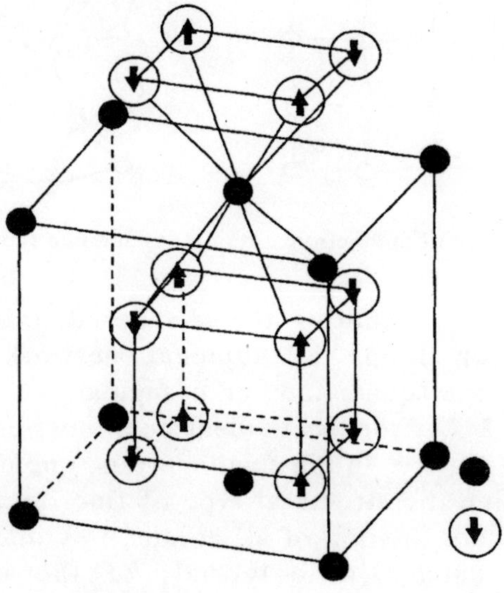

Figure 7.4. Tetragonal structure of ZrO_2.

located on them (Fig. 7.5*b*). The remaining anion networks, alternating with the former ones, are greatly distorted and their loops are represented by oblique squares with the double number of the triangles. As a result of monoclinic shift of the layers

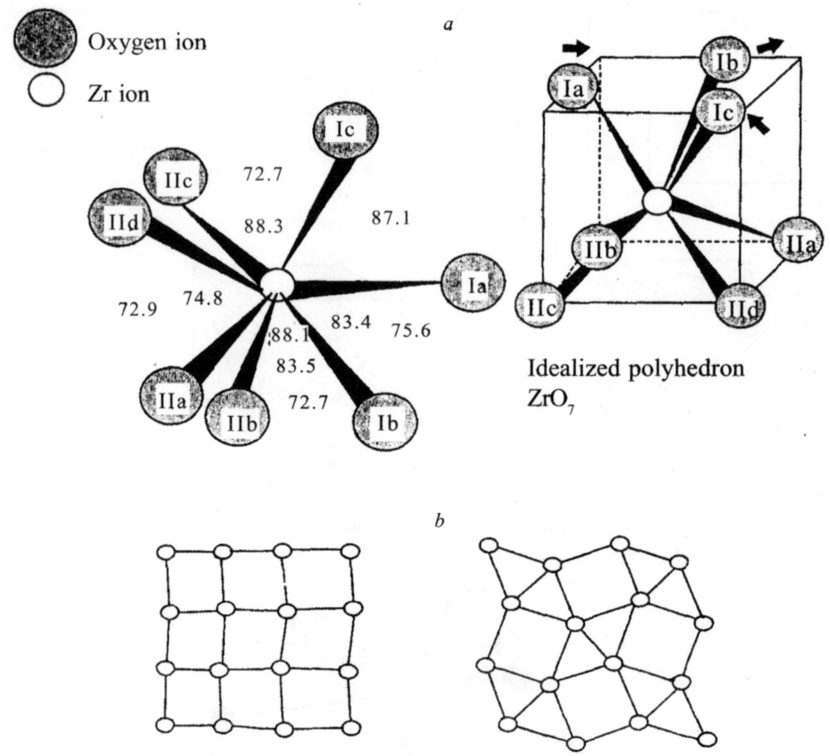

Figure 7.5. Structure of monoclinic zirconia (a) and two types of alternating anion networks (b).

zirconium atoms, retaining the checkered order of alternation, appear between loops of adjacent networks with different configuration: one square and one triangular. All the cations greatly deviate (by 0.2 Å) from their ideal positions, coinciding with the nodes of the F-lattice. In the final analysis, one of the main criteria which determine the structural type of fluorite and its derivatives – tetrahedral coordination of all anions, is violated. In the crystal structure of baddelyite only half of the oxygen atoms is characterised by quatenary coordination, whereas the remaining atoms have the coordination number 3.

The crystal structure of monoclinic ZrO_2 is characterised by very high stability, as indicated not only by the wide temperature range of existence of the monoclinic phase but also by the uncommon stability of baddelyite in the natural conditions. The monoclinic structure is characterised by twinning whose planes {100}*m* or {110}*m* are formed by the oxygen ions.

7.2. PHASE DIAGRAM OF THE SYSTEM ZrO_2–R_2O_3 (WHERE R – Y, Eu, Gd, Yb)

The equilibrium diagrams of the ZrO_2–R_2O_3 system, where R are the rare-earth elements (REE), are complicated systems of intermediate phases whose stability and length are not governed by simple relationships [11]. The allied structures of the phases, formed in these systems, change from the structure of the type of fluorite to the structure of the type Ti_2O_3 with the increase of the content of the REE oxide. This continuous transition is characteristic of high temperatures. At lower temperatures, there are phases of P-pyrochlore and H_1, H_2, H_3 and hexagonal phases. All these phases are derivatives of the structure of fluorite and their formation is the result of establishment of different degrees of structural order [12].

Some systems based on zirconia are characterised by a long history of investigations but, regardless of this, the results are highly contradicting. This is associated with a large number of difficulties resulting from the special features of the structure of zirconia and also from establishment of true equilibrium in these systems [13]. Firstly, at temperatures above 2000°C it is difficult to examine directly the crystal structure and, therefore, a large number of x-ray investigations were carried out at room temperature on specimens quenched from high temperatures. However, even with very fast cooling, there may be changes in the ratio of the phases and this complicates the interpretation of the results. Secondly, at temperatures below 1200°C, a reverse problem appears: the diffusion of ions is so slow that equilibrium is difficult to establish.

In these conditions, the process of construction of accurate equilibrium diagrams of several most important 'zirconium' systems has been very slow and the results have been contradicting. However, certain successes were achieved in the 60s and 70s of the previous century: the existence of a high-temperature cubic phase [14], discovery of the ordered phases in the ZrO_2–MgO [15,16], ZrO_2–CaO [17], ZrO_2–Y_2O_3 [18–20] systems, and the accurate determination of the field of stability.

The ZrO_2–Y_2O_3 system. In the system, the yttrium oxide is one of the main stabilisers of the cubic structure of ZrO_2. The system was investigated for the first time by Ruff and Ebert [21]. The system is characterised by the presence of wide areas of solid

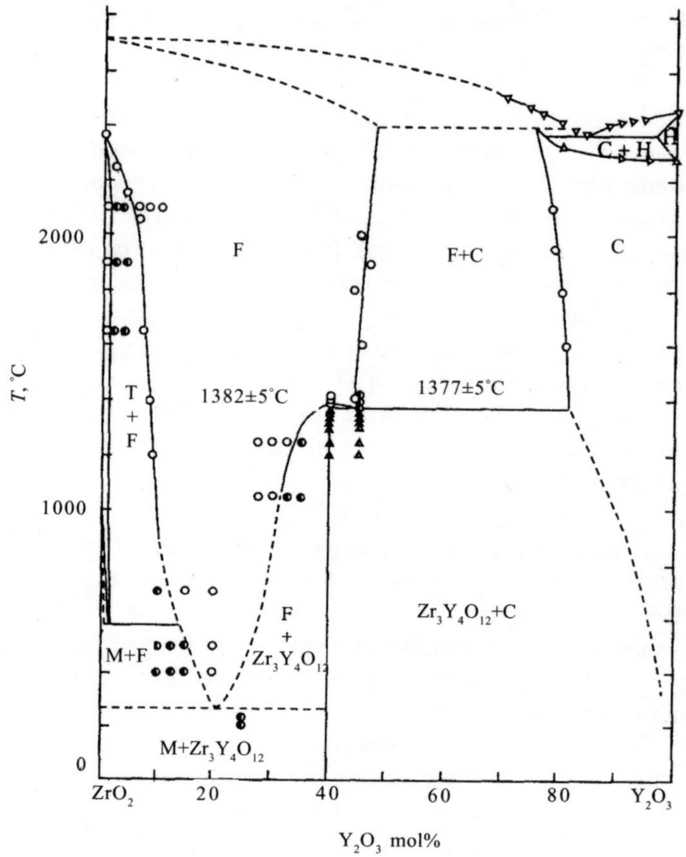

Figure 7.6. ZrO_2–Y_2O_3 equilibrium diagram [22].

solutions, both on the side of zirconia and on the side of yttrium oxide. Later, the equilibrium diagram of the zirconium systems were studied in detail by Stubican [22], Scott [23], Pascual and Duran [24] and the results are shown in Fig. 7.6.

All these diagrams show that in addition to the solid solutions, the system is characterised by the stable compound $Zr_3Y_4O_{12}$ which forms at a content of 40 mol.% Y_2O_3 and was synthesised for the first time and described by Ray and Stubican. Scott also detected a long area of ordering at 40 mol.% of yttrium oxide. According to Scott's data, $Zr_3Y_4O_{12}$ is stable below 1370°C and is characterised by a narrow region of homogeneity [23]. According to Stubican, the δ-phase forms as a result of the disruption of the order in the distribution of the ions in the cubic solid solution at temperatures

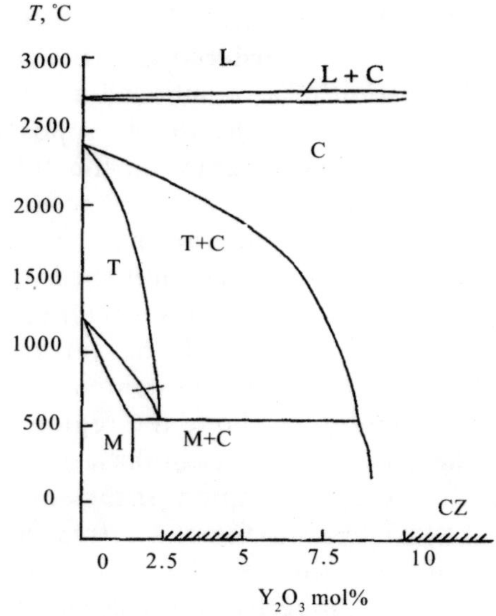

Figure 7.7. Part of the equilibrium diagram of the ZrO_2–Y_2O_3 system on the side rich in zirconium oxide [12].

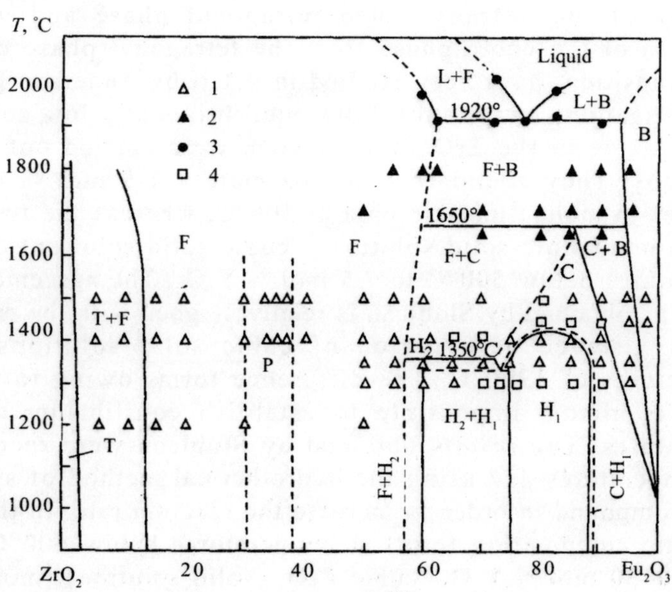

Figure 7.8. Diagram of the ZrO_2–Eu_2O_3 system [29]

above 1382+5°C [5]. The structure of the δ-phase is a derivative of the fluorite structure with ordered distribution of the oxygen vacancies and the formation of a rhombohedral cell. However, there are also contradictions between the three proposed diagrams.

Since the part of the phase diagram on the side of zirconium is very important from the practical viewpoint, the subsolidus equilibrium in this part of the diagram has been studied by many authors. The most detailed study was published by Scott (Fig. 7.7). The characteristic property of the phase diagram in this region is the decrease of the temperature of the tetragonal–monoclinic transition with increase of the yttrium content.

With increase of temperature above the region of stability of the monoclinic phase there is a narrow two-phase monoclinic-tetragonal region which is followed by the region of the so-called transformed tetragonal phase (*t*'). The transformed tetragonal solid solution which undergoes transformation to the monoclinic phase in cooling, exists in the composition range from 0 to 5 mol.% Y_2O_3. At higher yttrium oxide concentrations, there is a two-phase region of the non-transformed tetragonal (t) and cubic solid solutions. A further increase of the concentration of yttrium oxide results in the formation of the single-phase cubic solid solutions. The nature and property of the untransformed tetragonal phase and also the formation of the cubic phase from the tetragonal phase during a phase transition, have been studied in detail by Anderson [25].

Investigations of the subsolidus equilibrium at a low content of yttrium oxide in the ZrO_2–Y_2O_3 system were carried out by Ruh et al [26]. They found that approximately 1.5 mol.% of Y_2O_3 dissolves in monoclinic zirconia at 500°C, whereas the two-phase region (monoclinic solid solution + cubic solid solution) exists at temperatures below 500°C to 7.5 mol.% Y_2O_3. The agreement with the results obtained by Stubican is relatively good with the exception of the existence of a region of cubic solid solutions below temperatures of 1200°C. The difference forms owing to the fact that it is almost impossible to establish equilibrium at these temperatures. The results obtained by Stubican were recorded at low temperatures and using the hydrothermal method of synthesis of the compound in order to increase the reaction rate. In this case, a eutectic composition forms at temperatures below 400°C in the range 20–30 mol.% Y_2O_3: cubic ZrO_2 (solid solution) (monoclinic phase ZrO_2) +$Zr_3Y_4O_{12}$. It is assumed that the equilibrium phases below the eutectoid transformation temperatures are the monoclinic solid solution based on zirconia and $Zr_3Y_4O_{12}$ compound.

The regions of the two-phase mixture of the solid solutions based on zirconia and yttrium oxide are in the ranges: 64–76; 69–79; 73–86 mol.% Y_2O_3 at temperatures of 1400, 1700 and 2000°C, respectively.

In [27] in the investigations of the process of evaporation of the ZrO_2–Y_2O_3 solid solution it was found that the composition with the yttrium oxide content of approximately 20 mol.% behaves as an azeotrope.

The ZrO_2–Eu_2O_3 system

The phase transformations in the systems were investigated in [28, 29]. A solid solution with the cubic lattice of fluorite was identified in the ZrO_2–Eu_2O_3 phase diagram shown in Fig. 7.8. It is characterised by a wide range of homogeneity from 10 to 55 mol.% of Eu_2O_3. In the composition range 27–39 mol.% Eu_2O_3 a $Eu_2Zr_2O_7$ phase forms on the basis of the solid solution F. This phase is characterised by the ordered crystal structure of the pyrochlore type. The boundary of the region of homogeneity of this compound on the ZrO_2–rich side (T = 800–1400°C) is situated close to x = 1.48 mol.% Eu_2O_3 [29].

The ZrO_2–Gd_2O_3 system. The system was investigated in [30–33]. Specimens were synthesised by the solid phase method and a compound with the structure of pyrochlore was found in a wide homogeneity range (Fig. 7.9). Several types of solid solutions were also found in the system: 1 – the tetragonal solid solution based on zirconia in the subsolidus range (above 1000°C); 2 – the cubic solid solution of the fluorite type; 3 – the cubic solid solution of the Tl_2O_3 type; 4 – the solid solution with the pyrochlore structure (a = 10.527 Å with the composition $Gd_2Zr_2O_7$); 5 – the hexagonal phase existing in narrow ranges in the vicinity of 57 mol.% Gd_2O_3, with the lattice parameters: a = 10.002 Å, c = 18.644 Å; 6 – the hexagonal phase, stable below 1600°C and containing 77–78 mol.% of Gd_2O_3: a = 10.032 Å, c = 18.814 Å. The authors of [33] investigated the equilibrium diagram of the ZrO_2–Gd_2O_3 system with the gadolinium oxide content up to 10 mol.% (Fig. 7.9).

In the investigations of the system by the 'annealing and quenching' method it was established that there are solid solutions based on both Gd_2O_3 and ZrO_2 at a temperature of >2000°C. The wide region of the solid solutions based on zirconia is characterised by a continuous transition of the tetragonal solid solution to the

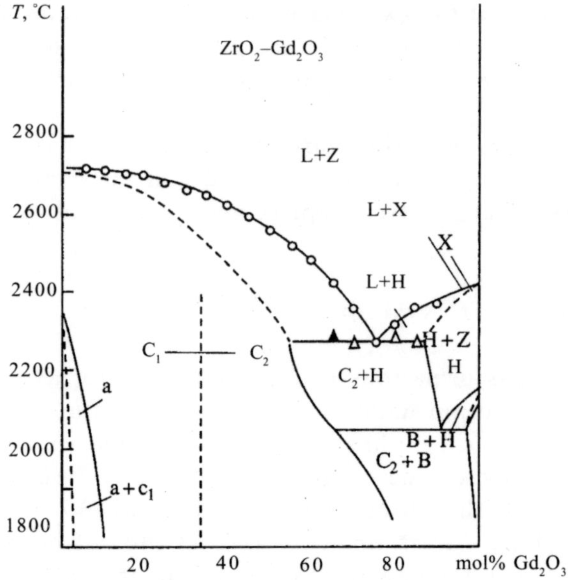

Figure 7.9. Diagram of the ZrO_2–Gd_2O_3 system [33].

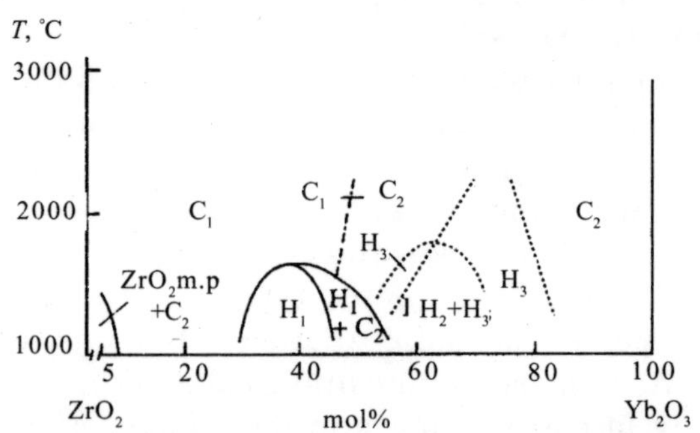

Figure 7.10. Diagram of the ZrO_2–Yb_2O_3 system [34].

cubic type of fluorite and of the latter to the cubic solid solution of the type Tl_2O_3.

The ZrO_2–Yb_2O_3 system

In the system (Fig. 7.10) investigations showed the existence of several types of solid solutions: 1) the solid solution based on tetragonal zirconia at temperatures above 1000°C; 2) the cubic solid solution of the fluorite type; 3) the cubic solid solutions of the Tl_2O_3 type. The latter should be regarded as solid solutions of zirconia in the cubic ytterbium oxide [31, 32]. The system is characterised by the presence of a rhombohedral phase (H) and two hexagonal phases. The parameters of the elementary cell were determined for the phase H. For the compositions with 33 mol.% Yb_2O_3: a = 8.178 Å, α = 72°32'30", and for the compositions with 40 mol.% Yb_2O_3: a = 8.259 Å, α = 72°40°20". The phase with the pyrochlore structure was not found in the system, but the existence of the $Zr_3Yb_4O_{12}$ compound was detected on the ZrO_2–Yb_2O_3 equilibrium diagram [34], Fig. 7.10.

Examination of the structural and phase transformations of nominally pure zirconia, stabilised by yttrium oxide, and also of rare-earth elements shows that there are three types of crystals, differing in the structure, physical and mechanical properties, which form three independent branches of development of new directions. These are the completely stabilised zirconia with the stable cubic structure whose crystals are referred to in Russia as fianites and are used efficiently in technology and jewellery industry. These are crystals with an incompletely stabilised cubic lattice, whose crystal structure depends on temperature, the nature and amount of the stabilising oxides. In this group of the crystals, it is necessary to mention the crystals with a partial phase transition from the cubic to tetragonal phase. This transformation results in the formation of domains of complicated geometry which fill the entire volume of the crystal. The high mechanical properties have been the main interesting points of these crystals in recent years: the strength and toughness of the material. The experimental results show that the increase of the concentration of the stabilising oxide above specific values, for example, for Y_2O_3>20 mol.%, results in a change of the local environment of the admixture cations; the crystals retain the cubic structure of the fluorite type. Thus, a new optical matrix, the C-OX single crystals, form. In doping with ions of the transition metals, the C-OX single crystals are characterised by greatly differing optical properties in comparison with the fianites. In the following chapters, all the three types of the crystals will be studied in detail.

The stability of the solid solutions based on zirconia depends on the nature and concentration of the stabilising oxide, temperature, annealing time and chemical purity of the initial reagents. The specimens, produced by crystallisation of the melt, as indicated by the equilibrium diagram, are not equilibrium at room temperature.

The stability of the solid solutions based on ZrO_2, produced by the directional solidification of the melt, was investigated in [35]. Crystalline specimens, containing 1–33 mol.% of the stabilising oxides, and powders prepared from them were annealed in air at a temperature of 1000°C for 10 days and at 1200°C for 45 days.

After annealing, no changes were detected in the x-ray diffraction patterns of the crystals. No turbidity was observed in the cubic single-phase crystals. Annealing of the powders of the same composition at 1000°C also caused no changes. In annealing for 45 days at 1200°C, examination showed the partial decomposition of the solid solutions of several compositions, shown in Table 7.1. A monoclinic phase appeared at low concentrations of the stabilising oxide: at a concentration of the stabilising oxide of 10–33 mol.%, the cubic phase in the solid solutions remained also after annealing.

Thus, the single crystals of the cubic solid solutions, produced by directional solidification of the melt, are stable at high temperatures. Annealing for 45 days at 1200°C of the crystals, refined into powder, show the formation of the monoclinic phase only for the composition the ZrO_2–8 mol.% Eu_2O_3. No monoclinic phase formed after annealing in the compositions ZrO_2–(8–33) mol.% R_2O_3, where R is Gd, Y, Yb and ZrO_2–(10–33) mol.% Eu_2O_3.

In [35], a series of experiments was also carried out with the growth of crystals with a low solidification rate (1–2 mm/h) followed by long-term annealing (14–40 *h*) by a smooth decrease of the power of the generator; the temperature in the crucible was varied from 2600 to 1200°C (Table 7.2).

The crystals with the composition ZrO_2–12 mol.% R_2O_3 (R–Y, Gd), and ZrO_2–8 mol.% the Gd_2O_3, grown in these conditions, were turbid, although no significant changes are detected on the x-

Table 7.1. Phase composition of powders after annealing at 1200°C for 45 days

Composition of crystals	R_2O_3 content, mol%							
	1	2	3	5	8	10	12-20	33
ZrO_2–Eu_2O_3	M	M+T	M+T	T+M	F+M	–	F	P
ZrO_2–Gd_2O_3	M	M+T	M+T	T+M	F	F	F	F
ZrO_2–Yb_2O_3	M	M+T	M+T	T+M	F	F	F	F

ray diffraction diagrams of these crystals. The examination by the Raman scattering method did not show the presence of a second phase in the crystals. Nevertheless, in investigations using a laser ultramicroscope, it was established that the scattering of the light in the crystals is caused by the particles of the second phase with the size of 1–0.1 μm. The crystals with the composition ZrO_2–R_2O_3 (R–Y, Yb) with the concentration of 14–33 mol.% and ZrO_2–16 mol.% Gd_2O_3, grown in the same conditions remained transparent. The appearance of the scattering particles may be associated both with the presence of secondary admixtures in the charge whose solubility decreases with decreasing temperature, and with the decomposition of the fluorite solid solution.

The direct determination of the composition of the second phase and of its structure is difficult because the content of this phase is very low and the particle size is very small. It should be mentioned that after heating the specimens to 2100°C (in vacuum) for 30 *h* and subsequent rapid cooling at a rate of 1000°C/h there was almost no scattering of the light in the crystals with the composition ZrO_2–12 mol.% Y_2O_3.

In slow crystallisation and subsequent slow cooling of the crystals with the composition ZrO_2–(8–12) mol.% R_2O_3 (R–Gd, Y, Yb a) a second phase appeared (Table 7.2), but the amount of this phase was small and its nature could not be determined by x-ray diffraction analysis. The single crystal nature of the specimens is not disrupted, but they intensively scatter the light. The crystals with the composition ZrO_2–(16–33) mol.% R_2O_3 (R–Gd, Y, Yb) do not show any changes in the same growth and annealing conditions. Thus, in directional crystallisation of the ZrO_2–R_2O_3 melts, depending on the type and concentration of the stabilising oxide and also on the growth regime, the process yields either non-transparent crystals, containing scattering particles of the second phase, or optically homogeneous single crystals (Table 7.2).

The phase composition of the crystals ZrO_2–R_2O_3 (R–Eu, Gd, Y, Yb), grown from the melt, is not thermodynamically equilibrium in relation to room temperature. For the crystals with the concentrations of the stabilising oxide of 1–5 mol.% R_2O_3, the characteristic feature is the presence of several phases. This problem will be examined in greater detail in Chapter 8, concerned with partially stabilised crystals of zirconia. At the concentration of the stabilising oxides 8–40 mol.% R_2O_3 (R–Eu, Gd, Y, Yb) the solid solution with the fluorite structure undergoes the order–disorder

Table 7.2. Effect of composition and growth conditions on the stability of solid solutions (growth rate 1–2 mm/h, cooling directly in crucible to 1000°C for 12–40 h) [35]

Composition	External appearance of crystals	XDA data on phase composition
ZrO_2–12 mol% Y_2O_3	Slightly scattering	Single-phase cubic solid solution with fluorite structure
ZrO_2–16 mol% Y_2O_3	No scattering	
ZrO_2–8 mol%Gd_2O_3	Strong scattering	
ZrO_2–12 mol%Gd_2O_3	Slight scattering	
ZrO_2–16 mol%Gd_2O_3	No scattering	
ZrO_2–14 mol%R_2O_3(Y,Yb)	No scattering	

transition with increase of the R_2O_3 concentration. It has been established that at the content of the stabilising oxide of > 8 mol.% R_2O_3, solid solutions are produced in the form of transparent single-phase single crystals, characterised by high phase stability at temperatures up to 1200°C in air for at least 45 days.

Examination of the process of decomposition of the solid solutions in high-temperature annealing shows that large single-phase, optically homogeneous single crystals can be produced only when using the conditions of slow solidification at a rate of 1–2 mm/h followed by a slow decrease of temperature to 1200°C (at a rate of 10°C/h) at the content of the stabilising oxides: R_2O_3 more than 14 mol.% for Y, Yb and more than 16 mol.% for R–Eu, Gd.

7.3. PRODUCTION OF FIANITES

As mentioned previously, the method of direct HF melting in a cold crucible is at present the only method of producing large single crystals of stabilised zirconia and hafnia–fianites.

The initial reagents for the growth of fianites are dioxides of zirconia, hafnium, yttrium, and also oxides of certain alkali-earth and rare-earth elements used as stabilising or colouring additions. In most cases, high-purity reagents are used.

Fianite technology includes several main stages. The first stage is the process of start melting. Start melting is usually realised by introducing, into the volume of the molten material, a metal included in the composition of the components of the material. For example, to produce ZrO_2–Y_2O_3 crystals, metallic yttrium is added to the charge, and for the ZrO_2–R_2O_3 (R–Eu, Gd, Yb) crystals it is metallic zirconium or the appropriate rare-earth element, for HfO_2–Yb_2O_4 it is metallic hafnium. Usually, the amount of starting metal is approximately 10–15 g per kilogram of the weight of the charge.

Heating of the metal and the surrounding charge in the process of start melting takes place as a result of the absorption of the energy of the HF field and was also the result of the exothermic reaction of oxidation of metal by atmospheric oxygen. In every specific case, the special feature of the process depends on the physical–chemical properties of the metal used. Another important factor in the conditions of the starting process: atmosphere, the depth of the starting metal in the charge, the degree of refining of the metal, dispersion of the charge. These factors determine to a certain degree the duration of the starting period. The physicochemical properties of the metal used for starting were published previously in Table 2.2.

In the fianite crystals with the composition ZrO_2–R_2O_3, where R is Eu, Gd, Yb, the concentration of the oxides of the rare-earth elements changes in the range 8–33 mol.% [33].

The main and stabilising oxides are mixed in plastic cuvettes. Thorough mixing of the charge, subsequent sintering and grinding of the mixture prior to melting has no significant effect on the quality of the ground crystal. Therefore, after mixing, the charge is directly loaded into the container. The melting of these compositions is carried out in a container with a diameter of 130 mm, height 160 mm. Equipment Donetsk-4 is used for melting and growth of the crystals. Zirconium oxide crystals, stabilised with yttrium oxide, were also grown in other systems, described in Chapter 2. Usually, the diameter of the crucible was 220 mm, height 300 mm. In growing ZrO_2–Y_2O_3 in large containers, the charge produced from powder reagents was used together with crystal waste of the same composition in the ratio of 4/3, and the total weight of the charge was 14 kg. The application of compacted material shortens the loading time and reduces the amount of waste. It should also be mentioned that directional crystallisation is accompanied by additional purification of the initial charge. The use of the crystalline material also improves the quality of grown crystals.

The loading of the container is the important stage of the technological process which controls the dimensions and optical homogeneity of the crystal. Prior to loading, a layer of the sintered charge several centimetres thick is placed on the bottom of the container and this is followed by loading the charge with the crystals. The shavings of starting metal are deposited at the top in the depths of 2–4 cm and compacted. The melting of the charge starts 5–20 min after activating the generator. Indication of melting is the high-intensity glow between the sections of the crucible. At the start of melting, the working parameters of the generator rapidly change. The power generated in the load in Donetsk-4 equipment is controlled using a calibrated constantan–copper thermocouple measuring the temperature of water at entry into and exit from the container. The temperature difference is proportional to the power removed from the container. The temperature of water is determined with the accuracy of ±0.1°C. The consumption of water, flowing through the container, is controlled by a contact pressure gauge. The power in the melt, is calculated from the equation:

$$p = V / t(T_2 - T_1)C, \tag{7.1}$$

where V is the volume of the water flowing through the container, m^3; t is time, s; $\Delta T = T_2 - T_1$ is the temperature difference of water at entry into and exit from the container, deg; C is the heat capacity of the water, J/(m^3 · K).

During melting, the power, absorbed by the melt increases as a result of the increase of the melt volume. Thus, by varying ΔT it is possible to control the melting process. For example, when the next portion of the charge is placed on the surface of the melt, the value ΔT rapidly decreased (Fig. 7.11), since part of the energy is used for heating the charge. This results in partial solidification of the melt, a decrease of the volume of the melt and, consequently, weakening of the inductive bond resulting in the instability of the parameters of the generator (U_a, I_a); with a subsequent gradually increase of the volume of the melt, the value of ΔT increases.

The charge is also added in the crucible to ensure that the volume of the melt reaches the required value. At the end of loading, the charge forms a cupola-shaped 'roof' above the melt which baked under the effect of intensive radiation of the melt. The 'roof' reduces the heat losses from the surface of the melt. The duration of starting up the container with the diameter of 220 mm with the charge is 1–1.5 h.

At the stable level of the supplied power the container with the

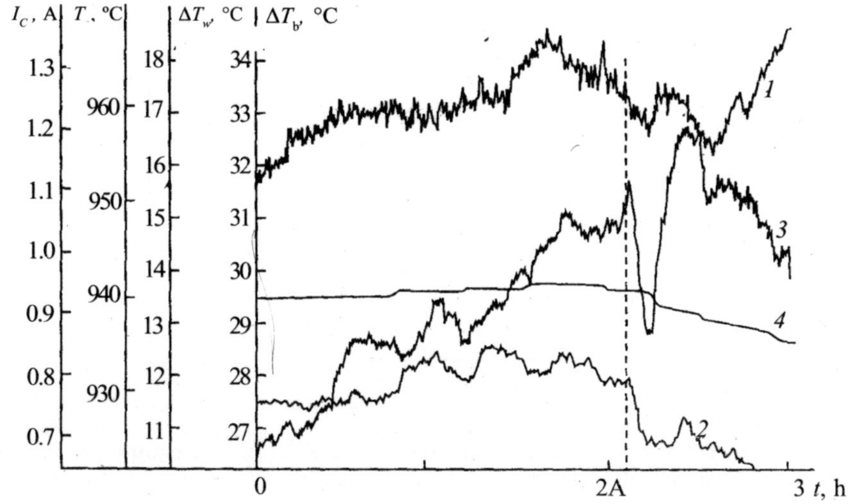

Figure 7.11. Time dependences of: 1) ΔT_b; 2) ΔT_w; 3) ΔT_p; 4) I_C [37].

melt is held for 30–40 min to establish phase equilibrium. Solidification is initiated by lowering the container with the melt in relation to the induction coil. With exit of the lower part of the container from the induction coil, the bonding of the melt with the induction coil decreases and, correspondingly, the supply of energy to the melt becomes smaller. As a result of the gradual cooling of the lower part of the container, the process of solidification of the melt starts spontaneously.

The anode (I_a) and grid (I_{gc}) currents are the parameters sensitive to the variation of the melt volume. Therefore, it is possible to carry out objective control of the phase equilibrium between the melt and the solid phase in the course of the process. Figure 7.12 shows the time dependences of I_{gc} and I_a in the processes of homogenisation, solidification of the melt, and annealing of the grown crystals. U_a was maintained constant.

In addition to the anode and grid currents, another sensitive parameter is the frequency of generation which is sensitive to the variation of the load. Donets-4, Fianit and Kristall-401 can be used for producing 1.5–6 kg of crystals in a single process.

To evaluate the effect of the conditions of directional solidification on the phase composition and optical homogeneity of the crystals, the rate of loading of the crucible was varied in the range 1–40 mm/h. The results show the increase of the size of the single crystals with a decrease of the lowering rate of the container

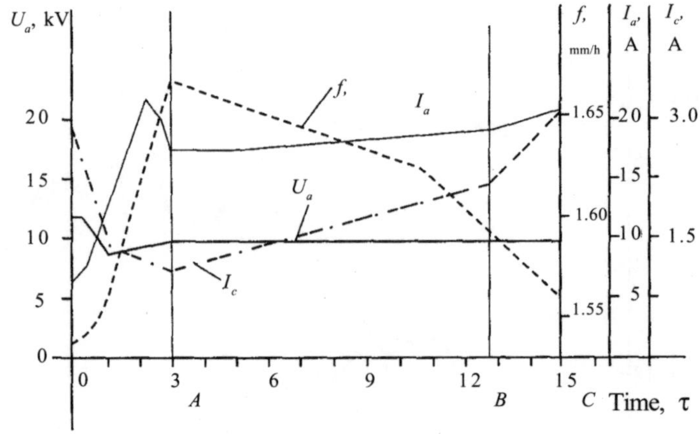

Figure 7.12. Drift of the electrical parameters of Kristall-401 equipment in real melting and solidification processes of cubic zirconia. I_a – anodic current; f – the frequency of HF generator; I_c – grid current; U_a – anodic voltage; (O–A) – process of homogenizing the melt; (A–B) – growrth of the crystal block; (B–C) – annealing.

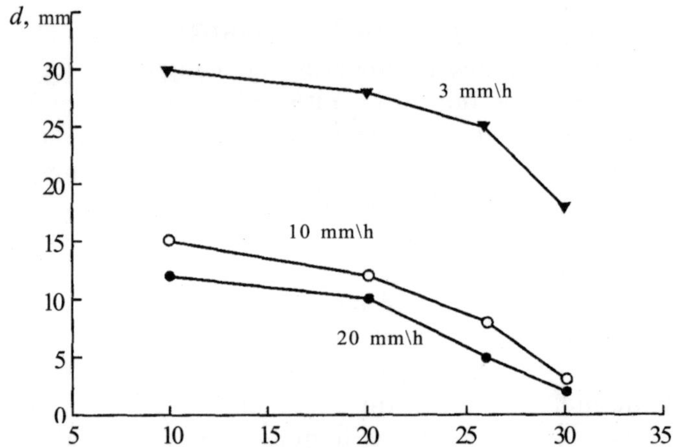

Figure 7.13. Dependence of the size of cubic zirconia crystals on the content of the stabilizing oxide Y_2O_3 and solidification rate (indicated at the curves).

[37, 38]. This is due to the fact that the polycrystalline layer of the solid shell contains a large number of crystallisation centres, and slow growth is sufficient for spatial selection as a result of which only the crystals with optimum orientation in the growth direction 'survive'. For example, at a container diameter of 130 mm and a growth rate of 2 mm/h, only 2–3 large crystals were grown. At a rate of 60 mm/h, the number of crystalline blocks was more than

20. The slow growth rates also resulted in the improvement of the quality of the crystals and increase of optical homogeneity. Figure 7.13 shows the dependence of the dimensions of the grown crystals on the rate of lowering of the container with the melt and the content of stabilising yttrium oxide. Table 7.3 shows the effect of rare-earth oxides on the yield and dimensions of the crystals which also makes it possible to predict the optimum technological parameters for obtaining the crystals of the required dimensions [39].

To examine the effect of mass transfer in the melt on the quality of crystals, investigations were also carried out into the effect of reverse rotation of the container. At a growth rate of 8–16 mm/h reverse rotation of the container has no significant effect on the quality of the grown crystals, but at a growth rate of 1–2 mm/h, the quality of crystals noticeably improves [40]. Evidently, reverse rotation results in the disruption of the diffusion layer ahead of the solidification front. The large crystals, grown at low rates and in

Table 7.3. Effect of composition and cooling rate on the dimensions of crystals (*d*) of zirconia ZrO_2 stabilized with R_2O_3 (R = Ce–Lu)

C, mol %	Crystal dimensions *d**, mm				
	R_2O_3 (R-Ce-Nd)			R_2O_3 (R-Eu–Lu)	
	V=16 mm/h	V=2 mm/h	V**=16 mm/h	V=16 mm/h	V=2 mm/h
0	2–3	–	10	2–3	1–2
0.1	2–4	–	–	3–5	–
0.3	10–12	–	15	10–15	–
1.0	10–12	–	–	10–15	–
5.0	4–5	8	5	12–15	15–20
10-20	2–3	4–5	2–3	12–15	20–40
25	–	–	–	8–10	12–15
33	1-2	–	–	2–3	8–12
40	–	–	–	2–3	8–10

*Crucible diameter 130 mm
**Thermal screens were used

small containers, usually crack. At a large number of small blocks the stresses are removed as a result of the formation of boundaries between blocks. However, the formation of a small number of large blocks results in the formation of high mechanical stresses in the volume of the crystals causing failure of the latter.

The experimental results show that the most justified method of increasing the dimensions of fianite crystals and their optical homogeneity is the increase of the volume of the crystal melt and a decrease of temperature fluctuations at the solidification front. The increase of the optical homogeneity is also supported by a decrease of the rate of growth of the crystals and by the application of intensive forced mixing of the melt. The decrease of the linear growth rate to 1 mm/h and the application of the system for reverse rotation of the container makes it possible to reduce the gradients of the refractive index of the crystals to 10^{-4} cm^{-1}. The increase of the volume of the solidifying melt makes it possible to produce large single crystals as a result of reducing temperature gradient and increasing heat capacity. During cooling in the temperature range 2800–1000°C the crystals still have sufficiently high electrical conductivity and continue to absorb the energy of the HF field and, consequently, the integrity of the crystals is not disrupted.

To produce colour crystals, it is necessary to use small additions (≈0.5 wt%) of rare-earth and transition metals to the melt. The technological scheme of the production of fianite crystals by the technology developed at the General Physics Institute of the Russian Academy of Sciences is described below.

The technology of the fianite crystals includes the following main stages: a) preparation of the charge; b) loading the charge into the container; c) melting; d) growth of single crystals by the method of directional solidification of the melt; e) extraction of the crystals from the crucible; f) machining and grading of the crystals; g) cleaning of crystal waste for subsequent melting and crystallisation. The diagram of production of the fianite crystals and equipment used at the General Physics Institute is shown in Figs. 7.14 and 7.15 [35].

The industrial technology of growth of zirconia-based crystals, used widely in many countries, is almost completely based on the results of investigations of Russian scientists obtained in the last 30 years [42, 43, 44–47]

The quality of the charge and the purity of the initial reagents

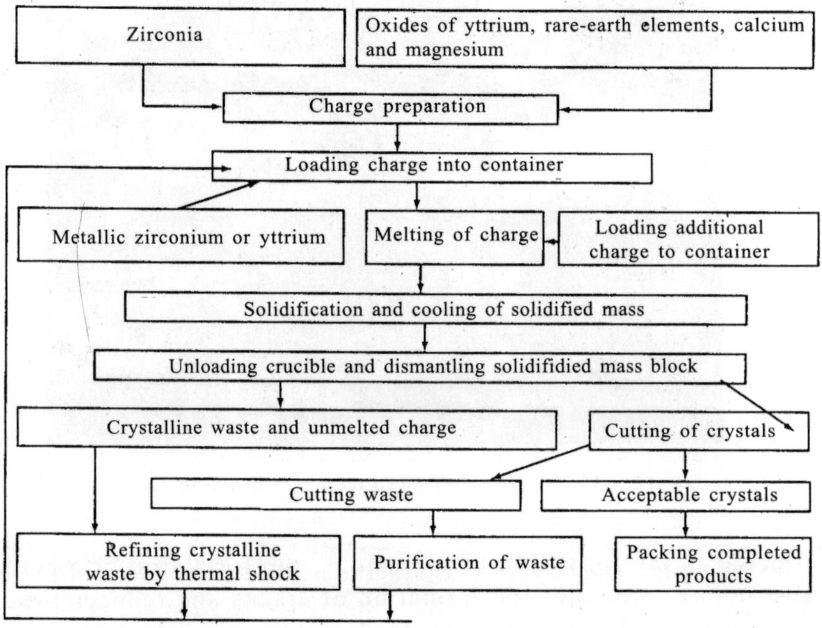

Figure 7.14. Flow chart of production of cubic zirconia crystals (fianites) at the General Physics Institute of the Russian Academy of Sciences [35].

have a controlling effect on the entire technological process of crystal production. The oxides in the proportion, corresponding to the given composition, are thoroughly mixed in a mechanical mixer and loaded into a container. Crystal waste of the same composition from the previous melts is added to the charge [43, 45]. This makes the technology almost completely waste-free and results in a considerable saving of expensive reagents, shortens the loading and melting time because the crystalline material has higher density. In directional solidification, there is also additional purification of the melt to remove a number of impurities and this improves the quality of the grown crystals.

The presence of a number of uncontrolled impurities in the initial oxides, used for the growth of single crystals by directional solidification, is highly undesirable. These impurities influence the both the growth process and the quality of the crystals. As shown later, the crystal growth is negatively affected mainly by the impurities such as Si, Al, Ti, Nb; the effect of Fe, Co, Ni, Mn, Cr, V, W, Mo is slightly less marked. The effect of these impurities is

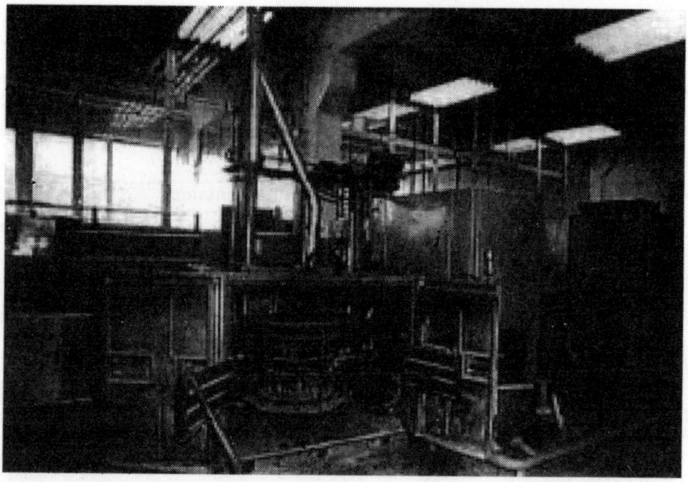

Figure 7.15. Kristall-403 equipment for growing fianites.

reflected in the formation of inclusions, the formation of a cellular substructure, promotes the formation of cracks and reduces the size of the single crystals. The growth defects, caused by the presence of microimpurities, reduce the optical homogeneity and strength characteristics of the crystals. In addition to this, there is a number of requirements on the purity of the initial material in relation to the application of the final product. For example, special requirements are imposed on the fianites used as substrates for the epitaxial growth of semiconductors (Si, Ge, $A^{III}B^{V}$, $A^{II}B^{VI}$, etc) as regards the content of the elements which may influence the electrophysical characteristics of the films. When producing a film-forming material based on zirconia, special attention is given to the impurities influencing radiation resistance, beam strength, and also impurities influencing the technological properties of the material in spraying (volatile impurities). Thus, these factors require careful analysis of both the components of the initial materials and the produced single crystals.

The composition of the impurities of the initial oxides and grown single crystals has been investigated by the method of atomic emission spectral analysis because the standard methods of atomic emission analysis cannot be used for zirconia because of its high thermal stability (Table 7.4).

Table 7.4 gives the content of microimpurities in the crystal grown from the given charge. The analysis results indicate that the

Table 7.4. Content of uncontrolled impurities in charge and in crystals [40]

Element	Impurity content, % $\times 10^{-5}$			
	In initial charge	In crystal		
		bottom	centre	top
Fe	2.0	<0.1	<0.5	1.0
Ca	3.0	1.1	1.6	1.8
Si	5.1	0.1	0.5	1.0
Mn	0.2	<0.1	<0.1	<0.1
Cu	1.0	<0.1	<0.1	<0.1
Mg	1.2	0.5	1.1	1.4
Al	0.5	<0.1	<0.1	<0.1
Nb	0.4	0.1	0.1	0.3
W	<0.1	<0.01		
Ti	0.5	<0.1	0.1	0.4
Be	<0.1	<0.01	<0.01	0.01
Sr	4.3	1.5	2.1	3.3

purity of the produced crystals is 5–100 times higher as regards the content of the impurities in comparison with the initial oxides. This effect is a direct consequence of the application of technology of directional solidification in the cold crucible because the crystals are not contaminated during growth which takes place in other methods of crystal growth characterised by contact of the melt with the crucible material. In the present case, the melt and the crystals are greatly purified as a result of the displacement of the microimpurities to the upper part of the crystal and evaporation of a number of impurities (As, Pb, Cu, etc.) as a result of the high temperature of the melt.

Filling of the crucible with the charge is an important stage of the technological process which controls the yield, dimensions and quality of crystals. The distribution of the material in the crucible determines the formation of the volume of the melt and the thermal conditions in which the melting of the material and solidification of the melt take place. Special features of loading of the cold crucible consist of the need for the formation of the lower bulk thermal screen in the form of rings produced from powder-like materials and crystal waste.

The lower thermal screen plays the following functions: 1– reduces heat losses through the bottom of the crucible; 2 – equalises the temperature fields in the diameter of the crucible; 3– reduces the axial gradients in the growing crystals; 4 – reduces the number of seed crystals and this increases the dimensions of the single crystals in the grown block.

The ring-shaped loading scheme has the following advantages: 1–saves the powder material, 2–reduces the duration of the process of starting melting; 3–supports the formation a high–density, sintered layer of the solid phase, 4–increases the stability of the melting process, 5– simplifies the process of loading large diameter crucibles.

7.4. CRYSTAL CHEMISTRY OF THE CUBIC PHASE OF SOLID SOLUTIONS BASED ON ZIRCONIA

The zirconia single crystals have a unique combination of the properties resulting in the special interest in this material. However, the presence of the monoclinic–tetragonal transition, accompanied by a sharpe change in the volume, results in the failure of the grown zirconia crystals and prevents their use in practice. In order to facilitate the production and used of zirconia crystals of the highest-temperature cubic modification, it is necessary to stabilise the modification in the entire temperature range, up to room temperature.

Examining the geometrical stability criteria, Goldsmith, Magnus and Pauling used the fluorite lattice as an example to shows that the critical parameter is the ratio of the ionic radii of the cation and the anion. At $R_c / R_a \geq 0.736$, the densest packing of the anions in the $Fm3m$ lattice is likely to exist. When this parameter is lower than this value, the ion crystals do not form the cubic lattice because at short distances between the anions the forces of electrostatic repulsion unavoidably distort the structure. For zirconia, this ratio is 0.66 [2].

The cubic structure may be stabilised by increasing the radius of the cation (*i.e.*, replacing zirconium by cations with a larger diameter) or by the formation of vacancies in the anion sublattice (*i.e.*, by introducing the cations with a smaller charge or by heterovalent substitution). The solid solutions, formed in this case, are denoted as follows: $Zr^{+4}_{1-x}R^{3+}_{x}O^{2-}_{2-x/2}(V^{O})_{x/2}$ where V^{O} is the vacancy in the oxygen sublattice.

To obtain the necessary increase of the ratio of the radii of the cation and the anion in the zirconia lattice, it is necessary to change the geometrical parameters by ~4% [48]. The lattice constants of stabilised zirconia are characterised by different values in relation to the nature and concentration of the stabilising oxide.

In a number of cases, stabilisation takes place during the introduction of the oxides whose cations have a smallert ion radius than the radius of zirconia. Consequently, the controlling factor for the formation of the cubic structure of ZrO_2 is not only the cation radius [49]. Thermodynamic considerations show that the crystal lattice is stable if its state is characterised by the minimum free energy. To reach this state, in addition to the radii of the cation and the anion, it is also important to consider the nature of interaction between the electron shells of the components of the lattice. According to the authors of [50, 51], the relationship between the secondary cation and oxygen should be more heteropolar than between the ions of zirconium and oxygen. In this case, the structure of fluorite is formed. However, in the establishment of the equilibrium state in the system, it is also necessary to take into account the kinetic factor.

In most cases, oxides structurally similar to zirconia are used. In this case, according to crystallochemical considerations, stable solid solutions may form. These oxides are, in particular, oxides of rare-earth elements, yttrium, and also the oxides of the group of alkali-earth elements.

Since the cubic solid solutions are stable in the entire temperature range from melting point, they are often referred to as the completely stabilised zirconia.

There is obviously a large number of intermediate states between the random and ordered distribution of cations and vacancies in the sublattice sites. The actual state has a very complicated nature and cannot be calculated [49].

The ZrO_2–R_2O_3 solid solutions with the general formula $Zr^{4+}_{1-x}R^{3+}_{x}O_{2-x/2}V^{O}_{x/2}$ in the vicinity of the composition 33 mol.% R_2O_3 (R–La–Gd) undergo the 'order-disorder' type transformation leading to the formation of a phase with the pyrochlore structure. Ordering takes place simultaneously in the cation and anion sublattices [52].

The structure of the pyrochlore mineral has the general formula $A_2B_2X_7$, where the position A may be replaced by the ions Na, Ca, Th, U, Ln; the position B by Nb, Ta, Ti, in positions X by O, OH, F. This structure is a derivative of the fluorite-type structure [53], whose spatial group is O_h^7–*Fd3m*, $Z = 8$, $a = 10.35–10.45$ Å. The structure is stable at the ratio of the ion radii $R_A/R_B = 1.2$. In transition of the fluorite structure to the pyrochlore structure, one of the eight anions, surrounding the cation, is removed from the nodes of the anion sublattice, and the other are displaced from their ideal positions. The displacement is characterised by parameter x [53–57]. Large ions A are coordinated by eight oxygen ions which surround them in the tips of the distorted cube. The ions B are distributed inside the oxygen cubes two tips of which situated at the ends of the spatial diagonal, remain free (coordination number 6), (Fig. 7.3*b*). Thus, the anion motive of the structure of pyrochlore is to some extent intermediate between the motive of the structure of fluorite where all polyhedrons are cubes and the motive of the structure of senarmontite where all polyhedrons are flattened octahedrons [57]. The breakdown of the polyhedrons into two types of components results in doubling of the identity period and in a change of the symmetry: the spatial group of symmetry Fd3m replaces *Fm3m*). The transition from the fluorite to pyrochlore structure is accompanied by the formation of superstructural reflections on the x-ray diffraction patterns corresponding to the establishment of an order in the distribution of the cations and distortion of oxygen polyhedrons. These reflections have odd indexes of the type (311), (331), (511), (531), etc, and low intensity whose value depends mainly on the difference of the values of the scattering capacity of the cations of the groups A and B of the structure. If this difference is small, the odd reflections (*hkl*) may not be detected. Thus, an indeterminacy forms in the identification of the structure by the method of x-ray diffraction analysis and it is necessary to examine the structure of pyrochlore by other methods.

The 'order–disorder' transformation in solid solutions takes place at critical temperature T_C [52,58]. At temperatures higher than T_C, the cations of zirconium and rare-earth elements are distributed randomly, like the anions and vacancies of oxygen. Below this temperature, the cations Zr^{4+} and Y^{3+} are distributed in an ordered manner. At temperature $T > T_C$ the fluorite structure forms and the

Table 7.5. Ordering temperatures in ZrO_2–R_2O systems [52]

System	r_{R3+}, A	T_C, °C
ZrO_2–La_2O_3	1.14	Order at all temperatures
ZrO_2–Nd_2O_3	1.04	As above
ZrO_2–Sm_2O_3	1.00	2400
ZrO_2–Gd_2O_3	0.97	1550
ZrO_2–Dy_2O_3	0.92	Disorder at all temperatures

pyrochlore structure forms at $T < T_C$. The transformation temperature increases with increase of the ion radius of the rare-earth element, as indicated by the data presented in Table 7.5 [52].

The composition of the non-stoichiometric phase is written in the general form: $R_{2\pm x}Zr_{2\pm x}O_{7\pm x/2}V^{O}_{1\pm x/2}$.

In addition to the ordered compounds of the type of pyrochlore, the system based on zirconia contains compounds of the type A_7O_{12} containing three- and four-valent cations, characterised by a chain of oxygen vacancies along the direction <111> of the initial structure [18–20, 34]. In the range 40 mol.% R_2O_3 the compounds $Zr_3Y_4O_{12}$ and $Zr_3Yb_4O_{12}$ with rhombohedral symmetry were found. The ordering process in the compound $Zr_3Y_4O_{12}$ takes place at T <1200°C [20]. For example, at a temperature of 1180°C, the ordered phase is formed only after annealing for 13 weeks.

The decrease of the amount of stabilising oxide is accompanied by the formation of a tetragonal solid solution based on zirconia. However, single-phase tetragonal specimens are produced only in strictly specific synthesis conditions [59–62]. For sintered ceramic specimens, the stability of the tetragonal structure of zirconia depends on various factors, such as density, composition, grain size and sintering temperature. Since the upper temperature boundary of phase stability of the solid solutions is the temperature of the phase tetragonal–cubic transition and it is lower than the melting point, these materials are referred to as partially stabilised zirconia (PSZ).

It has been established that there are two forms of tetragonal zirconia which may form in the ZrO_2–Y_2O_3 system [23, 63–65]: yttrium oxide-rich 'non-transformable' t'-ZrO_2 phase and the t-ZrO_2 phase depleted in Y_2O_3 which is capable of martensitic transformation to the monoclinic phase *m*. The *t*'-ZrO_2 phase is the main component in the rapidly quenched cubic solid solutions. It forms by a diffusion-less mechanism as a result of a shear phase transformation and is characterised by the presence of antiphase domain boundaries clearly visible in a transmission electron microscope.

ZrO_2 ceramics with a low content of the stabilising agent (2–9 mol.% Y_2O_3), produced by solid phase sintering, consists mainly of two (or three) phases, depending on the previously mentioned factors. The monoclinic tetragonal zirconium oxide exists in these materials in the form of coherent precipitates in the tetragonal or cubic matrix.

In directional solidification of the melt in the cold crucible, the optimum conditions of production of the specimens consisting only of tetragonal zirconia are: the yttrium oxide concentration 3 mol.% and rapid cooling of the melt (> 400°C/h) [65]. These results are in good agreement with the experimental conditions determined by Lefevre [30] for retaining the tetragonal phase. With a decrease of the cooling rate or when adding the sesquioxide oxide in the concentration of less than 3 mol.%, in addition to the tetragonal phase, the specimens also contains the monoclinic phase. The mixture of the cubic and tetragonal phases is obtained in the same experimental conditions, if the amount of the sesquioxide is more than 5 mol.%.

In pure zirconia, the cubic–tetragonal transition takes place at temperatures above 2300°C, depending on the partial pressure of oxygen. This transition is paraelastic ↔ferroelastic *(m*3*m* ↔ *F*4 */mmm)* [65]. In the phase diagram of the ZrO_2–Y_2O_3 systemt in the concentration range rich in zirconia (Fig. 7.7), the cubic and tetragonal phases are separated by the two-phase region (*c*+*t*) [12]. In growth of a crystal from the melt, the crystals with the cubic structure of the fluorite type are the first to grow, and the transition to the tetragonal phase takes place at a temperature below critical temperature T_C. These temperatures for a number of compositions were determined in [66]. The structural types of the zirconia crystals, partially stabilised with the yttrium oxide, are presented in Table 7.6. Figure 7.16*a* shows an elementary tetragonal cell whose nodes are denoted by the solid circles. The position ½, ½, ½ (empty circle) corresponds to the node equivalent to the fluorite lattice. The zirconium (or yttrium) cations, are surrounded by the oxygen anions that form the same coordination polyhedrons as in the case of the

Table 7.6. Structures of tetragonal zirconia (3 mol% Y_2O_3) [65]

	Tetragonal $Zr_{1.94}Y_{0.06}O_{1.94}$	Cubic fluorite MO_2
Spatial group	$P4_2/nmc$, $Z = 2$	$Fm3m$, $Z = 4$
Cell parameters	$A_t = 0.3610$ nm; $c_t = 0.5168$	$A_c = 0.5126$ nm
Atomic positions	2(a) Zr (or Y) 000B = 0.002 nm²	Z=0.025
	4(*d*) O (or vacancies) 01/2*Z*	
	$Z = 0.199$, $B = 0.025$ nm²	
Cation environment	4Zr–O 0.208 nm	8M–O 0.222 nm
	4Zr–O 0.238 nm	12M–M 0.362nm
	4Zr–O 0.361nm	
	4Zr–O 0.363 nm	
Anion environment	2O–Zr 0.208 nm	4O–M 0.222 nm
	2O–Zr 0..238 nm	6O–O 0.256 nm
	2O–O 0.258 nm	
	2O–O 0.261nm	

Comment. The cubic fluorite structure of MO_2 with the elementary cell of the same volume is given for comparison [68]

cubic lattice, but with the orientation different from 000 and ½, ½, ½ positions, as shown in Fig. 7.16.

The main difference between the tetragonal structure and the fluorite structure is that in the tetragonal zirconia there are two types of the metal–oxygen bond. The difference in the strength of the M–O bonds is determined by their lengths of 0.208 and 0.238 nm, which correspond to the binding forces of $s = 0.72$ and $s = 0.30$, respectively [65].

The consequence of this change in the structure is the transformation of cleavage planes in the tetragonal crystals on which preferential separation of the single crystals takes place. The cleavage planes for the fluorite structure belong to the family of {111} planes. The separation of the tetragonal crystals takes piace mostly along {001} planes, normal to the fourth-order axes [65].

The special features of the formation of phases and structure of

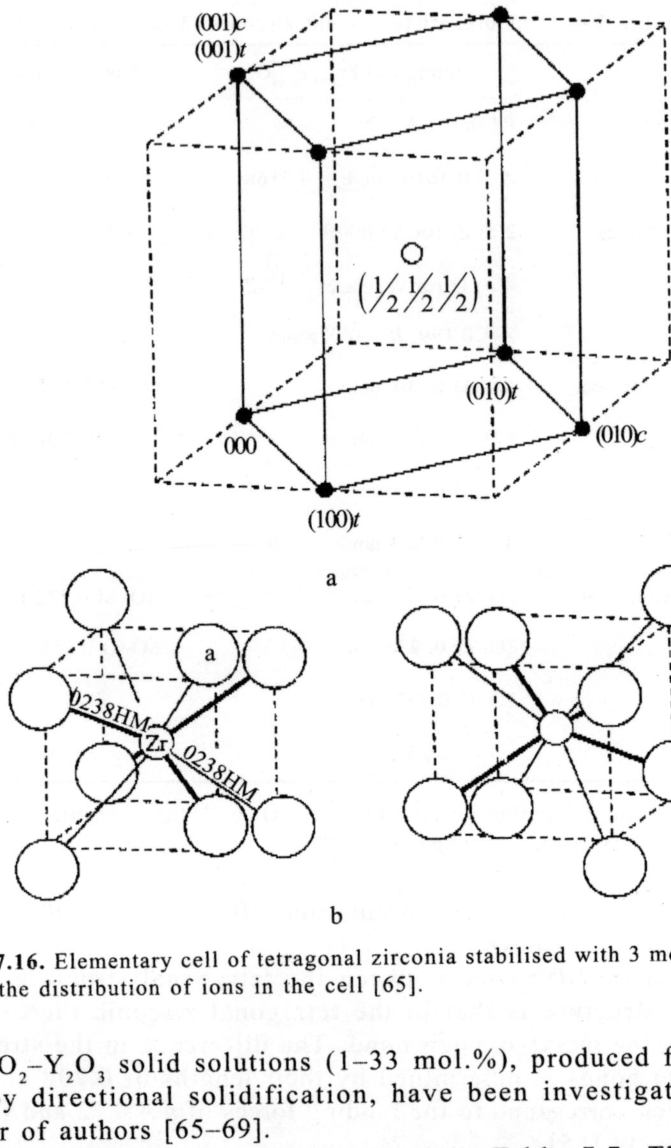

Figure 7.16. Elementary cell of tetragonal zirconia stabilised with 3 mol.% Y_2O_3 (a) and the distribution of ions in the cell [65].

the ZrO_2–Y_2O_3 solid solutions (1–33 mol.%), produced from the melt by directional solidification, have been investigated by a number of authors [65–69].

Experimental results are presented in Table 7.7. The solid solutions with the concentration of the yttrium oxide of 10–15 mol.% are characterised by some disordering which become smaller at higher and lower concentrations of the oxide [70].

The problem of the difference of the content of the cubic and tetragonal *t'* (pseudo-cubic) phases in the crystals of partially stabilised zirconia has not as yet been solved. For example, the

Table 7.7. Phase composition of crystals grown by directional solidification of the melt [65,68,70]

Mol% Y_2O_3	Number of phases	Composition of phases	Spatial group
1–2	2	Monoclinic +tetragonal	C^5_{2h}–$P2_1/c$ $P4_2/nmc$
3	1	Tetragonal	$P4_2/nmc$
3–8	2	Tetragonal + cubic	$P4_2/nmc$ O^5_h–$Fm3m$
8–35	1	Cubic	O^5_h–$Fm3m$

phase analysis of the crystals with 4.7 mol.% of Y_2O_3 shows that they consist of ~35 wt% of the trigonal *t*-phase and 65 wt.% of the pseudo-cubic *t*'-phase [71, 72].

The conclusions on the nature of the solid solution formed in solidification of the melt have been made on the basis of comparison of the x-ray density calculated from the crystal lattice parameter, and the density determined by experiments. Calculations were carried out usually using the model of the solid substitutional solution with oxygen vacancies which results in good agreement between the calculated and the experimental densities. When the experimental density was higher than the calculated values, it was assumed that a small fraction of interstitial ions may exist in the lattice of the investigated solid solutions [49, 73–77].

The results of phase analysis in these investigations slightly differ. The characteristic special feature of the single crystals, produced from the melt, is the existence, in the crystals, of the metastable phases at room temperature. The quantitative composition of the phases corresponds to the composition of the melt from which they were produced. The diffusion processes, taking place in cooling in the solid phase in accordance with the equilibrium diagram, even at the same synthesis conditions, may take place by different mechanisms depending on the technological conditions (for example, solidification rate, cooling rate, temperature gradients, additional heat treatment, etc). This may be the reason for the difference in the data obtained by different authors. Table 7.8 presents the comparative results of phase formation in the ZrO_2–R_2O_3 system.

Comparison of the results, obtained by the RS method, with the results of traditional methods of analysis of the phase composition

Table 7.8. Phase compositions of ZrO_2–R_2O_3 systems (R – Eu, Gd, Yb, Y) determined by the method of Raman scattering of light (RS), x-ray phase (XPA) and crystallo-optical analysis [35]

Composition of crystals, mol%	Phase composition		
	RS method	XPA	Crystallo-optical analysis
		ZrO_2–Eu_2O_3	
1	M + T	M + T	M + T (small amount)
3	M + T	M + T	M + T
5	T	M + T	M + T + K
8–10	T	F	T + K
10–12	F	F	K
15–20	DP	F	K
33	F	F	K
		ZrO_2–Gd_2O_3	
1	M + T	M	M + T (small amount)
3	M + T	M + T	M + T
5	T + M	M + T	T + M + K
8	T	F	T + K
10	K	F	K
12–33	F	F	K
		ZrO_2–Yb_2O_3	
1	M + T	M	M + T (small amount)
3	T	M + T	M + T
5	T	T	T + K + M
8–33	F	F	K
		ZrO_2–Y_2O_3	
6–8	T	F + T	F + T
10	F	F	F
12	DF	F	F
15–20	OF	F	F
33	DF	F	F

Notations: F – fluorite, DF – disordered fluorite, OF – ordered fluorite, DP – disordered pyrochlore. Phases: P – cubic pyrochlore, T – tetragonal, M – monoclinic, K – cubic show that the RS method provides the largest amount of information. For example, in investigation of the phase compositions and phase transitions in the materials with the low concentration of the stabilising oxides, investigations were carried out directly on the crystals, whereas powder x-ray diffraction analysis requires grinding of the specimens, activating the phase transformations in the specimens. In addition to this, in the investigations of the processes, associated with the ordering of the cubic solid solutions determined mainly by the ordering of the oxygen vacancies, formed in heterogeneous substitution, the sensitivity of x-ray diffraction analysis is low. Crystal-optical analysis is not capable of separating the cubic phases with different types of structure.

Only the complex approach is most reliable. In a number of cases (for example, for the inspection of the formation of the second phase in the range of cubic solid solutions), the sensitivity of the conventional x-ray diffraction and crystal-optical methods, is sufficiently high.

Parameters of the crystal lattice of fianites ZrO_2–Y_2O_3

Investigation of the structure of fianite single crystals is of both independent interest and of interest in connection with investigations of the polycrystalline stabilised zirconia synthesised by different methods: co-precipitation [78, 79], melting [49], plasma spraying [80] and sintering.

The authors of [81] carried out detailed x-ray diffraction examination of the ZrO_2–Y_2O_3 cubic phase. Single crystals were produced by directional solidification of the melt in air. The cubic structure was stabilised with Y_2O_3 in the amount of 6–40 mol.%.

X-ray diffraction analysis was carried out in a diffractometer with filtered CuK_α-radiation. The lattice constant was determined from the (531) peak with the accuracy of 0.001 Å. The degree of ordering of the single crystals was characterised by two quantities: the ratio of the intensity of the lines (600) and (200) to the normalised width of the peaks (531) at half height in relative units.

According to the results of x-ray diffraction analysis, all the single crystals were characterised by the stabilised high-temperature crystalline form of the cubic structure of the fluorite type. The spatial group $O_h^5 - Fm3m$. Figure 7.17*a* shows the dependence of the lattice parameter on the content of the stabilising oxide in the crystal. The equation for the calculation of the lattice parameter of

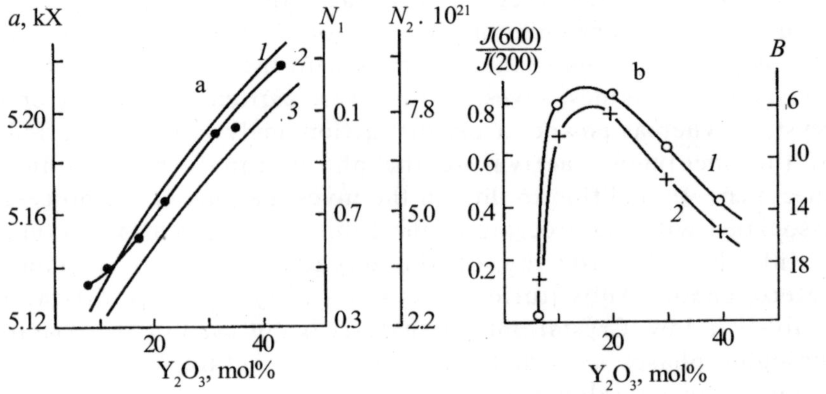

Figure 7.17. Variation of the lattice spacing and number of vacancies in relation to Y_2O_3 content [82]. a) 1 – lattice spacing calculated from equation (7.2); 2 – experimental values of the spacing; 3 – number of vacancies in the anion sublattice; N_1 – in the elementary cell; N_2 – in 1 cm³. b) disorder as a function of the Y_2O_3 concentration: 1 – half-width of the (531) line and 2 – intensity ratio of the (600) and (200) lines

zirconia, stabilised simultaneously by *n* different elements, taken at different ratios, was obtained on the basis of considerations of the spherical packing of the ions in the crystal structure [81]:

$$D = A\left[R_{Zr} + R_O + \sum_{k=1}^{n} P_k m_k \Delta R_k / 100 + \sum mk(P_k - 1)\right], \tag{7.2}$$

where $A = 2.31$; R_{Zr} and R_O are the radii of the ions of zirconium and oxygen; *n* is the number of various stabilising additions; P_k is the number of ions of each stabilising element per one molecule of the oxide; m_k is the amount of the *k*-th stabilising oxide, expressed in mol.%; ΔR_k is the difference of the radii of the ions of the *k*-th stabilising element and zirconium. For the binary system, in which $n = 1$ and $P = 2$,

$$D = 2.31\{R_{Zr} + R_O + 2m\Delta R/(100 + m)\}. \tag{7.3}$$

As shown by equation (7.3) and the graph (Fig. 7.17*a*), the dependence of the lattice spacing on the Y_2O_3 content is non-linear. The calculated and experimental curves have the same form and are in numerical agreement for the composition 10 mol.% Y_2O_3. A

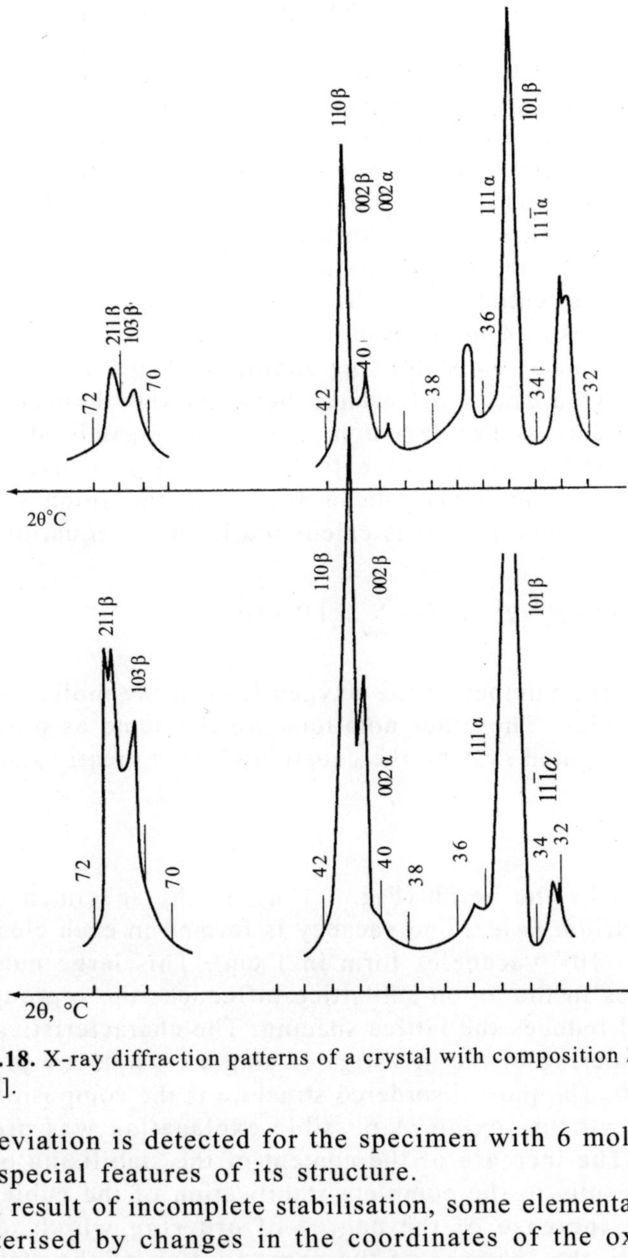

Figure 7.18. X-ray diffraction patterns of a crystal with composition ZrO_2–3 mol.% Y_2O_3 [82].

large deviation is detected for the specimen with 6 mol.% because of the special features of its structure.

As a result of incomplete stabilisation, some elementary cells are characterised by changes in the coordinates of the oxygen ions, leading to the formation of the tetragonal lattice. In this case, the parameter *a* of the cubic lattice decreases, and the size on the *c* axis greatly increases. The rearrangement of the structure results in the splitting of some lines and in the formation of doublets. The diffraction pattern of the specimen with 3 mol.% Y_2O_3 is shown in Fig. 7.18.

The increase of the lattice spacing may be associated with the presence in the crystal of two phases: cubic and distorted tetragonal. The main role of the stabilising addition in this case is played only by the cubic phase and this increases the yttrium concentration in this phase. The lattice spacing, equal to 5.136 A, corresponds to the specimen with the content of 8 mol.% Y_2O_3. The amount of the tetragonal phase in the investigated specimen is, according to the estimate, ~25%, which is in agreement with the intensity of peaks on the diffraction patterns.

At the content of the stabilising additions greater than 10 mol.% Y_2O_3, there is a small difference between the theoretical and experimental curves. The deviation of the experimentally determined spacings of the crystal lattice to low values is caused by the formation of vacancies. The mean number of the anion vacancies for a single elementary cell is calculated from the equation:

$$V = \sum 4P_k m_k (2 - P_{ok} / P_k) / [100 + \sum m_k (P_k - 1)], \quad (7.4)$$

where P_{ok} is the number of the oxygen ions in the molecule of the stabilising oxide. The other notations are the same as previously. For the investigated system, the equation (7.3) is greatly simplified:

$$V = 4m / 100 + m. \quad (7.5)$$

As indicated by the graph (Fig. 7.17*a*), in the specimen with 33 mol.% of yttrium oxide, one vacancy is formed in each elementary cell or $7.1 \cdot 10^{21}$ vacancies form in 1 cm^3. This large number of the vacancies in the anion sublattice influences the regrouping of the ions and reduces the lattice spacing. The characteristics of the degree of ordering of the investigated single crystals are presented in Fig. 7.17*b*. The most disordered structure is the composition with 6 mol.% of yttrium oxide. A possible explanation was presented previously. The increase of the content of the stabilising oxide to 10 mol.% results in the complete stabilisation of the cubic lattice and a large increase of the degree of ordering which tends to increase with the increase of the concentration of the stabilising oxide to 20 mol.%. A further increase of the content of the stabilising oxide results in a monotonic decrease of the degree of

Table 7.9. Density of zirconia single crystals in relation to the concentration of stabilizing oxide Y_2O_3 [82]

$[Y_2O_3]$, mol%	ρ_1, g/cm^3	ρ_2, g/cm^3	$[Y_2O_3]$, mol%	ρ_1, g/cm^3	ρ_2, g/cm^3
10.3	5.930	5.910	19.8	5.661	5.655
13.8	5.868	5.863	31.2	5.596	5.603
16.3	5.766	5.800	35.9	5.538	5.543

Comment. ρ_1 is the distance according to x-ray data; ρ_2 is the distance determined by hydrostatic weighing.

ordering as a result of the gradual transition from the lattice of zirconia to the lattice of the solid solution of oxides of zirconium and yttrium.

The calculation of the density of the substitutional solid solution taking into account the vacancies formed in the given composition is carried out using the equation:

$$\rho = [818.050 - 136.920m/100 + m]/d^3. \tag{7.6}$$

The values of the density of the single crystals with different yttrium content, calculated using equation (7.6), and the experimental results, obtained by the method of hydrostatic weighing, are presented in Table 7.9.

Figure 7.19 shows the calculated and experimental values of the density of the single crystal solid solutions of hafnium dioxide in

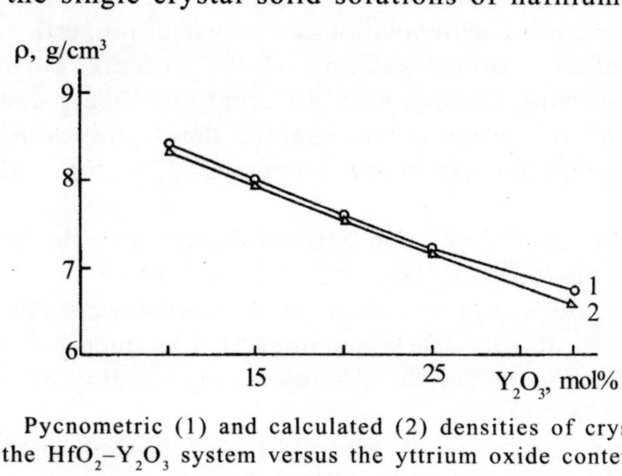

Figure 7.19. Pycnometric (1) and calculated (2) densities of crystals of solid solutions of the HfO_2–Y_2O_3 system versus the yttrium oxide content [83].

relation to the yttrium oxide content.

With the increase of the yttrium oxide content, the density of the crystals rapidly decreases. The main reasons for the decrease of the density is the increase of the volume of the elementary cell and the formation of anion vacancies. Small differences between the calculated and experimental values indicate the high perfection of the crystal structure.

7.5. DEFECTS IN THE STRUCTURE OF FIANITE CRYSTALS

Crystal structure defects include any deviation from the ideal periodicity of the atomic structure. Defects form in the solidification process under the effect of thermal, mechanical, electrical, and radiation influences and also when adding admixture elements to the melts.

Defectiveness influences many structure-sensitive properties of the crystal. The dislocations, vacancies, impurity atoms, phase precipitates and other elements of the real structure of the crystal determine all phenomena of plasticity and strength of the crystals. The detected optical absorption is mostly of the impurity type. The crystals, containing a large number of macroscopic ($\approx 10^{-4}$ cm or more) or even colloidal (10^{-6}–10^{-5} cm) inclusions and cracks are not suitable for practical application. Defects influence many physical and chemical properties of the crystals. The formation of defects is linked genetically with the phase transformations and the coefficients of distribution of the impurity atoms in the crystals. This circumstance makes it possible to restore the prior history of formation of the crystal on the basis of the morphology, form, dimensions, chemical composition and physical properties of crystals with the defects. Investigations of the defects, formed in the process of growing fianites with different stabilising additions in a wide range of the composition, enabled the definition of two more classes of compounds with the properties differing from those of the fianites.

The interest in the defects, formed during growth, is associated with the non-repeatability of the defective structure and, consequently, the variations of the structure-sensitive characteristics of the crystal not only from specimen to specimen but also within the limits of the same crystal. Therefore, high-quality crystals should be perfect from 'birth'.

The following growth defects are found most frequently:

–heterogeneous inclusions–inclusions of a second phase of the macroscopic dimensions from colloidal to several centimetres;

–point defects–impurity atoms and ions, vacancies and their aggregates;

–twins, stacking faults, dislocations, block boundaries;

–internal stresses;

–heterogeneities in the distribution of the previously mentioned defects–zonal, sectorial, 'pencil' and other structures [84].

Coefficients of distribution of impurities in the ZrO_2 crystals

The equilibrium coefficient of distribution of the impurities and components is one of the main parameters of the crystallisation process and it is also the parameter of the equilibrium between the solid and liquid phases. The distribution coefficient K is the ratio of its concentrations in the solid and liquid phases, having the common interfacial surface. In practice, it is necessary to consider the distribution coefficients determined for equilibrium (K_0) and non-equilibrium (K_{eff}) conditions. An objective relationship exists between the values of the equilibrium and effective coefficients.

The distribution coefficient for the equilibrium conditions K_0 characterises the ratios of the concentrations corresponding to the liquidus and solidus lines of the equilibrium diagram. The factor which determines the transition from the nonequilibrium to equilibrium conditions is the solidification rate.

The dependence of the effective distribution coefficient on the solidification rate is described by the Barton–Prim–Slichter formula which treats crystallisation as a heterogeneous process at the interface and the transfer of excess atoms of the second component (at $K_0<1$) or the main component (at $K_0> 1$) in the liquid phase [85].

If the solidification rate is high, the impurity component is displaced (at $K_{eff} <1$) or it may be trapped by the solid phase if the solidification rate is higher than diffusion rate in the volume of the liquid phase. A concentration gradient forms in the liquid phase ahead of the solidification front. This region, enriched at $K_{eff} < 1$ (or depleted at $K_{eff}> 1$), controls the process of implantation of the component in the solid phase in non-equilibrium crystallisation. If the liquid phase is mixed mechanically or by thermal convection, the concentration differences in the volume of the liquid are equalised. The flow in the liquid layer in the vicinity of the interface between the liquid and solid phases remains laminar. Diffusion in this layer remains the controlling factor and results in the distribution of the concentration of the component in the solid phase with the coefficient K_{eff} in the range $K_0 < K_{eff} < 1$ or $K_0 > K_{eff} > 1$. In a general case δ, the thickness of the layer in which the diffusion plays the controlling role, depends on the mixing rate. Assuming the purely diffusion transfer of the components in the melt and the stationary solidification conditions, the effective distribution coefficient is described by the equation:

$$K = \frac{K_0}{K_0 + (1 - K_0)\exp(-f \cdot \delta / D)}, \tag{7.7}$$

where D is the diffusion coefficient of the component in the liquid; f is the solidification rate.

This equation is used widely in describing the solidification processes. The value of K_{eff} varies continuously from $K_{eff} = K_0$ at low growth rates to $K_{eff} = 1$ at high growth rates.

In the initial stage of crystallisation, the conditions of removal of the impurity or its capture are unstable. In this stage, a diffusion layer with thickness δ forms in the liquid. The concentration of the impurity in the diffusion layer changes from the mean concentration in the liquid phase to some stable concentration in the solid phase [86]. On reaching the stationary crystallisation regime, the distribution of concentrations in the solid phase is described by the Gulliver–Pfann equation in which the equilibrium coefficient of distribution is replaced by the effective coefficient K_{eff} [87]:

$$C(g) = KC_0(1 - g)^{K-1}, \tag{7.8}$$

where g is the fraction of the solidified substance.

In the crystallisation process, a diffusion layer forms in the liquid phase. The diffusion layer is enriched with the low-melting component of the solution which has the temperature higher than the equilibrium temperature of crystallisation of the given system. The remaining mass of the liquid phase solidifies at this temperature. This phenomenon is referred to as the concentrational supercooling as a result of which the surface of the solidification front loses its stability.

The morphology of the interfacial surface must be taken into account in interpretation of the experimental data, characterising directional solidification. It is evident that different degrees of concentrational supercooling correspond to different effective distribution coefficients. The concentration dependence of the effective distribution coefficient in directional solidification may also be ambiguous. This is caused by the fact that with the increase of the concentration of the impurity component K_{eff} may approach unity in accordance with the variation of the equilibrium coefficient K_0 (isomorphous capture or displacement of the impurity), and also because of the development of concentrational supercooling (non-isomorphous capture or displacement). The ambiguity of K_{eff} may also be associated with the instability of the thermal conditions of directional solidification. Small variations of temperature at the solidification front result in a situation in which the main translational movement is supplemented by additional reciprocal movement. The variations of the growth rate result in the periodic heterogeneity of the composition of the crystals in the growth direction. The periodic or banded heterogeneity often forms in drawing the crystals from the melt as a result of the heterogeneities of the temperature field. Banded heterogeneity, like the cellular structure, displaces the effective coefficient to unity. Banded heterogeneity may be removed only by the stabilisation of the thermal, mechanical and hydrodynamic conditions of the solidification process and this is difficult to achieve. However, it

Table 7.10. Effective coefficients of distribution of Y_2O_3 in ZrO_2 crystals [35]

Y_2O_3 concentration, %	Rate of lowering cold crucible, mm/h	K_{eff}
5	16	1.08±0.05
8	16	1.06±0.05
10	1	1.12±0.03
10	5	1.05±0.02
10	8	1.02±0.02
10	30	1.07±0.02
15	16	1.02±0.20
20	16	1.00±0.01
30		1.01±0.02

Table 7.11. Composition of grown crystals and effective coefficient of distribution of components in zirconia [35]

Composition of crystals, $ZrO_2–R_2O_3–R'_2O_3–R''_2O_3$ wt.%				Effective distribution coefficient		
ZrO_2	R_2O_3	R'_2O_3		K_{R2O3}	$K_{R'2O3}$	$K_{R''2O3}$
$ZrO_2–Y_2O_3(V_r=16$ mm/h)						
89.49	10.51			1.07±0.02		
84.91	15.09			1.08±0.01		
81.29	18.21			1.6±0.03		
80.13	19.87			1.06±0.0.02		
77.63	22.37			1.03±0.01		
70.02	29.98			1.02±0.03		
53.93	47.09			1.01±0.02		
$ZrO_2–Gd_2O_3(V_r= 16$ mm/h)						
84.00	16.00			0.99±0.01		
79.62	20.38			0.98±0.04		
76.00	24.00			1.03±0.03		
72.05	27.95			1.02±0.02		
66.68	33.32			1.02±0.03		
60.96	39.04			1.01±0.04		
46.52	53.48			1.02±0.01		
$ZrO_2–Y_2O_3Nd_2O_3(V_r= 5$ mm/h)						
79.4	19.9	0.74		1.04±0.01	0.57±0.07	
77.9	19.7	2.44		1.08±0.01	0.60±0.02	
75.0	22.6	2.41		1.01±0.01	0.67±0.01	
73.0	22.3	4.74		1.04±0.01	0.59±0.01	
60.4	37.4	2.23		0.99±0.01	0.66±0.02	
58.7	36.9	4.40		1.01±0.01	0.64±0.01	
$ZrO_2–Y_2O_3–Nd_2O_3–Ce_2O_3(V_r= 5$ mm/h)						
76.1	17.8	4.83	1.24	1.09±0.01	0.51±0.02	0.39±0.02
76.2	16.2	6.03	1.54	1.09±0.01	0.52±0.01	0.42±0.01
77.2	16.3	5.30	1.24	1.12±0.01	0.55±0.02	0.39±0.03
78.2	16.4	4.28	1.13	1.09±0.01	0.66±0.01	0.40±0.01
78.2	16.4	4.28	1.13	1.07±0.01	00.71±0.01	0.47±0.02
79.0	14.8	4.90	1.25	1.08±0.01	0.60±0.03	0.49±0.04
$ZrO_2–Y_2O_3–Nd_2O_3–Ce_2O_3(V_r= 10$ mm/h)						
93.2	5.28	0.89	0.59	1.13±0.01	0.87±0.01	0.78±0.02

should be mentioned that the effect of banded heterogeneity on K_{eff} is far less marked than the effect of cellular or dendritic heterogeneity.

Diffusion-less crystallisation, i.e. crystallisation without diffusion displacement of matter leading to the separation of the components may take place at high cooling rates and high solidification rates with a steep temperature gradient [86].

The values of the concentrations of the stabilising oxides in the crystals of the ZrO_2–R_2O_3–R'_2O_3–R''_3O_3 system (where R is Gd, Y; R' – Nd; R'' – Ce) and effective coefficients of distribution for different growth conditions and compositions are presented in Tables 7.10 and 7.11.

Comparison of the distribution curves of yttrium for the ZrO_2–R_2O_3 single crystals, taken from different sections of the ingot, shows that the distribution of yttrium along the length of the crystals is periodic. The waves on the concentration curves taken from different parts of the ingot may not coincide and this indicates differences in the crystallisation conditions in the cross-section of the ingot (not a flat solidification front).

The variation of the solidification rate (1–16 mm/h) and the concentration of the yttrium oxide has only a slight effect on the form of the distribution curves. Figure 7.20 shows the characteristic curve of the distribution of yttrium oxide along the length of a single crystal with the ZrO_2–Y_2O_3 cubic structure. As indicated by

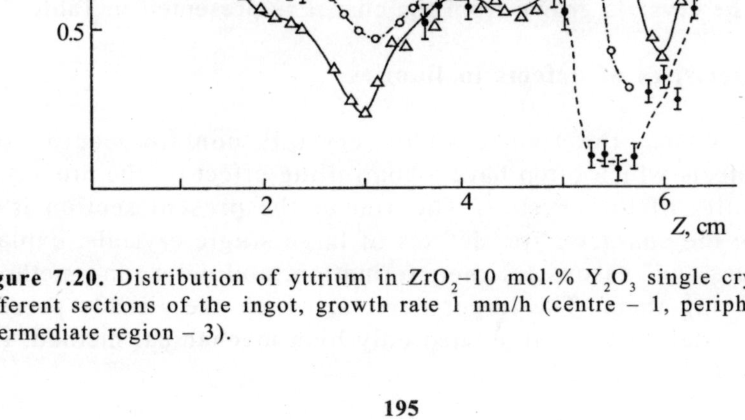

Figure 7.20. Distribution of yttrium in ZrO_2–10 mol.% Y_2O_3 single crystals in different sections of the ingot, growth rate 1 mm/h (centre – 1, periphery – 2, intermediate region – 3).

Tables 7.10 and 7.11, the effective distribution coefficients of the yttrium oxide, calculated using these data, are close to 1 and depend only slightly on the growth rate and the initial concentration of the yttrium oxide. For example, for the Y_2O_3 concentration of 10 mol.% (16.95 wt.%) the decrease of the growth rate increases the value of the distribution coefficient; at a rate of 16 mm/h, the increase of the concentration is accompanied by a small decrease of K_{eff}.

For the crystals ZrO_2–Gd_2O_3 at growth rates (V_P = 60 mm/h), the distribution coefficients of the stabilising oxide are also close to unity and slightly increase with increase of the oxide concentration.

The authors of [35] also determined the nature of penetration of oxides of rare-earth elements with high ion radii (Ce, Pr, Nd) into the crystal lattice of zirconia. The range of the compositions, which produce single-phase crystals in crystallisation of the melts, is very narrow as a result of the nature of phase formation in these systems, differing from ZrO_2–R_2O_3 system (R–Y, Gd, Yb). Therefore, the oxides of caesium, praseodymium and neodymium were added in the form of an additional mixture to the solid solutions ZrO_2–Y_2O_3. The distribution of the concentration of the ions of rare-earth elements along the length of the crystal greatly differs from the values calculated from the Gulliver–Pfann and Barton–Prim–Slichter equations.

In addition to admixtures, especially introduced into the melt to stabilise the crystal structure or activate laser generation, the crystal usually contains uncontrollable impurities from the charge. The content of the impurities varies in the range 10^{-3}–10^{6} wt.% and is determined by the purity of the initial reagents and the effective distribution coefficient. The amount of the impurities in the charge and in the crystals grown from the charge is presented in Table 7.4.

Characteristics of defects in fianites

In most cases, the fianite single crystals contain macro- and microdefects which often have a controlling effect on the properties and quality of the crystals. The aim of the present section is to describe the characteristic defects of large single crystals, explain the nature and reasons for their formation, and determine methods of removing them. The larger defects, which are already present in the crystal, can be eliminated only by a mechanical method, i.e.

by removing the defective part of the crystal.

Crystal technology should be such as to ensure that defects do not form during growth and cooling of the crystals. It should be mentioned that the defects, detected in fianites, are also found in other crystals, grown from a cold crucible and their formation mechanisms are similar.

Characteristic defects in fianites are growth bands, inclusions of secondary phases, the cellular substructure and also residual mechanical stresses. These defects cause the optical heterogeneity of the crystals. Undoubtedly, the formation of these and other defects is influenced by the impurities of secondary elements,

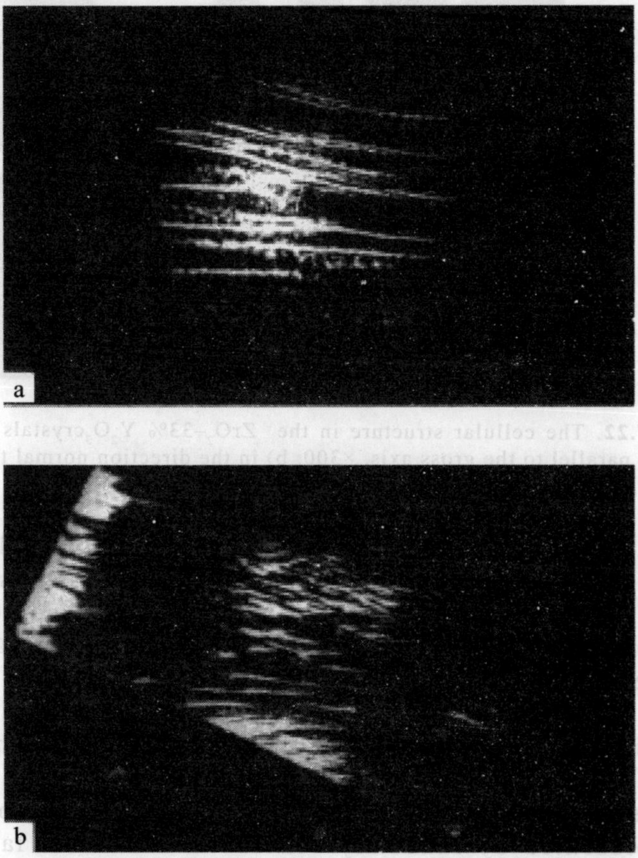

Figure 7.21. Growth bands: in the ZrO_2–33 mol.% Y_2O_3 single crystals, the growth rate 16 mm/h (a) and in single crystals grown in the 'lid' of the crystalline block (b) [40].

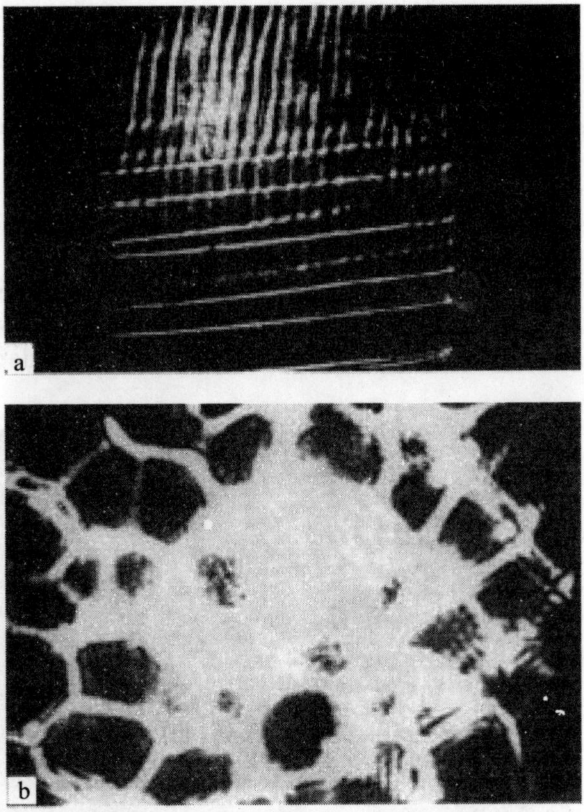

Figure 7.22. The cellular structure in the ZrO_2–33% Y_2O_3crystals [40], a) the direction parallel to the gross axis, ×300; b) in the direction normal to the growth direction, ×15.

present initially in the reagents. As indicated by Table 7.4, the content of the main impurities is relatively low but in some cases (for example, in drawing light fibre optics) the content of the impurities is outside the permissible range and the initial reagents must be additionally purified.

The growth bands are layers with different refractive indices found in all crystals. They are manifested in the form of bands normal to the direction of growth of crystal blocks (Fig. 7.21*a*). In the crystals, growing on the surface of the melt, the growth bands are represented by randomly distributed deformed layers (Fig. 7.21*b*).

The cellular structure is found in crystals in the central part of the ingot, usually in its upper part (Fig. 7.22). The upper end

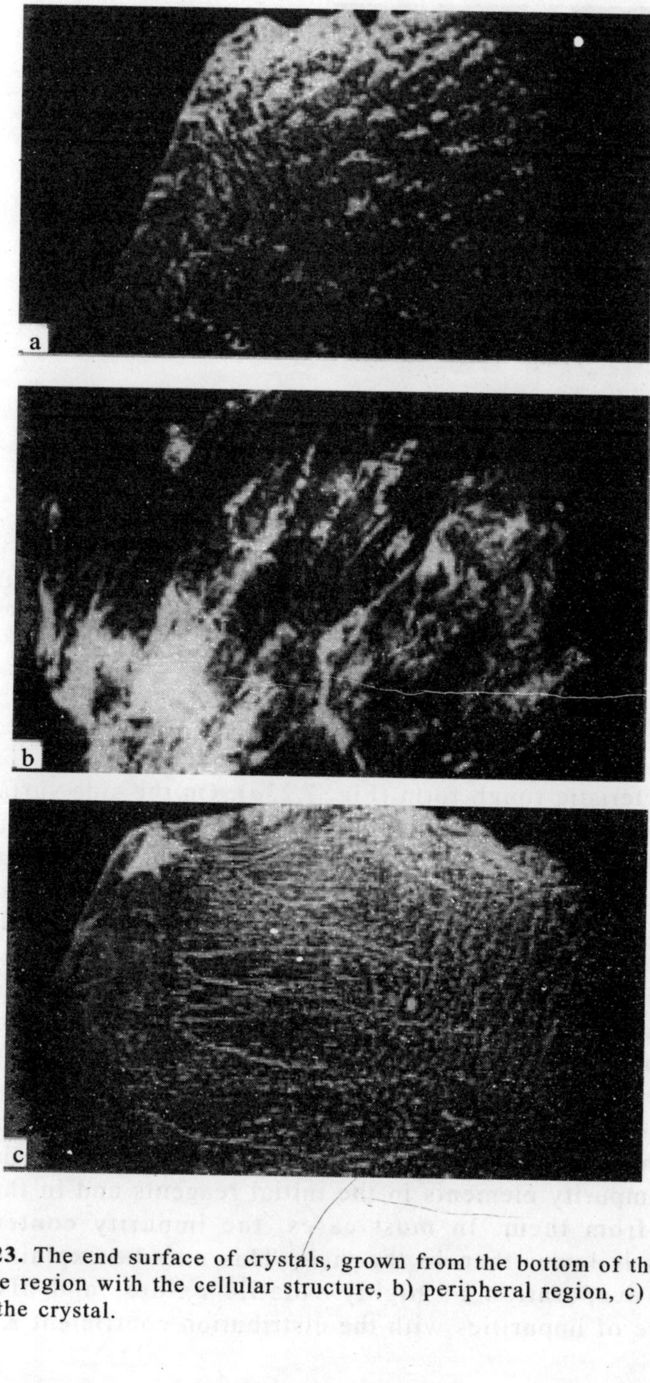

Figure 7.23. The end surface of crystals, grown from the bottom of the crucible [40]. a) the region with the cellular structure, b) peripheral region, c) transition region of the crystal.

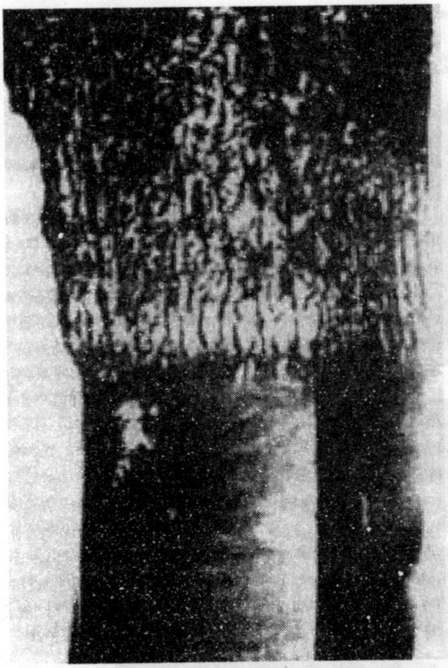

Figure 7.24. The side surface of the crystal with a highly developed cellular structure.

surface of the crystals in which the similar structure formed has a characteristic rough form (Fig. 7.23*a*). On the side surface of the structure there is a distinctive boundary between the smooth surface with strong shine and the uneven surface (Fig. 7.24). Figure 7.22*a* shows that the number of bands decreases with the formation of the cellular structure. The same crystal may contain alternating regions with distinctive bands and cellular structures. The dimensions of the region of the ingot with the cellular structure, extending to the surface of the ingot, reached 120 mm in diameter (using a crucible with a diameter of 220 mm and a high purity charge). The crystals distributed around the periphery are characterised by a smooth surface, and those in the transition region have a rough surface (Fig. 7.23). Table 7.4 gives the content of the impurity elements in the initial reagents and in the crystals grown from them. In most cases, the impurity content of the crystals is lower than in the melt. This may be explained by the partial evaporation of highly volatile oxides, and also by the presence of impurities with the distribution coefficient $K < 1$.

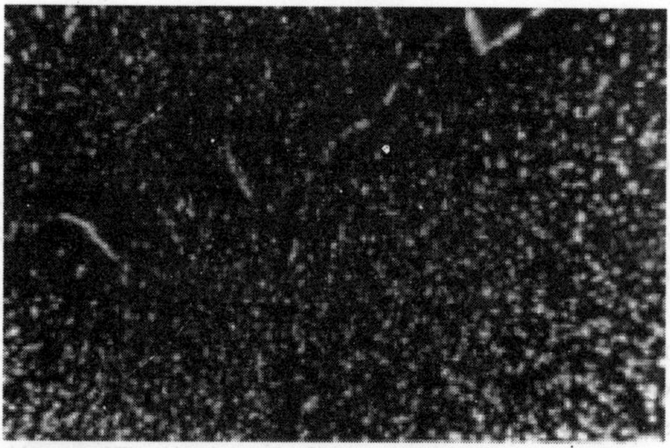

Figure 7.25. Inclusions of the second phase in the ZrO_2–Y_2O_3 10 mol.% + N_2O_3 0.5 mol% (electron microscopy) [40].

Inclusions of secondary phases

Large single inclusions of the solid phase and bubbles are found only seldom. The inclusions of the second phase in the form of particles with the size of 0.08–10 μm are found more frequently, cause scattering of light and can be detected visually (Fig. 7.25) or using a microscope. There are cases in which all the crystals of the ingot were 'turbid' to various degrees. The intensity of light scattering was the highest in the crystals taken from the central part of the ingot. However, the crystals contained areas with scattering particles in the form of a layer covering the entire block or its part.

Light scattering depends on the purity of initial reagents. Optical investigations with glowing points show that the scattering regions coincide with the regions of the cellular structure characterised by a large increase of the intensity of the glow of the Tindall cone. In addition to the inclusions in the volume of the crystal, the second phase is detected in the form of deposits on the surface of blocks. The individual sections, usually smooth with strong shine of the side surface, are covered with a matt deposit.

The formation of the inclusions causing light scattering in fianites, may take place both in the process of solidification of the melt at the phase boundary, at the boundaries of the cellular structure, and in cooling of the ingot in the process of high-temperature annealing of the crystals in the growth chamber with a decrease of the generator power [40].

Heterogeneities of chemical composition

The distribution of the stabilising oxide and the heterogeneity of the composition were investigated on the example of solidification of the solid solutions: ZrO_2–Y_2O_3 [7], ZrO_2–Y_2O_3–CaO [89]. The crystals were grown by directional solidification of the melt in air with lowering of the crucible with the melt at a rate of 1–20 mm/h. The initial charge was prepared by thorough mixing of the powders of zirconium and yttrium oxide. The concentration of the yttrium oxide was varied in the range 8–33 mol.%. In lowering the crucible, solidification of the melt started in the lower part of the crucible on crystalline grains of the skull. The solidified ingot consisted of several columnar single crystals. Solidification of the melt also started by means of a slow decrease of the generator power with the crucible in the stationary position. The rate of decrease of the generator power corresponded to the mean linear growth rate of 16–20 mm/h. In addition to the main solidification process from the bottom of the crucible, a 'roof', i.e. a single crystal layer, also formed on the surface of the melt, growing into the bulk of the melt as a result of high-intensity radiation. The solidified layer reached the maximum thickness (up to 3 cm) in solidification by decreasing the generator power.

The grown crystals were characterised by three types of optical heterogeneities: growth bands (ripples), light scattering and the cellular substructure. The ripples, alternating layers with the changed refractive index, were detected in all crystals. They are manifested in the form of straight bands, normal to the direction of growth of the crystals growing from the bottom of the crucible (Fig. 7.26*a, b*) or in the form of distorted bands in the 'roof'. The variation of the refractive index in the ripples is caused by the variation of the yttrium concentration determined by examination of the distribution of yttrium along the length of the crystals. Figure 7.20 shows the curves of distribution of yttrium $C(z)$ in the crystals taken from different parts of the ingot. The concentration of yttrium in the melt was 10 mol.%, the rate of growth of the crystals 1 mm/h. The figure shows that the distribution of yttrium $C(z)$ is wave-shaped. The waves in different parts of the crystals differ in amplitude and frequency, indicating different solidification conditions in the cross-section of the crystal. The concentration of yttrium at the end of the crystals is lower. This is characteristic of the impurity with the distribution coefficient K_{eff} higher than unity.

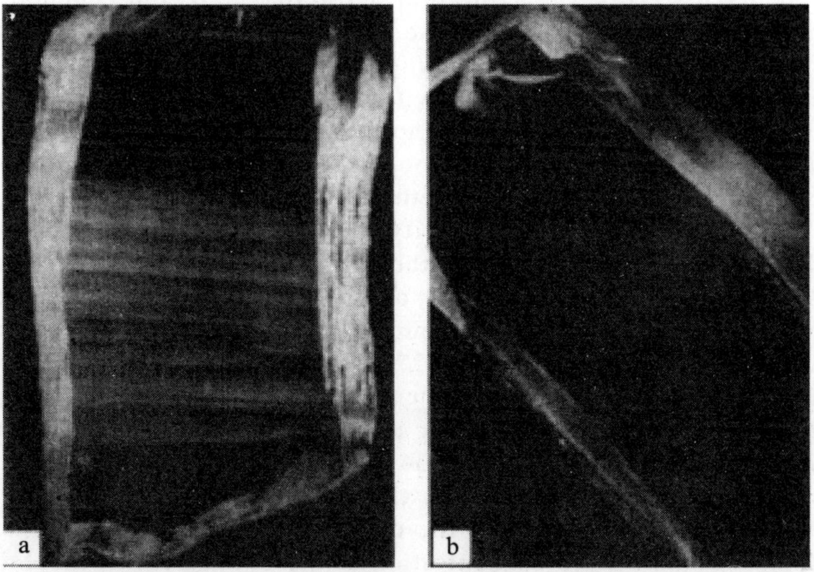

Figure 7.26. Scattering of light in ZrO_2 – 12% Y_2O_3 crystals in the region of the cellular structure [40]. The region includes the entire crystals; b) part of the crystal, ×3.

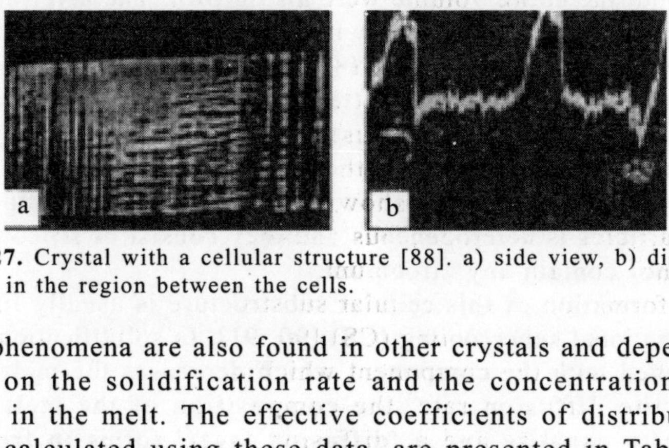

Figure 7.27. Crystal with a cellular structure [88]. a) side view, b) distribution of yttrium in the region between the cells.

Similar phenomena are also found in other crystals and depend only slightly on the solidification rate and the concentration of the impurity in the melt. The effective coefficients of distribution of yttrium, calculated using these data, are presented in Table 7.10. Comparison of Figs. 7.20 and 7.27, relating to the same crystal, shows that the presence of the ripples in the crystal is independent of the yttrium distribution $C(z)$. Evidently, the variation of the composition corresponding to the ripples is insufficient to be detected by the x-ray spectral method. The difference in the yttrium concentration between the sections with different refractive indices is smaller than 1%. In the crystals, growing from the bottom of the crucible, the number of ripples decreased with a decrease of the rate of lowering the container with the melt and when using reversing motion. The decrease of the growth rate also increases the size of the crystals in the ingot. At the rate of lowering of the crucible of 1 mm/h, an ingot with a diameter of 130 mm consisted of 4–5 blocks. Large blocks, grown at this rate, contained high stresses and cracked during cooling. After thermal annealing, the crystals remained without fracture.

Another type of heterogeneity is the cellular structure, usually detected in the upper part of the crystals, grown in the middle of the ingot at a lowering rate of the crucible with the melt of 60 mm/h and higher. An example of the crystal with a cellular structure is shown in Fig. 7.27*a*. The formation of the cellular structure is accompanied by the decrease of the ripples. The same crystal may contain alternating regions with ripples and a cellular substructure. The determination of the composition in the cross-section of the crystals with the cellular substructure shows that the concentration of yttrium in the centre of the cells is 2% higher in comparison with the concentration at the boundaries of the cells.

The particles, scattering the light, are grouped in ripples or at the boundaries of the cells, but crystals with uniform scattering throughout the entire volume were also grown. The layer distribution of the inclusions, coinciding with the banded structure, formed at a relatively low concentration (8–9%) of the yttrium oxide, and the crystals also showed light scattering.

One of the types of inclusions – large (~10 μm) particles, concentrate in the crystals at the boundaries of the cells. Local x-ray spectral microanalysis shows that the chemical composition of these particles is heterogeneous and they consist of silicon, yttrium, but do not contain any zirconium.

The formation of this cellular substructure is usually linked with concentrational supercooling (CS) [90, 91]. In solidification, the melt is enriched with the component which decreases the melting point. At a finite diffusion rate, the composition of the melt does not manage to equalise and a 'diffusion layer' forms in front of the solidification front. The solidification temperature in this layer increases in the direction into the bulk of the melt. The zone of the highest supercooling also travels into the bulk of the melt resulting in the instability of the flat solidification front: individual projections of the crystal, penetrating into the region of supercooling, grow at a higher rate than other regions of the solidification front.

Analysis of the distribution of temperatures in the melt shows that the maximum temperature is obtained on the level of the induction coil. The temperature in the upper and lower parts of the melt is reduced because of a less effective absorption of the energy of the high-frequency field and more intensive heat removal. This distribution of temperatures complicates thermal convection in the melt above the crystals, growing from the bottom of the crucible. The extent of mixing of the melt by the high-frequency field is very low. In these conditions, the mass transfer in the melt is difficult and this results in the formation of a wide diffusion layer ahead of the solidification front and in the development of concentrational supercooling for crystals growing from the bottom of the crucible.

As mentioned previously [92, 93], the formation of the banded structure in crystals may be caused by concentrational supercooling and oscillations of the growth rate in the diffusion layer, and also by the failure of the layer in rapid solidification of the supercooled region of the melt. The instability of the growth rate results in the oscillations of the effective coefficient of distribution of the impurity. The decrease of the rate of growth and high intensity of mixing of the melt result in the equalisation of the impurity concentration in the melt and this reduces the probability of formation of concentrational supercooling and, consequently, reduces the extent of the banded structure. This mechanism explains the formation of growth bands in the crystals growing from the bottom of the crucible. For the crystals, growing in the 'roof', the formation of concentrational supercooling is unlikely because growth takes place in the conditions of strong convective mixing and a highly superheated melt is situated below it. However, the formation of the banded structure in the crystals, formed in the 'roof', may be caused by irregular convection, resulting in the fluctuations of temperature and the solidification front, as indicated by disordering of the ripples in the 'roof'.

As indicated by previous considerations, the same mechanism (concentrational supercooling) is used to explain different types of defects: the cellular substructure and the formation of the banded structure. It may be assumed that the transition from the banded to cellular substructure is determined by dimensions of the region of concentrational supercooling, and the formation of cells results in a decrease in the amount of the banded structure.

In addition to the variation of the growth rate associated with concentrational supercooling, the variations of the composition may

Table 7.12. Composition of inclusions in ZrO_2–R_2O_3 crystals (here R – Y, Eu, Gd, Yb)

Composition of crystal		Composition of inclusions						
		Phase 1 (grey)				Phase 2 (black)		
ZrO_2, mol%	R_2O_3, mol%	ZrO_2, mol%	R_2O_3, mol%	SiO_2, mol%	CaO, mol%	ZrO_2, mol%	R_2O_3, mol%	SiO_2, mol%
90	$10Y_2O_3$	10.6	27.6	61.8	–	2.19	1.0	97.00
In the zone of cellular structure								
88	$12Y_2O_3$	8.15	28.14	62.69	–	10.06	7.65	97.23
End of the crystal								
88	$12Y_2O_3$	8.23	27.53	62.65	1.57	3.9	5.7	90.33
84	$16Y_2O_3$	11.5	25.36	63.08	–	1.48	8.55	89.96

be caused by 'apparatus' effects: instability of the generator power, variation of the relationship between the induction coil and the melt, and the nonuniformity of movement of the crystal. As shown by the measurements of the power absorbed in the melt from the variation of the temperature of the cooling water at entry into and exit from the container, the growth process is accompanied by variations of the temperature of the melt which may cause heterogeneity of the composition of the crystal. These external instabilities are not the main source of the growth bands and can be eliminated by modifying the working conditions of equipment. At the same time, they may be responsible for the relatively large oscillations of the composition reflected in the form of concentration waves $C(z)$.

Silicon inclusions form as a result of the displacement of the silicon impurity by the growing crystal into the region between the cells. Yttrium oxide builds up in the SiO_2 phase whose acidity is higher than that of ZrO_2. The heterogeneity of the inclusions is caused by the fact that they are trapped by the growing crystal in the form of melt droplets, and the segregation processes already take place in the crystal matrix. The appearance of the second phase in the crystal is possible in cooling as a result of polymorphous transformations if the concentration of the stabilising addition is insufficiently high. The variations of the composition in the bands being formed take place in the vicinity of the boundary of stability of the cubic phase.

In addition to the yttrium oxide, calcium oxide is also used as a stabilising addition. In [95], the crystals of the following systems were grown: ZrO_2–CaO, ZrO_2–Y_2O_3, ZrO_2–Y_2O_3–CaO. The distribution of Y_2O_3 and CaO in the charge, the crystals and in the residual melt was determined by chemical analysis. The results are presented in Table 7.12 which shows that the Y_2O_3 content of the crystal is always lower than in the initial materials; the content in the residual melt is lower than in the crystal. This indicates that Y_2O_3 may evaporate and the coefficient of its distribution in the crystal is lower than unity. At the Y_2O_3 content of the melt of 20 mol.%, the growth of macroscopic crystals is more stable but they are darker. At 12 mol.% of Y_2O_3, the crystals are almost colourless and transparent.

Analysis of the crystals containing the calcium oxide shows that the amount of this oxide in the crystal is considerably smaller in comparison with the initial material. A large part of the oxide remains in the residual melt, i.e. the coefficient of distribution of the calcium oxide in the crystals K_{Ca} is lower than 1. The crystals, containing 13 and 15 mol.% of CaO, are considerably smaller than the crystal stabilised by Y_2O_3; they are colourless and transparent but more brittle. The authors detected the same defects in the grown crystals and they will be described later. Etched specimens showed inclusions similar to pipes touching each other, and the form of the inclusions on the etched surfaces was tubular with a diameter of approximately 10 μm. The etching rate of the inclusions was considerably higher than that of the main matrix. The inclusions were concentrated at the ends of the crystal and removed by machining. The inclusions in the form of individual islands, detected in the centre of the crystal, were hexagonal. All the inclusions in the cubic crystals of stabilised zirconia had the same optical characteristics, but the refractive index and microhardness were clearly lower than those of the main matrix. Chemical analysis of the inclusions in the investigated crystal shows that the composition of the impurities is the same as that of the main matrix, and they consist mainly of a mixture of Si, Ca and Mg, with Si and Ca dominant. In polygonal inclusions in the form of polyhedrons, the content of Si and Ca was considerably higher, the content of Mg lower and that of Zr still lower. No silicon was found in the matrix of the crystal.

7.6. MECHANISMS OF FORMATION OF DEFECTS IN FIANITES

Growth bands

The appearance of the growth layers in fianites is caused by the presence of admixtures and, primarily, of the stabilising oxide. We have already mentioned an example of the distribution of the stabilising oxide along the length of the crystal (Fig. 7.20). Analysis of the results shows that the visually observed growth bands do not always correspond to the distribution waves of the concentration of the stabilising oxide. The discrepancy between the density of the concentration waves and visible growth bands are characteristic of the crystals with a high density of the layers. For low density of the layers (2–4 cm^{-1}), the agreement is quite good. Evidently, the amplitude of variations of the composition, corresponding to the growth bands, is below the sensitivity of the method of x-ray spectral analysis. In fact, the estimates of the composition made on the basis of the local measurements of the refractive index show that the difference between the absolute yttrium concentration in areas with different refractive indexes is usually approximately 1%. The density of the bands is unaffected by factors such as the concentration of the stabilising oxide, the rate of growth of the crystal, and the nature of mixing of the melt.

With increase in the concentration of the stabilising oxide, a large increase of the density of the bands was detected only in the range of the compositions close to 30 mol% R_2O_3 and a the lowering rate of the crucible of 16 mm/h. With a decrease of the lowering rate to 1mm/h, the density of the bands in the crystals decreased to 8 cm^{-1}. In forced mixing of the melt by reversing movement of the crucible and in the case of low growth rates, the density of the bands decreased to 1 cm^{-1}.

The growth layers are usually determined by two reasons: apparatus instability and special features of the mechanism of growth of crystals from the melt.

The determination of the heating power and the efficiency of removal of heat from the crucible with the melt shows that the maximum temperature is obtained on the level of the induction coil, and in the upper and lower volumes of the melt the temperature of the melt is usually lower because of the less extensive absorption of the energy of the high frequency field. Due to such a distribution of temperature, the temperature in the layer of the melt above the crystal, growing from the bottom of the crucible, increases with the height and, therefore, heat convection is more difficult. The decreased in the concentration of the yttrium ions in the vicinity of the solidification front in comparison with the concentration in the melt (since $K_{eff} > 1$) results in an increase of the density of the melt at the solidification front and this also prevents convective mixing. The electrodynamic mixing of the melt by the high frequency field in the investigated case is insignificant because the working frequency is relatively high (5.28 MHz) and the electrical conductivity of the melt is relatively low (10–100 ohm cm) [94]. In addition to this, the solidification front is positioned well below the upper level of the induction coil. In these conditions, mass exchange in the melt is complicated, resulting in the formation of a thick diffusion layer and in the development of concentrational supercooling. Evidently, this is accompanied by the formation of spontaneous oscillations of the growth rate during the formation of the diffusion layer depleted in Y_2O_3, but the rate of crystal growth decreases since the crystallisation temperature increases. When the crucible is lowered into a region of lower temperatures, the partial or complete failure of the diffusion layer takes place as a result of the rapid crystallisation of the supercooled melt. Nonuniform growth leads to variations of the effective coefficient of distribution of the admixture. As indicated by experience, a decrease of the growth rate and intensive mixing of the melt cause the equalisation of the concentration of the admixture in the melt leading to a decrease of the probability of occurrence of concentrational supercooling and to a decrease in the extent of formation of the banded structure in the crystals.

Possible reasons for the formation of growth bands may also include: instability of the power of the generator, variations of the consumption of the heat carrier in the cooling system of the crucible, nonuniform movement of the crucible with the melt. The effect of these factors has been investigated in experiments in [6]. The nonuniform movement of the crucible was recorded using an indicating device with a sensitivity of 0.01 mm. The maximum 'jumps' were 0.05–0.07 mm, at a rate of lowering of the crucible of 1 mm/h, and these variations may be ignored.

In the growth of single crystals with the composition ZrO_2–12 mol.% Y_2O_3 in Donets-4 equipment with the crucible lowering rate of 2 mm/h, the parameters of the generator (I_c, U_a, I_a), working in the conditions with stabilisation of U_a, were constantly controlled, together with the control of the power generated in the load and taken away by the cooling water throughout the entire growth process.

The recording of the heating power and the power, generated in the load, shows that the process of crystal growth is accompanied by variations of the melt temperature which may lead to heterogeneities in the composition of the crystals. Comparisons of the diagrams of temperature in the melt with the diagrams of distribution of yttrium along the length of the crystal at a growth rate of 1 mm/h shows that temperature jumps with time coincide with the jumps of the concentration of the stabilising oxide. Consequently, a mechanism of formation of concentration waves has been proposed. Disruption of the stability of the power of heating the melt leads to a disruption of the diffusion layer at the phase boundary and, consequently, to instability in the yttrium distribution. For the selected growth conditions, the number of bands formed throughout the entire growth process was 1–2.

Thus, the apparatus instability may lead to the formation of growth bands but it is not the main source of these bands. This position is confirmed by a decrease in the density of the growth bands with a decrease in the rate of lowering the crucible. The same relationships were also recorded for single crystals with other stabilising oxides (europium, gadolinium, ytterbium). It is important to note the anomalously low density of the growth bands in the crystals based on HfO_2. Evidently, the congruent crystallisation mechanism operates in this case.

For the crystals growing from the surface of the melt ('in the roof') the characteristic feature is the high density and considerable bending of the growth bands. The reasons for the formation of these bands differ from the reasons causing the formation of growth bands in the crystals growing from the bottom of the crucible. In these crystals, the formation of concentrational supercooling is unlikely because they grow in the conditions of strong convective mixing. The formation of the banded structure in these crystals is caused mainly by irregular convection, leading to the fluctuation of temperature at the solidification front. The irregular nature of convection flows is also the reason for the formation of the bent growth bands.

Thus, the formation of the banded structure in the fianite crystals is associated with small variations of the concentration of the stabilising oxide as a result of concentrational supercooling. Crystal defects may be removed by decreasing the rate of lowering the crucible and by forced mixing of the melt.

The cellular substructure

Another type of admixture heterogeneity, often found in fianite crystals, is the formation of the cellular substructure. The determination of the chemical composition in the cross-section of the ZrO_2–10 mol.% Y_2O_3 crystals with a cellular substructure shows that the yttrium concentration in the centre of the cells is 2% higher in comparison with the periphery. This corresponds to the coefficient of distribution of yttrium in the crystal higher than unity.

The formation of the cellular substructure is usually associated with the concentrational supercooling of the melt. The formation of this defect during growth of fianite crystals was substantiated previously. During crystallisation, the melt is enriched with the component decreasing the melting point (zirconia), and at a low diffusion rate the composition of the melt does not equalise in the entire volume. In this case, a 'diffusion' layer forms ahead of the solidification front. The concentration of Y_2O_3 and, consequently, the solidification temperature in this layer increase in the direction from the surface of the crystal into the bulk of the crystal. This results in the instability of the flat solidification front.

Thus, the same mechanism – concentrational supercooling – is responsible for the formation of both the cellular structure and growth bands in the fianite crystals. Concentrational supercooling (CS) is an essential but insufficient condition for the formation of the cellular structure. The cellular structure forms only at some critical value of concentrational supercooling. Cells appear mainly in the upper part of the crystal, at a high crystallisation rate and without mixing of the melt. It is evident that the values of CS, required for the formation of the cellular structure, should be higher than those required for the formation of growth bands.

The formation of the cellular structure may be associated not only with the stabilising oxides but also with foreign admixtures, present in the initial charge. As in many other crystals, the most detrimental impurity for the crystallisation of fianites is the silicon oxide with the distribution coefficient considerably lower than unity.

Inclusions of foreign phases

The light scattering in the fianite crystals, caused by the presence of the second phase, may be due to two reasons: a) decomposition of the cubic solid solutions; b) the presence of secondary phases in the initial reagents. The authors of [40] investigated the effect on the growth and perfection of single crystals of fianite of the most widely encountered admixtures in the form of oxides: SiO_2, TiO_2, Al_2O_5, WO_3, whose concentration in the initial reagents was approximately 10^{-2} wt.% and which had a significant effect on the properties of these crystals [101–103]. For example, it is well known that the presence of oxides of silicon, aluminium and titanium, found in the initial reagents, reduces the stability of the polycrystalline material. The SiO_2 impurity results in a large decrease of the strength and heat resistance of the crystals and decreases the softening temperature under load [103].

The formation of any specific type of admixture heterogeneity in the crystals is determined by the nature and concentration of the admixture. The previously mentioned admixtures are not isomorphous in relation to zirconia. In fact, the isomorphous interrelationship of the atoms (ions) in the crystals is determined by a number of factors, namely: 1) chemical indifference; 2) similar ion radii; 3) identical nature of ion bonding; 4) isostructural nature of the produced crystals. The isomorphism conditions are not fulfilled for these ions with zirconia atoms already in the case of the first two parameters. Thus, these oxides should be displaced by the growing crystals and, depending on the crystallisation conditions and the concentration of the given admixture, they will cause the formation of different admixture defects.

We have investigated crystals with the composition ZrO_2–R_2O_3, where R = Eu, Gd, Yb, Y, and crystals with intentionally introduced admixtures in the concentration from 0.05 to 1 wt.%. The admixtures of aluminium, titanium and tungsten were introduced into a charge with the composition ZrO_2–12 mol.%Y_2O_3 in the amount of 1 wt.% [93].

Effect of silicon dioxide on the crystallisation of fianites

The effect of the SiO_2 impurity on the nature of crystallisation and the optical homogeneity of the fianite crystals was investigated in [104]. Crystals of fianites with rare-earth stabilising oxides and a batch of crystals with 12 mol.% of yttrium oxide and with the intentionally introduced silicon dioxide in the amount of 0.05–0.5 wt.% were grown. In the crystals without intentional addition of SiO_2 into the melt, the blocks were characterised by the typical appearance and the dimensions, and also by the presence of typical defects examined in detail in the previous chapter.

Examination of the crystals in a polarisation microscope showed

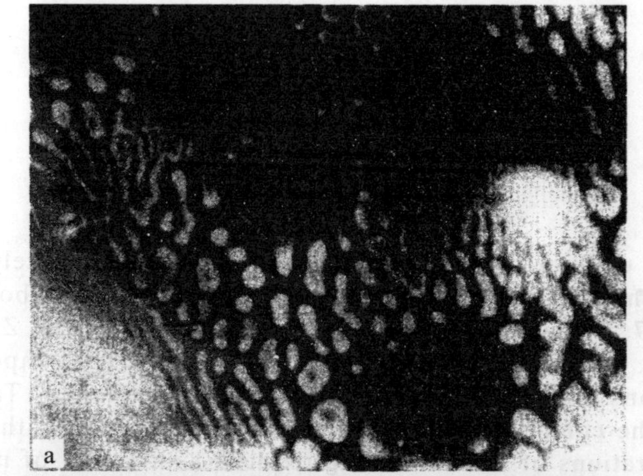

Figure 7.28. Deposits on the site surface, formed in the process of growth of $ZrO_2 - Y_2O_3$ 10 mol.% crystals [40]. a) the melt contained SiO_2 impurity, b) Al_2O_3 impurity.

Table 7.13. Composition of inclusions in crystals grown from a melt with the composition ZrO_2–12 mol% Y_2O_3 + 0.5 wt.% SiO_2 [40]

Examined areas of crystal*	Composition, mol%		
	ZrO_2	Y_2O_3	SiO_2
1	2.23	6.11	91.64
2	4.93	31.72	63.33
3	83.67	16.32	-
4	12.19	24.45	63.35
5	84.31	15.24	-

*Examined areas of crystal 1–5cm, in Table 7.14

that the cellular structure contained large (approximately 10 μm) inclusions of a second phase, distributed at the cell boundaries. Figure 7.25 shows a photograph of inclusions in the ZrO_2–Y_2O_3 crystals produced in scattered electrons. The shape of the inclusions was usually spherical or droplet shaped. Table 7.13 shows the composition of the inclusions of crystals with different concentrations of the stabilising oxide. Examination of matt areas of the surface of the crystals already showed the presence in them of individual clusters of phases with a high SiO_2 concentration (Fig. 7.28). With an increase in the silica concentration, the extent of light scattering increases, as indicated by an increase of the intensity and the diameter of the cross-section of the Tindall cone. Scattering was caused by small particles of the second phase with a size of ~0.1 μm. At an SiO_2 content in the melt of 0.2 wt.%, the crystals contained a columnar structure and growth bands. With an increase in the content of the admixture, this complicated structure of the defects was transformed into a cellular substructure. At 0.3 wt.% SiO_2, the crystals in the centre of the block were cellular and the blocks were polycrystalline. The periphery of the block was characterised by the growth of large single crystals with a high level of light scattering. In solidification of the melt, containing 0.5 wt.% SiO_2, examination showed the formation of a completely polycrystalline block. The phases, containing silicon dioxide in the polycrystals, were characterised by the form of continuous vertical columns passing through the entire length of the block. X-ray diffraction analysis showed that the crystals with a SiO_2

concentration of 0.05–0.5 wt.% have the cubic structure of the fluorite type.

According to the equilibrium diagram, the ZrO_2–SiO_2 system contains the $ZrSiO_4$ compound which decomposes in the solid phase into ZrO_2 and SiO_2 at a temperature of 1950 K, below the eutectic temperature. In directional crystallisation of the melt, the formation of a compound consisting of ZrO_2 and SiO_2 in the process of growth of crystals is unlikely to take place. Silicon dioxide, present in the investigated system, with a distribution coefficient lower than unity $K_{eff} < 1$, should be displaced during the growth of the crystal of the solid solution of SiO_2 with ZrO_2. The region of the phase diagram rich in ZrO_2 does not contain the $ZrSiO_4$ compound (the SiO_2 concentration in the crystal does not exceed 1.3 10^{-2} [105, 106].

During growth of the crystals, the concentration of SiO_2 ahead of the solidification front increases because at a high growth rate the diffusion does not manage to reduce the concentration of SiO_2 because of the segregation process. This results in the formation of initially fine inclusions of the second phase and, subsequently, the formation of various morphological (columnar and cellular) structures forming large inclusions. The spherical shape and heterogeneity of the composition of large inclusions indicate that the capture of the admixtures takes place in the liquid state. Possibly, the unstable growth, taking place during the formation of the cellular structure, is accompanied by the capture of melt droplets from the boundary layer enriched with the silica. As a result of the lower (in comparison with ZrO_2) melting point, these droplets are enclosed in the hard matrix of the crystal. Thus, this phenomenon is associated with the concentrational supercooling [106]. The composition of the inclusions depends on the type and concentration of the stabilising oxide and also on the growth conditions of the crystal. The SiO_2 concentration in these droplets reaches more than 60 mol.%, and the droplets are characterised by the higher concentration of the stabilising oxides of rare-earth elements. Possibly, the R_2O_3 oxides, having the alkali nature, concentrate mainly in the droplets in which SiO_2 generates a more acid medium in comparison with the zirconia. The process of chemical differentiation takes place either during the formation of the droplet or during cooling of the droplet as a result of diffusion from the solid matrix. In order to explain the sequence of the distribution of the admixture in the droplet, it is convenient to use the ZrO_2–SiO_2 and SiO_2–R_2O_3 binary diagrams. The solidification of the droplet takes place from the periphery to the centre and in accordance with the SiO_2–Y_2O_3 diagram, the first phase to form in the temperature range 1933–1973 K is the solid phase containing mainly SiO_2; the centre of the droplet is characterised by the crystallisation of the eutectic mixture of the solid phases SiO_2+Y_2O_3 $2SiO_2$, together with the ZrO_2 component [107]. The composition of the inclusions in the ZrO_2–12mol.%Y_2O_3+0.5 wt.% SiO_2 crystals is presented in Table 7.13.

It is natural to expect that the composition and structure of the inclusions in crystals with other stabilising elements should be the same because these elements have the phase diagram is similar to SiO_2. However, Table 7.13 shows that this is not so in reality: the nature and compositions of the inclusions differ for different stabilising elements. Possibly, these differences are caused by the composition of trapped droplets or by the extent of exchange of components between the droplet and the solid matrix. It should be mentioned that the conditions of growth of the crystals, stabilised with the oxides of europium, gadolinium and ytterbium, differ from the conditions of growth of the crystal stabilised with yttrium (the volume of the crucible with the melt was considerably smaller). The high content of the stabilising oxide in the inclusions makes it possible to assume that in the case of relatively high concentrations of $SiO\text{-}_2$ and a low content of the stabilising oxide, the silica may induce the decomposition of cubic solid solutions and the formation of the monoclinic phase ZrO_2, present in the polycrystalline materials [102].

Thus, the production of zirconia single crystals requires efficient purification of the components to remove silica, the content of silica in the melt should not exceed ~5 × 10^{-3} wt.%.

The presence of an admixture of aluminium oxide in the amount of 1 wt.% (1.2 mol.%) in the ZrO_2–2mol.% Y_2O_3 crystals at a growth rate of 10 mm/h did not disrupt the single crystal nature of the individual blocks, but the side surface of the crystals was characterised by the formation of a matt deposit. The crystals, growing from the surface of the melt ('the roof') were polycrystalline. Examination of samples in an x-ray microanalyser showed the presence of Al_2O_3 in the volume of the crystals and on the side surface (Fig. 7.28*b*). The distribution of oxides of zirconium, yttrium and aluminium in the cross-section of a single crystal block is shown in Fig. 7.29. The matt deposit on the side surface of the crystal blocks has the following composition: 85

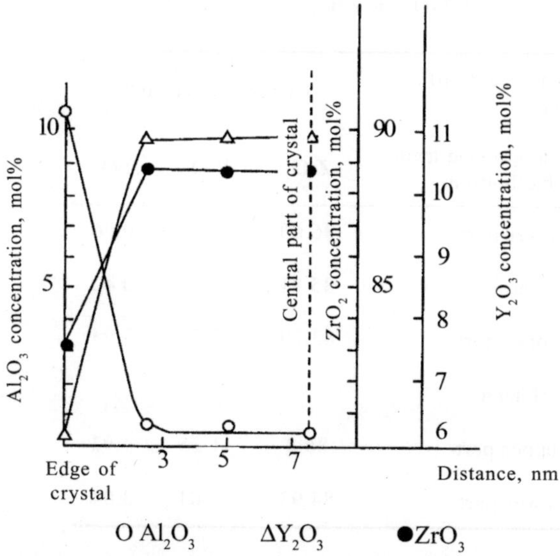

Figure 7.29. The radial distribution of the oxides of Zr, Y, Al in the ZrO_2+Y_2O_3+Al_2O_3 single crystals [40].

mol.% Zr_2O_3, 12.61 mol.% ZrO_2, 2.20 mol.% Y_2O_3. In the 'roof', the aluminium oxide precipitated between the blocks in the form of a second phase (Fig. 6.9*b*) with the composition (86.02±2.19)Al_2O_3; (11.83±1.72)ZrO_2; (0.14±0.5)Y_2O_3. The Debye patterns of the crystals did not contain additional lines. Thus, the aluminium oxide was rapidly displaced in the process of growth and accumulated mainly in the crystals which grew at the very end of the crystallisation process.

The admixtures of the oxides of titanium and tungsten (TiO_2 and WO_3) in the amount of 1 wt.% (1.5 and 0.5 mol%, respectively) do not disrupt the single crystal nature during the growth of crystals. After adding WO_3, the 'roof' of the ingot in the lower part formed in the form of a polycrystalline dense material. The single crystals in the upper part were characterised by slight colouring which, in the case of titanium, was yellow and for tungsten oxide it was green, with nonuniform distribution in the cross-section of the crystal. The crystals with titanium grown at a rate of 10 mm/h contained an area with a cellular structure formed at the end of growth and characterised by more intensive colouring. According to the results of x-ray diffraction microexamination, the content of tungsten in the volume of the crystals was considerably smaller than

Table 7.14. Distribution of TiO_2 in ZrO_2–12% Y_2O_3+1 wt.% TiO_2 crystals [40]

Examined area of crystal	Concentration,mol%		
Crystal growing from crucible bottom:	ZrO_2	Y_2O_3	TiO_2
1 – lower part	86.00	13.92	0.06
2 – centre	85.81	13.76	0.42
3 – upper part	83.29	12.79	3.95
Top of ingot:			
4 – upper part	86.2	11.35	1.93
5 – lower part	84.93	12.01	3.04

the amount added to the melt, ~0.2 mol.%. The chemical composition of the single crystal part of the 'roof' of the ingot was: ZrO_2 85.6 mol.%, Y_2O_3 14.4 mol.%, WO_3 0.2 mol.%, and the polycrystalline part was characterised by the presence of large inclusions with a size of 6–8 μm of irregular shape and containing large quantities of tungsten: ZrO_2 34.2 mol.%, Y_2O_3 23.0 mol.%, WO_3 42.8 mol.%. The concentration of TiO_2 in the single crystals changed along the length with a gradual increase towards the end of growth of the crystal (Table 7.14). In the crystals of the 'roof', the TiO_2 content increased in the direction from top to bottom.

Polarisation–optical and electron microscopic examination of the regions with the cellular structure in fianite with the additions of titanium oxide showed that the cell boundaries are characterised by the presence of small (approximately 1 μm or smaller) spherical inclusions, consisting mainly of titanium, silicon, zirconium and yttrium. The titanium and silicon content of these inclusions was high but the quantitative composition could not be determined because of the small size of the inclusions.

The examination of the effect of the oxides of aluminium, titanium and tungsten shows that these oxides, like silicon dioxide, cause the formation of defects disturb the optical homogeneity and other properties of crystals. However, this requires higher concentrations than in the case of silicon.

Thus, of the investigated secondary impurities, the most

detrimental effect on the growth and perfection of the finite crystals is exerted by silicon. A small amount of silica (0.05 wt.%) disrupts the growth of crystals and increases the density of various defects and admixture heterogeneity. The experimental results show that all the previously mentioned admixtures are displaced by the front of the growing crystal during crystal growth. Therefore, a reduction in the rate of growth to 10 mm/h and the use of forced mixing of the melt increase the efficiency of the process of displacement of the admixtures and improve the optical homogeneity of the crystals.

Formation of the microzonal structure

The authors of [103] investigated the conditions of formation of a microzonal structure in refractory oxide crystals based on zirconia (fianite). As shown in previous chapters, the crystals contain growth bands, slight light scattering by submicron inclusions of the second phase, a cellular substructure and other defects [104, 105]. Crystals for the investigations were produced by the method of direct RF melting in a 'cold' crucible. The concentration of the stabilising oxide Y_2O_3 in the charge was 12 mol.%, the content of the alloying admixture of titanium oxide was 0.01 wt.%.

Figure 7.30 shows the topography of the microzonal distribution of the admixtures of titanium in these crystals. The distribution was determined by the method of activation radiography, based on recording secondary β-radiation. The identification of radionucleotides, the evaluation of their content and the drop in activity were determined on the basis of γ-spectra of Ti^{47}. The radiographs were interpreted by the method of digital image processing using an interactive system for image processing. The digital processing programs were prepared on the basis of the methodology of cluster analysis, used widely in object recognition problems.

As indicated by Fig. 7.30, the distribution of titanium is characterised by steep local gradients. Analysis of the topographs shows the following:

1. Examination showed the existence of a periodic microzonal structure (zones *1–3*) with a thickness of 300–400 μm, and with increase of the size of the crystal, the size of the microzones increases both with respect to the area and the total content.

2. The presence of a stochastic cellular structure was confirmed (zones *5–6*). The observed large decrease in the concentration of

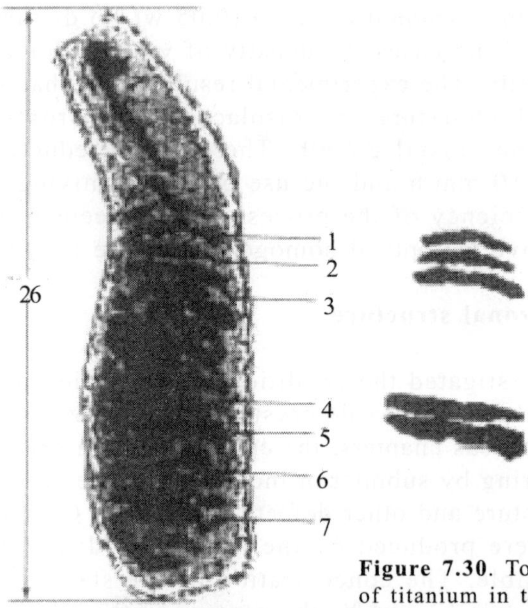

Figure 7.30. Topography of the distribution of titanium in the ZrO_2+Y_2O_3 crystal [96].

the admixture in the region of the cell boundaries is characteristic of an admixture with the effective distribution coefficient $K_{eff} > 1$.

The effective distribution coefficient of titanium calculated as the ratio of its total content in the single and polycrystalline regions, determined by the γ-spectra, is $K_{eff} = c_m/c_p = 2.2$,

3. A transition was found between the regions with the microzonal and cellular structures (zones *4, 7*).

The banded microzonal structure is described using various mechanisms based on: the variation of temperature at the crystallisation front, the variation of the speed of displacement of the front, and concentrational supercooling [106]. However, these models are restricted to diffusion–kinetic or thermodynamic approaches, determining the conditions of formation of an instability at the crystal–liquid interface [106]. The majority of the models, describing the formation of the banded microzonal structure, were developed for single-component crystals. Using these models, it is not possible to take into account correctly the kinetics of the chemical interaction between the admixture and the crystal material and describe the transition from the banded to cellular structure. It is important to take into account the kinetics of chemical interaction because in the majority of cases the process of growth takes place at high temperatures and with participation of catalytically active components of the crystallising material.

The authors of [103] investigated the mechanisms of formation of the microzonal structure and its transition to a cellular structure which takes place as a result of the diffusion–transport catalytic reactions of oxidation of titanium on the surface of the crystal by the reaction of the type:

$$O+E \leftrightarrow EO; \quad Ti+E \leftrightarrow ETi; \quad ETi+O \rightarrow TiO_2+E,$$

where E is the catalyst (in the present case, the role of the catalyst may be played by Y, Zr). The reactions of this type are well known in the case of fermentative catalysis and belong to the class of substrate–inhibitor reactions [108].

The mathematical analysis of the problem shows that there is a smooth transition from the unidimensional microzonal structure in the thin end of the crystal to the cellular structure of the microzones in the vicinity of the thicker part of the crystal. The length l at which there is a transition is determined mainly by the geometrical factor (the ratio of the radii and the length of the modelling cone r, r_0 and l), and also by the diffusion and reaction-capable properties of the interacting components. The external appearance of the detected microzonal structure resembles domain structures investigated previously.

7.7. MECHANICAL STRESSES AND THERMAL ANNEALING OF FIANITES

The examination of the crystals of the ZrO_2–Y_2O_3 and HfO_2– Y_2O_3 solid solutions under a microscope in crossed polarisers showed a slight heterogeneity of extinction, caused by the presence of volume stresses in the crystals. Conoscopic images show that under the effect of the residual stresses, the isotropic crystals become biaxial, optically positive. The angle between the optical axes is close to 90°. The optical indicatrix of the crystals has the form of a slightly deformed sphere, elongated in the direction of the growth axis of the crystals [001]. The birefringence of the crystals is $\Delta n = n_2 - n_1 = 0.000707$ [109].

In order to relieve the internal stresses, the fianite crystals are subjected to high-temperature annealing in vacuum or nitrogen. Annealing in an oxidation atmosphere is carried out in gas flame or silit-rod furnaces.

In order to determine the optimum annealing conditions of the

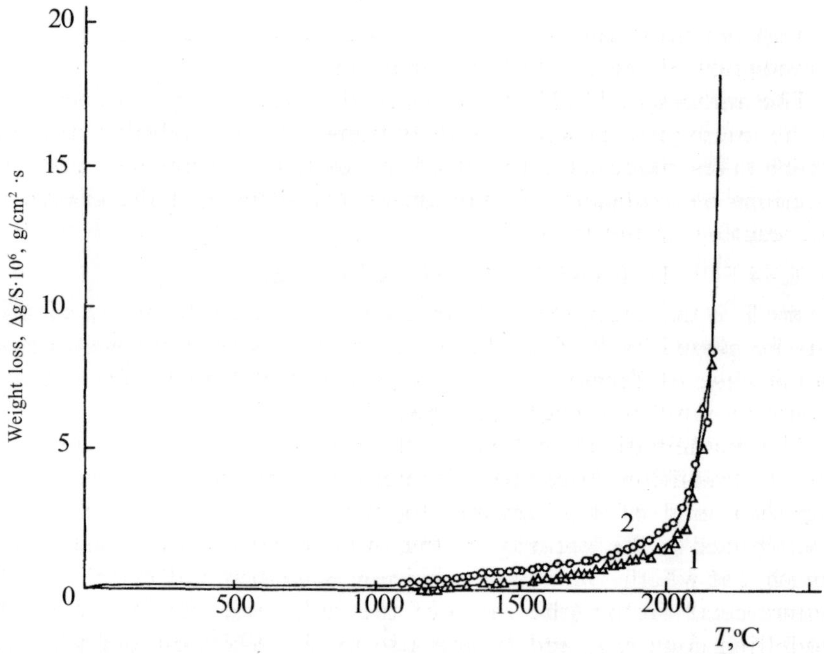

Figure 7.31. Curves of evaporation of the crystals of the $ZrO_2+Y_2O_3$ (1) and $HfO_2+Y_2O_3$ (2) solid solutions [101].

single crystals of $ZrO_2–Y_2O_3$ and $HfO_2–Y_2O_3$, the rate of evaporation of the crystals in vacuum was determined in advance. Evaporation was carried out in a vacuum of $10^{-4}–10^{-5}$ mm Hg on specimens in the form of disks with a diameter of 10–12 mm and a height of 1–2 mm. The extent of evaporation from the open surface was determined by continuous weighing. The minimum value of the recorded rate of evaporation was 10^{-8} g/cm^2 s. Figure 7.31 shows the temperature dependences of the weight loss of the investigated crystals in continuous heating at a rate of 8–10 deg/min. Extensive evaporation starts at a temperature of 2100°C. The rate of evaporation of the crystals of ZrO_2 and HfO_2 stabilised with 10 mol.% Y_2O_3 and measured at 2100 °C was respectively (7–8) and (3–5) 10^{-6} g/cm^2 s. Thus, vacuum annealing of the crystals–fianites should be carried out at temperatures not exceeding 2100°C.

The residual stresses were determined by the polarisation–optical method on the basis of birefringence. Figure 7.32 shows the interference patterns of the crystals in the crossed polarisers of the microscope prior to and after annealing. Annealing in the optimum

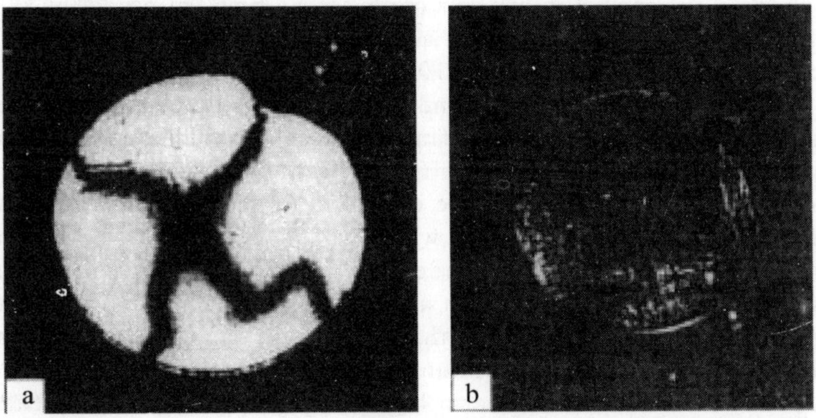

Figure 7.32. Interference patterns of the crystals of ZrO_2+Y_2O_3solid solutions with thermal stresses in the cross polarizers of the microscope (a) prior to and (b) after annealing [101].

conditions (T_{max} = 2100°C, cooling rate 85 deg/h) almost completely eliminated the stresses. The light spots observed in the annealed crystals in crossed polarisers are associated with the inhomogeneous with the impurities present in the crystal.

The stresses in the crystal are also determined by the formation of dislocations, producing slip planes and the boundaries of mosaic blocks. It is interesting to investigate the effect of high-temperature annealing on the density of dislocations. The dislocation structure was developed by chemical etching. An efficient etching solvent for the fianite crystals is a mixture of nitric and hydrofluoric acids at a ratio of 2:3. Etch pits are triangular, with nonuniform distribution. The density of the etch pits greatly increased in the direction to the edges of the crystal, where thermal stresses evidently greatly increased. Annealing reduced the density of dislocations by approximately an order of magnitude.

The experimental results show that vacuum annealing at 2000°C and subsequent rapid cooling result in the blackening of transparent and colourless crystals, stabilised with 10 mol.% Y_2O_3. This phenomenon does not take place at low cooling rates. When heating dark specimens in air top 1200°C they lost colour. The darkening of the specimens is associated with the loss of oxygen at high temperatures and with quenching phenomena caused by a rapid decrease of temperature in vacuum. Vacuum annealing of the crystals reduces tetravalent cerium, praseodymium and terbium to the trivalent condition and europium to the bivalent condition [110]. The annealing atmosphere also affects the microhardness of crystals which increases by 10–12%.

The optical heterogeneity and distribution of defects in the volume of crystals of Zr- and Hf-fianites was investigated in a laser ultramicroscope. The scattering of light was observed under an angle of 90° in relation to the direction of the incident light beam. The crystals of ZrO_2–Y_2O_3 with oxides of erbium, cerium and neodymium clearly show under the microscope microscopic and submicroscopic particles with a density of 10^6 mm^{-3} and a diameter of several microns. Possibly, the particles of the admixture decorate the structural defects, the boundaries of the blocks and slip layers. Under the effect of long-term high-temperature vacuum annealing, the size of the particles greatly increased and the density of the particles dropped. In the nominally clean (unalloyed) fianites, the scattered light was detected in the form of the Tindall cone and the intensity of light scattering depended greatly on the purity of the initial materials and the technological growth conditions. The scattering of light by the crystals is caused by the particles with the size of the order of 0.1 μm. In block crystals and in crystals with volume stresses, the admixture is susceptible to aggregation. The concentration of the Y_2O_3 stabiliser in the range 10–20 mol% has almost no effect on the extent of light scattering. In the presence of a third admixture, scattering centres were detected in many cases.

Attention should be given in greater detail to the mechanism of high-temperature reduction of the crystals of fianite because this is of practical importance for application of these crystals.

High-temperature and electrolytic reduction of the polycrystalline sintered materials based on ZrO_2 was investigated in [104–107]. The experimental results show that the ions of O^{2-} are removed from the crystal lattice and a phase nonstoichiometric in relation to oxygen forms:

$$Zr_{1-y}Y_yO_{(2-0.5y)} \rightarrow Zr_{1-y}Y_yO_{(2-0.5y)-x} + x/2O_2 \uparrow .$$

As a result of extensive reduction, the process may continue and lead to the precipitation of a metal-like phase, a solid solution of oxygen in α-Zr, suboxides ZrO_6 and Zr_3O and metallic zirconium [113].

High-temperature reduction of the crystals was carried out in a

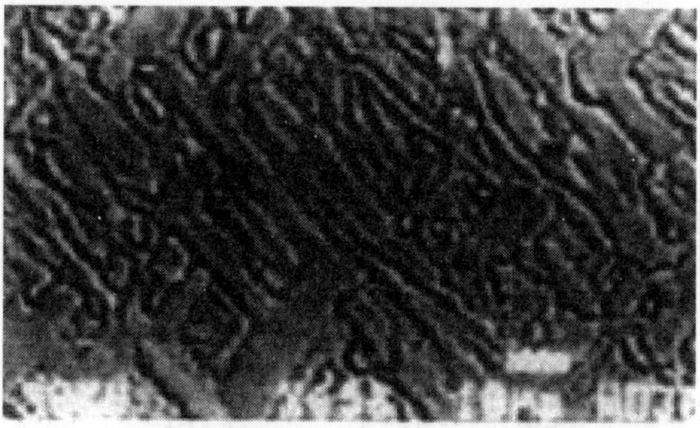

Figure 7.33. Electron micrograph of the surface of ZrO_2+Y_2O_3 crystals after thermal annealing.

vacuum furnace with a tungsten heating element at a temperature of 1400–2000°C for 10–30 hours, in a vacuum of 1.33 × 10^{-2} Pa. Cooling was carried out at a rate of 20–100 deg/min [111]. After vacuum annealing the crystals became black. In heating in air, the reduced crystals lost their colour but the single crystal nature was not disrupted. In cooling at high rates, the surface of the crystals was characterised by the formation of a network of shallow cracks and thermal etch pits, Fig. 7.33. According to the results of x-ray diffraction analysis, a new phase does not appear in the material, but the parameter of the elementary cell decreases. In heating in air of the reduced specimens, examination showed small exothermic peaks in the range (200–500)°C. The formation of these peaks is associated with the oxidation process. The dependence of the degree of reduction on the composition of the solid solution is presented in Fig. 7.34. The maximum value of $\Delta P/P$ of the crystals corresponds to the two-phase region of the solid solutions, consisting of the monoclinic and tetragonal phases, at a content of the stabilising oxide of (3–6) mol.% [115, 116]. In the region of the cubic solid solutions, the degree of reduction decreases with increasing concentration of the stabilising oxide from 8 to 33 mol.%. The deficit of oxygen x in the crystal $Zr_{1-y}Y_yO_{(2-0.5y)-x}$ changes in relation to the degree of reduction from 0.0003 × 10^{-2} to 2.606 × 10^{-2}.

The amount of generated oxygen reaches the value $\Delta n_0/n_0$ = 1.32% for the crystals containing 3 mol.% of Y_2O_3. In the range of the cubic solid solutions, this value decreases and equals ≈0.4%. The density of the crystals increases during the reduction. An

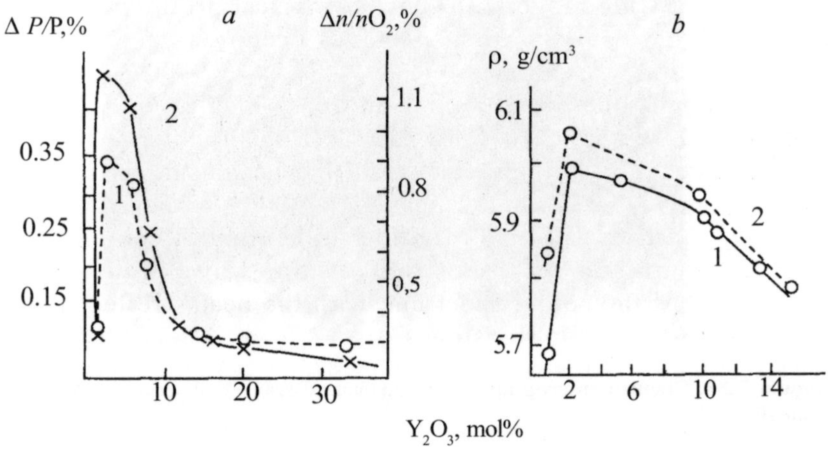

Figure 7.34. Dependence of the degree of restoration of the crystals of the $ZrO_2+Y_2O_3$ solid solution on the Y_2O_3 content after vacuum annealing (1400°C, 30 h) [104]. a: 1 – the fraction of lost oxygen, 2 – relative weight loss, b: – variation of the density of crystals: 1 – prior to annealing, 2 – after annealing.

increase in the amount of the generated oxygen (< 0.1%) [114] results in the compression of the crystal lattice as a result of the formation of vacancies. Possibly, at high temperatures and long holding times the metal may be transferred to the interstitial positions and shear structures may form as a result of the compression of the structure along some crystallographic planes during the loss of oxygen. An increase in the density of the crystals depends on the composition of the solid solution. The largest changes were detected in two-phase crystals containing from 1 to 6 mol.% of the stabilising oxide. The variation of the parameters of the elementary cell of the fianite single crystals is presented in Table 7.15.

A decrease in the parameter of the elementary cell and, correspondingly, increase in the density of the solid solutions increase the microhardness of the crystals after annealing. The values of microhardness of the crystals with the composition ZrO_2 (88 mol.%) $+Y_2O_3$ (12 mol.%) were as follows: (1360 + 60) prior to vacuum annealing, (1550 + 60) Pa for the reduced black crystals and (1390 + 50)Pa for the oxidised crystal.

Examination of the optical properties of the black fianites of the previously mentioned compositions shows that the edge of the absorption band is displaced to the longwave region of the spectrum and depends on the density of the reduced layer of the crystal. It

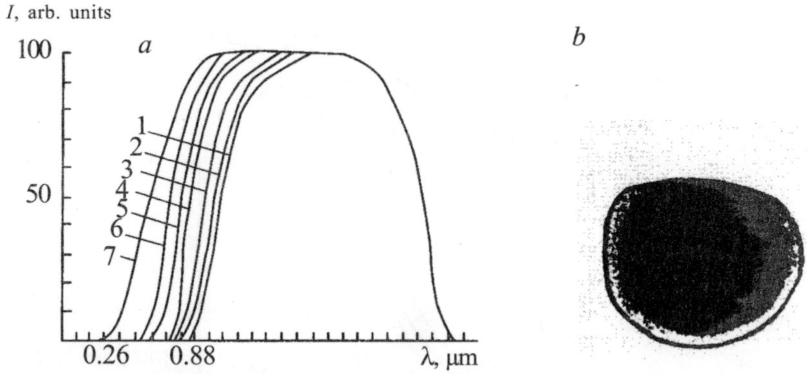

Figure 7.35. The transmission spectra of a sheet of black fianite in relation to the reduction conditions [104]. 1) annealing in a vacuum, $1.33 \cdot 10^{-2}$ Pa, 30 h, 2000°C; 2–7) in air at 350°C from 0.5 to 3 h (b).

Table 7.15. Variation of the parameters of the crystal lattice of single crystals of fianites ZrO_2–Y_2O_3 after vacuum annealing (vacuum $1.33 \cdot 10^{-2}$ Pa, annealing temperature 2000°C, time 30 h) [104]

$C_{Zr}O_2$	C_{Y2O3}	*a*, Å Prior to annealing	*a*, Å After annealing	Δ*a*/*a*,%
89.70	10.30	5.1432	5.1390	0.08
87.20	12.80	5.152	5.143	0.17
84.8	16.2	5.158	5.154	0.11

may be assumed that the loss of oxygen by the crystal during reduction is accompanied by the formation of different electronic dye centres whose absorption spectrum overlaps the entire visible range of the spectrum. Figure 7.35*a* show that the process of oxidation of the single crystal starts on the surface and is caused by the diffusion of oxygen whose rate is determined by the annealing temperature. Figure 7.35*b* shows the transmission spectra of fianites for different degrees of reduction.

The partially reduced 'black' fianite is of considerable interest. In heating of the colourless fianite in vacuum at a temperature of 1800°C, the zirconia is reduced and, during this process, the crystal loses part of oxygen and the Zr^{4+} ions transfer to the condition with lower degrees of oxidation. This process is accompanied by the formation of electronic dye centres whose absorption spectrum

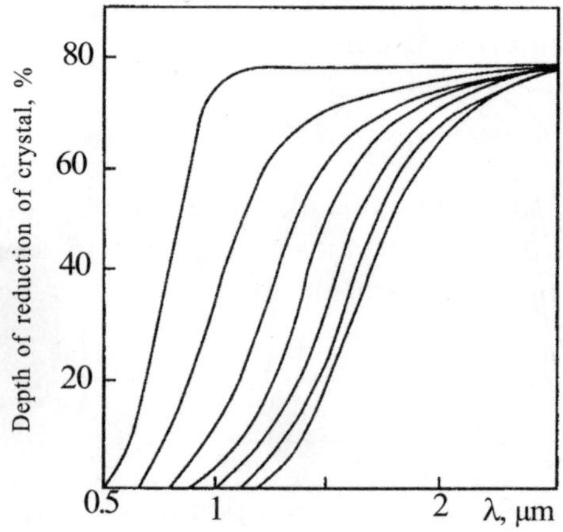

Figure 7.36. Long-wave edge of the absorption spectrum of 'black fianite (d = 1.2 mm) in relation to the depth of reduction of the crystals.

overlaps the entire visible range of the spectrum.

Figure 7.36 shows the long-wave edge of the absorption spectrum of black fianites, corresponding to different depths of reduction. It may be seen that the absorption spectrum can be varied in a wide range and, consequently, the black fianite may be used for infrared filters together with other well-known semiconductor materials, such as germanium and silicon.

In 'moderate'reduction of fianites, for example, in heating in a vacuum of 10^{-4} mm Hg at 1700°C for 30 hours, and subsequent cooling at a rate of 120 deg/h, the changes taking place in the crystals are fully reversible: in subsequent annealing in ar at 1000°C, the crystals are again colourless and optically homogeneous. This fact on its own is remarkable: regardless of the fact that the loss of oxygen in reduction, according to experimental data, reaches 0.02 wt.%, the phase homogeneity of the crystals is not disrupted. This results in the formation of a stable solid solution whose composition includes two or more valency forms of zirconium [117].

Figure 7.37 shows the absorption spectrum of the crystals prior to and after annealing in oxygen at 2000°C with subsequent quenching in air and in vacuum at 1250 and 1550 K at v_{cool} = 6 deg/ min. Annealing in air increases the absorption in the range λ< 350 nm (see also 7.37b), soft annealing (T ~1400 K) in vacuum results in clarification in this range. Vacuum annealing at high temperatures increases absorption in the entire examined range of the spectrum

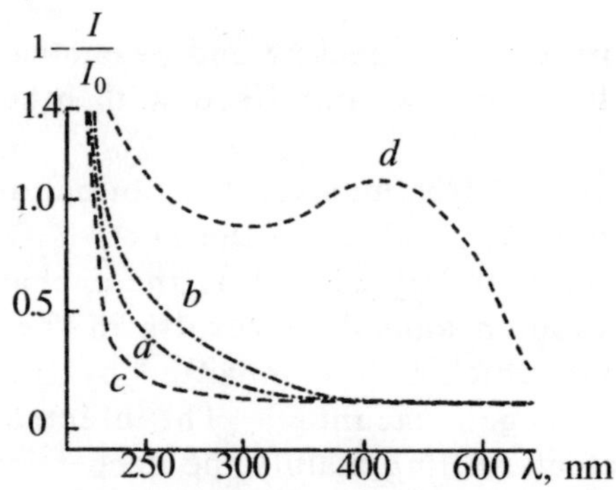

Figure 7.37. Absorption spectra of HfO_2–Y_2O_3 crystals prior to annealing (a) and after annealing: in air at 2000 K (b), in vacuum at 1200 K (c) and at 1550 K (d).

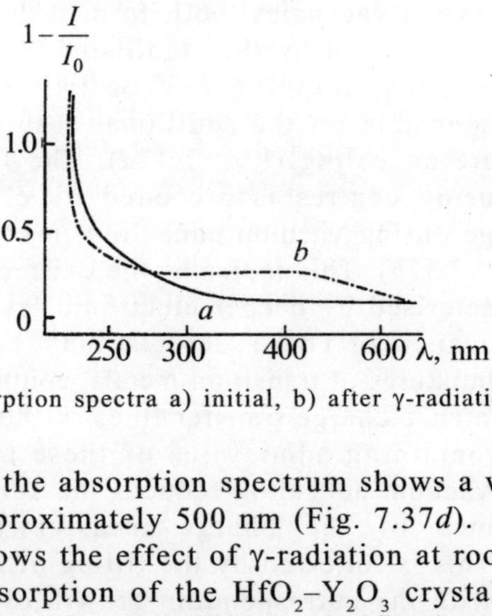

Figure 7.38. Absorption spectra a) initial, b) after γ-radiation, dose 10^6 R.

230–800 nm, and the absorption spectrum shows a wide band with a maximum at approximately 500 nm (Fig. 7.37*d*).

Figure 7.38 shows the effect of γ-radiation at room temperature on the optical absorption of the HfO_2–Y_2O_3 crystals. In the non-annealed crystals, γ-radiation results in clarification in the range $\lambda <$ 260 nm and in the appearance of additional absorption in the longer wave length range. The intensity of additional absorption increases with increasing dose and reaches saturation at a dose of 10^5–10^6 roentgens. Qualitatively, these effects are caused by γ-radiation in the annealed crystals, but preliminary annealing affects the intensity of additional absorption and clarification. In the vacuum-annealed crystals, the intensity of radiation-induced additional absorption reaches saturation at higher doses and on a higher level, in comparison with the non-annealed crystals. Annealing in air with quenching has the opposite effect. Identical dependences of the spectra of absorption on annealing and γ–gamma radiation were also detected in HfO_2 crystals, stabilised with other oxides R_2O_3 and also CaO.

The effect of heat treatment on the optical absorption of HfO_2–Y_2O_3 crystals is associated, as in the ZrO_2–Y_2O_3 crystals, with the variations of stoichiometry caused by the exchange of oxygen with the medium. Vacuum annealing results in the release of oxygen from the crystal which is accompanied by the formation of free electrons and oxygen vacancies. The electrons are distributed between the defects, filling mainly the deeper levels. Thus, vacuum annealing of the HfO_2–Y_2O_3 crystals is identical with the additive colouring of the alkali–halide crystals in the metal vapours. The main type of defects, capable of trapping the electrons in HfO_2–Y_2O_3 are the oxygen vacancies, both formed during the departure of oxygen and introduced by the stabilising Y_2O_3 admixture. The vacancies of oxygen, trapping one or two electrons, may be regarded as responsible for the additional absorption formed after high-temperature annealing (Fig. 7.37*d*). The formation of these intrinsic colouring centres is preceded by clarification in the ultraviolet range during vacuum annealing in comparatively 'soft' conditions (Fig. 7.37*b*). This indicates the existence of another type of defect characterised by deeper electronic levels in comparison with anion vacancies. These defects may be represented by uncontrolled admixtures of transition metals, containing in the higher valency state intense charge transfer lines in the ultraviolet range. In a crystal, containing admixtures of these metals in a higher valency state, vacuum annealing reduces the admixtures, resulting in disappearance of the charge transfer band (ultraviolet clarification). This is followed by the filling of less deep levels of the vacancies with the corresponding growth of the absorption of the intrinsic colouring centres. This is also clear from Fig. 7.37*c,d*. In annealing in air, the crystal absorbed oxygen and, consequently, the absorption of the admixture in the higher valency state increased (Fig. 7.37*a, b*).

The presence of two competing centres of capture of electrons – intrinsic, associated with vacancies, and admixture – may also be used to explain the nature of changes in the spectra caused by ionising radiation. In this case, additional absorption may be associated with the formation of the intrinsic colouring centres during capture of the electrons by anion vacancies, and the clarification in the ultraviolet range may be attributed to the reduction of the uncontrolled admixtures of transition metals. In contrast to vacuum annealing in which the reduction of the admixtures and the formation of the intrinsic dye centres take place successively in the order of decreasing depth of the level, in γ-irradiation, both processes are parallel because the distribution of electrons between the levels of capture is regulated in this case not by the depth but by the kinetic parameters and concentration of the centres. Therefore, in the case of γ-irradiation, additional absorption increases together with the clarification in the ultraviolet range (Fig. 7.38) [118]. The charge state of the admixture ions and the appropriate absorption spectrum, formed after annealing crystals in different conditions, will be investigated in Chapter 8.

The crystals based on ZrO_2 are characterised, at elevated temperatures, by almost completely pure ionic conductivity. In the reduced condition, the crystals have a high fraction of electronic conductivity; in oxidation of the crystal and approaching the stoichiometric composition, the electronic component of electrical conductivity decreases [113, 119].

7.8. PHYSICOCHEMICAL AND MECHANICAL PROPERTIES OF FIANITES

Investigations of the fianites of different chemical composition show that they have properties not found in other available natural or synthetic crystals. It is important to note the physicochemical stability of fianites at high temperatures. They are characterised by a low degree of evaporation in vacuum, described by the value $7\cdot10^{-6}$ g/cm^2 and 2100°C. In air, the fianites do not change their chemical composition and structure up to the melting point (2700–2900°C). In this respect, the fianites are inferior only to the thorium dioxide which cannot at present be produced in the form of large single crystals and, in addition to this, the radioactivity of thorium must also be taken into account.

At room temperature, the fianites do not interact with alkali and acids, including hydrochloric, hydrofluoric and sulphuric acids. The chemical stability of fianites at high temperatures is characterised by, for example, the fact that they do not interact with melts of various corrosive metals, such as sodium, and melts of many semiconductor compounds. Especially high chemical resistance is typical of the crystals based on HfO_2. Not having porosity and disruptions in the structure of the surface, the single crystals are greatly superior in chemical resistance to the sintered ceramics of identical composition. The hardness of the fianites is 8.5 on the Moos scale, and of the currently available natural crystals, only diamond and corundum are harder than fianites.

The electrical properties of the fianites are very interesting. At room temperature, the fianites are typical dielectrics. However, already at 300°C they show oxygen–ionic electrical conductivity which increases with increasing temperature and reaches 1 ohm^{-1} cm^{-1} at 2000°C.

As already mentioned, the fianites are solid solutions with a wide range of phase homogeneity. In addition to the oxide-stabilising agents, compounds of other elements, modifying the physical properties of the crystals, may be introduced into the composition of the fianites. The number of these compounds–modification agents is very large, and the concentration in which they can be added is several percent. The possibility of changing the chemical composition of the fianites in a wide range enables the physical properties of the fianites to be varied in a wide range.

The density of fianites

Density is one of the most important characteristics of the material and controls the majority of the mechanical, thermal, optical and other properties. Density is a physical constant of an individual substance. It is determined by the structure of the crystals, in particular, the density of packing of the atoms in the crystal lattice. The density of large crystals was measured by hydrostatic weighing based on the determination of the volume of the crystal on the basis of the mass of the liquid displaced by the crystal. The accuracy of measurement can be increased using high-density liquids. The results of measurement of the density of fianites are presented in Table 7.16 and Fig. 7.39. The presented data show that the density of the fianites depends on the concentration and nature of the stabilising oxide. The inflection point on the dependence of density on the concentration of the stabilising oxide corresponds to the phase transition from the region of two-phase solid solutions to the region of cubic solid solutions. The lattice parameters of the cubic solid solutions were used to determine the x-ray density which was compared with the density determined by hydrostatic weighing of the crystals.

The x-ray density was calculated from the equation: ρ = 1.66020

Table 7.16. Density of cubic solid solutions ZrO_2–R_2O_3 (R – Eu, Gd, Yb) [35]

Stabilizing oxide	$RO_{1.5}$ concentration	Structure of solid solution	Lattice parameter	Density, kg/m³ ·10⁻³	
				experimental	calculated
Eu_2O_3	19.02	Fluorite	5.172	6.415	6.411
	23.82	As above	5.185	6.471	6.461
	29.86	"	5.198	6.573	6.563
	37.04	"	5.219	6.675	6.662
	50.66	Pyrochlore	10.518	6.855	6.830
Gd_2O_3	17.65	Fluorite	5.149	6.483	6.483
	20.84	As above	5.155	6.541	6.548
	25.32	"	5.162	6.651	6.617
	30.29	"	5.181	6.702	6.693
	43.82	"	5.186	6.906	6.893
Yb_2O_3	16.82	"	5.132	6.659	6.655
	19.93	"	5.134	6.765	6.759
	22.79	"	5.136	6.867	6.856
	27.88	"	5.138	7.039	6.988
	35.62	"	5.174	7.167	7.161
	49.85	"	5.179	7.645	7.641

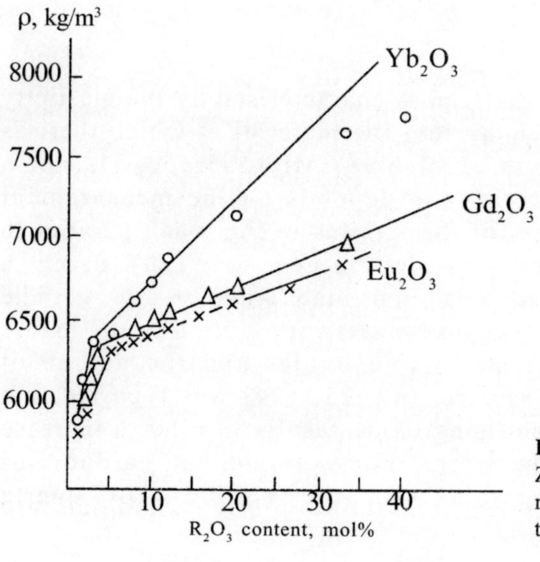

Figure 7.39. The density of ZrO_2–R_2O_3 fianite crystals in relation to the concentration of the stabilising oxide [40].

($\Sigma Z/V$), where ρ is density, ΣZ is the sum of the molecular weight of all molecules in the elementary cell; V is the volume of the elementary cell. When calculating ΣZ, it was assumed that the heterovalent substitution of ZrO_2 by the oxides of rare-earth elements is accompanied by the formation of vacancies in the anion sublattice [121, 122], and the cation subsystem was assumed to be ideal. The following equation was used in the calculations: $\Sigma Z = xZrO_2 + yRO_{1.5}$, where x and y are the number of moles of ZrO_2 and $RO_{1.5}$ per the elementary cell of the structure, on the condition: $x + y = 4$; x/y = mol.% ZrO_2/mol.% $RO_{1.5}$. According to the Debye patterns, the composition 33 mol.% Eu_2O_3 has a pyrochlore structure. X-ray density was determined for the $Eu_2Zr_2O_7$ compound. The results of the calculations are presented in Table 7.16, and the difference between the experimental and calculated values does not exceed 0.5%. However, in most cases, ρ_{exp} is higher than ρ_{cal} and this may be caused by the presence of a small number of interstitial atoms in the lattice of solid solutions.

It should be mentioned that for low concentrations of the stabilising oxide, the grinding of the crystals in an agate mortar into powder for x-ray diffraction examination results, due to the metastable state at room temperature, in the occurrence of phase transitions, and the powder x-ray diffraction patterns do not reflect the actual phase composition of the crystals. This problem will be investigated in greater detail in the chapter concerned with the crystals of partially stabilised zirconia (PSZ, Chapter 8).

Microhardness of fianites

The microhardness of the material is characterised by the elasticity of indentation of another body into the material at which there is no residual deformation or scratching. Microhardness is not a physical constant of material and depends on the measurement method. The microhardness of the fianites in the plane parallel to the direction of growth was determined in [120] using a microhardness meter. The optimum load was 70–100 g. The microhardness of fianites was investigated in relation to the nature and concentration of the stabilising oxides and the degree of ordering of the crystal structure. In [123, 124] was reported that a small amount of the stabilising oxide results in a large increase of the hardness of the specimens. For example, the hardness of pure zirconia is comparable with the hardness of quartz (approximately 7 on the Moos scale or 700 kg/mm²), and the hardness of stabilised zirconia is comparable with that of topaz (8 on the Moos scale or 1200 kg/mm² according to Vickers). The microhardness of the tetragonal solid solutions (ZrO_2–8 mol.% Gd_2O_3) is considerably higher than that of cubic solid solutions (20 mol.% Gd_2O_3). In addition, the experimental results show that the microhardness depends only slightly on the nature of the stabilising rare-earth oxide [125].

The value of microhardness in the series of solid solutions reflects the effect of the change in the spacing of the ions as a result of including the ions of the stabiliser in the composition of these solutions and also the disruption of the ordering of the structure. The solid solutions with the composition $R_2Zr_2O_7$ (33 mol.% R_2O_3) undergo the order–disorder transformation, and the temperature of this transition decreases with an increase in the atomic number of the lanthanide [126]. The microhardness of the solid solutions decreases in the same sequence as a result of disordering of the structure. The dependence of the microhardness of the specimens of ZrO_2–R_2O_3, where R = Eu, Gd, Yb, on the R_2O_3 content is shown in Fig. 7.40. The microhardness is maximum in the concentration range 5–7 mol.% R_2O_3, corresponding to the transition of systems from the single-phase cubic solutions to multiphase ones.

The large increase of the strength of the crystals in this range of the compositions is evidently determined by the mechanical inhibition of movement of the dislocations [127, 128, 129]. The crystals with these compositions are semi-transparent and greatly scatter the light because they consist of a mixture of the tetragonal

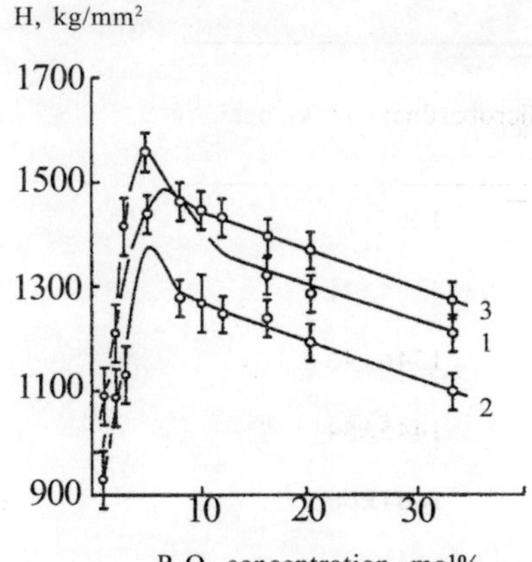

Fig. 7.40. Microhardness of fianite crystals in relation to the concentration of the stabilising oxide [40]. The number at the curves corresponds to the oxides: 1) Y_2O_3, 2) Gd_2O_3, 3) Eu_2O_3.

and monoclinic phases. Hardening may be associated with the microscopic stress fields formed around particles during precipitation of the phase in the fine-dispersed condition. The subsequent decrease in the microhardness of the single crystals of cubic solid solutions with an increase in the content of the stabilising oxide is associated with the ordering of the structure of the solid solutions. An increase in the concentration of the stabilising oxide decreases the microhardness most markedly in the case of the Eu_2O_3 crystals and, to a lesser degree, in the crystals with Gd_2O_3 and Yb_2O_3. Attention should be given to the high microhardness values of the composition ZrO_2–8 mol.% Gd_2O_3 which may be associated with the presence of high stresses in these crystals.

The polished samples produced from the grown fianites were annealed in air at a temperature of 1200 °C for 10 hours. This heat treatment does not reduce the residual volume stresses, formed in the process of crystallisation, but eliminates the local stresses in the

Table 7.17. Microhardness of fianites ZrO_2–Y_2O_3 in directions of axes [111], [110], [100] (load 200 g) [40]

Composition of crystals, mol%		Microhardness *H*, kg/mm^2		
ZrO_2	R_2O_3	[111]	[110]	[100]
88	$12Y_2O_3$	1455±47	1532±60	1694±54
92	$8Gd_2O_3$	1350±28	1472±30	1602±49

Table 7.18. Microhardness of fianites HfO_2–Yb_2O_3 in growth direction (load 100 g) [40]

Composition of crystals, mol%		Microhardness *H*, kg/mm^2
HfO_2	Yb_2O_3	
99	1	1067±36
98	2	1174±71
97.5	2.5	1346±46
95	5	1445±54
93	7	1631±60
88	12	1439±38

surface layer, formed during mechanical treatment of the crystals and increases their microhardness/.

The specimens of HfO_2 crystals, containing 1–7 mol.% of Y_2O_3, were not single-phased. The microhardness of these crystals increased with an increase in the content of the stabilising oxide to 7 mol.%, and in transition to the range of single-phase cubic solutions the microhardness decrease. The microhardness of the fianites based on the zirconia and hafnium dioxide, is presented in Tables 7.17 and 7.18. Attention should be given to the high brittleness of the crystals, containing small quantities of silicon oxide. The introduction of even as little as 0.05 wt.% of SiO_2 greatly decreases the microhardness.

Thermal expansion of fianites

For technical applications of components produced from crystals of solid solutions based on zirconia, especially at high temperatures, it is important to know the coefficient of thermal expansion of these materials. These data for several compositions in the temperature range 20–900°C can be found in [35,40]

7.9. OPTICAL AND SPECTROSCOPIC PROPERTIES OF FIANITES

The optical properties of the zirconia crystals, stabilised by the oxides of yttrium and the rare-earth elements, are of considerable interest for optics and jewellery industries.

Refractive index

The optical properties are extremely important characteristics of the cubic single crystals of zirconia: the refractive index, the dispersion in the visible range of the spectrum, the transmission band. The values of the refractive indices for the case of application of various oxides as stabilising agents are presented in Table 7.19 [40].

For the ZrO_2–Y_2O_3 cubic crystals, the dispersion of the refractive index is in the range 0.060–0.063 and the transmission band 0.26–7.5μm.

For the 11 spectral lines, the refractive indices of the transparent cubic crystals were measured by the prism method [120]. The values of the refractive index are presented in Fig. 7.41 *a–c* and

Table 7.19. Experimental and calculated parameters of refraction of single crystals of solid solutions based on zirconia [40] (λ = 6328 Å)

Stabilizer	Concentration, mol%	n_{Dexp}	n_{Dcalc}
MgO	10	2.178	2.178
CaO	16	2.165	2.164
	24	2.104	2.116
MnO	10	2.21	2.20
	15	2.18	2.18
Y_2O_3	10	2.173	2.178
	15	2.153	2.147
	20	2.121	2.129
CeO	25	2.244	2.194
	30	2.212	2.200

Table 7.20. Dispersion of the refractive indices of zirconia crystals [118]

λ, μm	Refractive index, *n*			
	$[Y_2O_3] = 9$	$[Y_2O_3] = 11$	$[Y_2O_3] = 15$	$[Y_2O_3] = 17$ mol%
0.4358	2.2224	2.2101	2.1783	2.1232
0.5461	2.1856	2.1737	2.1434	2.0949
0.5779	2.1790	2.1669	2.1372	2.0897
0.5893	2.1768	2.1653	2.1352	2.0881
0.6328	2.1718	2.1581	2.1287	2.0826

in Table 7.20. The resultant dispersion curves are measured in the wavelength range 400–650 nm. The reference value is represented by the refractive index of a crystal with the ZrO_2–12 mol%Y_2O_3 composition at a temperature of 77 K at a wavelength of λ = 6328 Å, n = 2.1546, the error of the measurement is +0.01 [35].

The refractive indices depend on the density of the crystals. The presence of oxygen vacancies decreases the density of crystals and,

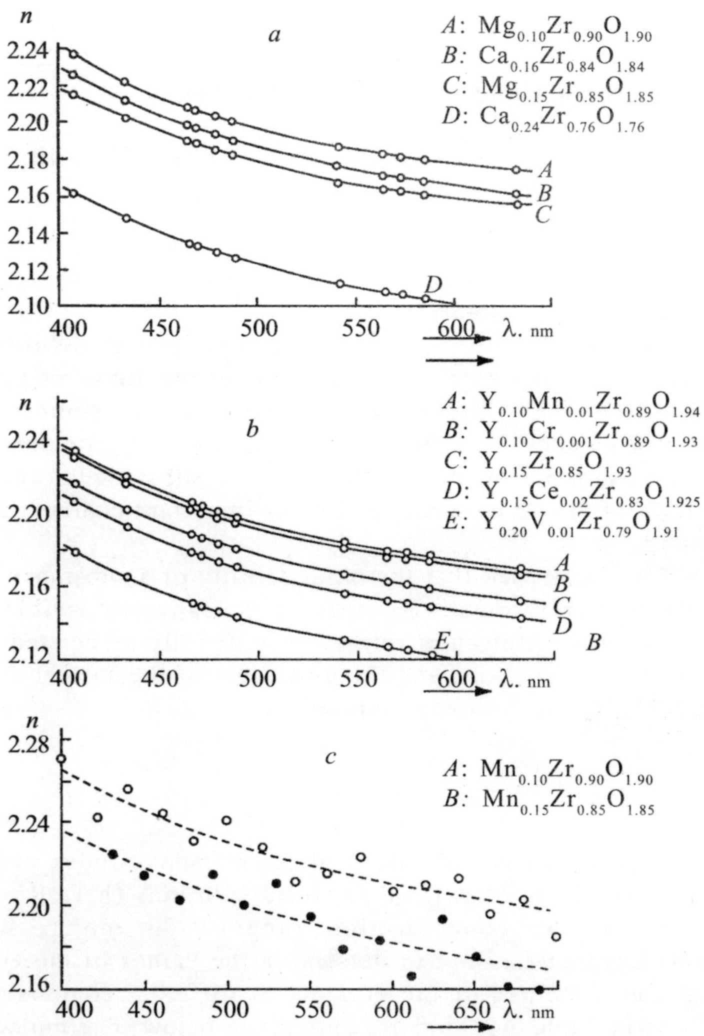

Figure 7.41. Dispersion curves of zirconia, stabilised with the oxides of magnesium and calcium in the wavelength range 400–650 nm (a); with yttrium oxide in the range 400–650 nm (b); with manganese oxide in the range 400–700 nm (c) [119].

consequently, the refractive index also decreases. The additions (Ce, Mn, Cr and V) in the crystal stabilised with the oxides of magnesium, calcium and yttrium, have only a small effect on the refractive index values. The concentration of the main stabilising agent and its effective distribution in the crystal always have a considerably stronger effect.

Molecular refraction and dispersion of fianites

Molecular refraction (R) links the electronic polarisability of a substance (α_{el}) with its refractive index (n) [115, 120]

$$R = [(n^2 - 1)/(n^2 + 2)](M/\rho) = 4/3\pi N_A \alpha_{el}, \quad (7.7)$$

Here M is the molecular mass of the substance; ρ is the density, N_A is the Avogadro constant.

The molecular refraction is suitable for practical applications – it may be regarded as a sum of 'refractions' of the atoms or groups of atoms, forming the molecule of a complex substance. The additivity of R enables the efficient application of the refractometric methods for examination of the structure of substances, and the determination of dipole moments and other important characteristics of compounds.

It should be mentioned that the polarisability of a substance may be regarded as purely electronic only in the range of visible and ultraviolet radiation. Molecular refraction is directly associated with the polarisability of the ions and the molecules of the substance and is calculated from the following equation:

$$R = 4/3\pi N_A \alpha_{el} = 2.52 \cdot 10^{24} \alpha_{el}. \quad (7.8)$$

On the basis of the measured values of the refractive index and the density of the crystals of ZrO_2–R_2O_3 solid solutions (here R = Eu, Gd, Yb, Y) in the concentration range 9–33 mol.% R_2O_3, calculations were carried out to determine the values of molecular refraction. The densities of the crystals of different compositions are presented in Table 7.21 and are also given below. The molecular mass of mixed crystals was calculated on the basis of the assumption that solid solutions are characterised by a filled cation sublattice with vacancies in the anion sublattice. The values of R and α for the previously mentioned compositions are presented in Table 7.21 which shows that the values of the refractive index are determined by the nature and concentration of the stabilising oxide. An increase in the content of the stabilising oxide (Eu, Gd, Yb) increases the density of the crystals and the refractive index. In the case of stabilisation with the yttrium oxide, an increase of the

Table 7.21. Molecular refraction R and electronic polarizability of single crystals of ZrO_2–R_2O_3 (R – Eu, Gd and Yb) and HfO_2–Y_2O_3.

R_2O_3, mol%	Molecular mass	Density. kg/m^3·10^{-3}	Refraction index	Molecular refraction, R	Electronic polarizability, α_{el} 10^{39} C·m^2/V
			ZrO_2–Eu_2O_3		
11.06	133.560	6.4153	2.13	11.26	0.496
13.52	135.627	6.4712	2.14	11.40	0.502
17.55	138.821	6.5726	2.15	11.55	0.509
22.73	142.637	6.6750	2.18	11.87	0.523
33.92	149.807	6.8546	2.20	12.15	0.535
			ZrO_2–Gd_2O_3		
9.68	133.273	6.4913	2.11	10.99	0.485
11.63	135.079	6.5413	2.12	11.11	0.490
14.50	137.686	6.6510	2.14	11.26	0.496
28.01	148.421	6.9058	2.18	11.94	0.526
			ZrO_2–Yb_2O_3		
9.18	135.461	6.6588	2.03	10.37	0.457
11.07	137.749	6.7648	2.05	10.51	0.463
12.86	139.880	6.8669	2.11	10.90	0.481
16.20	143.630	7.0386	2.14	11.10	0.490
21.67	149.370	7.1674	2.16	11.46	0.505
33.20	159.850	7.6481	2.17	11.55	0.509
			ZrO_2–Y_2O_3		
8.00	121.510	5.9919	2.194	11.35	0.500
9.00	121.360	5.9385	2.1768	11.34	0.500
11.00	121.024	5.8990	2.1653	11.34	0.498
17.00	120.084	5.8204	2.1352	11.20	0.493
			HfO_2–Y_2O_3		
10.00	192.770	9.6500	2.0830	37.92	1.672

concentration of this oxide reduces the density and refractive index of the crystals. In the series Eu, Gd, Yb, the refractive index slightly decreases and this may be associated with a decrease in the polarisability of rare-earth elements included in the composition of the solid solution.

Colouring additions have no significant effect on the refractive index because of their low concentration in comparison with the main stabilising oxide. However, in certain cases, the nature of the third admixture affects the refractive index, for example, in the series mentioned below the following inequalities apply: $n(Y+Nd) > n(Y+Ho) > n(Y+Er)$, $n(Y+Co^{2+}) < n(Y+Co^{3+})$. The calculated values of the electronic polarisability of the ZrO_2–Y_2O_3 system are close to the experimental data. The main contribution to the polarisability comes from the oxygen ions. An increase in the fraction of the oxygen vacancies with increasing concentration of the Y_2O_3 (yttrium oxide) results in a larger decrease of α_{el} in comparison with the increase of the concentration of the yttrium ions because of the low value of α_y. The difference between the experimental and calculated values of the polarisability is evidently caused by a decrease of the averaged-out polarisability of the ions under the effect of the electric fields of the adjacent ions and the formation of oxygen vacancies.

For single crystals of the ZrO_2–Y_2O_3 and HfO_2–Y_2O_3 systems, investigations were carried out to determine the dispersion of refraction and to calculate the value of the dispersion coefficient (Abbe number):

$$\nu = (n_D - 1)/(n_F - n_C), \qquad (7.9)$$

where n_F, n_C are the refractive indices of the spectral lines F and C, n_D is the reference.

It should be mentioned that in the visible range of the spectrum, the dispersion has the normal form, *i.e.* the refractive index increases with a decrease in the wavelength, and the dispersion value is relatively high. In order to describe the dispersion in this wavelength range, it is sufficient to use the semiempirical Zelmayer relationship:

$$n^2 = n_\infty^2 + B/(\lambda^2 - \lambda_0^2). \qquad (7.10)$$

The constant of the Zelmayer equation for the fianites with different contents of the oxide of yttrium and other activating elements are presented in Table 7.22.

The dispersion of the refractive index and the dispersion coefficient (Abbe number) of these crystals, calculated using the Zelmayer equation, are presented in Table 7.23. The temperature dependence of the refractive index of the fianites, stabilised by the yttrium oxide, in the temperature range 20–1000°C was investigated in [40]. The error of measurements was $\pm 2 \times 10^{-4}$ /deg, Fig. 7.42. The values of the refractive index change with temperature because of two reasons [135]. Firstly, because of the displacement of the

Table 7.22. Constants of Zellmayer equation for calculating the dispersion of refractive indices of the fianites [40]

Composition of crystals	n_∞	B	λ_0, Å
ZrO_2–9 mol% Y_2O_3	2.137±0.06	4.82±2.45	0.24
ZrO_2–11 mol% Y_2O_3	2.114±0.03	6.98±1	0.22
ZrO_2–17 mol% Y_2O_3	2.088±0.002	6.36±0.1	0.20
ZrO_2–10 mol% Y_2O_3	2.13	4.92	0.20
ZrO_2–10 mol% Y_2O_3+0.5 wt% Nd_2O_3	2.12	8.03	0.22
ZrO_2–10 mol% Y_2O_3+0.5 wt% Er_2O_3	2.11	6.58	0.19
ZrO_2–10 mol% Y_2O_3+0.5 wt% CoO	2.14	3.99	0.27
ZrO_2–10 mol% Y_2O_3+0.5 wt% Ho_2O_3	2.12	7.34	0.20

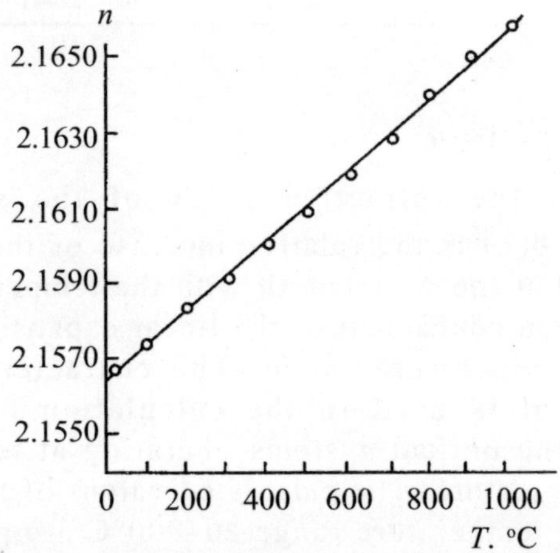

Figure 7.42. The temperature dependence of the refractive index of crystals with the composition ZrO_2–12 mol.% Y_2O_3, wavelength λ = 6328 Å [40].

absorption bands into the ultraviolet range of the spectrum, resulting in the displacement of the entire dispersion curve and in an increase of the refractive index; secondly, as a result of a decrease in the density of the substance with increasing temperature.

The thermo-optical constant is one of the important characteristics of the material and is determined by the following equation:

Table 7.23. Dispersion of the refractive indices of single crystals ZrO_2–Y_2O_3, ZrO_2–10 mol% Y_2O_3–0.5 wt.% La_2O_3, HfO_2–15 mol% Y_2O_3

Composition	$n(\lambda)$, Å					Dispersion coefficient
	4358	5461	5779	5893	6328	
ZrO_2–9 mol% Y_2O_3	2.2224	2.1856	2.1790	2.1768	2.1718	31.50
ZrO_2–11 mol% Y_2O_3	2.2101	2.1737	2.1669	2.1653	2.1581	30.72
ZrO_2–17 mol% Y_2O_3	2.1783	2.1434	2.1372	2.1352	2.1287	33.89
ZrO_2–9 mol% Y_2O_3–0.5 wt% CeO_2	2.2285	2.1844	2.1778	2.1757	2.1691	33.88
ZrO_2–9 mol% Y_2O_3–0.5 wt% Nd_2O_3	2.2285	2.1916	2.1848	2.1759	2.1750	30.97
ZrO_2–9 mol% Y_2O_3–0.5 wt% Ho_2O_3	2.2189	2.1813	2.1744	2.1727	2.1658	34.80
ZrO_2–9 mol% Y_2O_3–0.5 wt% Er_2O_3	2.2061	2.1699	2.1634	2.1613	2.1547	34.87
ZrO2–9 mol% Y_2O_3 –0.5 wt% CoO	2.2177	2.1804	2.1740	2.1722	2.1675	36.29
ZrO2–9 mol% Y_2O_3–0.5 wt% CoO (annealed)	2.2201	2.1835	2.1769	2.1748	2.1678	34.55
HfO_2–15 mol% Y_2O_3	2.2132	2.0949	2.0897	2.0881	2.0826	39.14

$$\nu(\lambda)=\beta(\lambda)/[n(\lambda)-1)]-\alpha, \tag{7.11}$$

where $n(\lambda)$ is the refractive index of the substance at the wavelength λ, $\beta(\lambda)$ is the relative increase of the refractive index of the material at the wavelength with the temperature changed by 1°; α is the mean coefficient of the linear expansion of the material in the same temperature range. The characteristic $\nu(\lambda)$ of the optical material is used in the calculation of thermo-optical aberration of the optical systems operating at temperatures other than room temperature. The calculated values of the thermo-optical constant in the temperature range 20–900°C are presented in Table 7.24.

The spectroscopic properties of fianites

The special feature of stabilised crystals of ZrO_2 and HfO_2 is that their crystal lattice may contain a large number of different impurities in high concentrations, reaching 10 mol.% whilst retaining the phase homogeneity. Rare-earth elements, elements of the iron group – chromium, vanadium, titanium and others, may be introduced into the fianites. Many impurity elements may exist in the solid solutions based on dioxides of zirconium and hafnium in different valency states. In principle, the solid solutions represent an entire class of substances which, retaining their microstructure

Table 7.24. Thermo-optical constants of the fianites λ = 6328 Å, n = 2.1567, T = 20°C [40]

Temperature range	$\beta \cdot 10^{-6}$	$\alpha \cdot 10^{6}$	$\frac{\bar{\beta}_\lambda}{n_\lambda - 1} \cdot 10^{6}$	$\nu_\lambda \cdot 10^{6}$
20–100	8.75	8.48	7.56	–0.92
20–200	8.89	9.44	7.73	–1.75
20–300	8.57	9.66	7.41	–2.25
20–400	6.68	10.03	7.50	–2.52
20–500	8.75	10.24	7.56	–2.67
20–600	8.96	10.38	7.75	–2.63
20–700	9.12	10.63	7.88	–2.59
20–800	9.36	10.68	8.09	–2.53
20–900	9.43	10.68	8.15	–2.62

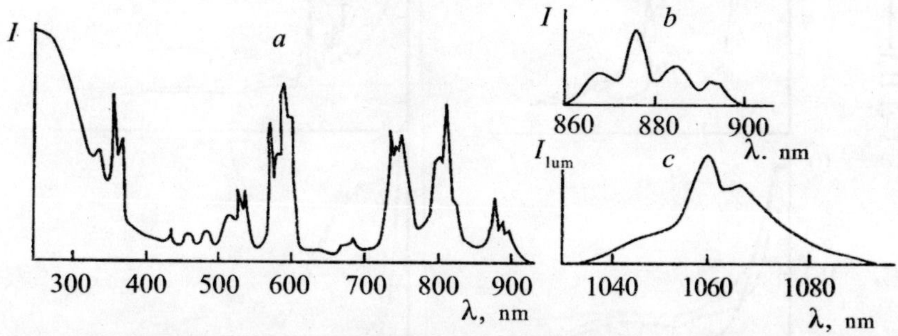

Figure 7.43. Optical spectra of HfO_2–Y_2O_3–Nd^{3+} crystals [118]. a) the absorption spectrum of HfO_2–Y_2O_3–Nd^{3+} crystals at 300 K; b) the absorption band of Nd^{3+} corresponding to the $^4I_{9/2} \rightarrow {}^4F_{3/2}$ transition; c) the luminescence spectrum for the $^4I_{9/2} \rightarrow {}^4F_{3/2}$ transition at 300 K.

and phase homogeneity, may change their physical properties in a wide range. The spectroscopic properties of the solid solutions based on zirconium and hafnium dioxide with different impurities, illustrating the special features of these systems in comparison with other impurity crystals, used in optics and quantum electronics, have been investigated in [136].

Colour fianites

Prior to discussing important technical aspects of the application of fianites, it is important to mention the application of fianites in the jewellry industry for which the main bulk of the crystals is produced [137]. As already mentioned, the crystal structure of the fianites is such as that a large number of different additional elements, in

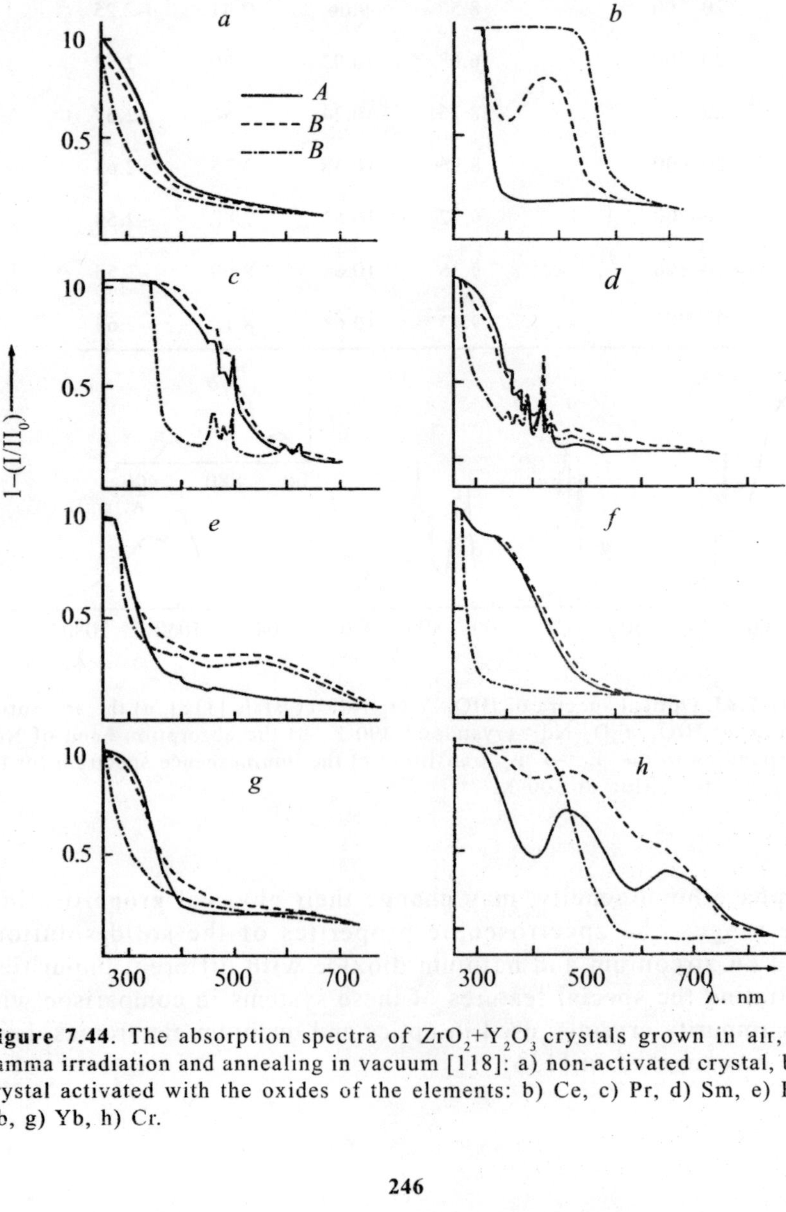

Figure 7.44. The absorption spectra of ZrO_2+Y_2O_3 crystals grown in air, after gamma irradiation and annealing in vacuum [118]: a) non-activated crystal, b) the crystal activated with the oxides of the elements: b) Ce, c) Pr, d) Sm, e) Eu, f) Tb, g) Yb, h) Cr.

particular, rare-earth elements, elements of the group of iron, chromium, vanadium, titanium and other chromophores, may be added to the crystal structure in the isomorphous manner and in large amounts. The introduction of stabilising additions such as the oxides of terbium, holmium, europium and thullium, colours the crystals correspondingly to the brown-green, yellow-green and white green colours. A small addition (0.1%) of the iron oxide results in a yellow shade of the crystal stabilised with the gadolinium oxide, and the cobalt oxide (0.01–0.5%) colours the crystals to dark violet colour, the nickel oxide (0.1–2%) to light brown colour, the vanadium oxide (0.1–1%) to green colour, the chromium oxide (0.1–2% cycle) to brown-purple colour, and the copper oxide (0.1–2%) makes the crystals to acquire yellow colour. The yellow-orange colour is typical of the fianites containing the addition of tetravalent cerium, and the trivalent cerium makes the crystals light red [137]. This wide range of the colours and shades enables many natural stones to be imitated, including diamonds and emeralds. Table 7.25 presents the elements whose oxides, added to the melt, result in the corresponding colour of the fianites.

7.10. ELECTROPHYSICAL, ACOUSTIC, ELASTIC AND PHOTOELASTIC PROPERTIES OF FIANITES

The authors of [126] established that the sintered ceramics of ZrO_2, stabilised with oxides CaO, SrO, Y_2O_3, La_2O_3, is characterised by ionic conductivity. The temperature dependences of the specific resistance in the temperature range 500–1300°C are presented in [127]. Figure 7.45 shows the temperature dependence of the specific resistance of single crystals of ZrO_2–Y_2O_3 (10 mol.%) and HfO_2–Y_2O_3 (10 mol.%), measured by the three-probe method [128]. In the low temperature range, the electrical conductivity is extrapolated by the equations: $\rho = 1.6\cdot10^{-4}\exp(1.20/kT)$ for HfO_2–Y_2O_3 (10 mol.%) crystals, and $\rho = 3.1\cdot10^{-5}\exp(1.12/kT)$ for ZrO_2–Y_2O_3 (10 mol.%), where kT is expressed in eV, ρ in ohm/cm.

The specific resistance of different specimens of zirconia single crystals, stabilised with yttrium oxide, of the same nominal composition greatly differs in the low temperature range. Evidently, this is associated with different concentration of the defects in the crystals, grown in different conditions. If an electrical field is applied to a crystal, heated to a temperature of 300°C or higher,

Table 7.25. Colours of cubic zirconium oxide after adding admixtures

Admixture	Chemical symbol	Colour
Cerium	Ce	Yellow–orange–red
Chromium	Cr	Olive
Cobalt	Co	Lilac
Copper	Cu	Yellow
Erbium	Er	Rose
Europium	Eu	Rose
Holmium	Ho	Rose
Iron	Fe	Yellow
Manganese	Mn	Brown–violet
Neodymium	Nd	Lilac
Nickel	Ni*	Yellow–brown
Praseodymium	Pr	Amber
Thullium	Tm	Green
Titanium	Ti	Yellow–brown
Vanadium	V	Green

*Typical colour for several tenths of a per cent; the colour may depend on the degree of oxidation of the admixture.

the negative electrode is characterised by the formation of a coloured region, spreading to the anode. In the absorption spectrum, a wide band appears in the vicinity of the fundamental absorption edge, determined by oxygen vacancies, neutralised by free oxides.

Investigations of the single crystals of zirconia in the temperature range 800–1900°C in vacuum and in air showed that the electrical conductivity of single crystals is 1.5 times higher in comparison with the ceramic specimens. When a second addition in the series of rare-earth elements was made, electrical conductivity increased in

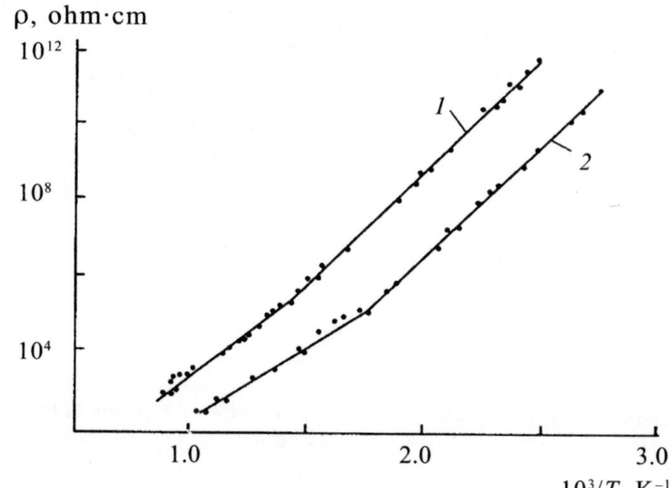

Figure 7.45. Temperature dependence of the specific resistivity of fianites ZrO_2–10 mol.% Y_2O_3 (curve 1) and HfO_2–10 mol.% (2) [127].

the direction from Ce to Pr, decreased in the case of Nd and again increased in the direction to Tb.

The frequency dependences of ε and tg δ are presented in Fig. 7.46. The relatively low dispersion of dielectric permeability indicates that the polarisation of the investigated crystals is of the relaxation type and, possibly, is caused by the displacement of weakly bonded oxygen vacancies or admixture ions.

The temperature dependences of tg δ and ε of the compositions HfO_2–10 mol.% Y_2O_3 and ZrO_2–10 mol.% Y_2O_3, presented in Fig. 7.47, are also of the relaxation type.

The authors of [129] investigated the elastic and photoelastic properties of single crystals of the ZrO_2–Y_2O_3 and HfO_2–Y_2O_3 solid solutions which depend on the concentration of the stabilising oxide Y_2O_3 and are characterised by the molecular scattering of light.

The investigations of the elastic properties of ZrO_2–Y_2O_3 crystals with 8 and 12 mol.% of Y_2O_3 by ultrasound methods were carried out in [130, 131]. The elastic and photoelastic properties of the crystals of ZrO_2–Y_2O_3 and HfO_2–Y_2O_3, containing 10 mol.% of Y_2O_3, were investigated by the Mandel'shtam–Brillouin methods and the patterns of molecular scattering of light by these crystals were determined. The investigations of the elastic properties of

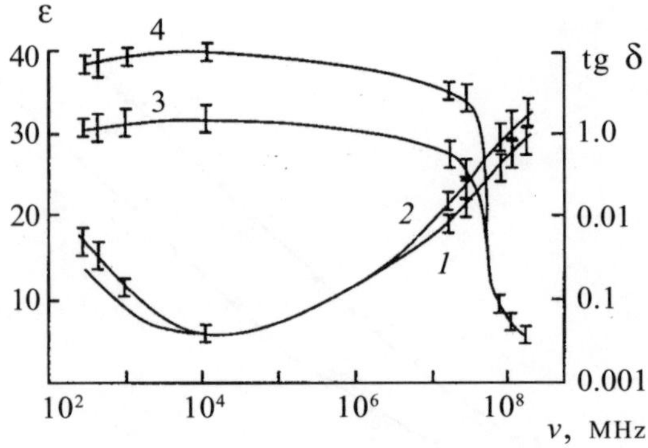

Figure 7.46. Frequency dependence of tg δ and ε of fianites [118]: the curves 1 and 3 – HfO_2–10 mol.% Y_2O_3; curves 2 and 4 – ZrO_2 – 10 mol.% Y_2O_3.

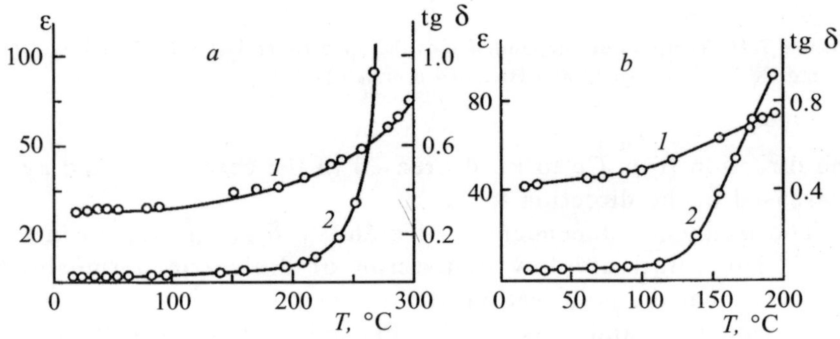

Figure 7.47. Temperature dependences of tg δ and ε of fianites [118].

ZrO_2–Y_2O_3 crystals as a function of the Y_2O_3 concentration, were carried out in [134].

Preliminary experiments on a large number of single crystal specimens of solid solutions showed that the results are highly sensitive to the accuracy of orientation of the crystal planes in relation to the crystallographic direction. The accuracy of the measurement methods was improved using the same method of light scattering on crystal specimens with accurately oriented planes. These results, together with other data [130–134], made it possible to improve the accuracy of determination of several properties of the dynamics of the lattice of ZrO_2–Y_2O_3 and HfO_2–Y_2O_3 crystals.

The speed of sound waves and the values of the elasticity constant, calculated from the speed of sound waves, show that in the ZrO_2–Y_2O_3 crystals, the speed of sound is considerably greater than that detected in the HfO_2–Y_2O_3 crystals. In both crystals, the speed of longitudinal and transverse sound waves depends greatly on the direction of propagation of sound. Figure 7.48 shows the speed of the longitudinal and transverse sound waves in the ZrO_2–Y_2O_3 crystals as a function of the Y_2O_3 concentration. In order to facilitate comparison, the speeds of ultrasound waves V_{L1}, V_{T1}, V_{T2}, measured for the [110] direction, were reduced to the direction [100] and [11 $\sqrt{2}$] for which the speed of ultrasound waves is available. Reduction was carried out using the relationships derived by Borcherds [135]. Figure 7.48 shows that the ultra- and supersonic speeds are similar and change only slighty with the Y_2O_3 concentration.

The speeds of sound v were used to calculate the elastic constants using equations [135]. The results are presented in Fig. 7.49

The photoelastic constants $P_{i,j}$ were calculated from the Mandel'shtam–Brillouin spectrum using the ratio of intensities and equations (7.11) and (7.12)

$$Z_1 = (p_{44} / P_{12})^2 c_{11} / (2c_{44}) \tag{7.12}$$

$$Z_2 = 2P_{44} / (P_{11} - P_{12})^2 c_{11} / (2c_{44}). \tag{7.13}$$

Here Z_1 and Z_2 are the ratios of the intensities of the transverse components to the longitudinal components for the $V\Sigma$ and $H\Sigma$ spectra, respectively, for the case of scattering of light on the

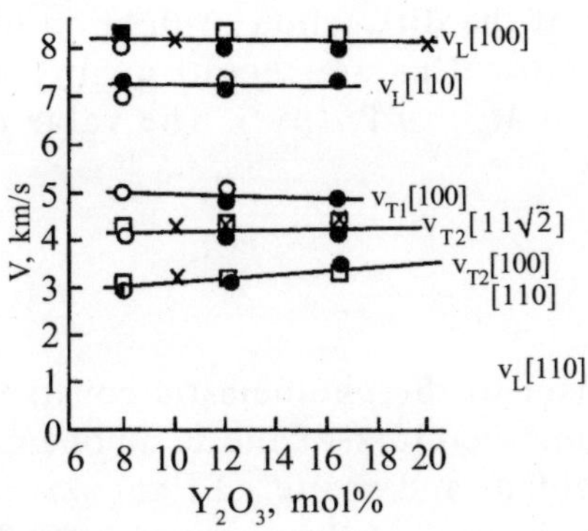

Figure 7.48. The speed of sound in $ZrO_2+Y_2O_3$crystals in relation to the concentration of Y_2O_3 for different crystallographic directions [130, 131]. • – ultrasound measurements; O – ultrasound measurements for the direction [110] and the speed reduced to the direction [100]; □× – the hypersonic measurements (for explanation see the text).

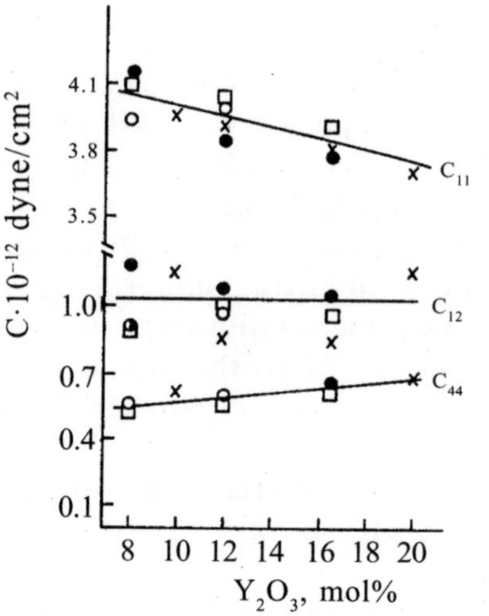

Figure 7.49. Elastic constants of the $ZrO_2+Y_2O_3$crystals in relation to the Y_2O_3 content. The data are in complete agreement with Fig. 7.48.

photons passing along the [100] axis. The equations (7.11) and (7.12) form four combinations of the ratios P_{11}, P_{12}, and P_{44}. The authors selected orientations, providing the best agreement between the calculated and experimental spectra for the 2–4 orientation (Fig. 7.50). For the ZrO_2–Y_2O_3 crystals with 0.08, 0.1 and 0.156 mol.% Y_2O_3, only the parameters of photoelastic anisotropy $(P_{11}-P_{12})(2P_{44})$ were determined. The resultant relationships of the photoelastic constants and the parameters of photoelastic anisotropy are presented in Fig. 7.51.

The absolute values of the photoelastic constants were obtained in the examination of the diffraction of light on ultrasound. Directly measured acousto-optical parameters of quality M_2 are linked with P using the equation $M_2 = n^6P^2/(\rho v^3)$. The value of P is determined by the expression:

$$p = p_{ijkl}\alpha_i\beta_j\chi_k\gamma_l. \tag{7.14}$$

Here P_{ijkl} is the tensor of the photoelastic constants of the material, α_i and β_j are the unit verctors of the components of the electrical field (E) of the incident and scattered light waves, respectively; χ_k and γ_l and the unit vectors of the components of the direction and of polarisation of sound, and v is the speed of sound. For a longitudinal sound wave:

$$P_{11}, \text{for } \alpha \text{ along } [100] \tag{7.15}$$

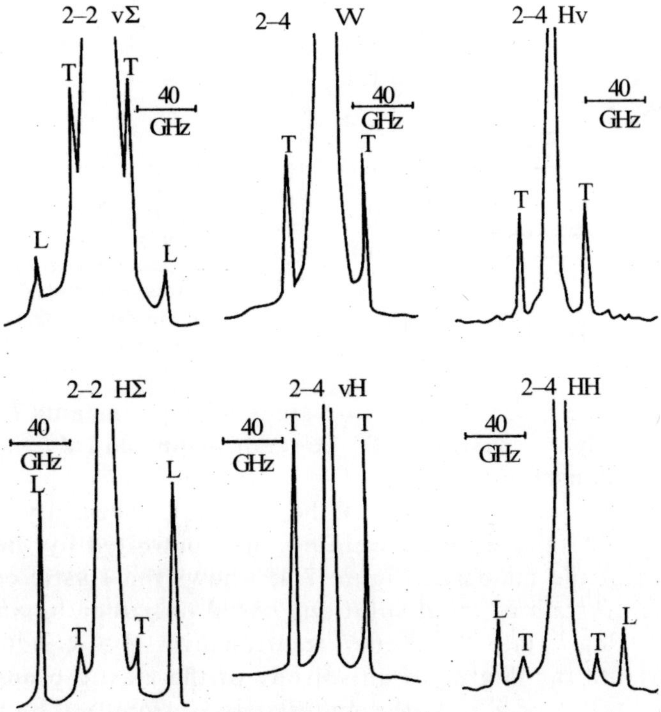

Figure 7.50. Mandel'shtam–Brillouin spectra for ZrO_2+Y_2O_3 crystals containing 20 mol.% of Y_2O_3.

$$P_{12}, \text{for } \alpha \text{ along } [001]\ \chi \text{ along} [100] \tag{7.16}$$

$$0.5(P_{11}+P_{12})+P_{44} \text{ for } \alpha \text{ along} [100]\ \chi \text{ along} [110] \tag{7.17}$$

$$0.5(P_{11}+P_{12})-P_{44}, \text{ for } \alpha \text{ along } [110]. \tag{7.18}$$

For the transverse sound wave, propagating along the [100] direction, the photoelastic constant P is equal to P_{44} and is independent of the orientation of the field E in accordance with the vectors of propagation of light and sound

Figure 7.51 shows the photoelastic constants of the ZrO_2–Y_2O_3 crystals as a function of the Y_2O_3 concentration. In order to determine the dependence of the absolute values of the photoelastic constants on the Y_2O_3 concentration, the data of ultrasound measurements were used. Figure 7.51 also shows the parameters of the anisotropy $P_{12}/2P_{44}$ in relation to the Y_2O_3 concentration. The

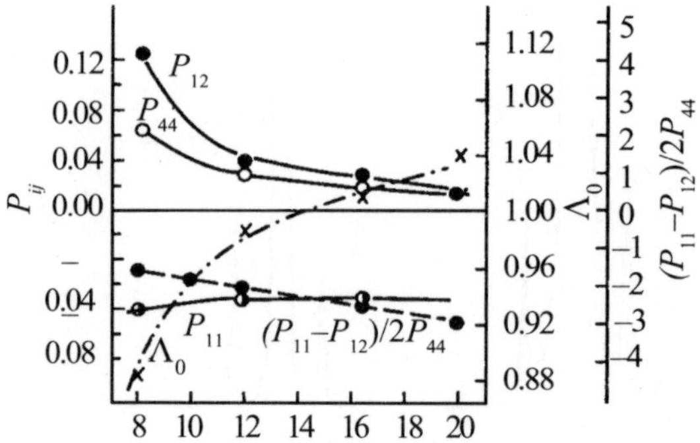

Figure 7.51. Photoelastic constants and the parameters of photoelastic anisotropy in relation to the Y_2O_3 content of the ZrO_2+Y_2O_3 crystals (for explanation see the text).

graphs indicate that the absolute values of the constants P_1 and P_2 decrease with increasing Y_2O_3 concentration, and P_{11} is almost completely constant.

The mechanical properties of the material, especially cracking resistance and failure under welding, are controlled by the elastic properties of the material. Figure 7.49 shows the elastic constants for single crystals of solid solutions based on zirconia, which can be used to calculate the Zener relationships $Z = C_{44}/(C_{11}-C_{12})$, characterising the degree of anisotropy of the elastic properties in actual crystals (for $Z = 1$, the material is isotropic). The value of this ratio also determines the direction of the extreme value of the Young modulus: at $Z < 1$, the maximum value of the Young modulus is in the <100> direction, and at $Z > 1$ its maximum is found in the <111> direction [136].

Analysis of the results indicates the following:

1. Depending on the concentration of the yttrium oxide in the entire range of the compositions, including the multiphase composition, the constant C_{11} monotonically decreases, C-$_{44}$ monotonically increases, C_{12} remains constant; the direction [100] is the high-modulus direction, E ~360 GPa, and {111} is the low-modulus direction, E ~ 170 GPa [136].

2. In the solid solutions ZrO_2–Y_2O_3 and HfO_2–Y_2O_3 there are directions of the exciting and scattered light for which the longitudinal waves provide almost no contribution to the scattering of light. Identical results were obtained previously for non-cubic crystals of sapphire and lithium niobate [136].

3. The speed of the longitudinal and transverse sound waves in the ZrO_2–Y_2O_3 and HfO_2–Y_2O_3 solutions depends strongly on the direction of propagation of the waves, *i.e.* the fianites are characterised by a strong anisotropy of the speed of sound.

4. The elastic and photoelastic anisotropy of the fianites are high and change in relation to the content of the stabilising oxide. The main physical properties of the fianites are presented in Table 7.26.

7.11. JEWELLERY QUALITIES OF THE FIANITES

The unique combination of the physical and physical–mechanical properties of the fianites is the reason why they are included in the class of valuable technical crystals. For the efficient application of these crystals in technology, it was necessary to develop methods of physical–chemical and physical–mechanical treatment: cutting, grinding and polishing to the 14th class of surface finish, drilling, optical clarification, brazing with metals, ceramics and glass, precision fractioning. The development of these methods of treatment has made it possible to produce components of complicated shapes from fianites and also join these crystals with other materials in sections of different devices and systems. The physical properties of the fianites are presented in Table 7.26.

The fianites are used on a large-scale in the jewellery industry. In particular, the fianites are regarded as an excellent imitation of diamonds.

The precious stones have been the subject of reproduction for many years. However, only at the end of the 19th century, the advances in chemistry and physics made it possible to develop synthetic stones whose properties do not differ at all from those of the natural precious stones and, in many cases, are higher.

A turning point in the production of synthetic precious stones was achieved by the French engineer V.A. Verneuil who in 1892 invented a method of production of synthetic ruby from a melt. His method was subsequently used for the growth of other synthetic crystals: sapphire, spinel, alexandrite and other ‘precious’ stones. In the period since Verneuil’s invention, many methods have been developed for the growth of crystals and new types of crystals have been produced which have no analogues in nature, with the fianites occupying an important position [42, 43].

All synthetic crystals may be classified on the basis of the following features: 1. Synthetic jewellery stones, having no natural analogues, which include ruby, sapphire, spinel, rutile, diamond, isumrud, quartz, alexandrite, opal; 2. synthetic materials with no natural analogues, with suitable examples being strontium titanate, fabulite, lithium niobate, yttrium–aluminium garnet, fianite and other

Table 7.26. Physical properties of fianites [108]

Characteristic	Value	Comment
Melting point, °C	2700–2850	Depends on composition
Density, g/cm³	5.85–9.65	As above
Refractive indes, n_D	2.2	ZrO_2–Y_2O_3–10 mol%
Mean dispersion of refractive index $n_{F-}n_c$	0.038	"
Transmission band	0.26–7.6	"
Abbe number	30	"
Hardness (mineralogical scale)	8.5	"
Young modulus, kg/mm²	3930	"
Shear modulus, kg/mm²	10.87	"
Poisson coefficient	0.374	"
Resistivity, ohm·cm^{-1}		"
at 20 °C	$1 . 10^{12}$	"
at 300 °C	$1 . 10^{4}$	"
at 2000 °C	1	"
Heat conductivity, kcal/m·h·deg	2.2	"
Coefficient of linear thermal expansion at 20–1800°C, deg^{-1}	$10.5 \cdot 10^{-6}$	"
Evaporation in vacuum, g/cm²·s	$(7–9) \cdot 10^{-6}$	"
(10^{-4} mm Hg at 2100 °C)	$(3–5) \cdot 10^{-6}$	HfO_2-Y_2O_3–10 mol%

crystals; 3. imitations of the jewellery stones: glass, doublets and triplets. A special commission in 1970 published more accurate definitions of synthetic stones: 'Synthetic stones are crystallised produced completely or partially by the work of human hands. Their chemical composition, crystal structure and physical properties coincide in a wide range with those of natural prototypes' [137].

The jewellery inserts from diamonds occupy a special position because of four properties:

1. Diamond (non-faceted) has an exceptionally high refractive index (n = 2.42) which enables the faceting in which a large part of the incident light is refracted, resulting in unusual sparkling.

2. Diamond is characterised by a high dispersion of the refractive index (d = 0.044) resulting in the play of colours formed as a result of the decomposition of the white light into the components of the colour spectrum, with the beams reflected under different angles inside the stones.

3. Diamond has the highest hardness (h = 10) and, consequently, after polishing with diamond powder, it is not scratched by other minerals (although the tendency to cleaving makes it susceptible to impacts).

4. The colourless diamond is characterised by the complete absence of colouring and this additionally increases the optical effect.

Any diamond substituent should be evaluated on the basis of these four properties. Although some synthetic and natural substituents are very close to diamond in respect of some properties, in other aspects they cannot be compared with diamond. For example, the dispersion of light in crystals such as strontium titanate (d = 0.190) and rutile (d = 0.280) is considerably higher than in diamond. However, these materials are characterised by low hardness (h = 5.5–6.5) and, consequently, have insufficient wear resistance. However, even if they were harder, these stones could be easily identified due to their excessive sparkling ('fire') because their dispersion is considerably greater than that of the diamond.

More recent substituents of the diamond, aluminium–yttrium amd gadolinium–gallium garnets, have the hardness similar to that of the diamond (h = 7.5–8.5) and intermediate values of the refractive index and dispersion (Table 7.27) and, consequently, they are characterised by a relatively high shine and sparkling ('fire'). Regardless of these properties, other characteristics of these stones are greatly inferior to those of the diamond.

At present, it is possible to grow fianite crystals weighing up to 5 kg. The colouring of the fianites and their density are determined by chemical composition. Small amounts of the impurities of rare-earth and transition metals result in different colours and shapes of the fianites: red, rose, violet, blue, yellow, etc. As regards the colour range, the fianite is comparable with amethyst, garnet and zircon, and in the attractiveness it is compared with diamond.

The high refractive index of the fianites, which is close to that of the diamond, and the high dispersion result in special light effects. These properties, combined with different colouring, make it possible to simulate natural precious stones and also produce new ones with original colours. In the ultraviolet beams, the fianites in relation to the impurities may sparkle by blue, yellow, violet and other colours.

The fianites can be machined only in specific crystallographic directions. The fianites are relatively difficult to process, they easily crack and chip. The yield in the faceting of the crystals is usually not higher than 10–15%. In the production of juvenile inserts, the height of the lower part of the crystal should be deeper in order to improve its colour 'play', and the upper area should be wider in

Table 7.27. Diagnostic properties of diamond and its imitations [134]

Name	Characteristic formula	Crystal system	Refractive index	Birefringence
Diamond	C	Cubic	2.417	Often anomalous
Fianite	(Zr, Hf)O_2	"	2.10–2.20	None
Yttrium–aluminium garnet (YAG)	$Y_3Al_5O_{12}$	"	1.8834	None
Synthetic rutile	TiO_2	Tetragonal	2.61–2.90	0.287
Strontium titanate	$SrTiO_3$	Cubic	2.41	None
Lithium niobate	$LiNbO_3$	Trigonal	2.20–2.30	None
Zircon	Zr[SiO_4]	Tetragonal	1.920–2.010	0.036–0.059
Scheelite		Trigonal		
natural	Ca(WO_4)	Tetragonal	1.918–1.934	0.016
synthetic				
Sphalerite	ZnS	Cubic	2.37	None
Corundum				
natural	Al_2O_3	Trigonal	1.757–1.776	0.008–0.009
synthetic				
Spinel, synthetic	$MgAl_2O_4$	Cubic	1.728	Often anomalous
Evklaz	$Al_2Be_2[SiO4]_2(OH)_2$	Monoclinic	1.652–1.672	0.019–0.020
Phenakite	$Be_2[SiO_4]$	Trigonal	1.654–1.670	0.019–0.020
Tourmaline, achroite	Na(Mg, Fe)$_3$(Al, Fe)$_6$[BO_3]$_3$[Si_6O_{18}](OH)$_4$		1.614-1.634	0.016–0.024
Topaz	$Al_2[SiO_4](F_2\ OH)_2$	Rhombic	1.607-1.637	0.010
Beryll, colourless	$Be_3Al_2[Si_6O_{18}]$	Hexagonal	1.566-1.600	0.004–0.010
Mine crystal	SiO_2	Trigonal	1.544-1.553	0.009

comparison with the diamond. The faces of the inserts of the fianites are slightly rounded and this is another feature by which they differ from the diamonds.

The faceted stones produced from the fianites immediately attracted the attention of jewellers. As already mentioned, the initial batches of the fianite crystals were produced by the Lebedev Institute of Physics, Russian Academy of Sciences. Later, the production of similar crystals started in Switzerland by the company Grand Dzewahirdzhan which supplied the crystals to jewellery markets under the name djevalite. In the United States of America, the company Serez Corporation Corporation synthesized the material 'diamonesc' whose properties are very similar to those of the fianites. In Austria, Swarowski and Co produces 'zircon' using a Russian license [43] .

Name	Dispersion	Shine	Density, g/cm^3	Bonding	Hardness	Luminiscence colour in UV rays
Diamond	0.063 (0.044)	Diamond-like	3.52	Perfect	10	Blue, yellow, green, rose, orange, colourless
Fianite	0.060	"	6–10	"	7.5–8	Green, blue, yellow, violet
Yttrium–aluminium garnet (YAG)	0.028–0.038	"	4.57–7.09	Imperfect	6.5–8.5	Green, blue, violet, yellow
Rutile, synthetic	0.180–0.300	"	4.25–4.32	Perfect	6–6.5	No luminiscence
Strontium titanate	0.10–0.20	"	5.13–5.15	"	5.5–6.5	No luminiscence
Lithium niobate	0.120	"	4.64		5.5	No luminiscence
Zircon	0.039	"	3.95–4.80	Imperfect	6.5–7.5	Yellow
Scheelite						
natural	0.026	"	6.1	Mean	5	Blue
synthetic						
Sphalerite	0.156	"	4.09	Highly perfect	3.5	No luminiscence
Corundum						
natural	0.018	Glassy	3.99–4.05	Imperfect	9	Sometimes slightly rosy
synthetic						
Spinel., synthetic	0.020	"	3.63	Perfect	8	Green, blue
Evklaz	0.016	"	3.02–3.10	"	7.5	
Phenakite	0.015	"	2.95–3.00	Clear	7.5–8	
Tourmaline-achroite	0.017	"	2.90–3.31	Highly imperfect	7–7.5	
Topaz	0.014	"	3.50–3.57	Perfect	8	
Beryll, colourless	0.014	"	2.50–2.90	Imperfect	7.5–8	
Mine crystal	0.013	"	2.65	"	7	

Diagnostic properties of diamond and its imitations

It should be stressed that all possible imitations of the diamonds have been used in the jewellery industry for many years: transparent, colourless minerals: zircon, corundum, evklaz, phenakite, tourmaline–achroite, topaz, beryll, rock crystal, etc., and also various glasses with a strong dispersion of the refractive index because of various additions (mainly lead oxide). At present, the most effective imitations of the diamonds are the synthetic crystals such as yttrium–aluminium garnet (YAG), strontium titanate (fabulite), fianite, rutile, spinel, corundum, lithium niobate, etc. (Table 7.27) Some imitations are characterised by high dispersion, exceeding that of the diamond. These imitations differ from the diamonds by a number of properties. The main difference is the lower hardness. Up to now, there is no mineral or synthetic material with such a high hardness (10 on the Moss scale) like the diamond.

A significant role in the identification of the diamonds is played by the optical properties. Diamond is characterised by a high refractive index which cannot be measured by conventional optical methods. Some imitations with lower refractive indexes, presented in the table, can be easily distinguished from diamond on conventional refractometers. The minerals with high refractive indexes are characterised by a strong diamond shine, identical with that of the diamond; the shine of the minerals with low diffraction indexes is of the glassy type. This is especially clearly visible on the faceted stones. The recognition on the basis of the shine is carried out in a refractometer or in a 'magic eye' device used for distinguishing the diamonds from the crystals, characterised by less marked shine.

The property of birefringence of the crystal is often utilised in diagnostics. The presence of distinctive birefringence in the crystal makes it possible to separate diamond from the crystals of zircon, rutile, corundum, evklaz, phenakite, tourmaline, beryll, topaz, rock crystal. However, it must be remembered that the diamond may also sometimes be characterised by anomalous birefringence.

In the identification of the stones not secured in a holder, it is recommended to use the measurements of the density of the stone. Using a liquid with a density of 3.5 g/cm^2 (for example, the diluted Clerichi liquid) it is quite simple to separate the diamonds. All the imitations are characterised by higher or lower density. Density is also determined by hydrostatic weighing. Only the density of topaz is close to the density of the diamond. The densities of the most widely used imitations are also presented in table 7.14. The density of the glasses changes in relation to the lead content of the latter, but the hardness of the glass is many times lower than that of the diamond and, consequently, they can be easily distinguished from the diamond. Additional diagnostic properties include the presence of the cleavage planes, luminescence and inclusions.

The universal method of identification of the diamond is to illuminate the insert by the x-rays. In the x-rays, the diamond is transparent, and the majority of minerals and glasses are non-transparent. The x-ray method can be used for the inserts, secured in the components. Recently, 'samplers' have been used widely for the identification of diamonds. The method is based on the measurement of the heat conductivity of diamond which differs by two orders of magnitude from the heat conductivity of the majority of imitations (Table 7.28). The method based on wetting the surface, with the determination of the contact wetting angle, is also used. Special black dyes are also used; they leave a continuous line on the diamond and a dotted line on other stones with high (more than 1.8) refractive indexes [121].

Because of the lower hardness and also as a result of less efficient polishing of less valuable materials, the substitutes of the diamonds contain unavoidably the faces which are not ideally flat and rounded edges. Consequently, the edges themselves are not so sharp and clear as in the diamond [45]. The diagnostic features of the diamond and of its imitations are presented in Table 7.27.

Table 7.28. Heat conductivity of diamond imitations

Material	Heat conductivity (approximate) V/cm °K
Diamond	10–26*
Corundum	0.35
Yttrium–aluminium garnet	0.11
Gallium–adolinium garnet	<0.1
Strontium titanate	<0.1
Rutile	0.08
Cubic zirconia (fianite)	0.02

*Depends on the amount of the admixture.

References

1. J. McCullough and K. Trueblood, *Acta Crystallogr.*, 18, 6, 983–991,1965.
2. G.B. Bokii, Crystallochemistry, Nauka, Moscow, 1971.
3. A.A. Voronkov, et al., Crystallochemistry of minerals of zirconium and their artificial analogues, Nauka, Moscow, 1978.
4. G. Teufer, *Acta Crystallogr.*, 15, 11, 1187, 1962.
5. Yu.M. Polezhaev, *Zh. Fiz. Khimii,* 41, 11, 2958–2959, 1967.
6. C.T. Lunch, et al., *J. Amer. Ceram. Soc.*, 44, 3, 143–147, 1961.
7. L.M. Komissarova, et al., *Zh.Neorg. Khimii,* 5, 7, 1413-1415, 1960.
8. R. Cupres and R. Wollast, *Ber. Deut. Keram. Ges.*, 40, 9, 527–532, 1963.
9. D.K. Smith and H.W. Newkirk, Acta Crystallogr., 18, 6, 983–991, 1965.
10. N.V. Belov, Kristallografiya, 5, 3, 460–461, 1960.
11. M. Perec-y-Jorba, High-temperature studies, Mir, Moscow, 1967.
12. H.G. Scott, *J.Mater. Sci.*, 10, 1527–1535, 1975.
13. V.S. Stubican and J.R. Hellman, Phase Equilibria in Some Zirconia System, pp.25–37, in: Advances in Ceramics, vol.3. Edited by A.H. Heuer and L.W. Hobbs, The American Ceramic Society, Columbus, OH, 1981.
14. D.K. Smith and C.F. Cline, *J. Am. Ceram. Soc.*, 45, 5, 249–50, 1962.
15. C. Delamarre, *C.R. Acad. Sci.*, C269, 113–115 (1965).
16. H.J. Rossell and R.H.J. Hannink, Phase $Mg_2Zr_5O_{12}$ in MgO Partially Stabilised Zirconia, pp.139–51, in: Advances in Ceramics, Vol.12. Edited by N. Claussen, M. Ruhle, and A.H. Heuer, The American Ceramics Society, Columbus, OH, 1984.
17. J.R. Hellman and V.S. Stubican, *Mater. Res. Bull.*, 17, 459–65 (1982).
18. R. Ray and V.S. Stubican, *Mater. Res. Bull.*, 12, 5, 549–556, 1977.
19. H.G. Scott, *Acta Crystallogr.*, B33, 1, 281–282 (1977).
20. V.S. Stubican, et al., *J. Am. Cer. Soc.*, 61, 1–2, 17–21, 1978.
21. O. Ruff and F. Ebert, *Z. Annorg. Allgem. Chem.*, 180, 19–41, 1929.
22. V.S. Stubican, et al., Phase Relationships in Some ZrO_2 System, pp.96–106 in: Advances in Ceramics, Vol.12, Edited by N. Claussen, M. Ruhle, and A.H. Heuer, The American Ceramics Society, Columbus, OH, 1984.
23. H.G. Scott, *J. Mater. Sci.*, 12, 2, 311–316 (1977).
24. J. Pascual and P. Duran, *J. Amer. Ceram. Soc.*, 66, 1, 23–27, 1983.
25. C. Anderson, et al., Diffusionless Transformations in Zirconia Alloys, p.78–85, Advances in Ceramics. Vol.12, Science and Technology of Zirconia II, The American Ceramics Society, Ohio, 1983.
26. P. Ruh, et al., *J. Amer. Ceram. Soc.*, 67, 9, A, 190–192, 1984.
27. A.I. Belov and T.A. Semenov, *Izv. AN SSSR, Neorg. Mater.*, 13, 10, 1817–1821, 1977.
28. I.A. Davtyan et al., *Izv. AN SSSR, Neorg. Mater.*, 5, 890–895, 1966.
29. S.S. Kiparsov, et al., *Izv. AN SSSR, Neorg. Mater.*, 12, 9, 1693–1694, 1976.
30. J. Lefevre, *Ann. Chem.*, 8, 1–2, 117–149, 1963.
31. M. Perez-y-Jorba, *Ann. Chem.*, 7, 479–511, 1962.
32. M. Perez-y-Jorba, et al., *C.R. Acad. Sci.*, 219, 22, 2329, 1959.
33. F. Sibieude and A. Rovanet, *C.R. Acad. Sci.*, 205, 459–468, 1972.
34. M.R. Thornberm, et al., *J. Solid State Chem.*, 11, 3–4, 536–544, 1970.
35. E.E, Lomonova, Dissertation, General Physics Institute, Moscow, 2001.
36. V.I. Aleksandrov, et al., *Pribory Tekh. Eksper.*, 5, 222–225, 1970..
37. V.I. Aleksandrov, et al., *Vestnik AN SSSR*, 1973, No. 12, 29–39.

38. B. Scott and H. Rawson, *Glass. Technol.*, 1973. Vol.14. ! 5. P. 115-124.
39. V.I. Aleksandrov, et al., Proc. Indian National Sci. Acad., V. 57, No 2,1991, 133–144.
40. Lomonova E.E., Production and examination of single crystals of solid solutions based on ZrO_2 and HfO_2, Disseration, FIAN, Moscow, 1980.
41. H.J. Rossell, Ordering in Anion-Deficient Fluorite-Related Oxides, pp. 47-63, in: Advances in Ceramics, v.3, edited by A.H. Heuer and L.W. Hobbs The American Ceramic Society, Columbus, OH, 1981.
42. V.I. Aleksandrov, et al., A method of production of single crystals based on solid solutions of refractory oxides, Author's Cert. No. 1048857, 15.06.1083.
43. V.I. Aleksandrov, et al., A method of production of single crystals based on stabilised zirconia, Author's Cert. No. 1365739, 08.09.1987.
44. V.I. Aleksandrov, et al., A method of production of coloured crystals of oxides (its variants), Author's Cert. No. 1086833, 23.12.1983.
45. V.I. Aleksandrov, et al., A method of production of dielectric crystals, Author's Cert. No. 1549123, 08.11.1989.
46. V.I. Aleksandrov, et al., A method of production of single crystals based on stabilised zirconia, Author's Cert. No. 1568686, 01.02.1990.
47. V.I. Aleksandrov, et al., A method of production of coloured crystals of refractory oxides (its variants), Author's Cert. No. 1112821, 08.05.1984.
48. Mazdiysni K.S. and Brown L.M., *Journal Amer. Ceram. Soc.*, 54, 1, 43-45, 1970.
49. V.P. Gorelov, et al., *Izv. AN SSSR, Neorg. Mater.*, 13, 1, 181–182, 1977.
50. H.-H. Mobius, *Z.Chemie,* 4, 3, 81–94, 1964.
51. I.I. Vishnevskii, et al., Nauch. Tr. Ukr. Nauch. Issled. In-ta Ogneuporov, 53, 6, 74, 1952
52. R. Collongues, *Ann. Chem.*, 8, 7–8, 395–408, 1963.
53. R.S. Roht, *J. Res. Nat. Bureau*, 56, 17–25, 1956.
54. Yu.A. Pyatenko, *Kristallografiya*, 4, 2, 204–208, 1959.
55. Yu.A. Pyatenko, *Kristallografiya*, 4, 5, 708–713, 1963.
56. E. Alechin, R.Roy, *J. Amer. Ceram. Soc.*, 45, 1, 18–25, 1962.
57. N.V. Belov, *Sb. L'vov. Geol. Ob-va*, 4, 21, 1950.
58. R. Kollong, Nonstoichiometry, Mir, Moscow, 58–88, 1974.
59. T.K. Gupta, et al., *J. Mat. Sci.*, 12, 2421, 1977.
60. T.K. Gupta, *Science of Sintering*, 10, 205, 1978.
61. C.A. Andersen and T.K. Gupta, Phase Stabi!ity and Transformation Toughening in Zirconia, in: Advances in Ceramics, V. 3, Sience and Technology of Zirconia, The Amer. Ceram. Soc., Columbus, Ohio, p.184, 1981.
62. F.F. Lange, *J. Mater. Sci.,* 17, 10, 225–34, 1982.
63. V. Lantery, et al., Tetragonal Phase in the System ZrO_2–Y_2O_3, pp. 118–130, in: Advances in Ceramics, v. 12, Sci. and Techn. II, The Amer. Ceram. Soc., Columbus, OH, 1984.
64. T. Yagi, et al., *J. Amer. Ceram. Soc.*, 69, 1, A 3–4, 1986.
65. D. Michel, et al., J.Mater. Sci., 18, 2618–2628, 1983.
66. Yu.K. Voron'ko, et al., *Neorg. Mater.*, 34, 4, 439–443, 1998.
67. D. Michel, et al., *J. Cryst. Growth*, 43, 4, 546–48, 1978.
68. R.H. Ingel, et al., Physical, Microstructural and Thermomechanical Properties of ZrO_2 Single Crystals, in: Adv. in Ceram., V. 12, Science and Technology of Zirconia II, Ed. by N. Claussen, M. Ruhle and A.H. Heuer, Am. Cer. Soc., Columbus, OH, p.408–414, 1989.

69. V.I. Aleksandrov., et al., *Fiz. Tverd. Tela*, 20, 2, 528–534, 1978.
70. V.V. Osiko, et al., Spectroscopic Investigation of Defects Structure and Structure Transformations in Ionic Crystals, Crystals, Springer, 10, 37–86, 1984.
71. J. Martinez-Fernandez, et al., *Acta Metall. Mater.*, 43, 2, 593–601, 1995.
72. T. Noma, et al., *J. Ceram. Soc. Jpn.*, 94, 8, 887–90, 1986.
73. V.I. Aleksandrov, et al., *Fiz. Tverd. Tela*, 16, 8, 1456–1459, 1975.
74. V.I. Aleksandrov, et al., *Kr. Soob. Fiz.*, 3, 21–27, 1975.
75. I.L. Chistyi, *J. Raman Spectroscopy*, 6, 4, 183–192, 1977.
76. R.P. Ingel, D. Lewis III, *J. Am. Cer. Soc.*, 69, 4, 325–32, 1986.
77. R.P. Ingel, D. Lewis III, *J. Am. Cer. Soc.*, 71, 4, 265–71, 1988.
78. A.I. Ioffe, et al., *Dokl. AN SSSR*, 203, 657, 1973.
79. V.N. Filatova, et al., *Zh. Neorg. Mater.*, V. 11, 1427, 1975.
80. P.P. Budnikov, et al., Chemical technology of ceramics and refractories Stroiizdat, Moscow, 1972.
81. A.B. Ivanov and Yu.L. Krasulin, All-Union Conference on Chemistry, Technology and Application of Zirconium, Hafnium and their Compounds, Giredmet, Moscow, 1972.
83. J.M. Farley, *J. Mater. Sci.*, 7, 475 (1972).
84 Modern Crystallography, Vol. 3, Nauka, Moscow, 1980.
85. J. Barton, et al., in: Germanium, Izdatinlit, Moscow, 1955, 74–81.
86. V.M. Vigdorovich, et al., Directional Crystallisation and Physico-Chemical Analysis, Khimiya, Moscow, 1976.
87. G.. Pfann, Zone Melting, Mir, Moscow, 1970.
88. V.I. Aleksandrov, et al., *Neorg. Mater.*, 16, 1, 99–105, 1980.
89. D.B. Zhang. *J. Crystal Growth*, 79, 336–340, 1968.
90. M.G. Mil'viskii and V.B. Osvenskii, Problems of Modern Crystallography, Nauka, Moscow, 1975.
91. L.S. Milevskii, *Kristallografiya*, 6, 2, 249–255, 1961.
92. D.A. Petrov and B.A. Kolachev, Crystal Growth, Vol. 1, 159–169, 1957.
93. L.V. Tir and N.M. Smirnov, Proc. 4th Conference on Application of Magnetic Hydrodynamics in Industry, Donetsk, 1969.
94. E.K. Keler and A.B. Andreeva, *Ogneupory*, 4, 184–192, 1964.
95. G.P. Kalliga and L.A. Lyutsareva, *Ogneupory*, 9, 412–417, 1964.
96. M.Kh. Ashurov, et al., *DAN SSSR*, 313, 3, 590–593, 1990.
97. V.V. Voronov, et al., *Neorg. Mater.*, 21, 10, 1710–1714, 1988.
98. A.A. Chernov, Problems of Modern Crystallography, Vol. 3, Nauka,Moscow, 1980.
99. A.A. Chernov, *Usp. Fiz. Nauk*, 100, 2, 277–328, 1979.
100. F.F. Seelig, *Z. Naturforsch.*, 31, 731–738, 1976.
101. V.I. Aleksandrov, et al., *Neorg. Mater.*, 13, 2, 2192, 1971.
102. L.A. Krasikov, et al., *Fiz. Tverd. Tela*, 18, 12, 304, 1976.
103. L.A. Krasikov, et al., *Zavod. Lab.*, 46, 3, 242, 1980.
104. V.I. Aleksandrov, et al., *Neorg. Mater.*, 19, 1, 99, 1983.
105. P.A. Tikhonov, et al., *Zh. Fiz. Khimii*, 48, 3, 643, 1979.
106. A.K. Kuznetsov, et al., in: Electronic Structure and Physical Properties of Solids, part 2, Naukova Dumka, Kiev, 1972.
107. V.B. Glushkova, et al., *Izv. AN SSSR, Neorg. Mater.*, 13, 12, 2197–2201, 1977.
108. V.I. Aleksandrov, et al., *Vestnik Akad. Nauk SSSR*, 6, 65–72, 1980.
109. M.Kh. Ashurov, et al., *Dokl. Akad. Nauk SSSR*, 296, 1, 121–122, 1987.

110. F. Hund, *Z. Physikal. Chem.*, 199, 1/3, 142–151, 1952.
111. M. Spizbergen and P.M. Houpt, The British Ceram. Soc., 1968. Vol. 4. P. 247.
112. A.M. Leyus and M.R. Collounques, *C.R. Sci. Paris*, 1969. Vol. 269, 20, 1198–1200.
113. V.F. Kitaeva, et al., Experimental examination of the properties of ZrO_2–Y_2O_3 solid solution, Preprint No. 136, FIAN, Moscow, 1976.
114. B.F. Ormont, Introduction to Physical Chemistry and Crystal Chemistry of Semiconductors, Vysshaya Shkola, Moscow, 1973.
115. B.I. Ioffe, X-ray Diffraction Methods in Chemistry, Khimiya, Leningrad, 1976.
116. A. Anthony, *Rev. Hautes Temp. Refract.*, 2, 147, 1966.
117. D.W. Strikler and W.G. Carlson, *J. Amer. Ceram. Soc.*, 47, 122, 1966.
118. V.I. Aleksandrov, et al., *Usp. Khimii*, 47, 3, 385–427, 1978.
119. A. Baerman, et al., *J. Crystal Growth*, 79, 331–335, 1986.
120. L.I. Demkina (editor), Physical Fundamentals of Production of Optical Glass, Khimiya, Leningrad, 1976.
121. K. Nassau, *The Lapidary Journal*, 35, 6, 1194–1200, 1981.
122. K. Nassau, *The Lapidary Journal*, 35, 6, 1210–1214. 1981.
123. V.I. Aleksandrov, et al., Brief Comments on Physics, FIAN, Moscow, 2, 17, 1973.
124. V.I. Aleksandrov, et al., *Izv. AN SSSR, Neorg. Mater.*, 13, 12, 2192–2196, 1977.
125. V.I. Aleksandrov, et al., *Fiz. Tverd. Tela,* 26, 5, 1313–1318, 1984.
126. F. Krieger, Chemistry of Imperfect Crystals, Mir, Moscow, 1969.
127. V.I. Aleksandrov, et al., Brief Comments on Physics, FIAN, Moscow, 3, 3, 1975.
128. V.I. Aleksandrov, et al., Brief Comments on Physics, FIAN, Moscow, 6, 10, 1976.
129. I.L. Chistyi, *Journal of Raman Spectroscopy*, 6, 4, 183–192, 1977.
130. N.G. Pace, et al., *J. Mater. Sci.* 4 1106, 1969.
131. V.I. Aleksandrov, et al., *Kristallografiya*, 18, 1085, 1973.
132. P.H. Borcherds, *Opt. Acta*, 20, 147 (1973).
133. N.I. Kornilov and Yu.P. Solodova, Jewellery Stones, Nedra, Moscow, 1983.
134. V.S. Balitskii and E.E. Lisitrsyna, Synthetic Analogues and Simulation of Natural Precious Stones, Nedra, Moscow, 1981.

8

Partially stabilised zirconia (PSZ) crystals

One of the current problems of materials science is the development of new structural nonmetallic materials in which the strength, low brittleness and chemical inertness are maintained in the oxidation of atmosphere in a wide temperature range.

These materials include partially stabilised zirconia (PSZ) which was previously produced only in the form of ceramic polycrystalline specimens. The method of direct high-frequency melting of dielectrics in the cold crucible has made it possible to produce for the first time partially stabilised zirconia in the form of nanostructured crystals and carry out investigations exposing the excellent properties of these materials.

8.1. PHASE TRANSFORMATIONS IN CRYSTALS OF PARTIALLY STABILISED ZIRCONIA

At present, it is regarded as established that the phase transitions in zirconia have the form of martensitic transformations, characterised by a number of features:

– diffusionless-shear mechanism of phase transition;

– the presence of a hysteresis

– the athermal nature of the process, i.e. the absence of the activation energy;

– quenching is not possible [1–6].

This nature of transition has been confirmed on single crystals of zirconia grown at low temperatures from the solution in the melt and by the hydrothermal method [6]. It has been shown that the temperature range of the $m \leftrightarrow t$ transitions depends on the chemical composition of the specimen and its prior thermal history, in particular, on the synthesis method.

Martensitic transformations in zirconia

The martensitic transitions take place in the solid state, in the absence of local changes in the chemical composition, i.e. without diffusion, and take place by a collective mechanism.

The development of the theory of martensitic transformations is based on the concepts proposed by G.V. Kurdyumov [4] on the regular nature of the rearrangement of the lattice and on the coherence of coexisting phases.

The $m \leftrightarrow t$ and $t \leftrightarrow c$ phase transitions in pure zirconia are martensitic transformations with athermal kinetics [1–12]. This means that the transition takes place with the speed of sound (the zero energy of activation) as soon as a viable nucleus of the new phase appears. This mechanism can be efficiently described using the model of the 'floating domino'. The process with the athermal kinetics cannot be prevented by rapid cooling (quenching) even at the temperature of liquid nitrogen or helium. However, the absolutely athermal mechanism is possible only in the presence of ideal interphase boundaries with no potential barriers, preventing the movement of the boundary. The situation completely changes even in the presence of point defects. In this case, the displacement of the interphase boundary is complicated by the barriers resulting from the presence of structural defects. The resultant energy of thermal activation differs from zero, the rate of the process rapidly decreases, and the kinetics of the process becomes isothermal. This circumstance indicates that it is possible to stabilise the metastable high-temperature phase by rapid cooling from the temperature of the equilibrium phase transition to a lower temperature. At cooling rates greatly exceeding the rate of transformation, the phase transition has no time to occur. From the thermodynamic viewpoint, the high-temperature phase is metastable and the absolute loss of stability may take place at some critical fluctuation of temperature. From the kinetic viewpoint, the fluctuation is the formation of a viable nucleus of the low-temperature phase. However, the probability of fluctuation formation of the nucleus with the critical volume decreases with decreasing temperature. At some relatively low temperature, this probability is actually equal to zero and the metastable high-temperature phase may exist indefinitely long in the absence of external factors causing nucleation (for example, mechanical loading). These considerations qualitatively explain the process kinetics. Lattice defects lead to the formation of a potential barrier on the path of movement of the interphase boundaries. This is possible, for example, in the case of formation of the oxygen vacancies during the stabilisation of zirconia by the oxides of metals with the valency of the cations lower than that of zirconia. For example, with addition of yttrium oxide, one oxygen vacancy forms per two Y^{3+} ions:

$$(1-x)ZrO_2 + xY_2O_3 \leftrightarrow Zr_{1-x}Y_xO_{2-x/2} + V^{o}_{x/2}$$

For better understanding of the processes of fully amorphous phase transformations in the solid solutions based on zirconia and the processes of phase formation in synthesis from melts of materials based on zirconia, it is essential to examine certain special features of the martensitic phase transitions.

Crystallography and energetics of martensitic transformations

The crystallographic characteristics of the martensitic transition are the strict orientation relationships between two lattices: the habit form of the martensitic lath and the macroscopic form of the transformed part of the crystal [9–13]. In a martensitic process, the lattices are conjugated via coherent or semicoherent interphase boundaries.

Conjugation of the phases with different crystal lattices should provide for the mutual accommodation of the lattices [4,5]. The accommodation may result from elastic displacements of the atoms from equilibrium positions and from inelastic displacements due to misfit dislocations and vacancies condensed at the phase boundaries. Only elastic displacements are sources of internal stresses. The magnitude of these displacements is determined by two factors:

1. The mismatch of the crystal lattices of both phases in the contact plane
2. The mechanism of conjugation of the phases, determining the compensation of the mismatch.

A completely coherent phase conjugation is the one in which all crystal planes of one phase change continuously into the planes of the another phase, thus forming the closed Burgers contour. This conjugation results in the maximum elastic stresses in the crystal.

A semi-coherent boundary forms when the lattices of the contacting crystals are elastically deformed in such a manner that some individual areas are characterised by coherence and other areas by the gradual buildup of the mismatch which is periodically eliminated by resultant breaks. The Burgers contour becomes open. The closing vector *b* is the Burgers vector of the mismatch dislocation. The mismatch dislocations consequently decrease the level of internal stresses [11].

The coherent conjugation forms in an early stage of the transition when the stored elastic energy is still insufficient for the formation of a mismatch dislocation. The increase in the size of the inclusion creates suitable conditions for the semi-coherent conjugation mechanism.

Polycrystalline materials and fine-grained ceramics (and also materials produced by means of nanotechnologies) may be characterised by the presence of completely non-coherent conjugation in which there is no friction between the matrix and the inclusion resulting in the failure of the material by the mechanism of intergranular sliding.

Khachaturyan [5] developed a theory which makes it possible to predict, on the basis of crystallographic and energetics relationships, the morphology of particles of the second phase, formed in the habit plane of the crystal matrix. The optimum form of the inclusion is determined by the opposite effects of two components of energy: volume and surface. In a general case, the form of the inclusion is determined by competition between the energy of elastic distortions, which is minimum for the infinitely thin and elongated plate, and the surface energy, minimum for the equilibrium form. In cases in which the contribution of surface tension is dominant, the equilibrium form of the plate in the habit plane is a circle, if the elastic energy is dominant, the plate is needle-shaped, representing in fact an infinitely prolate ellipse. In intermediate cases, inclusions have the form of a less prolate ellipse.

Hysteresis of the martensitic phase transition

The change of the volume during the $m \leftrightarrow t$ transition ($m \leftrightarrow t$ – compression, $t \leftrightarrow m$ –expansion) indicates that this transformation is the transition of the first type. In this case, the classic theory

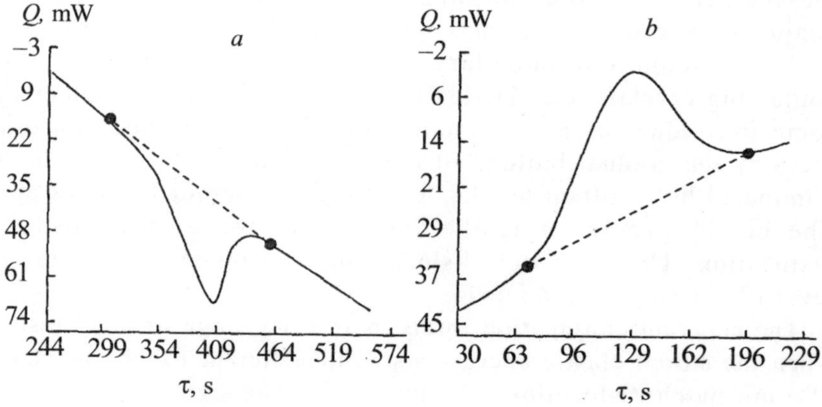

Fig. 8.1. Curves of differential scanning calorimetry of direct (a) and reversed (b) transirions of ZrO_2 crystals [19]

of allotropic transformations does not permit the existence of a transition with a temperature hysteresis.

The reason for hysteresis is assumed to be the change in the free energy of the crystal during the $m \leftrightarrow t$ transition [14–16]. From this viewpoint, the most probable case of a hysteresis is the stresses formed by a sudden change of the volume during the phase transition. For example, the introduction of the energy of the residual stresses into the equation for the internal energy of the phase decreases the temperature of the direct transition and increases that of the reverse transition.

The transformation temperature depends to a greater degree on the surface energy. The size of the particles (~100 Å), below which the transformation to the tetragonal modification is supported by the difference if the free energy of two polymorphous modifications, has been determined in [17]. In the determination of the critical size of the particles the potential energy of elastic deformation was not taken into account because it was assumed to be very small in comparison with the surface energy.

It was also assumed that the hysteresis is determined by different degrees of perfection of the structure of individual crystals present in the specimen [18, 19]. However, the existence of a hysteresis in single crystals shows that this cannot be the main reason.

Kinetics of the martensitic phase transition

The majority of martensitic transformations are characterised by athermal kinetics. The degree of transformation in this case is determined only by temperature and is almost completely independent of time [6, 12–18].

The athermal kinetics is evidently determined by the expression which does not include the time t:

$$\xi = A \cdot \exp\left(-\Delta\Gamma_A / RT\right), \qquad (8.1)$$

Here ξ is the degree of transformation, A is the pre-exponential multiplier, ΔG_A is activation energy (for the case of a non-ideal interphase boundary).

These kinetic special features of the martensitic transformation are valid for the ideal crystal. In the real conditions, the transition is accompanied by plastic deformation or failure: martensite crystals are stress concentrators and areas of plastic failure. Some differences in the description of the experimental kinetics of the martensitic transformation are associated with the difficulties of fulfilling the isothermal conditions in the examination of the transition and with the effect of the inhomogeneous distribution of components in the crystals.

As a result of the martensitic phase transformation, the material is hardened. During the phase transition, the stored energy (elastic, surface, energy of defects) increases and the effect of dissipative forces, associated with the irreversible scattering of energy during the movement of the interphase boundary and dislocations in a medium with defects, becomes stronger. To overcome this 'friction' and for further development of the transition, it is necessary to increase constantly the thermodynamic driving force. This results in hardening of the material.

There are two qualitatively different types of kinetics of phase transformations in single-component crystalline solids. Depending on the external conditions, the same phase transition may take place at a high rate (of the order the speed of sound) which depends on temperature only slightly (for the majority of martensitic transformations) or exponentially. In the latter case, the transformation is limited by the diffusion kinetics of 'settling' of the vacancies at the interphase boundaries and by their reversed dissolution. The diffusion kinetics is observed in cases in which the crystal initially contains a large amount of the martensitic phase. The self-diffusion kinetics of the transformation does not exclude the crystallographic and morphological relationship of the initial and new phases, in particular, a transformation may be accompanied by macrodeformation and the formation of a relief.

Attention should be given to the resultant contradiction between the diffusionless nature of the martensitic transformation and the presence of the two-phase region on the *T–x* diagram, where the composition of the phases should change in accordance with the lever rule. This is associated with the fact that the phase diagram is plotted without considering elastic stresses. A metastable (from the viewpoint of equilibrium thermodynamics) phase is actually stable. This may be accompanied by the formation of intermediate phases which are not reflected on the *T–x* diagram. This fact is confirmed by the existence of a large number of contradicting data on the phase composition of the ceramics and single crystalline materials based on zirconia and its dependence on the synthesis conditions. For example, the existence in single crystals of a 'frozen' tetragonal phase at concentrations of ~(1–2) mol% Y_2O_3 was explained by the fact that the rearrangement of the structure in the single crystal is more difficult in comparison with that in fine-dispersion powder in which the chain of the atoms, undergoing the transformation, breaks up on the surface of the particle. Therefore, the stresses do not reach high values, and the volume of the individual grains increases as a result of a decrease of their porosity. In contrast to the powders, the single crystals split up during transformation because the simultaneous breaking of a large number of atomic bonds may take place only at high energy losses $\Delta G_{t \to m}$. For example, the latter phenomenon takes place in the case of strong supercooling of a substance in relation to the phase transition temperature. As the crystal becomes more perfect and larger, the strength of its chemical bonds increases and the phase transition requires very high supercooling up to the complete freezing of the structure.

The data on the kinetics and thermodynamics of the phase transitions in PSZ crystals, grown from the melt, are very scarce. The kinetics of the $m \to t$ transition has a significant effect on the degree of stabilisation of the tetragonal phase because the possibility of quenching of this phase depends on the synthesis conditions. The authors of [19] investigated the $m \to t$ phase transitions in the crystals of PSZ in relation to the concentration of the stabilising oxide. The curves of differential-scanning calorimetry (DSC) for direct and reverse transitions of pure zirconia are shown in Fig. 8.1 [19] and were used to calculate the enthalpy of phase

Table 8.1. Thermal effects of the $m \rightarrow t$-phase transition and the content of the monoclinic phase in crystals*

x	ΔH, J/g	Content of the monoclinic phase, wt.%	
		DSC	XPA
0.000	39.16	100	100
0.0086	-	59	100
0.0178	7.28	18.6	36
0.020	4.74	11.5	28
0.0234	3.03	7.5	21

*Evaluation on the basis of the results of DSC and XPA analysis

transitions (Table 8.1). In the case of zirconia crystals, the calorific effect of the direct and reverse transitions differs only slightly ($\Delta H_{m \rightarrow t}$ = 4.92 kJ/mol), being in the range of the experimental error and in good agreement with the literature data: 4.75 and 5.95 kJ/mol [20].

The experimental DSCdata on the calorific effects of transitions were used for the quantitative evaluation of the phase composition of specimens with different contents of the stabilising agent.

The amount of the monoclinic phase in the specimen of the given composition is determined by the $C_M = \Delta H / \Delta H_{ZrO2}$ relationship, where ΔH_{ZrO2} is the enthalpy of transition in pure zirconia (100% of the monoclinic phase), ΔH is the enthalpy of transition in the given specimen. The calculated concentrations are presented in Table 8.1. [19] This Table also gives the results of the quantitative evaluation of the phase composition of the specimens obtained by x-ray diffraction analysis of powders of refined crystals. Incomplete agreement in the data on the phase composition of the specimens evidently indicates that the refining of the crystals increases the content of the monoclinic phase in the specimen.

The characteristic feature of partially stabilised zirconia crystals produced from the melt is the formation of the cubic crystallographic system during crystallisation. Subsequent phase transitions take place in the solid phase. The initial composition, the rate of growth and cooling of the crystal and the annealing conditions determine the mechanisms of formation, the quantitative ratio of these phases and the degree of deviation from the thermodynamically equilibrium state, corresponding to the phase diagram.

With an increase in the content of Y_2O_3 in the crystals, the temperatures of the $m \rightarrow t$ phase transitions decrease but do not coincide with the temperature range of the two-phase region (m + t) of the corresponding composition on the equilibrium diagram. This may be associated both with changes of the specimen composition and with the metastable state of the specimens: in transition through the two-phase range (t + c) in the process of synthesis (cooling) of the crystal, diffusion processes are incomplete and strongly depend on the cooling rate. This results in a change in the content of Y_2O_3 in the tetragonal phase (a decrease in comparison with the initial value) and the temperature of the phase transitions becomes higher than the temperatures corresponding to the initial composition on the equilibrium diagram.

On the basis of kinetic data, one can propose the following model of the process taking into account analysis of the phase transformations. The process may be regarded as a two-state process, consisting of the nucleation of a new phase and movement of the interphase boundaries. The limiting stage of the direct transition in pure zirconia is the nucleation stage. The initial parts of the kinetic curves are flat. During the formation of nuclei of the tetragonal phase, the interphase boundaries start to move at a high speed. This rapid movement is possible only within the limits of a specific zone of the crystal where the conjugation of the phases is coherent. A new act of nucleation is required in areas with a mismatch dislocation or a microcrack. This pattern is usually observed on a crystal with a low content of Y_2O_3 (x = 0.005) where the relief associated with the phase transition forms in different parts of the crystal and rapidly propagates within the limits of these areas. The rate of phase transition in the crystal is higher in comparison with the powder. This is associated with the fact that the crystal requires a smaller number of nuclei in comparison with that in the powder because in the powder one nucleus may cause a transition only within one grain.

It is evident that after completion of the $m \rightarrow t$ transition, the specimens consist of 100% of the tetragonal phase and in the reverse transition the nucleation process is always homogeneous and slow. In addition, mismatch dislocations and microcracks appear. The movement of the interphase boundary during the reversed transition is of the dissipative nature. Therefore, the reversed transition is slow in all specimens.

With an increase in the content of the stabilising oxide x (mol) the concentration of vacancies increases and the rate of transition drops. The effect of oxygen vacancies on the kinetics of direct and reverse transitions is confirmed by the data on the kinetics of phase transitions in the specimens in which vacuum annealing results in the formation of additional vacancies. The slowdown of the direct transition is very marked, whereas in the case of the reversed transition it is less pronounced. As a result of annealing the hysteresis range becomes smaller and the same takes place with an increase in the content of the stabilising agent.

According to the theory of martensitic transformations, the main dissipative process during the displacement of the boundaries is the diffusion of vacancies. Transformation is limited by the 'settling' of vacancies (dissolved in the crystal) on the interphase boundaries and by the 'dissolution of the deposit' reversed process. If the crystal contains a large number of dislocations, acting as sources and sinks of the vacancies, there is no significant deceleration in the process and equilibrium is established very rapidly. However, if prior to the start of the phase transition the crystal contains a large amount of the new phase (for example, in specimens with $x = 0.02$ or $x = 0.0234$), the role of the dislocations in the establishment of equilibrium becomes less and less important, and the kinetics is controlled by self-diffusion interaction between the particles of the new tetragonal phase through the monoclinic phase. The self-diffusion process is far slower in comparison with the conservative movement of the boundary, limited by the removal of heat. Of all the examined specimens, only the specimen with $x = 0.0234$ could be quenched because with an increase in the content of the yttrium oxide the number of vacancies increases and the rate of the $t \rightarrow m$ transition greatly decreases.

The results of examination of the phase transitions in the ZrO_2–Y_2O_3 crystals show that:

1. The increase in the concentration of the stabilising oxide decreases the temperatures of the $m \rightarrow t$ and $t \rightarrow m$ transitions and the magnitude of the temperature hysteresis;

2. The kinetics of the $m \rightarrow t$ phase transformations in the crystals of partially stabilised zirconia depend on the concentration of the stabilising oxide and may be limited by both the stage of nucleation (for the pure zirconia crystal) and by the stage of movement of the interphase boundary (at $x > 0.02$); the maximum rate of transformation is recorded for the crystals with the composition $x = 0.0086$;

3. The $t \rightarrow m$ transition is characterised by homogeneous nucleation which also determines the rate of the transition;

4. At $x > 0.0234$, the crystals of partially stabilised zirconia did not contain any monoclinic phase, as established by the differential–thermal analysis method.

Differential–thermal and dilatometric analysis methods we used for the determination of the temperature ranges of transformations in the ZrO_2–R_2O_3 crystals (where R is Eu, Gd, Yd), Table 8.2 [21].

8.2. THE PHASE COMPOSITION OF THE CRYSTALS OF PARTIALLY STABILISED ZIRCONIA

The results of phase analysis of the synthesised specimens of the crystals of partially stabilised zirconia by the methods of x-ray diffraction analysis, crystallo-optical method and the Raman scattering method were presented previously in Chapter 7. It should be mentioned that in grinding the specimens, the phase composition of the specimens changed and the monotonic phase appeared in the specimens with 3 mol% R_2O_3, indicating the metastability of these materials and the presence in them of a transformed tetragonal phase [22].

The x-ray diffraction examination of the phase composition of the crystals of partially stabilised zirconia shows that the monoclinic phase is present in the specimens with the concentration of the stabilising oxide to ~2–2.3 mol% Y_2O_3; the crystals containing 3–4 mol% Y_2O_3 contained mainly the tetragonal phase. The structural characteristics of the crystals of partially stabilised zirconia are presented in Table 8.3.

The monoclinic phase formed only in certain cases in the upper part of the crystals of this composition. Evidently, at concentrations of < 1.5 mol% Y_2O_3, the tetragonal phase, detected at room temperature, was represented by the t'-phase, formed from the cubic phase, and the M-phase, formed from the t-phase. At a concentration of ~3–4 mol% Y_2O_3 and higher, the breakdown of the cubic phase, accompanied by the diffusion of the cations did not lead to the formation in the crystals of the t-phase with a low concentration of Y_2O_3, where further cooling would have been accompanied by the $t \rightarrow m$ transition. Only in individual upper regions of the crystals, absorption of the yttrium oxide by the inclusions of the second phase resulted in a decrease of the local concentration of Y_2O_3 to the limit at which both transitions $c \rightarrow t'$ and $t \rightarrow m$ take place.

Table 8.2. Temperature ranges of phase transitions in ZrO_2–R_2O_3 crystals (R – Eu, Gd, Yb)

Composition of crystal	Content of stabilizing oxide			Transition temperature heating			Transition temperature cooling
			Differential thermal analysis				
ZrO_2–Eu_2O_3		1		800			730
		2		670			600
		3		640			460
ZrO_2–Gd_2O_3		1		880			800
		2		650			520
		3		520			400
ZrO_2–Yb_2O_3		1		900			780
		2		620			470
			Dilatometric analysis				
	mol %		$m \rightarrow t$			$m \rightarrow t$	
		t_s, °C	t_f, °C	Δt	t_s, °C	t_f, °C	Δt
ZrO_2–Eu_2O_3	1	780	900	120	900	400	320
	2	640	740	120	620	300	320
	3	520	660	140	1460	200	200
ZrO_2–Gd_2O_3	1	890	900	–	900	800	–
	2	550	650	100	520	250	270
	3	480	600	120	400	280	120
ZrO_2–Yb_2O_3	1	820	900		780	500	280
	2	480	620	140	470	300	240

When examining the phase formation in the ZrO_2–Ln_2O_3 systems (where Ln = Y, Eu, Gd, Yb), quenched from the melt and crystalline specimens in the range of concentration of the stabilising agents 0–8mol% by the Raman scattering method in [22], the authors

Table 8.3. Characteristics of crystals of partially stabilized zirconia (PSZ)

Composition ZrO_2–3 mol %Y_2O_3–0.3 mol % CeO_2		Colour	Heat treatment after growth		Growth direction [*hkl*] of phase *T*	Phase composition		Parameters of phase *T*, Å		Size of mosaic blocks, nm	
Y_2O_3	CeO_2		*T* °C	τ, h		T	M	*a*	*c*	T	M
3	0.3	Red			111	57	43	5.091	5.167	200	115
3	0.3	Crimson			111	71	29	5.091	5.167	81	78
3	0.3	Gray	1600	2	111			5.105	1.152	260	>500
3		White			331	61	39	5.090	5.166	67	243
3		Black	2200	2	111			5.108	5.157	>500	263

proposed two mechanisms of phase formation, depending on the synthesis conditions:

1. The tetragonal T-phase forms in the volume of the initial cubic solid solution as a result of the high mobility of the ions in accordance with equilibrium diagram of the ZrO_2–Ln_2O_3 systems. In the nuclei of the T-phase, the concentration of the stabilising oxide should change rapidly in relation to the initial concentration and may greatly differ from it. The remaining volume is enriched with Ln_2O_3 which may stabilise the cubic structure of the solid solution up to room temperature. If the increase of the concentration of Ln_2O_3 cannot prevent the $c \rightarrow t$ phase transition, two tetragonal phases *t* and *t'* form in the specimens, and the impurity concentration in the *t'*-phase is higher than the initial concentration and is characterised by a low value of the tetragonal shift in comparison with the t phase. Thus, the t-phase forms in the initial stage, and the *t'*-phase forms with a further decrease in temperature. The probability of this mechanism of phase formation decreases when the temperature is reduced to 1000–1200°C, because a large decrease in the diffusion rate complicates the nucleation process of the T-phase.

2. The formation of phases at temperatures below 1200°C (i.e. below the temperature of the $c \rightarrow t$ phase transition for the concentration of the stabilising oxide of 0–10 mol%) results in the formation of domains of the tetragonal phase without participation of the diffusion processes. The concentration of the stabilising agent in this case is close to the initial concentration. The concentration of Ln_2O_3 and the dimensions of the domains may vary subsequently during annealing and depend on the temperature and duration of the process.

On the basis of the examined phase formation processes it may be concluded that both mechanisms operate in directional solidification of the melt in different stages of synthesis. For example, the mechanism of type 2 is controlling in growth from small volumes of the melt at high cooling rates of the crystals of ~400°C/min in the ZnO_2–Ln_2O_3 system. For the ZrO_2–(7–9)mol% Gd_2O_3 (Eu_2O_3) compositions, this mechanism results in the synthesis of transparent single crystals. The crystals contain *t*-domains, strictly oriented along the axis of the fourth order, with the concentration of the stabilising oxide of 2.5 mol% Gd_2O_3 (Eu_2O_3). These crystals are characterised by birefringence. This effect is not detected in the crystals of ZrO_2–Ln_2O_3 (Ln – Y, Yb, Lu) with the same concentration of Ln_2O_3, in which the T-phase disappears at the concentration of the stabilising oxide of ~6 mol%. This difference may be associated with different temperature ranges of existence of the phases in the crystals with different stabilising agents.

8.3. PRODUCTION OF CRYSTALS OF PARTIALLY STABILISED ZIRCONIA

The crystals were grown in equipment of series Kristall from large cold crucibles. The rate of growth of the single crystals was varied in the range (1–50) mm/h, the cooling rates, depending on the size of the crystal, were 4–250 °C/h [19].

For applications in practice, it is necessary to prepare large homogeneous single blocks of material without cracks. For this purpose, investigations were carried out into the effect of the initial composition and growth conditions on the physicomechanical properties and the size of the crystals. The range of the concentrations of the stabilising oxide (yttrium) characterised by the minimum number of cracks, formed after completion of crystal growth, was determined.

For example, at Y_2O_3 concentrations of ~0.5–2 mol%, the crystals were mat, oblique, with a large number of microcracks resulting from the tetragonal–monoclinic transition during cooling of the crystal after growth. Examination also showed the presence of a large number of macrocracks, formed as a result of thermal shocks during rapid stoppage of heating of the melt.

When the concentration of Y_2O_3 was increased from 2.5 to 5 mol%, the crystals were of the white milky colour, with the shiny or slightly mat surface; cracks formed mainly as a result of thermal shocks and growth residual stresses.

The increase of the yttrium oxide content above 5 mol% supported the formation of semi—transparent crystals with strong opalescence and a large number of growth cracks. Evidently, this is associated with an increase of the content of the cubic phase which does not have high strength characteristics. Inclusions of the resultant tetragonal phase caused high stresses in the cubic lattice.

At Y_2O_3 concentrations of (7.5–8) mol%, the number of cracks again decreased to the value typical of this method, the crystals were slightly opalescent, and examination in crossed polarisers showed the conoscopic pattern characteristic of strong stresses in the crystal (Chapter 6). Evidently, this is associated with a small content of the 2nd (tetragonal–pseudocubic) phase and the fact that the parameters of this phase were similar to those of the cubic lattice. The number of cracks was also affected by additional dopin with third impurities (Ce, Nd, Tb, Co, etc.).

The technological conditions of production of large crystals of partially stabilised zirconia were selected taking into account the dependence of the size of the crystals on the initial composition of the melt (Chapter 4). In the growth of crystals with low concentrations of the stabilising oxide, a decrease in the rate of crystallisation has almost no effect on the size of the crystals. To obtain large crystals, it is necessary to increase the distance between the solidification front and the front of the phase transformation. This was achieved by using a lower thermal screen which decreases the vertical temperature gradient in the growing crystal (Chapter 4). A second variant of technology is also possible in which the phase transformations take place only after completion of the growth process.

Large crystals of partially stabilised zirconia were grown using a seed crystal in the central part of the ingot. They showed no cracking during cooling. This technology was used to produce large single crystal blocks with a size of up to 40 mm in the cross-section and up to 120 mm long. The crystals of partially stabilised zirconia are presented in the insert.

Distribution of the stabilising oxide in the crystals of partially stabilised the zirconia

Many mechanical and physical properties of partially stabilised zirconia depend on the chemical composition of the material. The heterogeneity of the distribution of the components of the solid solution may result in the heterogeneity of the properties in the volume of the crystal. Therefore, the determination of the chemical composition and the distribution of the main and stabilising oxides along the length of the crystal are important characteristics of the material.

The composition was determined by x-ray spectrometric microanalysis. Figure 8.2 [19] shows the curves of distribution of yttrium oxide in the crystals of ZrO_2–(1.5–4)mol%Y_2O_3 along the growth axis.

There are several special features in the distribution of stabilising oxide:

–the characteristic decrease in the concentration of yttrium oxide towards the end of growth by 25–40%;

–the presence, on the distribution curves, of 'concentration waves' associated with the instability of the process of crystallisation, caused by concentration supercooling; the amplitude of these oscillations reaches 4–5% of the mean concentration in the crystal;

–the local (in areas ~100 μm in size) oscillations of the composition of the crystals with a low concentration of yttrium; at a concentration of 1.5 mol% Y_2O_3 in the initial charge, local deviations from the mean values may amount to ~25% and they are concentrated in the upper part of the crystal, corresponding to the final stage of growth (Fig. 8.2). Such deviation may be related to the phase transformations taking place during cooling of the crystals in the process of growth. At high temperatures of phase transformations (~2000°C) diffusion processes of the redistribution of yttrium between the cubic and tetragonal phases can occur in

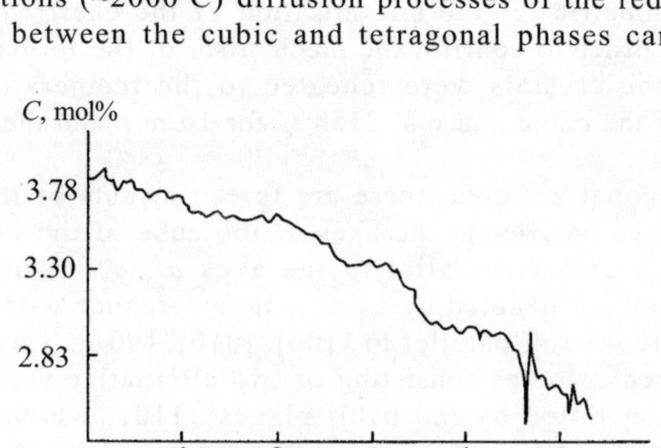

Fig. 8.2. Distribution of yttrium oxide in ZrO_2–3 mol.% Y_2O_3 crystals.

the solid phase. At low yttrium concentrations, the distribution coefficient is higher than unity and, consequently, the upper part of the crystal is depleted in this component and this increases the temperature of phase transformations. A decrease in the yttrium concentration is also supported by the presence in the melt of several impurities with a high chemical affinity for the yttrium oxide, such as: SiO_2, Al_2O_3, TiO_2, etc.

8.4. THE DOMAIN STRUCTURE OF CRYSTALS OF PARTIALLY STABILISED ZIRCONIA

The transition of the cubic to the tetragonal phase is accompanied by the formation of the domain structure in the crystal [23, 24]. The structure of the resultant domains are determined by the symmetry of the interacting phases, the initial cubic and ferroelastic tetragonal phases (*t*'). The domains in the crystals with 3 mol% M_2O_3 are distributed along the <001> direction in the {001} habit plane and their minimum size is 50 nm [23].

A detailed examination of the domain structure of the tetragonal phase was carried out in [25]. The results show that the individual domains form colonies distributed in the irregular structures in the form of 'fir trees' [26]. The spatial distribution of the colonies of the domains in tetragonal zirconia differ from the distribution in the partially stabilised zirconia where they are implanted in the uniform cubic matrix [27]. In the *t*'-phase of zirconia, the domains fill the entire volume and the spatial distribution of the domains should be highly symmetric for the minimisation of the energy of coherent strains. In order to confirm the mechanism of the formation of the *t*' phase, the crystals were reheated to the temperatures of the stability of the cubic phase at 2150°C for 10 min and then quenched [25].

In tetragonal zirconia, there are three variants of the domains with the axes parallel to the axes of the cube of the initial phase [28]. The c axis is parallel to the axes a_1, a_2, a_3 of the cubic matrix. They are denoted by t_1, t_2, t_3 in accordance with the c axes of the cubic matrix, parallel to [100], [010], [001]. These domains form ordered colonies consisting of two alternative variants of the domains, separated by the habit planes {110}, which are at the same time the twinning planes of both variants. Each colony consists of the pairs t_1/t_2, t_1/t_3 or t_2/t_3. The appropriate designation of these colonies is c_1, c_2, c_3. For each pair (for example t_1/t_2) there are

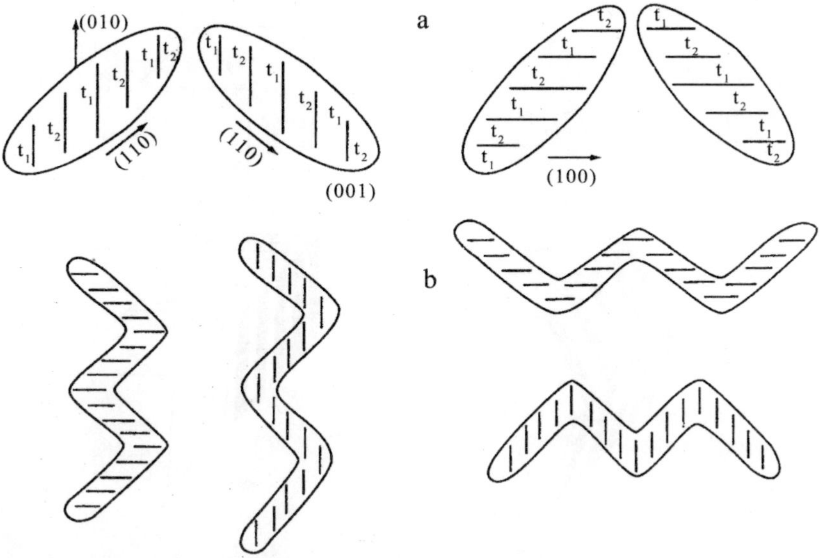

Fig. 8.3. Scheme of four variants of colonies in the (001) plane formed by domains t_1 and t_2 (a) and variants of 'fibres' of inclusions formed by four variants of colonies (b) [28].

four possible variants of colonies, including two twinning planes {110}, as shown in Fig. 8.3a.

Each pair of the domains in combination with two types of the twinning planes may form four types of colonies. Six twinning planes with three variants of the domains form 12 types of colonies.

In the *t*'-material, the colonies occupy the entire volume, are close to each other and form a regular structure. All colonies are arranged in the same manner, namely, they border with each other only by the {110} planes and are elongated along the <111> directions which are lines of intersection between the boundaries of the colonies. Three external planes of the colonies are situated under an angle of 60°. In all cases, one type of the domains of the colony at the boundary borders with approximately the same type of colony (for example, t_1), whereas other two types of domains (for example, t_2, t_3) border on the {110} plane which is the twinning plane (Fig. 8.4).

Different types of the structure of the colonies in the cross sections are produced by the single microstructure. The colonies may be connected along the <111> edge, completely filling the volume of the crystal. This geometry is shown in Fig. 8.5 (1). For the single variant of the domain structure to extend to the adjacent colony, and to ensure that in the following layer the domains border

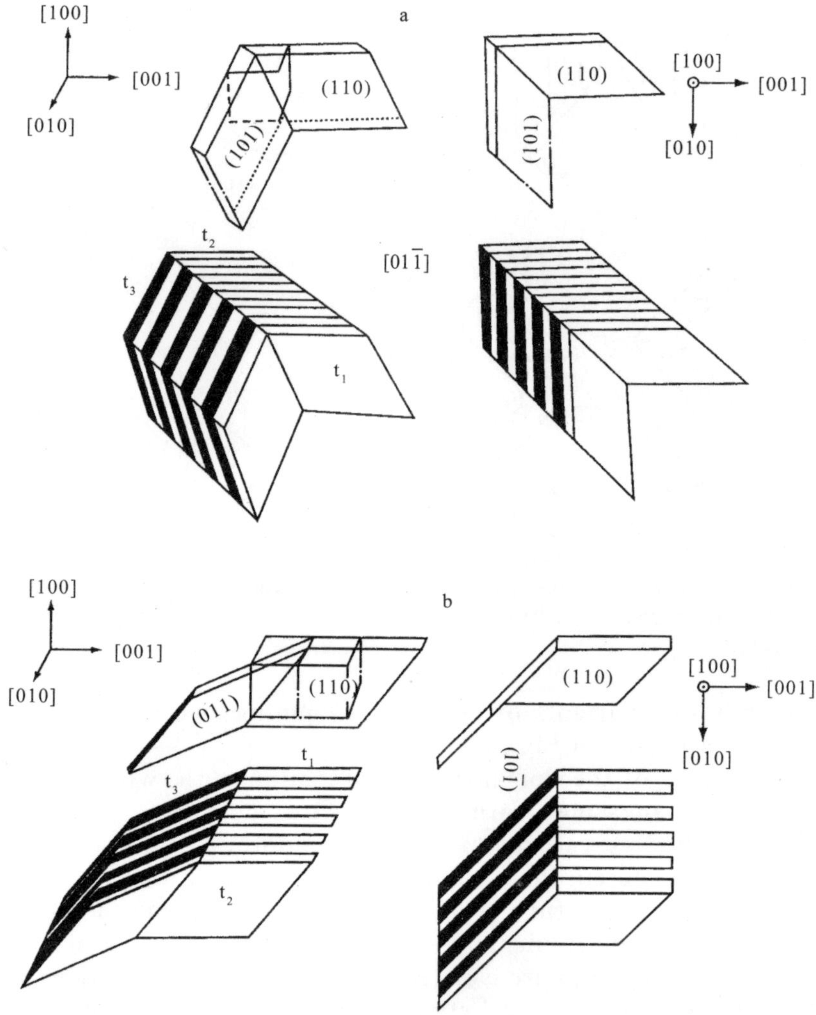

Fig. 8.4. Morphology of three variants of plate-shaped domains inclined in relation to the longitudinal axis of the colonies [28].

with the twinning planes and form the three-dimensional structure, the colonies should be linked in a unique manner. Analysis of the structure shows that three colonies are not in accurate contact along the <111> direction, and, as shown in Fig. 8.5 (2) *c*, *d*, the colonies c_1 and c_3 are shifted in relation to c_2 in the central region. This shift is caused by the shift of the domain t_3 of colony c_3, in the upward direction. If this shift is equal to the domain distance in the direction [111], the domain t_3 in c_3 of the lower level, Fig.

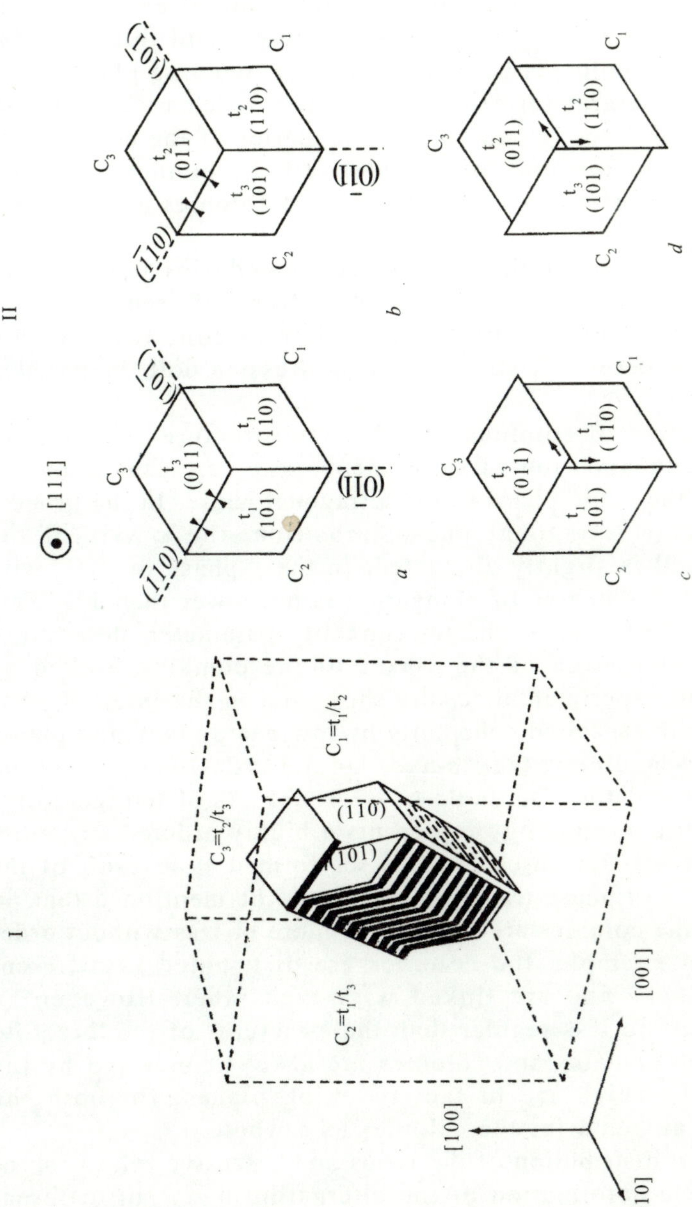

Fig. 8.5. Filling of the volume of a tetragonal crystal with domains [29].

8.5 (2)*c*, meets t_3 from c_2 of the upper level, Fig. 8.5 (2)*c*, and forms a helical structure around the longitudinal axis [111]. In subsequent rotation, t_2 from c_3 meets t_1 from c_3. In this case, the boundary of the colonies coincides with the twinning plane (110). The helical structural formations are shown schematically in Fig. 8.5 (1). Low stresses remain at the boundaries of the colonies. In accordance with the proposed model, all the boundary planes, namely the boundaries of the domains and colonies are twinning planes.

The examination of thin anti-phase boundaries between the domains shows that they form arbitrarily from different minimum domains, generating the structure similar to colonies but on a different scale because a structure of the Russian doll 'matryshka' [29].

The formation of colonies and their substructure provides for optimum accommodation of spontaneous stresses, caused by the transition of the cubic phase to the tetragonal phase. In the process of the cubic to tetragonal phase transition, the *c* axis of the elementary cell is slightly elongated. In the *t*'-phase of ZrO_2 with 3 mol% Y_2O_3, the degree of elongation is not lower than 1%. This elongation, referred to as the tetragonality parameter, determines the mutual orientation of the *c*-axes of the domains, linked by twinning. The experimental results show that in the internal parts the domains are separated coherently by low-energy twinning planes {110}. Correspondingly, their *c*-axes are not orthogonal in relation to each other and form an angle of 89.4° for 1% of tetragonality.

This specific morphology represents a highly ordered structure, which effectively decreases the stresses formed as a result of the ferroelastic $c \rightarrow t$ phase transition. It should be mentioned that, in many cases, the colonies are distributed more or less without order in space. For example, the colonies are distributed in different directions <111> and are linked with each other. However, in structures that are less regular than the structures of the threefold symmetry, the domains and colonies are always connected by the {110} planes, which are in fact twinning planes, for both the adjacent domains and for the colonies as a whole.

This mutual distribution of the twins and extensive relaxation of the ferroelastic deformation of the alternating layers of different domains and colonies are the basis for the coherence of large macroscopic polydomain crystals.

As mentioned previously, if the concentration of the stabilising oxide differs from 3 mol% then the crystals contain several phases. For example, it was shown in [30] that crystals with the composition 2.56–6.38 mol% Y_2O_3, grown by directional solidification of the melt, represent a matrix — a cubic solid solution with an inclusion of the tetragonal phase whose volume increases with a decrease in the content of yttrium oxide: 30 vol% at 4.2 mol% of Y_2O_3 and to 50 vol% at 2.56 mol% Y_2O_3.

The authors of [31] investigated specimens containing 4.5 mol% of Y_2O_3, produced by crystallisation in the cold crucible, which consisted of the t'-ZrO_2 (matrix) with a small number of disc-shaped inclusions of t-ZrO_2 (depleted in the yttrium oxide). These inclusions form during the period of cooling of the crystals after growth. In the course of additional annealing at 1600°C, the matrix continues diffusion separation. This is accompanied by the formation of the inclusions of t-ZrO_2 with a concentration of the order of 2 mol% Y_2O_3, whereas the c-ZrO_2 matrix is gradually enriched with the yttrium oxide to the equilibrium concentration of ~7 mol%. In subsequent cooling, the matrix transforms to t'-ZrO_2. The temperature of the $c \rightarrow t'$ transition differs for different authors [7, 32, 33], possibly as a result of complicated measurements. It is assumed that $T_{c-t} < 200°C$ for the composition ZrO_2–3 mol%Y_2O_3.

The development of the structure during annealing at 1600°C is determined by the minimum of the deformation energy in the formation of inclusions and by the energy of interaction between the inclusion and the matrix. The inclusions of t'–ZrO_2 form a structure of plate-shaped 'colonies', where each colony consists of two twinned t-ZrO_2 variants with the c-axis close to 90° and separated by coherent and low-energy habit planes {101}. This specific structure of the inclusions, which effectively minimises the energy of the stresses associated with the formation of inclusions, was investigated by high-resolution electron microscopy [13, 29, 34, 35]. During annealing, the dimensions of the colonies and the space occupied by the plates increased at a rate dependent on the diffusion kinetics. Examination of the annealing process showed that the kinetics of enlargement is controlled by the conventional diffusion-controlled ageing. In this case, the size of the plates increased as a function of $(t_{ann})^3$. The mean size of the plates after annealing at 1600°C for 2–150 hours changed from 15–20 nm to 60–70 nm.

The microstructure of the crystals with the composition ZrO_2–4.5 mol%Y_2O_3 may be represented as a mutually penetrating grid of the fibres with a mean radius of 300–500 nm and with a length

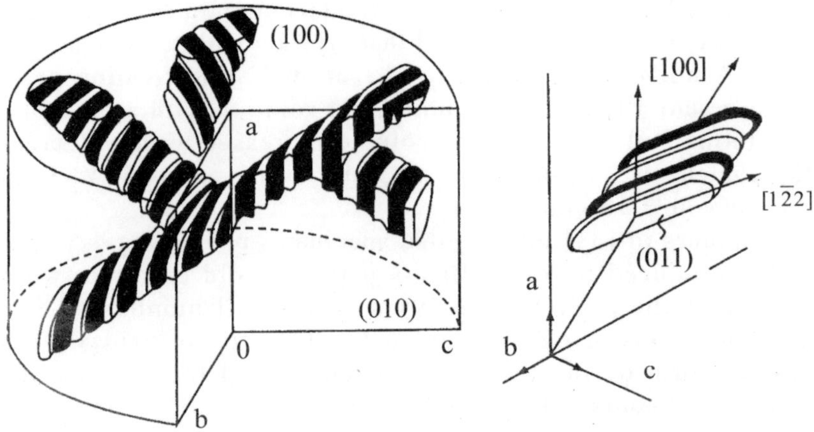

Fig. 8.6. Structural model of inclusions in a crystal of partially stabilised zirconia [36].

of several microns, elongated along the three directions <100>. Each fibre consists of continuously bending colonies in relation to the {110} planes.

A similar examination of the structure of the crystals with the composition ZrO_2–4.5 mol.% Y_2O_3, grown by the method of the floating zone was carried out in [36]. The crystals were grown on a sead crystal oriented in the [100] direction. After growth, the crystals were annealed at 1700°C. Analysis of the experimental results showed the three-dimensional structural model of the inclusions. The main structural unit of the inclusions is a thin sheet with a tetragonal symmetry and approximate dimensions of 3 × 0.5 × 0.07 μm. The structural model is presented in Fig. 8.6.

The concentration of yttrium oxide in the inclusions and the matrix was 1.9 and 4.7 mol%, respectively. The volume of the fractions of the inclusions in the crystal was approximately 70%. This value is in agreement with the estimate obtained from the equilibrium diagram.

The crystals grown in the [100] direction by the floating zone method have a simpler microstructure. The method is characterised by high temperatures gradient along the axis of growth, and in the radial direction, the gradient has the axial symmetry in relation to the growth axis. All the *c*-axes of the tetragonal inclusions were distributed in the plane normal to the growth direction. There were no tetragonal inclusions with the *c* axis parallel to the axis of growth. The field of stresses, generated by these gradients, probably decreases the number of variants of the microstructure.

The cubic–tetragonal phase transition affects the stress field in the crystal. The main axis of the tetragonal phase can rotate in relation to the axis of the cubic matrix in order to decrease the energy of elastic stresses, associated with the transition. The angle of rotation is 0.48° in relation to the tetragonal axis *a* which is parallel, with the limit of the experimental error, to the growth direction [34]. The direction of growth also corresponds to the axis *a* of the cubic prototype.

8.5. THE MICROSTRUCTURE OF PARTIALLY STABILISED ZIRCONIA CRYSTALS

It is well known that the properties of the crystals, including mechanical properties, are closely linked with the real structure of the crystals. The literature contains a large number of studies concerned with the examination of the microstructure of sintered polycrystalline and single crystals of partially stabilised zirconia. Like many other characteristics of this material, the microstructure depends strongly on the composition, synthesis conditions and subsequent treatment conditions. In this chapter, the microstructures of these materials have already been examined. The experimental results show that the material is characterised by a nanosized domain twinned structure. If the specimens consist only of the non-transformed tetragonal phase, the domains with the distribution close to orthogonal form a pseudo-cubic crystal. A matrix phase with inclusion forms in the case of the multiphase composition of the crystal, Because of the small size of the domains and the multilayer nature of the structure, the structure of this type is difficult to examine. Relatively detailed investigations can be carried out only on oriented crystals of partially oriented zirconia using high-resolution transmission electron microscopy. Information on the structure, in particular, on the size of the domains, is also provided by x-ray diffraction analysis.

The morphology of the crystal surface

It is well-known that the morphology of the crystal depends on the crystal structure of the given substance and reflects both the solidification conditions and the conditions of subsequent stages of formation of the actual structure of the crystal.

The authors of [19] investigated the morphology of crystals of partially stabilised zirconia by scanning and optical electron microscopy.

In the selection of the specimens for the investigations, special attention was given to the nature of temperature distribution in the diameter of the cold crucible, associated with the special features of the generation of heat in the melt, placed in the high frequency field. This distribution resulted in different conditions of growth of the crystals in the central zone, in the zone adjacent to the wall of the crucible and in the intermediate zone. The cooling conditions are different for dirrerent positions of the crystal in the volume of the melt, cspecially in the case of the crystals of partially stabilised zirconia from the viewpoint of the formation of its phase composition and microstructure as a result of phase transformations in the solid phase. The smaller the crucible diameter the stronger the conditions of formation of the actual structure of the crystal depend on its position inside the cold crucible. Since this material is of considerable interest for the practical application of the material, the crystals for investigations were grown in large diameter crucibles (> 200 mm). For comparison, investigations were also carried out on blocks taken from the central, intermediate and near-wall parts of a crystal boule.

The examination of the surface of the crystals after growth showed that the crystal morphology depends on the concentration of the stabilising oxide, the presence of third impurities and growth conditions. As an example, Fig. 8.7 shows the growth surface of crystals with different concentrations of the stabilising oxide.

The surface of the growth faces of the specimens of ZrO_2–0.5 mol%Y_2O_3 after the tetragonal–monoclinic transition shows the typical pattern of twinning of the monoclinic phase with a distinctive irregular network of microcracks at the boundaries of the regions (Fig. 8.7*a*). The growth of surfaces (Fig. 8.7 *b,c*) [19] of the ZrO_2–(1.0–2.5) mol%Y_2O_3 specimens shows clearly microcracks, and the structure contains elements of both the previous specimen and the resultant 'tweed' pattern (considerably finer), characteristic of high yttrium oxide concentrations. For the most relevant region of the yttrium oxide concentration of 2.5–4 mol%, examination showed two characteristic types (type 1 and 2) of the structure of the growth surface (Fig. 8.8) [19].

The surface of any of these types is characteristic either of all faces of the crystal block or of only several faces, or only of a part

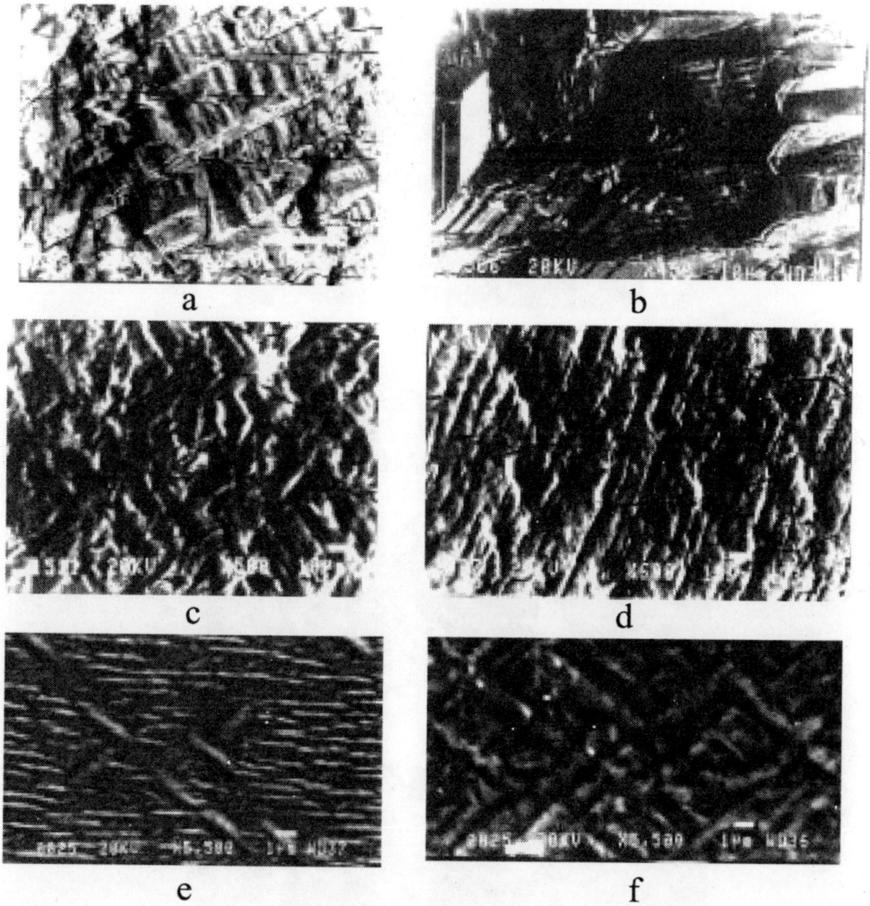

Fig. 8.7. Micrographs (a–c) and schemes (d) of three main types of deformation of the surface typical of crystals $(1-x)ZrO_2 \cdot x\ Y_2O_3$, x = 0.0005 (type I) (a); x = 0.010 (type II) (b); x = 0.015 (type III) (c). Magnification: a) 850; b) 500; c)1000.

of one of the faces. The surface of the first type (smooth) is shown in Fig. 8.8*a*, and represents a regular structure similar to the 'tweed' structure. The structure consists of fine (length 1–10 mm, width 0.5–3 μm) parallel 'rods', distributed in groups. The groups themselves are oriented in relation to each other either under the right angle or under an angle close to 45°.

The structure of the surface of the second type (Fig. 8.8b) is characterised by a more pronounced relief and consists of larger convex 'rods', distributed in parallel (parallelepipeds), with elements

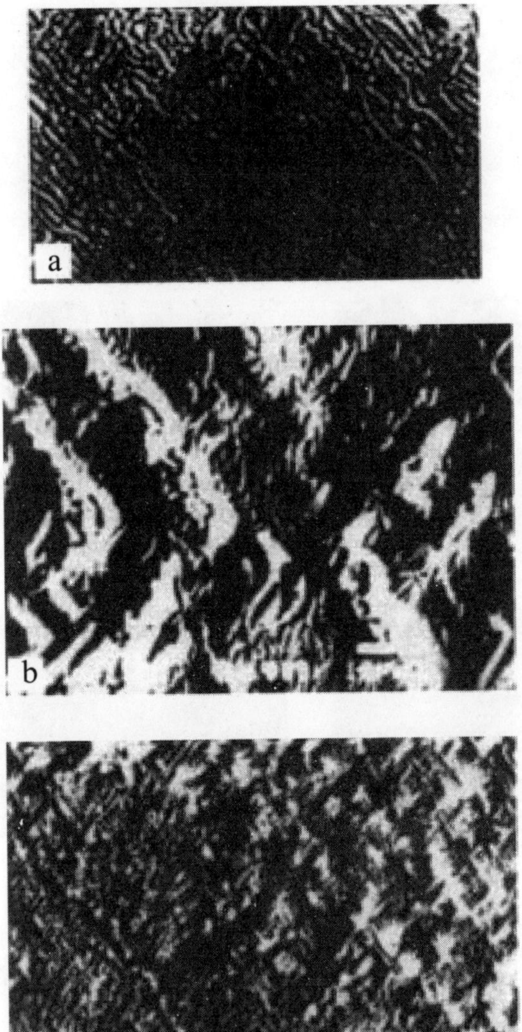

Fig. 8.8. Surace of growth faces of ZrO_2–2.5–4 mol% Y_2O_3 crystals [19]. a: type I – smooth faces; b: type 2 – matt faces; c: boundary of areas with smooth and matt faces.

of up to 50 μm long and more than 5 μm wide. The relief of the structure results in the mat appearance of the faces.

Similar surfaces are often found in the central and lower areas of the grown crystals, and the appearance of these surfaces is determined by the volume of the melt and the length of the crystals.

In this case, the single crystalline blocks often grow into each other and cannot be separated without breaking up the crystal themselves. The nature of the surface is also affected by thermal postgrowth annealing which determines the phase composition and formation of the microstructure.

It is important to note the effect of certain admixtures on the nature of the growth surface of the crystals. With the introduction of the admixtures Tb, Ce, Nb into the crystals of partially stabilised zirconia with an yttrium concentration of approximately 3 mol%, the characteristic appearance of the smooth faces is retained but the dimensions of the structural elements ('rods') greatly decrease. The nature of the structure of the mat faces varies greatly: the structure is characterised by intersections under an angle of 45° with the almost complete absence of perpendicular intersections. The surfaces are characterised by the presence of defects in the form of depressions resembling cracks. The relationship of the microstructure with the mechanical properties of these crystals will be discussed later.

The mat crystals grow when the initial melt contains a relatively large amount of admixtures, with the distribution coefficient considerably smaller than 1. However, in contrast to the pattern discussed previously, the surfaces of these crystals contain a characteristic deposit of the admixtures displaced by the crystal, which is clearly visible in the electron microscope, with the composition identified by the method of x-ray spectral analysis.

In most cases, the microstructure and the defects of partially stabilised zirconia are determined by the method of selective chemical etching [100, 101]. The rate of dissolution depends on the nature of the crystal, the crystallographic orientation of the surface, the presence of defects in the crystal, the composition of the etching agent and the etching conditions. These factors determine the shape of etch pits, the profile of the dissolved surface and also the rate of etching in the areas with defects. For water-insoluble crystals, such as the solid solutions of zirconia, suitable etching agents are acids and alkali.

Hot phosphoric acid is a contrast etching agent for partially stabilised zirconia so that it is possible to examine the microstructure of specimens in a scanning electron microscope.

Etching agents for the detection of the microstructure of the crystals of partially stabilised zirconia include solutions of HF, KOH, H_3PO_4. The results of action of these etching agents are

Table 8.4. Etching agents for PSZ crystals, etching time and temperature

Composition of solid solution crystal	Etching agent	Etching temperature and time
PSZ	Concentrated HF	20 °C, 5 min.
ZrO_2–9.4 mol% Y_2O_3	KOH	500–550 °C, 10–15 min
ZrO_2–2–14 mol% Y_2O_3	H_3PO_4	100–300 °C
PSZ	HF	20 °C, 4 min

presented in Table 8.4 [19]. For selective etching, the best results have been obtained using phosphoric acid. It should be mentioned that after etching a deposit appears on the surface and it can be removed by the effect of HF acid during period of 40 s at room temperature.

The microstructure of the specimens was developed by etching crystals of partially stabilise zirconia in phosphoric acid at a temperature of 280°C for 10–30 min. Figure 8.9 shows the microstructure developed by chemical etching: in the volume of the crystal (a,b), in the cleavage plane (c) and on the growth surface (d). An increase in temperature to 310°C and an increase in etching time result in the formation of chemically polished areas on the surface of the specimens.

When discussing the hardening mechanisms, it was noted that significant factors affecting the strength characteristics of partially stabilised zirconia are the dimensions and morphology of the domains. It has also been noted that these characteristics of partially stabilised zirconia depend on the nature, composition of the solid solutiona and the synthesis and heat treatment conditions.

In order to determine the disorientation of the crystal block, the structure was examined in a two-crystal x-ray diffractometer by the 'tilting' method. The size of the region of coherent scattering was determined on the basis of the broadening of the diffraction maxima: (111), (221) for m- and (111), (222) for the t-phases. The results of the experiments are presented in Table 8.5 [22].

The presented results show that the crystals are characterised by a nanodomain crystalline structure. The dimensions and orientation of the domains depend on the composition, growth and heat treatment conditions.

Examination of the microstructure of the crystals by scanning

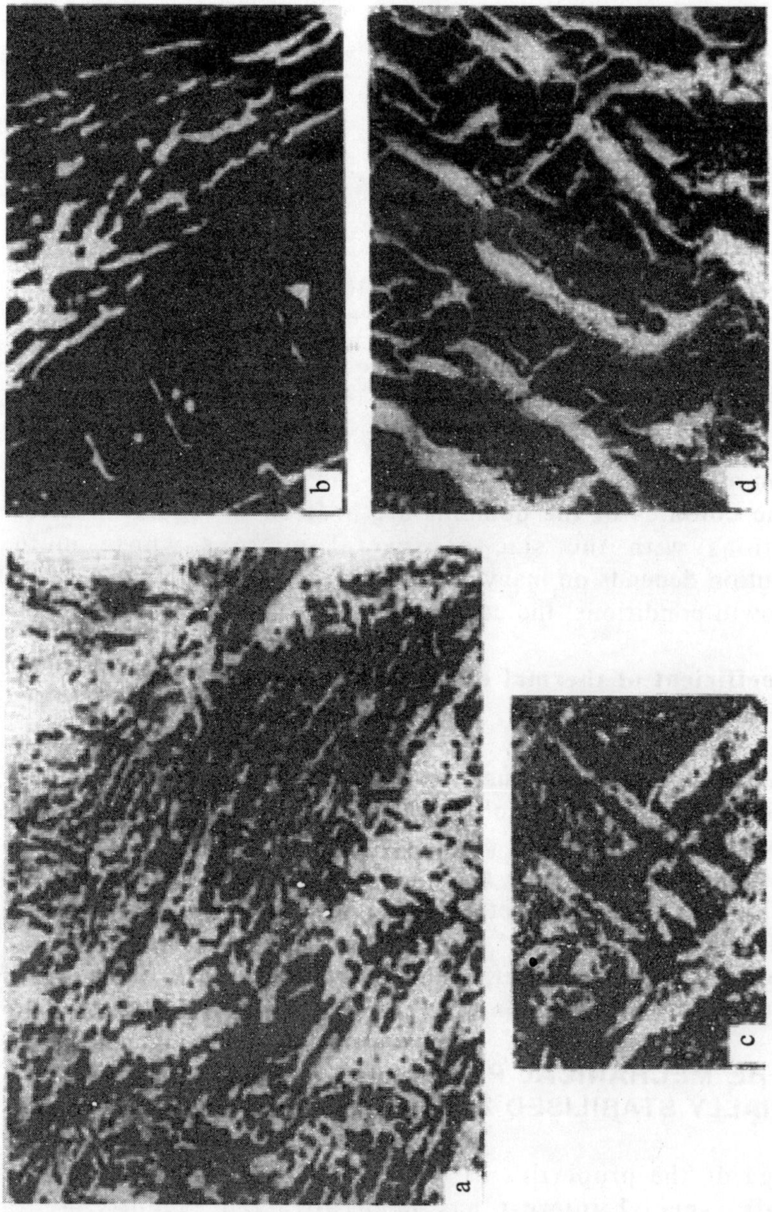

Fig. 8.9. Microstructure of crystals of partially stabilised zirconia in the temperature range 200–900°C

Table 8.5. Crystallographic characteristics of PSZ materials [22]

Composition of PSZ crystal	*T*, °C	Growth direction	α, deg	*D*, nm
ZrO_2–3 mol% Y_2O_3	–	[331]	1–2	6.7
ZrO_2–2.5 mol% Y_2O_3–0.2 mol% Ce_2O_3		[111]	2–8	20.0
ZrO_2–3.5 mol% Y_2O_3–0.1 mol% Ce_2O_3		[111]	2–8	8.1
ZrO_2–3 mol% Y_2O_3	2100	[111]	1–2	50.0
ZrO_2–3 mol% Y_2O_3–0.15 mol% Ce_2O_3	2100	[111]	1–2	26.0

Comment. T is the temperature of postgrowth vacuum treatment (10^{-3}–10^{-4} mmHg for 2–10 h); α is the mean misorientation angle in relation to the growth direction; *D* is the size of regions of coherent scattering.

electron microscopy, chemical etching, and optical microscopy shows that the colonies of the domains are ordered in regular structural formations with the size of several microns whose mutual distribution depends on many factors: the composition of the crystal, the growth conditions, the cooling conditions, thermal stresses, etc.

The coefficient of thermal expansion of the crystals of partially stabilised zirconia

The majority of the previously examined characteristics of partially stabilised zirconia were obtained on polycrystalline materials. However, the development of the technology of synthesis of single crystal materials from the melt has expanded the possibilities of examination of the structure, nature and properties of these materials.

Table 8.6 shows the coefficients of thermal expansion of crystal of partially stabilised zirconia.

8.6. THE MECHANICAL PROPERTIES OF CRYSTALS OF PARTIALLY STABILISED ZIRCONIA

As regards the properties of the crystals of partially stabilised zirconia, special interest has been attracted recently by the mechanical properties. This is associated with the high strength characteristics of these crystals, in particular, with high cracking resistance of the material (with high fracture toughness). Tables 8.6

Table 8.6 Coefficients of thermal expansion of PSZ crystals in the temperature range 200–900°C

Composition of crystal		l_0, nm	$\alpha \cdot 10^6$ deg^{-1}							
R_2O_3	mol%		200°	300°	400°	500°	600°	700°	800°	900°
Eu_2O_3	1	15.05	6.10	6.80	7.13	7.11	6.78	5.7	–6.41	–13.8
	2	47.00	7.38	8.05	8.44	7.63	5.72	–27.4	–28.3	–23.8
	3	12.2	7.49	8.94	8.05	6.38	-1.61	–0.92	0.55	1.8
	5	17.4	8.32	8.94	9.23	9.51	9.69	9.81	9.84	10.04
Gd_2O_3	1	19.00	7.78	8.23	8.59	8.68	8.66	8.64	8.63	8.28
	2	42.2	6.1	6.62	6.02	4.74	0.21	–9.33	–6.79	–4.90
	3	21.9	9.16	9.66	9.89	9.92	6.92	8.74	7.85	8.28
	5	29.15	9.16	9.48	9.63	9.72	10.03	10.42	10.42	1.55
	8	32.8	8.6	9.12	9.23	9.40	9.68	9.96	10.04	0.20
Yb_2O_3	1	18.8	9.4	9.85	10.0	9.7	9.7	9.6	8.65	–21.5
	2	37.00	8.13	13.45	11.89	5.12	–7.6	–7.75	–7.1	–5.21
	3	21.1	9.44	10.01	10.02	9.61	8.82	8.61	9.08	9.36
	5	13.7	9.99	10.37	10.68	10.72	10.81	10.84	10.87	10.78

[19] and 8.7 *a,b* [30] show the coefficients of thermal expansion of the crystals of partiallt stabilised zirconia. Table 8.7 *a,b* gives several physical and mechanical properties of the crystals of partially stabilised zirconia (a) and of sintered ceramics of the same composition (b).

The hardness of different compositions decreases with an increase of the content of the tetragonal and monoclinic phases, and catastrophically decreases with an increase of the load on the indentor applied to the crystal. The decrease of hardness with increasing content suggests that the inclusions of the phase reduce the stresses in the vicinity of the indentor due to the phase transition and twinning [37]. The hardness of Mg-FSZ and Ca-FSZ is 14.4 and 17.2 GPa, respectively, and in the case of Y-FSZ it is 13.6 GPa. Fracture toughness was maximum at 2.56 mol% of Y_2O_3. Fracture toughness increases with increasing volume of the fraction of the tetragonal inclusions, i.e. with a decrease in the content of the stabilising oxide.

Table 8.7. Comparison on physical and mechanical properties of PSZ crystals and sintered ceramics of the same composition based on ZrO_2 [30]

(a)

Crystal	Composition, wt%	Inclusions, vol%	Density, kg/m³	Hardness, GPa	Young modulus, GPa	Bending strength, MPa	Fracture toughness, MPa·m$^{1/2}$
Mg-FSZ	2.8	48	5790	14.4	200	685±48	4.82±0.56
Ca-FSZ	4.0	38	5850	17.1	210	661±33	3.97±0.46
Ca-PSZ	9	0	5680	17.2	210	241±28	2.54±0.13
Y-FSZ	5	52	6080	13.6	233	1384±80	6.92±0.14
Y-PSZ	20	0	5910	16.1	233	356±58	1.91±0.16

(b)

Polycrystalline ZrO_2	Young modulus, GPa	Bending strength, MPa	Fracture toughness, MPa·m$^{1/2}$
Mg-FSZ	200	430–700	4.7–15
Ca-FSZ	200–217	400–650	5.0–9.6
Y-FSZ	210–238	696–980	5.8–9.0

Comment. Fully (FSZ) and partially (PSZ) stabilized zirconia

In heating the crystals of Y-FSZ and Y-PSZ, the strength and fracture toughness decrease with increasing temperature. However, the fracture toughness of Y-FSZ material is approximately twice that of Y-PSZ [38].

The elastic properties

The mechanical properties of the material, especially cracking resistance and fracture under loading, are determined to a large extent by the elastic properties of the material. Table 8.8 [39] shows the elastic constants of the single crystals of solid solutions based on zirconia and the Zener calculated ratio: $Z = Cu_{44}/(C_{11}-C_{12})$, which determines the degree of anisotropy of the elastic properties in the cubic crystals (for $Z = 1$ the material is isotropic). The value of this ratio also determines the direction of the extreme value of the Young modulus: for example, at $Z < 1$, the maximum value of the Young modulus is oriented in the direction <100>, at $Z > 1$ in the direction <111> [39].

Depending on the concentration of yttrium oxide in the entire range of composition, including the multiphase compositions, constant C_{11} monotonically decreases, C_{44} monotonically increases,

Table 8.8. Elasticity constants of single crystals of solid solutions of PSZ

Y_2O_3 mol%	Density, kg/m³	C_{11}, GPa	C_{12}, GPa	C_{44}, GPa	Zener ratio $Z = C_{44}/(C_{11-}C_{12})$
20.00	5760	373.0	123.0	65.0	0.520
	5760	372.0	117.0	68.0	0.533
17.90	5826	390.4	110.8	69.1	0.494
16.50	5810	381.0	80.0	63.0	0.419
	5810	390.0	96.0	61.0	0.415
		370.0	106.0	67.0	0.508
		380.0	74.0	62.0	0.405
15.50	5868	397.6	108.6	65.8	0.455
12.10	5932	405.1	105.3	61.8	0.412
12.00	5890	403.0	100.0	57.0	0.376
		385.0	108.0	59.0	0.426
		391.0	71.0	60.0	0.375
		388.0	70.0	56.0	0.352
	5890	391.0	120.0	60.0	0.443
	5910	403.0	100.0	46.0	0.304
	5910	400.0	113.0	56.0	0.397
		413.0	115.0	61.0	0.409
	5908	404.0		62.0	
		413.0	112.0	61.0	0.405
11.10	5945	403.5	102.4	59.9	0.398
10.30	5910	403.0	83.0	58.0	0.363
10.00	5950	403.0	103.0	58.0	0.387
	5950	395.0	117.0	61.0	0.439
9.00	5980	380.0	130.0	59.0	0.472

Table 8.8. continued

Y_2O_3 mol%	Density, kg/m³	C_{11}, GPa	C_{12}, GPa	C_{44}, GPa	Zener ratio $Z = C_{44}/(C_{11}-C_{12})$
20.00	5760	373.0	123.0	65.0	0.520
	5760	372.0	117.0	68.0	0.533
17.90	5826	390.4	110.8	69.1	0.494
16.50	5810	381.0	80.0	63.0	0.419
	5810	390.0	96.0	61.0	0.415
		370.0	106.0	67.0	0.508
		380.0	74.0	62.0	0.405
15.50	5868	397.6	108.6	65.8	0.455
12.10	5932	405.1	105.3	61.8	0.412
12.00	5890	403.0	100.0	57.0	0.376
		385.0	108.0	59.0	0.426
		391.0	71.0	60.0	0.375
		388.0	70.0	56.0	0.352
	5890	391.0	120.0	60.0	0.443
	5910	403.0	100.0	46.0	0.304
	5910	400.0	113.0	56.0	0.397
		413.0	115.0	61.0	0.409
	5908	404.0		62.0	
		413.0	112.0	61.0	0.405
11.10	5945	403.5	102.4	59.9	0.398
10.30	5910	403.0	83.0	58.0	0.363
10.00	5950	403.0	103.0	58.0	0.387
	5950	395.0	117.0	61.0	0.439
9.00	5980	380.0	130.0	59.0	0.472
		475.0	144.0	61.0	0.369
8.50	5960	430.0	94.0	64.0	0.381
8.10	6011	401.8	95.2	55.0	0.359
8.00	5990	394.0	91.0	56.0	0.370
	6010	410.0	90.0	53.0	0.331
		415.0	119.0	57.0	0.385
6.90	6008	405.0	110.0	58.0	0.393
		410.0	103.0	58.0	0.378
3.90	6049	412.0		60.0	
		412.0	110.0	55.0	0.364
3.40	6075	406.0	105.0	55.0	0.365
		407.0	101.0	55.0	0.359
2.80	6085	425.0		56.0	
		410.0	125.0	52.0	0.365
		410.0		52.0	
		414.0		56.0	
2.20	6044	390.5		64.4	
1.7	5842	365.4		73.6	

and C_{12} remains constant. Direction [100] corresponds to the high value of the modulus $E \sim 360$ GPa, direction {111}to the minimum value of $E \sim 170$ GPa [36].

As already mentioned, the crystals of partially stabilised zirconia are characterised by high strength properties which, in combination with chemical inertia and high melting point, enable these materials to be used in different areas of science and technology. Many materials in the single crystal form and having the macroscopic dimensions have been obtained for the first time. The mechanical properties of partially stabilised zirconia depend greatly on the conditions of synthesis and subsequent heat treatment. Consequently, it is necessary to estimate the effect on the mechanical properties of the technological factors, composition and nature of the stabilising oxide. Investigations were carried out into various characteristics of the materials, such as: microhardness, cracking resistance, compression strength, bending strength, the elastic properties of crystals, the temperature dependences of certain mechanical properties.

In most cases, the mechanical characteristics of low-formability materials, such as ceramics, are determined by the bend test with a relatively simple loading procedure which has been substantiated theoretically and tested in practice [40–53].

The method is based on recording the deformation diagram in loading of the specimen and determination, on the basis of these diagrams, of ultimate tensile strength, limiting deformation, the static modulus of elasticity and many other characteristics of the material. The deformation diagrams reflect the relationship between the stresses and strains, formed in the specimen during loading. Pure bending is preferred, but the method of bending the specimen with the application of a force concentrated in the centre of the specimen (three-point bending) is also used widely.

The measurements of specimens were taken in [45–53]. The specimens were in the form of prisms, with the dimensions 3.5 × 5 × 50 mm. The dynamic modulus of elasticity (E_d) was determined from the values of the longitudinal speed of sound in specimens, the static modulus of elasticity (E_s) was determined on the basis of the dependences of the deflection of the load, applied to the specimen (0.5–0.15 of the limiting load. The ultimate strength was calculated in accordance with the classic Nadai equation [54], establishing the relationship between the stresses and strains in the process of loading of specimens.

The condition for the growth of brittle cracks is the violation of equilibrium between the released energy of elastic deformation and the increment of total surface energy. The brittle strength of the elements with a crack is inversely proportional to $l^{1/2}$, where l is the half length of the crack.

The measure of brittleness χ is the energetic measure of the deviation from the linear stress–strain dependence and is calculated from the results of determination of the strength of single crystals in four-point bending. The stress intensity coefficients, characterising the crack resistance of the material, were determined in both the method of bending of prisms with a notch and by the microindentation method. The *R*-curves were plotted in cyclic loading of specimens with a sharp crack [55].

Investigations were carried out on crystals of partially stabilised zirconia and, for comparison, on single-phase crystals with a cubic structure. Specimens were cut out in the direction parallel to the axis of growth, which was close to the direction <101> for the cubic single crystals, and to <111> for crystals of partially stabilised zirconia. The anisotropy of the mechanical properties was investigated on the specimens oriented along the specific crystallographic directions. The compression strength characteristics of the specimens were determined in an 1251 Instron machine (Great Britain). Tests were carried out on cylindrical specimens with a diameter of 10 + 0.5 mm and 10 mm long.

Microhardness was determined by a microhardness meter with a diamond Vickers pyramid as an indentor and was calculated using the equation

$$H = 1.854P / d^2 \tag{8.1}$$

where P is the load on the indentor, d is the diameter of the indentation.

The deformation and strength of crystals of partially stabilised zirconia were determined by the three-point method. Regardless of the difference in the dynamic moduli of elasticity and crystallographic orientation of the specimens, the values of the ultimate strength in three-bending were quite similar. The characteristics of the crystals are presented in Table 8.9 [56].

It should be mentioned that the moduli of elasticity of the specimens with the composition ZrO_2–3 mol%Y_2O_3 differ for different crystallographic directions. The static moduli of elasticity were slightly lower than the dynamic moduli of elasticity. This is characteristic of ceramic materials. In all experiments, examination showed a good agreement between the values of the dynamic and static moduli of elasticity. Figure 8.10 [50] shows the characteristic diagrams of deformation of the specimens with the composition ZrO_2–3 mol%$Y_{-2}O_3$ with the direction of the axis corresponding to <111>, the temperature range from –140 to 1100°C.

It should be mentioned that when the temperature is reduced –140°C, the ultimate strength in bending rapidly increases (to 70%). With an increase of temperature, the deformation diagram for specimens with the composition ZrO_2–3 mol%Y_2O_3 (<111>) remains linear for up to 800°C, and for the composition ZrO_2–3mol%Y_2O_3 (<001>) up to 1000°C.

Table 8.9. Dependence of the mechanical properties of ZrO_2–xY_2O_3 crystals on the content of stabilizing oxide x mol% (room temperature)

Composition, mol% Y_2O_3	3	3	10	20
Density, 10^3 kg/m^3	6.04	6.03	5.85	5.75
Direction of axis of specimen	<111>	<001>	<101>	<101>
Speed of ultrasound oscillations, m/s	5357	7058	5965	6107
Dynamic modulus of elasticity, E_d, GN/m^2	173	301	210	215
Static modulus of elasticity, E_{st}, GN/m^2	149	397	180	185
Ultimate strength in three-point bending, MN/m^2	862	738	224	158
Ultimate strength in four-point bending, MN/m^2	642	506	135	139
Critical stress intensity factor, N/mm$^{3/2}$	19	13.4	2.01	1.41
Limiting strain ε, 10^{-4} m/m	41.4		7.5	6.6

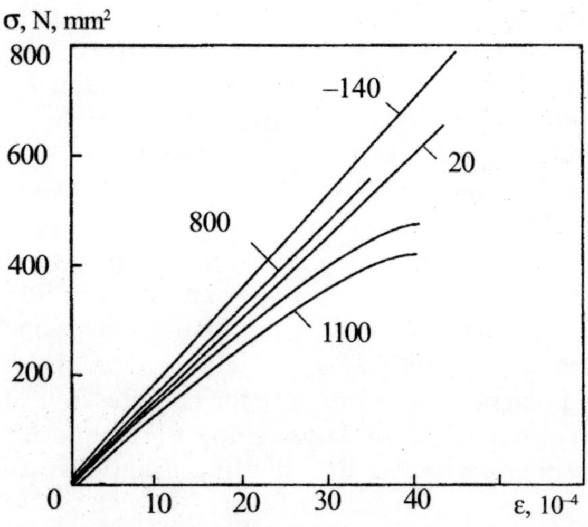

Fig. 8.10. Characteristic diagrams of deformation of specimens with the composition ZrO_2–3 mol% Y_2O_3 in the temperature range –140 to 1100°C. Direction of the growth axis of the crystals <111>.

Room-temperature tests show the linear–elastic nature of deformation of crystals up to breakdown, i.e. they belong to the group of brittle materials. The characteristics of these materials are sensitive to the rate of deformation at temperatures of inelastic deformation [50]. The interest in these materials is determined by the possibility of using them as structural materials useable at high temperatures and in oxidation media and, therefore, investigations were carried out at different loading rates at a temperature of 1400°C, which is half as high as the melting point ($T_{test} \approx 0.5\ T_m$).

For comparison, single crystals of completely stabilised cubic solid solutions based on zirconia were investigated under the same conditions. These investigations are interesting because of the fact that their behaviour may reflect quite accurately the special features of deformation typical of the cubic matrix in partially stabilised zirconia.

The comparison of strength values at temperatures of up to 1400°C shows that the crystalline materials of partially stabilised zirconia are superior to sintered ceramics based on zirconia with identical composition and also to other similar materials [50]. The mean values of the mechanical characteristics of the cubic and tetragonal crystals annealed at a temperature of 1400°C are presented in Table 8.10 [50].

Analysing the behaviour of the mechanical properties of the materials, it should be noted that a temperature of 1400°C is characterised by the formation of high residual strains in the single crystals of cubic solid solutions, and also that the mechanical characteristics are highly sensitive to the strain rate. These factors may be explained by the result which shows that a decrease in the strain rate may lead to the activation of diffusion processes and formation in the specimens of an additional slip systems [56, 57] attributed to the appearance of the upper and lower yield limit, i.e. the characteristic peak of loading on the loading diagram (Fig. 8.11) [50]. The deformation diagrams of the single crystals of the cubic solid solutions, presented in [53,58,59] for compression of miniature cylindrical specimens are in almost complete quantitative and qualitative agreement with the deformation diagrams of crystalline specimens.

With an increase of the concentration of the stabilising oxide in the cubic solutions from 10 to 20 mol% Y_2O_3, the deformation diagrams show an increase in the difference (ΔP) of the load corresponding to the upper and lower yield limits (Fig. 8.11) ($\Delta P_{20}/\Delta P_{10}$) ≈ 2.8 at 1400°C and V_{crack}= 0.005 mm/min. Evidently, this is associated with a higher degree of ordering of the cubic solid solution with the composition ZrO_2–20 mol%Y_2O_3 in comparison with the composition ZrO_2–10 mol%Y_2O_3.

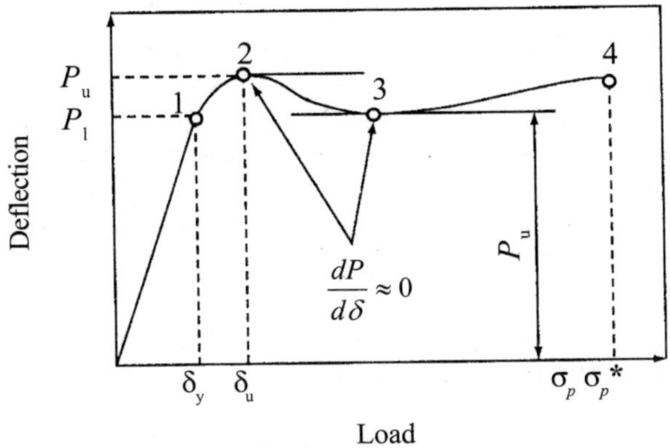

Fig. 8.11. Characteristic points of the deformation diagram. 1 – force $\mathbf{P}_y$ and deflection δ_y corresponding to the upper yield limit; 3 – force P_l corresponds to the lower yield limit; 4 – deflection δ_p at the moment of fracture of the specimen, δ_p* – at the moment of unloading the specimen.

Table 8.10. Effect of chemical composition and annealing on mechanical characteristics of single crystals of cubic solid solutions based on zirconia (V_{crack} = 0.5 mm/min, T_{test} = 20°C)

Material	ZrO_2–10 mol% Y_2O_3		ZrO_2–20 mol% Y_2O_3	
	prior to annealing	after annealing*	prior to annealing	after annealing*
Ultimate strength, MN/m²	134	172	144	216
Limiting strain, 10^{-4} m/m	7.4	10.2	7.7	11.7
Statistical elasticity modulus E_{st}, GN/m²	180	169	187	185

The crystals of partially stabilised zirconia are characterised by distinctive blocking of the slip systems in the crystals, presumably as a result of the presence of the domain structure with the size expressed in nanometres and the possible presence of inclusions of a second phase.

As regards the character of failure during deformation, the crystals of partially stabilised zirconia occupy an intermediate position between sintered ceramics and cubic crystals [50]. At all loading rates, the crystals break up into a smaller number of fragments in comparison with the specimens of the cubic crystals and their surface contains fine cracks near the main crack. The residual strain in the specimens was almost insignificant in comparison with cubic crystals which indicates a low creep rate.

An increase in the strength of the single crystals with the cubic structure at 1400°C at all V_{crack} rates in comparison with the strength at room temperature evidently is evidently caused by the a decrease of the degree of damage of the crystals during annealing as a result of mechanical treatment. This has been confirmed by testing specimens at room temperature after additional annealing at a temperature of ~1400°C for 15 min.

8.7. THE STRENGTH CHARACTERISTICS OF CRYSTALS OF PARTIALLY STABILISED ZIRCONIA

The results of examination of the compression strength characteristics are presented in Table 8.11 [19]. The Table shows that the maximum compression strength is shown by the materials with an yttrium oxide content of ~(2.5–3) mol%. The addition of other admixtures in small quantities has a strong effect on the mechanical characteristics of these crystals. The effect of the nature and concentration of a specific admixture on the properties of single crystals is specific and requires special investigations.

The microhardness of crystals of partially stabilised zirconia

The microhardness usually depends on the load at low loading and is independent at high loads. In [60–68] it is reported that in the case of ion crystals, the formation of an indentation is associated with the mobility and multiplication of dislocations in the field of concentrated loading.

The deviation of the shape of the indentation from the ideal shape in anisotropic materials results in a situation in which the diagonals of the same indentation differ. The values of microhardness, calculated on the basis of the length of the diagonals, may also differ. Thus, the microhardness may be anisotropic. To determine the anisotropy, the indentor is often represented by a Knoop pyramid in which the long diagonal is seven times longer than the short diagonal, with minimum cracking of the specimen. Microhardness is calculated from the following equation:

$$H_K = P / 0.07028d^2, \tag{8.2}$$

where d is the length of the long diagonal.

Table 8.11. Strength characteristics of PSZ crystals

Composition of crystals			Loading rate, mm/min	Compression strength σ_B, MPa	Strain Ψ	Elasticity modulus E, GPa
ZrO_2, mol%	Y_2O_3 mol%	Additional admixture (0.1–0.3 wt.%)				
99.5	0.5	–	2	793.1	0.062	125
88.5	1.5	–	0.2	1510	0.0072	222.5
88.0	2.0	–	0.2	1556	0.00213	369.2
87.5	2.5	–	0.2	2133	0.0028	360.0
87.0	3.0	–	0.2	1729	0.0065	264
86	4	–	0.2	1552	0.00808	196
85.3	4.7	–	0.2	1305	0.0064	196
87.5	2.5	Er	2	1811.0	0.087	188
87.5	2.5	Co	2	2741	0.11	214
87.5	2.5	Cr	0.2	1323	0.081	151
87.5	2.5	Tb	0.2	1832.0	0.086	205

According to Griffith's fracture theory, the stress (σ_f), required for brittle fracture of material containing a crack is [60, 61]:

$$\sigma_f = (1/Y)\sqrt{2E'\Gamma/a}, \tag{8.3}$$

Here Γ is the surface energy of fracture, a is the length of the crack, Y is a dimensionless parameter which depends on the shape of the crack; E' is a coefficient which depends on Young's modulus (E) and Poisson's coefficient (ν). At plane stress

$$E' = E, \tag{8.4}$$

and plane strain

$$E' = E/(1-\nu^2). \tag{8.5}$$

The methods of deformation of the material using an indentor have been used for investigation of measurement of some mechanical properties of materials (hardness, microhardness, toughness, brittleness, etc). The anisotropy of microhardness is determined mainly by the structure of the lattice and primary slip systems, acting during indentation. The anisotropy of the elastic properties of the crystals of ZrO_2–Y_2O_3 with an yttrium oxide content of 1.7–20.0 mol% has been investigated in [39] on the basis of the speed of propagation of ultrasound. The experimental results show that cubic modification single crystals are characterised by the highest anisotropy. The presence of the tetragonal phase in the crystals decreases the anisotropy of the elastic properties.

The anisotropy of the microhardness of cubic zirconia crystals has been studied in greater detail using Knoop's indentor. No cracks were found in the vicinity of the indentations on the (100) plane in the load range of 0.5–2.0 N. The microhardness of the specimens changed depending on the orientation of the long diagonal of the indentor being maximum along the direction <100> and minimum along <110> crosses at an angle of 45° [63, 64]. The maximum and minimum values of microhardness were: 19.0 and 14.5 GPa for a load of 0.5 N; 15.7 and 13.7 GPa for a load of 1 N; 13.8 and 12.6 GPa for a load of 2 N.

The hardness anisotropy coefficient H_{max}/H_{min} decreased from 1.3 to 1.11 with the load increasing from 0.5 to 2.0 N. The similar character of the effect of the load on the coefficient of microhardness was observed for the hardness of semiconductors and is explained by a strict restriction of the slip systems for small loads and by the activation of other slip systems at higher loads [65].

For monoclinic zirconia, the anisotropy of Knoop's hardness at a load higher than 1 N increases and the ratio H_{max}/H_{min} is 1.62 at the absolute values of 9.4 GPa for the orientation <101>{010} and 5.8 GPa for <010>{100) [66]. No anisotropy of Vickers hardness was detected on the (100) plane of cubic crystals. With temperature increasing from 25 to 1300°C, microhardness decreased for the planes {100} and {111} from 14 to 2 GPa, for {110} from 16 to 2 GPa. Owing to the fact that no cracks formed during indentation at temperatures above 800 °C, this temperature is the temperature of the brittle–ductile transition [67].

The plastic deformation of cubic zirconia, stabilised with yttrium oxide, in indentation in the temperature range 200–800°C takes place as a result of dislocation slip and crack formation. Dislocation slip takes place through the primary slip systems {001} <110> and $\{111\}\langle 1\bar{1}0\rangle$. The resultant geometry of the zone of plasticity is characterised by strong anisotropy. At a temperature of > 400°C, the mobility of edge dislocations is higher than that of screw dislocations.

In single crystals of cubic zirconia, stabilised with 9.4 mol% of yttrium oxide, special attention was given to the dislocation structure around indentations of the indentor on the planes {111}, {011} at temperatures of 200–800°C. Investigations were carried out in an electron microscope with selective etching [67]. Indentation on the (001) plane results in the formation of radial cracks propagating along the <110> direction. The results show that the formation and propagation of these cracks is a function of temperature and load. Examination of a number of sections under an indentor shows that cracks propagated along the {110} planes. This is in agreement with the results obtained for indentation at room temperature [68]. No lateral cracks were found in the examined temperature and load ranges.

The indentation of partially stabilised materials based on zirconia is characterised by a stress-initiated martensitic transition of metastable tetragonal zirconia to the stable monoclinic phase [67, 69]. The compressive stresses, formed around the indentation, improve the mechanical properties of tetragonal zirconia in comparison with the completely stabilised zirconia (cubic) (CSZ). The expansion, associated with phase transformation, also results in the swelling of the surface around indentation, which was detected in examination under an optical microscope by the phase contrast method [68]. A second possible mechanism of increasing the toughness of tetragonal zirconia may be the stress-induced reorientation of ferroelastic domains [23, 70]. The majority of investigations of these materials were carried out on ceramics and polycrystals. The mechanical properties of single crystals of tetragonal zirconia have been studied less extensively.

The anisotropy of failure in the crystals of partially stabilised zirconia is clearly manifested in the orientation of the diagonal of the indentor along <110> at which only radial cracks form, and radial and lateral cracks appear at the orientation along <100> [71, 72]. The indentation on the (111) surface results in the formation of radial cracks along the directions $\langle 2\bar{1}\bar{1}\rangle$, $\langle \bar{1}2\bar{1}\rangle$ and $\langle \bar{1}\bar{1}2\rangle$. At higher loads or low temperatures, fine cracks formed along the direction $[\bar{2}11]$. A series of sections also confirms that these cracks propagated along the {110} planes. Lateral cracks developed at higher loads or low temperatures.

The microhardness of the specimens with the composition ZrO_2–3 mol% Y_2O_3 at a load of (50–100) N is 14–15 GPa and is independent of the orientation of the indented polished surface, i.e. at higher loads the anisotropy of microhardness disappears.

Crack resistance of the crystals of partially stabilised zirconia

The stress intensity factors, characterising the cracking resistance of the material, were determined by both the methods of bending crystalline parallelepipeds and by the microindentation method. The *R*-curves were plotted in cyclic loading of specimens with a sharp crack.

The stress field in the vicinity of the tip of the crack is characterised by a stress intensity factor. Catastrophic failure takes place when the intensity of the stress field at the tip of the crack reaches a limiting value. Deformation in this region can be divided into three types: I – tensile loading and opening of the crack; II – slip inside the crack surface; III – slip outside the crack surface. The deformation of the first type is most significant. The stress intensity factor for this deformation is determined by the equation:

$$K_1 = \sigma Y\sqrt{a}. \tag{8.6}$$

When the value of K_I reaches the critical value $K_I = K_{IC}$, fracture starts to take place (uncontrolled crack growth) and equation (8.6) takes the following form:

$$K_{1C} = \sigma_f Y\sqrt{a}. \tag{8.7}$$

This equation, 'criterion-1' of brittle fracture has been established from the conditions of the stress state at of the crack tip and is in fact identical to the Griffith criterion derived from the law of conservation of energy in this process:

$$\sigma_f = (1/Y)\sqrt{2}E'\Gamma/a = K_{1C}/Y\sqrt{a}. \tag{8.8}$$

The limiting value K_{1C} is a parameter of the material referred to as the stress intensity factor which characterises the brittle fracture resistance of material. The method of measuring K_{1C} for brittle materials is the subject of discussion in many studies [73–75].

The crack formation resistance of the crystals of partially stabilised zirconia is determined by the method of three-point bending of notched specimens and by indentation of polished surfaces.

Examination of the cracking resistance of crystals with the composition ZrO_2–3 mol%Y_2O_3 shows that this material, characterised by a high elasticity modulus in the direction <001>, has a higher resistance to tensile loading of the cracks in this direction in comparison with the <111> direction (see Table 8.12). The maximum value of the stress intensity factor K_{1C} for the specimens is 16 MPa $m^{1/2}$. The results obtained in the examination of the *R*-curves, i.e., the dependence of the stress intensity factor on the length of the sharp crack, are presented in Fig. 8.12 [44].

The cracking resistance (stress intensity factor) depends on the load and orientation of the surface of the specimen (Table 8.12).

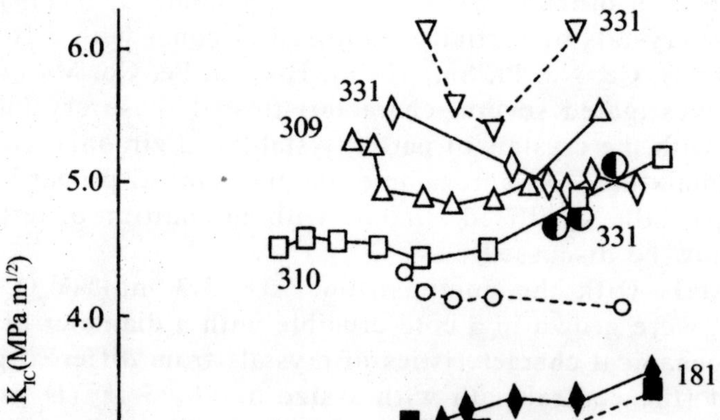

Fig. 8.12. Dependence of the stress intensity factor K_{IC} of ZrO_2–3 mol%Y_2O_3 crystals on the length of a sharp crack l_{crack} (numbers at the curves are the dynamic moduli of elasticity E_d, GPa).

Table 8.12. Critical coefficients of stress intensity of ZrO_2–3 mol% Y_2O_3 crystals (10^3 N/mm$^{3/2}$) [50]

Load, N	Surface orientation	
	(111)	(001)
50	8.7	10.2
100	5.3/3.9	7.7/5.7
300	4.2/2.3	6.8/4.6

The effect of admixtures on the mechanical properties of crystals of partially stabilised zirconia

It is well-known that the addition of a small amount (<0.5 mol%) of rare-earth and transition elements to the composition of solid solutions based on zirconia with a stabilising admixture within a specific range of concentrations (away from the concentration boundaries of the phase fields) does not change the phase composition of the crystals but results in changes of certain characteristics. In order to examine these phenomena, in [19] the authors grew crystals of partially stabilised zirconia with different third admixtures, Ce, Nd, Pr, Sm, Tb, Er, Ho, Cu, Fe, Co, Mn (Table 8.11) and investigated several characteristics of these crystals in comparison with the crystals of partially stabilised zirconia without additional admixtures. As an example, the mechanical properties of crystals of partially stabilised zirconia with an addition of terbium oxide will now be discussed.

The crystals with the composition ZrO_2–3.3 mol%Y_2O_3–0.3 mol% Tb_2O_3 were grown in a cold crucible with a diameter of 400 mm. The mechanical characteristics of crystals from different parts of the ingot (the central path with a size of ~0.25 D_c (D_c is the crucible diameter), peripheral, adjacent to the walls of the crucible ~0.1–0.15 D_c, and intermediate) were investigated on specimens cut out along the axis of growth without any preliminary crystallographic orientation. The mechanical characteristics of the investigated crystals are presented in Table 8.13.

The experimental results show that the values of the dynamic elasticity moduli of this material are slightly lower in comparison with the crystals of partially stabilised zirconia without additional admixtures, and are independent of the position of the crystal block in the ingot. The values of the moduli did not exceed 220–230 kN/

Table 8.13. Physical–mechanical characteristics of ZrO_2–3.3 mol% Y_2O_3–0.3 mol% Tb_2O_3 crystals

Composition, mol% Y_2O_3	ZrO_2–3.3 mol% Y_2O_3–0.3 mol% Tb_2O_3				
	1 (centre)			2	3
Crystal growth zone	After growth	Annealing (vacuum 10^{-3}, 1400 °C, 2 h) prior to	Annealing (vacuum 10^{-3}, 1400 °C, 2 h) after	Intermediate region	Periphery
Density, 10^3 kg/mm^3			6.06		
Dynamic elasticity modulus, E_d, kN/mm^2	213	180	196	195	176
Static elasticity modulus, E_{st}, kN/mm^2	163	141	155	155	136
Ultimate strength in four-point bending, σ_y, N/mm^2	354	369	448	416	398
Critical stress intensity factor, K_{IC}, N/mm$^{3/2}$	10.7	10.5	6.0	6.8	7.0

Table 8.14. Mechanical characteristics of oriented crystals of PSZ, activated with Ce and Tb

Crystal (sample oriented in [100])	Density, g/cm^3	Speed of propagation of UW, m/s	Dynamic elasticity modulus, GPa	Bending strength, MPa Three-point	Bending strength, MPa Four-point	K_{IC}, MPa·m$^{0.5}$
Y-Ce-PSZ	6.07	7769	366	1378	1250	11.43
Y-Tb-PSZ	6.06	7730	352	1611	-	9.96

mm^2, and the elasticity moduli for other crystals of partially stabilised zirconia were in certain cases higher than 300 and even 350 kN/mm^2. Previously, it was reported that the elastic properties of the crystals of partially stabilised zirconia are anisotropic and the [100] direction is the 'high-modulus' direction with E ~360 GPa, and {111} is the low-modulus direction with E ~170 GPa. Therefore, specimens of crystals whose surfaces coincided with the {100} plane of the pseudocubic crystal system were prepared. The bend tests yielded the ultimate strength and critical stress intensity factors of the specimens of ZrO_2–3.3 mol%Y_2O_3–0.3 mol%Tb_2O_3 presented in Table 8.14.

Comparison of Tables 8.13 and 8.14 shows clear differences in the values obtained on the oriented and non-oriented specimens.

Vacuum annealing is characterised by a change of the valency state of terbium ion: $Tb^{4+} \rightarrow Tb^{3+}$. Recovery annealing increases the ultimate strength and decreases the cracking resistance of the material.

Tests carried out at a low temperature (–150°C) showed that the strength of the material increased by approximately 30% in comparison with the strength at room temperature. Microhardness depends only slightly on the load on the indentor and the crystallographic orientation of the plane on which the tests were carried out. For example, the hardness of crystals at a load of 10 N was 12.8–13.8 GPa and at a load of 200 N it was 12.2–13.0 GPa.

Thus, small variations of the concentration of admixtures within 1 mol% result in a large scatter of the mechanical characteristics of the crystals of partially stabilised zirconia. The experimental results show that in the synthesis of high-strength and crack-resistant crystals of partially stabilising zirconia by the method of directional crystallisation of the melt it is important to take into account many factors: the effective coefficient of distribution of the admixtures, the technological conditions of incorporation of admixtures (the rate of crystallisation, temperature gradient, etc), the effect of the type and concentration of the stabilising oxide and the third admixture, the thermal conditions of annealing and cooling of the crystals, containing ions of variable valency.

Thus, the following conclusions may be drawn: with increasing concentration of the stabilising oxide the coefficients of thermal expansion (CTE) of the material increases (Tables 8.1, 8.2 and 8.6), the phase transition temperature decreases, the magnitude of changes of the volume of the specimen during phase transitions becomes smaller and the hysteresis loop disappears. In stabilisation with the europium and gadolinium oxides, this transformations take place at a content higher than 3 mol%, and in the case of the ytterbium oxide the hysteresis loop disappears almost completely at 3 mol%. The shape and size of the loop are greatly affected by the synthesis conditions. In directional solidification of the melt, the specimens have zero porosity and are characterised by the nonuniform distribution of the phases along the length of the crystal. It is possible that this is the reason why the shape of the hysteresis loop is smooth and the inflection point on the curve corresponds to the temperature at which the rate of phase transformation is higher than expansion or compression of the tetragonal phase present in the specimen. Therefore, the range of transformation becomes blurred and wide. A short measurement time and low temperatures resulted in the situation in which the metastable tetragonal phase (T'), present in the specimen, does not change its composition and does not undergo transformations, and changes take place only in the monoclinic phase, formed during crystal growth.

8.8. CURRENT VIEWS REGARDING THE MECHANISM OF HARDENING OF CRYSTALS OF PARTIALLY STABILISED ZIRCONIA

In order to explain the anomalously high mechanical properties of the crystals of partially stabilised zirconia, investigators have proposed several hardening mechanisms. The following mechanisms have been developed most extensively: 1) 'transformation'; 2) 'dispersion'; 3) ferroelastic.

The high strength and toughness of partially stabilised zirconia in the initial stage of development of theory were explained exclusively by transformation hardening in which the propagating microcrack results in the martensitic $t \leftrightarrow m$ transition which absorbs the energy of stresses, transforms this energy into thermal energy, and also constrains and arrests the propagating microcrack [76].

Subsequently, attention was given to the fact that the high strength characteristics of the single crystals of partially stabilised zirconia are also retained in the high-temperature range (higher than 1000°C) at which there is no $t \leftrightarrow m$ phase transition. In addition to this, in fracturing a crystal with the composition ZrO_2–4.5 mol%Y_2O_3 produced by directional crystallisation, there were no traces of the monoclinic phase on the fracture surface [28]. Regardless of this, the specimen were characterised by high strength and toughness. Examination of the microstructure and plastic deformation of these materials has resulted in a proposal for a new mechanism of hardening, i.e. dispersion hardening.

The experimental results show that the crystals contain inclusions of the disc-shaped tetragonal phase (*T*). In annealing of crystals (1400–1600°C), the inclusions gradually acquire the equilibrium shape and dimensions, determined by the minimum sum of the volume and surface energy. The annealing temperature and time have a significant effect of the mechanical properties of crystals.

For example, the yield limit rapidly increases with increasing annealing time because the density of the inclusions of the tetragonal phase is determined by the diffusion kinetics. An increase of the yield limit in annealing is directly proportional to the width of the inclusions and the distance between the colonies in which they are grouped.

Loading results in the formation of dislocations in the cubic matrix. According to the authors of [28], the dispersion hardening mechanism is associated with inhibition of the movement of matrix dislocations through a network of the inclusions of the second phase. For example, there were dislocations with the Burgers vector $b = 1/2[110]$, which travelled along the (001) plane.

As shown previously, 12 variants of tetragonal inclusions may form in the cubic matrix. It may be shown that only for one of the 12 variants, the [110] vector represents the natural translation vector of the tetragonal lattice. When a sliding dislocation meets any of the remaining 11 variants of the inclusions, the dislocation is inhibited in its movement because it cannot enter the tetragonal inclusions without energy losses associated with the rotation of the Burgers vector. For a dislocation to enter the tetragonal inclusions, the dislocation reaction of the type $1/2[102] \leftrightarrow [101] + 1/2[\bar{1}01]$ is required, i.e. the initial dislocation generates a dislocation pair. One dislocation enters the ([101]) tetragonal inclusions, the other one remains at the interphase boundary forming an Orowan's kink. This 'sessile' dislocation (identical with the well–known Lomer–Cottrell dislocation) is a barrier to the movement of the next dislocation inside the inclusions. A cluster of dislocations forms around the interphase boundaries. Each subsequent dislocation is arrested at larger and larger distances from the interface.

It is well-known that the dislocations, like a new phase, may nucleate homogeneously or heterogeneously (on dislocation clusters during closure of a vacancy disc, on the Frank–Read sources). The Lomer–Cottrell barriers around the interface suppress the generation of dislocations by means of the Frank–Read source. Homogeneous nucleation of dislocations requires high stresses increasing the yield and tensile strength, i.e. increasing the mechanical characteristics. This is the principle of the proposed hardening mechanism.

A number of authors [23, 25, 27, 29] explain the mechanism of hardening of the crystals of partially stabilised zirconia by a ferroelastic transformation associated with the transition of the cubic phase into a tetragonal phase. It is well-known that the ferroelastic properties are shown by substances having two or more stable orientation states and capable of transferring from one state to another under mechanical loading. The ferroelastic transition differs from the martensitic transition induced by mechanical stresses by the fact that the process takes place by means of the reorientation of ferroelastic domains without any change of the crystal structure. The ferroelastic properties of the tetragonal zirconia form as a result of the t'-phase transition which decreases the symmetry of the initial cubic phase to the tetragonal phase. This results in the formation of three orientation states nonequivalent from the energy viewpoint which are referred to as variants of twins or domains. With the mutual distribution similar to orthogonal, these domains form pseudocubic crystals. During loading, the domains may reorient their axes in order to accommodate the stresses. The details of this effect were investigated by the authors of [70, 76–80]. They depend on the direction and the type of loading and also the orientation of the crystal and domain switching in the ferroelastic materials. X-ray diffraction and ultrasound examination of the crystals of partially stabilised zirconia under loading show that when loading with 1.5 GPa, the tetragonal axis c changes its direction by 90°, i.e. the tetragonal axis becomes the a axis. When the load is removed, a reversed transition takes place [78–81].

By analogy with the ferromagnetic transition, the ferroelastic transition has a hysteresis at loads below the ultimate strength. In ultrasonic examination, the hysteresis loop in the speed of sound–stress coordinate was clearly visible. At low loads of ~0.4 GPa, there was also a jump in the speed of ultrasound explained by the tetragonal–monoclinic transition. However, this jump is irreversible because at low temperatures the monoclinic phase cannot transfer to tetragonal. The hardening and high toughness of the ferroelastic material are explained by the absorption of the energy of the load in transition.

Thus, it may be assumed that there are at least three different mechanisms of hardening of partially stabilised zirconia. Evidently, one of the mechanisms may be dominant in the materials of partially stabilised zirconia of different composition and in different loading conditions, or several mechanisms may operate simultaneously. For example, the main contribution in the tetragonal partially stabilised zirconia should come from the ferroelasticity mechanism. If the material contains several phases, several mechanisms may act simultaneously, depending on composition. In this case, the dispersion hardening mechanism will be controlling. For specimens containing the transformed *t*-phase, the controlling contribution is provided by the mechanism of transformation hardening, and for partially stabilised zirconia consisting of a mixture of cubic and nontransformed tetragonal *t'*-phases, the main role will be played by the mechanism of dispersion and ferroelastic hardening.

Since the composition, synthesis conditions and heat treatment parameters have the controlling effect on the degree of stabilisation of partially stabilised zirconia, its phase composition and microstructure, the mechanical properties of these materials also greatly depend on the above factors.

8.9 MOST PROMISING AREAS OF APPLICATION OF CRYSTALS OF PARTIALLY STABILISED ZIRCONIA

Recently, intensive research has been carried out into the application of ceramics of zirconia in many technical applications. The practical application of these materials is determined by a number of various characteristics such as: high melting point (~2800°C), the high-strength and cracking resistance (σ = 600–1200 MPa; K_{Ic} = 6–16 MPa m$^{1/2}$); high hardness (8.5 according to Moose, 1200–1500 kg/mm^2); low friction coefficient (0.04 on steel, 0.08 on tungsten–cobalt alloys); high resistance to abrasive wear; high coefficient of thermal expansion close to steel (10–11·10^{-6} deg^{-1}); chemical resistance to corrosive media (acids, alkali, melts of metals, oxidation atmosphere at high temperatures); good heat resistance (280°C); at low temperatures, partially stabilised zirconia has good electrical properties (at 20°C 2.2·10^{10} ohm·cm). Other typical features of partially stabilised zirconia are bioinertia; retention of high mechanical properties in a wide temperature range (–140 to 1400°C).

The practical application of the materials of partially stabilised zirconia is being developed in the following directions:

1. Structural ceramics utilising high strength, wear resistance and low friction coefficient, and also heat insulating properties and corrosion resistance. These ceramics are promising for fabrication of:

–coatings on the components of internal combustion engines [82], increasing the efficiency of operation, the reliability and durability of the engine, decreasing the fuel consumption, etc;

–the components of pumps for chemical production (crystals, sleeves of cylinders, sleeves of valves, valves, based on links, etc) [83];

–bearings and various friction couples.

2. High hardness, cracking resistance and heat resistance make this material promising for fabrication of various cutting and processing tools. For example, dies made of ceramics based on partially stabilised zirconia in the drawing of copper wire showed durability values twice those of wires of any hard alloy [82].

3. Because of chemical and biological inertness, strength and high cracking resistance, this material is regarded as one of the promising materials for use in medicine for implants (bone replacement, joints, teeth) [48, 85, 86].

Single crystals of partially stabilised zirconia have a large number of valuable properties by which they differ from both metallic and, in particular, high-strength dielectric materials (including ceramics produced from partially stabilised zirconia). Because of the absence of the grain boundaries, the partially stabilised zirconia crystals have high strength (comparable with the strength of metals), fracture toughness and hardness. They are characterised by increased resistance to acids, alcohol, water vapours, high resistance to abrasive wear and are also distinguished by a low friction coefficient. The degradation of the mechanical properties at high temperatures (at up to 1400°C) in the oxidation media in single crystals of partially stabilised zirconia is considerably smaller in comparison with metals where the majority are subjected to high-intensity oxidation and structural ceramics undergoing recrystallisation processes. In addition to this, when the temperature is reduced to –140°C, the mechanical properties of the crystals decrease by 70%.

The crystals of partially stabilised zirconia may be used for the fabrication of components of mechanisms operating in extreme conditions of high mechanical loading, chemically corrosive media, increased temperatures, in the absence of lubrication. Because of the existence of the nanodomain structure in the crystals of partially stabilised zirconia and because of their high mechanical strength, it is possible to produce from them cutting tools with extremely sharp cutting edges, for both high-precision machining of different materials (metals, wood, glass, crystals, etc.) and for high quality medical tools.

The biological inertness [85, 86] of this material makes it

promising for application in medicine for implants. In recent years, mechanical and medical–biological comparative tests were carried out on partially stabilised crystals of zirconia, single crystals of cubic zirconia, tetragonal polycrystalline partially stabilised ceramics and also single crystals of aluminium oxide [48]. The single crystals of partially stabilised zirconia show highrt strength and failure resistance (toughness) in comparison with other materials. The experimental results show that the investigated implants of partially stabilised zirconia do not cause any inflammatory reactions in the living tissue and do not lead to pathological changes in the internal organs of animals. On the basis of the experimental results, the crystals of partially stabilised zirconia may be recommended as new materials for preparation of various implants.

Thus, the most promising areas of application of the crystals of partially stabilised zirconia are:

1. Wear resistant structural components for application in the conditions with increased temperatures, load and corrosive media: components of pumps for chemical production (pistons, the cylinder lining, valves sleeves, valves, piston rings, etc); components for mills (ball, roller), crushers, etc; filament guides for textile production; bearings (thrust, sliding), and different friction couples; prisms for precision devices; solid electrolytes, fuel elements, heaters; sensors of the oxygen content of exhaust gases, in water vapour and in molten metal; heat shielding screens, corrosion-resisting components, wear-resistant components for drilling equipment.

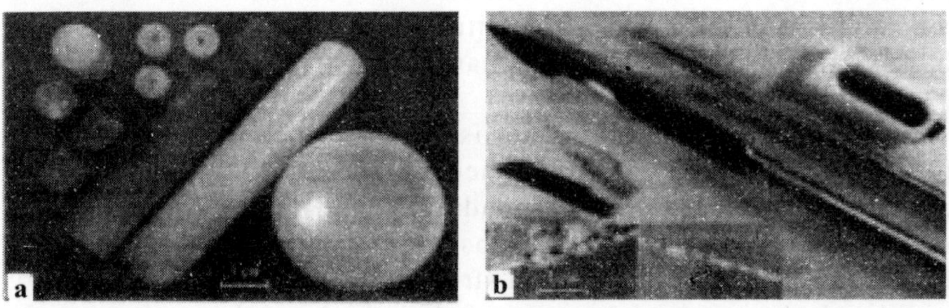

Fig. 8.13. Components made of PSZ crystals; a) medical scalpels, b) cutting edge of PSZ crystals.

2. Cutting and processing tools for machining the surface of components to the mirror-like state with a high degree of accuracy; in drawing wire; rolling rolls for the thinnest strips and foil

(platinum, gold, copper, aluminium), cutting tools for high precision machining of different materials (metal, wood, glass, crystals, etc); guide devices weaving machines; attachments for the formation of jets of high-pressure liquids; extruders and dies for the production of optical fibres for light guides; dies for the production of glass and basalt fibres, replacing platinum.

3. High-quality medical tools: scalpels with a large sharp cutting edge (~0.1 μm) for heart surgery, neurosurgery, vessels surgery, cosmetology, and opthalmology.

4. Bioinert implants with high fatigue strength for dental and orthopaedic surgery.

5. Goods of mass application: scissors, knives, components of meat-processing (knives, grates), golf clubs, links, jewellery components. Figure 8.13 shows several components produced from crystals of partially stabilised zirconia.

References

1. Y. Sakka, Y. Oishi, K. Ando, *Bull.Chem. Soc. Jpn.*, 55, 420–422, 1982.
2. N. Yoshikawa, H. Sato, *J. Jap. Inst. Metals*, v.50, 108–113, 1986.
3. R.A. Miller, J. L. Smilek, R. G. Garlich, Phase Stability in Plasma-sprayed Partially Stabilized Zirconia-Yttriya, Adv. in Cer. V.3., Ed. By A.H. Heuer et. al, Columbus The American Ceramic Society, 1082113, 1981.
4. G.V. Kurdyumov, Imperfections of the Crystal Structre and Martensitic Transformations, Nauka, Moscow, 1972.
5. A.G. Khachaturyan, Theory of Phase Transformations and Structure of Solid Solutions, Nauka, Moscow, 1974.
6. T. Mitsuhashi, Journal of American Ceramic Society, 48, 493, 1973.
7. C.A. Ànderson, J, Greggi and T.K. Gupta, Diffusionless Transformations in Zirconia Alloys, p.78-85, Advances in Ceramics Vol.12, Science and Technology of Zirconia II., The American Ceramic Society, Columbus, Ohio, 1983.
8. H.-H. Mobius, *Z. Chemie,* 4,3, 81–94, 1964.
9. A.G. Kachaturyan, Theory of Structural Transformation in Solids, Chapter 8, Wiley, New York 1983.
10. B.Ya. Sukharevskii, B.G. Alapin, A.M. Gavrish, *Dokl. Akad. Nauk SSSR,* 156, 3, 677–680, 1964.
11. Yu.K. Sirotin, M.P. Shaskol'skaya, Fundamentals of Crystallography, Nauka, Moscow, 1979.
12. H. Varlimont, Martensitic Transformations in Alloys based on Copper, Silver and Gold, Nauka, Moscow, 1974.
13. V. Lantery, T.E. Mitchell, and A.N. Heuer, *Journal of American Ceramic Society*, 69, 7, 564–569, 1986.
14. J. Lefevre, *Ann. Chem.*, 8, 1-2, 117-149, 1963.
15. B.Ya. Sukharevskii, A.M. Gavrish, *Dokl. Akad. Nauk SSSR*, 155, 2, 438-441, 1964.

16. H.S. Maiti, K.V.G.K. Gokhale and E.C. Subbarao, Journal fo the American Ceramic Society, 55, 6, 317–322, 1972.
17. I.T. Dailey, D. Levis, Z.M. Librantand, L.I. Porter, *Trans. J. Brit. Ceram. Soc.*, 71, 1, 25–30, 1972.
18. V. Ali, S.G.H. Bukhari, K.F. Shoaib and F.H. Hashmi, *Phys. Stat. Sol.* (a), 35, 181–184, 1976.
19. E.E. Lomonova, Technology, Properties and Application of Crystals based on Zirconia, Dissertation, General Physics Institute, Moscow, 2001.
20. G.V. Samsonov(editor), Physicochemical Properties of Oxides, Metallurgiya Moscow, 1978.
21. E.E. Lomonova, Growth and Examination of Single Crystals of Solid Solutions based on Zirconia and Hafnium, Dissertation, FIAN, Moscow, 1980.
22. A.F. Shchurov, I.F. Perevozchikov, *Neorg. Mater.*, 33, 9, 1087–1092, 1997.
23. D. Michel, L. Mazerolles, M. Perez y Jorba, J. Mater. Sci., 18, 2618–2628, 1983.
24. Yu.K. Voron'ko, et al., *Optika i Spektroskopiya*, 51, 4, 569–571, 1981.
25. D. Baither, B. Baufeld, U. Messerschmidt, F.H. Foitzik, M. Ruhle, *Journal of the American Ceramic Society*, 80, 1691–98, 1997.
26. K.M. Prettyman, J.-F. Jue, A.V. Virkar, C.R. Hubbard, O.B. Cavin, and M.K. Ferber, *J. Mater. Sci.*, 27, 4167–74, 1992.
27. D. Baither, B. Baufeld and U. Messerschmidt, *Phys. Status Solidi* (a) 137, 569-76, 1993.
28. A.H. Heuer, V. Lanteri, A.Dominguez-Rodriguez, Journal of the American Ceramic Society, 69, 3, 285–287, 1986.
29. B. Baufeld, D. Baither, U. Messerschmidt, and L. Bartsch, *Physica Status Solidi* (a), 150, 1297–306, 1995.
30. R.H. Ingel, D. Lewis, B.A. Bender, R.W. Rice, in: Advances in Ceramics, V.12, Science and Technology of Zirconia II, Ed. by N. Claussen, M. Ruhle and A. H. Heuer, American Ceramic Society, Columbus, OH, p.408–414, 1984.
31. A.N. Heuer, V. Lantery, A. Dominguez-Rodriguez, *Acta Metall.*, 37, 2, 559-567, 1989.
32. Yu.K. Voron'ko, et al., *Neorg. Mater.*, 34, 4, 439–443, 1998.
33. A.N. Heuer, R.Chaim, and V. Lantery, *Acta Metall.*, 35, 3, 661–666, 1987.
34. N. Ishizava, A. Saiki, T. Yagi, N. Mizutani, and M. Kato, Twin-Related Tetragonal Variants in Yttria Partially Stabilized Zirconia, *J.Amer.Ceram.Soc.*, 69, C-18 – C-20, 1986.
35. M. Stadtwald-Klenke, Elektronenmikroskopische Untersuchungen an Inneren Grenzflechen in Verschidenen mit Y_2O_3 Dotierten ZrO_2-Keramiken, PhD Dissertation, Stuttgart University, 1994.
36. N. Ishizava, A. Saiki, and M. Kato in: Advances in Ceramics, Vol. 24: Sience and Technology of Zirconia, The American Ceramic Society, Inc. p.479-484, 1988.
30. R.H. Ingel D. Lewis, B.A. Bender, R.W. Rice, in: Advances in Ceramics, V.12, Science and Technology of Zirconia II, Ed. by N. Claussen, M. Ruhle and A. H. Heuer, American Ceramic Society, Columbus, OH, p.408-414, 1984.
37. R.H.J. Hannink, K.A. Johnston, R.T. Pascoe, and R.C. Garvie, in: Advances In Ceramics, v.3, Ed. By A.H Heuer and L.W. Hobbs, The American Ceramic Society, Columbus, OH, 1981.
38. R.P. Ingel, D. Lewis, B.A. Bender, P.W. Rice, Comm. Amer. Ceram. Soc.,

65, 9, C150-C152, 1982.

39. R.P. Ingel, D. Lewis, Journal of the American Ceramic Society, 71, 4, 265-71, 1988.
40. R.G. Hoagland, C.W. Marshall, W.M. Duckworth, Journal of the American Ceramic Society, 59, 5/6, 189-192, 1976.
41. F.J. Baretta, In: Methods for assessing the structural reliability of brittle materials, ed. S.Freimann and C.Hudson, ASTM STP 844, 194-222, 1984.
42. G.D. Quinn, F.J. Baretta., J.H. Conway, Commentary of Army Standart Test Method for Flexural stregth of High Performance Ceramics at Ambient Temperature, US Army Materials and Mechanics Research Center, AMMRG TR 85-21, August 11, 1985.
43. G.D. Quinn, F.J. Baretta, Flexure test: can it be used for ceramics part design? MTS System Corporation, 1986, (Reprinted from Adv. Mater. And Processes, December 1985).
44. G.A. Gogotsi, et al., *Ogneupory*, 8, 14–17, 1991.
45. G.A. Gogotsi, et al., *Ogneupory*, 3, 15–19, 1992.
46. G.A. Gogotsi, et al., *Ogneupory*, 6, 2–8, 1993.
47. G.A. Gogotsi, E.E. Lomonova, V.G. Pejchev, *J. Eur. Cer. Soc.*, 11, 123-132, 1993.
48. G.A. Gogotsi, E.E. Lomonova, Yu. Furmanova, and I.M. Savitskaya, *Ceramics Internationale*, 20, 343–348, 1994.
49. S.N. Dub, G.A. Gogotsi, E.E. Lomonova, *J. Mat. Sci. Lett.*, 1, 446–49, 1995.
50. D.Yu. Ostrovoi, Deformation and strength of ceramics at different tem-erature-rate effects, Dissertation, Kiev, 1993.
51. G.A. Gogotsi, *Vestnik Mashinostroeniya*, 27–33, 1989.
52. G.A. Gogotsi, et al., *Problemy Prochnosti*, 1, 73–77, 1991.
53. A.Domingues-Rodriguez, K.P.D. Lagerlof, A.H. Heuer, *Journal of the American Ceramic Society*, 69, 3, 281-284, 1986.
54. G.D. Quinn, *British Ceram. Trans. J.*, 88–94, 1989.
55. G.A. Gogotsi, Problemy Prochnosti, 1, 77–87, 1977.
56. N.N. Malinin, Applied Theory of Plasticity and Creep, Mashinostroenie, Moscow, 1968.
57. K.G. Phillips, Fracture, Vol. 7, ed. H. Liebowitz, Mir, Moscow, 19–26, 1976.
58. E. Fries, F. Guiberteau, A. Dominguez-Rodriguez, *Philosofical Magazine* A, 60, 1, 107–121, 1989.
59. J. Lankford, R.A. Page, L. Rabenberg, J. Mater. Sci., 23, 4144–4156, 1988.
60. Yu.S. Boyarskaya and O.Z. Grabno, Physics of Microindentation Processes, Shtinda, Kishinev, 1–170, 1986.
61. K. Miyata, *Ceramikuusu,* 20, 1, 3–11, 1985.
62. Yu.A. Kontsevoi, et al., Physics and Strength of Semiconductor Materials and Structures, Radio i Svyaz', Moscow, 1982.
63. A.Saiki, N. Jshizawa, N. Mizutani, K. Masanori, *J. Ceram. Soc. Jap.*, Int. Edn., 89, 1, 41–46, 1989.
64. A. Pajares, F. Guiberteau, A. Dominguez-Rodriguez, A. H. Heuer, *Journal of the American Ceramic Society*, 71, 7, 332–333, 1988.
65. V. I. Karban' and Yu.I. Borzakov, Processing of Crystals in Microelectronics, Radio i Svyaz', Moscow, 1988.
66. Z. Li, Y. Li, Y. Fang, M.V. Nevitt, Microhardness Anisotropy of m-ZrO_2 Single Crystal., 11th Int. Ceram. Sci. and Technol. Congr., Anaheim, Oct. 31, v.3, 45, 1989.

67. D. Holmes, A.H. Heuer and P. Pirouz, *Philosophical Magazine*, A, 67, 2, 325-342, 1993.
68. A. Pajares, F. Guiberteau, A. Dominduez-Rodriguez, A.H. Heuer, *Journal of the American Ceramic Society*, 74, 4, 859-862, 1991.
69. J. Martines-Fernandez, M. Jimenez-Melendo, A. Dominguez-Rodriguez and A H. Heuer, *Journal of the American Ceramic Society*, 74, 5, 1071–1081, 1991.
70. C.J. Chan, F.F. Lange, M. Ruhle, J.F. Jue, A.V. Virkar, *Journal of the American Ceramic Society*, 74, 8, 807–813, 1991.
71. G.A. Gogotsi, et al., *Ogneupory*, 3, 2–12, 1996.
72. G.N. Mosrcher, P. Pirouz, A.H. Heuer, *Journal of the American Ceramic Society*, 74, 3, 491–500, 1991.
73. G.P. Cherepanov, Brittle Fracture Mechanics, Nauka, Moscow, 1974.
74. V.I. Trefilov, et al., *Izv. AN SSSR, Neorg. Mater.*, 20, 6, 958–966, 1984.
75. G.A. Gogotsi and A.V. Bashta, *Problemy Prochnosti*, 9, 49–55, 1990.
76. R.H.J. Hannink, P.V. Kelly, B.C. Muddle, *Journal of the American Ceramic Society,* 83, 461–487 (2000).
77. F. Sanchez-Bajo, et al., *Journal of the American Ceramic Society*, 80, 1, 232-36 (1997).
78. A.V. Virkar, et al., Ferroelasticity and Related Properties of Crystals, pp.3-103, in: Phase Transitions, vol.3, Gordon and Breach Science Publishers, New York, 1982.
79. J.F. Jue, A.V. Vikar, Fabrication, *Journal of the American Ceramic Society*, 73, 3650–57, 1990.
80. A.V. Virkar, *Engineering Materials*, 1998, v.153–154, p.183–210.
81. M.G. Gain, M.H. Lewis, *Mat.Lett.*, v.9, 9, 309–312, 1990.
82. J. Huber, J. Heinrich, Ceramics in internal combustion engines, in: 2nd Eur. Symp. Eng. Ceram., London, 23-24 Nov., Proc. London; New York, 1989, 203–228, 1987.
83. K. Wolf, Th-Holschujer, Zirconoxid ein neuer Werkstoff fur den Pumpenbau in der chemischer Industrie., Chem-Techn. (BDR), 17, 8, 24–25, 1988.
84. V.G. Peichev, S.Y. Pliner, Ogneupory, 2, 12–14, 1987.
85. T. Motoxupo,Ceramikuusu, Ser. Jap. 23, 6, 25-26, 1988.
86. P. Christei, A. Meunir, M. Heller, Y.P. Torre, C.N. Peille, *J. Biomed. Mater. Res.*, 23, 1, 45–61, 1989.

9

C–OX crystals

9.1. PRODUCTION OF C–OX CRYSTALS

In previous chapters, special attention has been paid to the examination of the conditions of growth and the possibilities of producing large single-phase cubic crystals in the ZrO_2–R_2O_3 systems (R=Eu, Gd, Yb, Y), i.e. fianites. The ranges of the concentration of the stabilizing oxide have been determined: 10–20 mol% R_2O_3 for R = Gd, Eu, and 8–20 mol% R_2O_3 for R =Yb, Y at the optimum growth rates of 10–20 mm/h. An increase in the concentration of the stabilising oxide above these limits results in a large decrease in the crystal size.

The difficulties in the growth of large single-phase single crystals ZrO_2– R_2O_3, R = Ce, Nd, Pr (elements of the beginning of the lanthanum series) are associated with the existence of relatively narrow ranges of the cubic solid solutions and with phase transformations in these systems at temperatures quite close to the crystallisation temperature. This is reflected in their phase homogeneity. The crystals grow either small (1–3 mm) or are turbid with a large number of cracks. Figure 9.1 presents the phase diagrams of the ZrO_2–CeO_2 and ZrO_2–Nd_2O_3 systems.

As shown in Chapter 7, the introduction into cubic fianite crystals of large quantities of oxides of cerium, praseodymium and neodymium in the form of additional admixtures, results in considerable difficulties because the distribution coefficients of these elements are considerably smaller than unity.

It was shown in Chapter 4 that the increase of the concentration of the stabilizing oxide requires a decrease in the rate of growth,

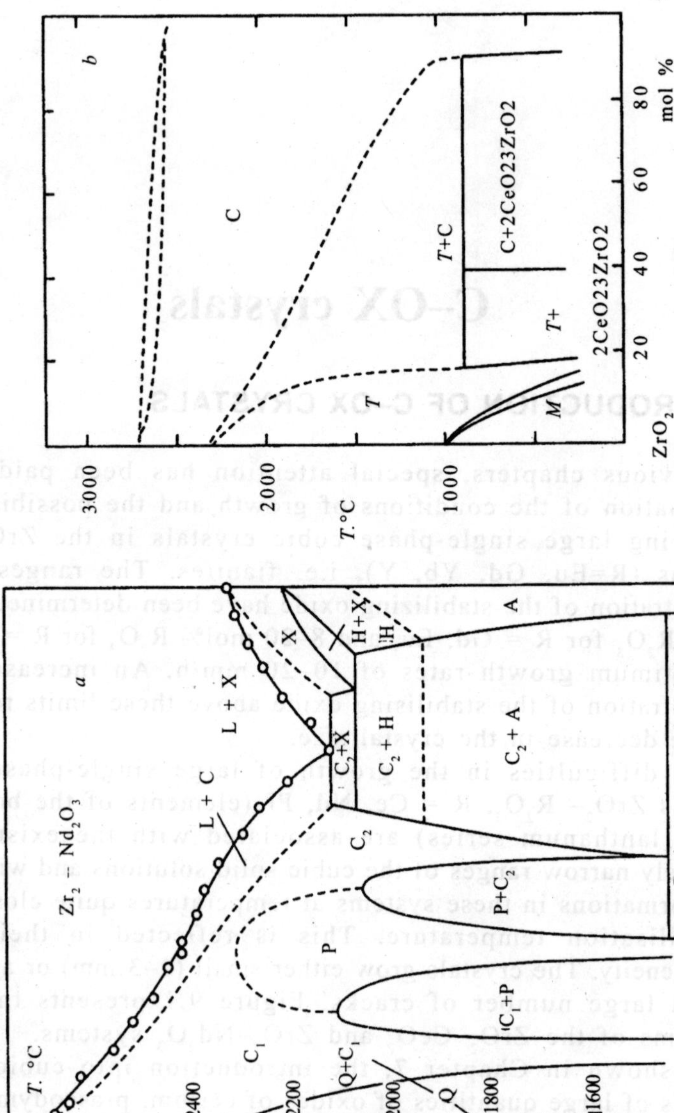

Figure 9.1. Equilibrium diagrams of the systems: ZrO_2–Nd_2O_3 (a) and ZrO_2–CeO_2 (b).

and the occurrence of the phase transformations, close to the crystallisation temperature, requires a decrease in the temperature gradient in the solid phase.

By taking into account the special features of crystallisation of solid solutions based on ZrO_2 in relation to the type and concentration of the stabilizing oxide, it has been possible to produce new large refractory oxide single crystals of cubic solid solutions based on zirconia and oxides of the rare-earth elements from the left half of the periodic table with a high content of stabilising oxides referred to as cubic oxides (C–OX).

Examination of the structure and optical properties of the C–OX crystals, activated by the ions of transition metals, as well as investigations of the effect of the heat treatment conditions on these crystals have shown large differences in the properties of C–OX crystals in comparison with the fianites which contain, as stabilizing oxides, rare-earth oxides of the right half of the series with the concentration not exceeding 20 mol%.

The families of the ZrO_2–R_2O_3–Pr_2O_3 (R=Y, Yb, Lu) crystals

As shown in Chapter 4, the amount and dimensions of the crystals, produced by directional solidification in the cold crucible depend on many technological parameters: growth rate, the size of the crucible, thermal conditions, the purity of the initial reagents, the type and concentration of the stabilizing oxide, etc. Figure 9.2 shows the dependences of the mean size of ZrO_2–R_2O_3–Pr_2O_3 crystals on the concentration of the oxides of yttrium and praseodymium at a growth rate of 5 mm/h and a cold crucible diameter of 130 mm.

The presence of praseodymium oxide results in a decrease in the size of single crystals. This is most distinctive in the case of low concentrations of the praseodymium oxide. An addition of 5–8 mol% of Pr_2O_3 to crystals with 10 mol% Y_2O_3 results in the formation of an ingot consisting of small crystals. However, when the yttrium concentration is increased, the size of the crystals increases from 2–3 mm at 10 mol% to 10–15 mm at 20–30 mol% Y_2O_3. Thus, by combining the rare-earth elements from the left and the right hand parts of the rare-earth series, it is possible to produce cubic structure crystals containing high concentrations of 'light' rare-earth elements (Fig. 9.2).

It should be mentioned that in the ZrO_2–R_2O_3 systems, in the

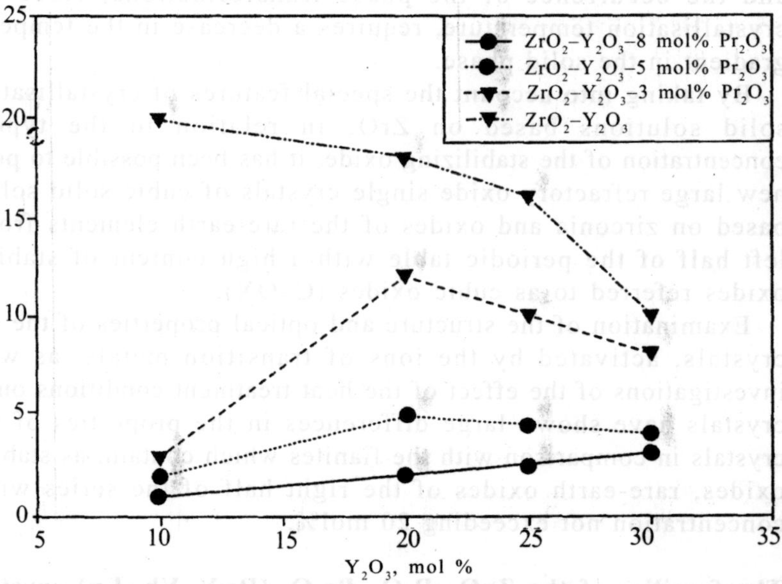

Figure 9.2. Dependence of the mean size of ZrO_2–Y_2O_3–Pr_2O_3 crystals on the concentration of yttrium and praseodymium oxides at a growth rate of 5 mm/h and the diameter of the cold crucible of 130 mm [1]

formation of substitution solid solutions, the dimensional factor of the rare-earth ions plays a significant role, for example, it affects the temperature of the eutectoid transformation and the width of the range of existence of the single-phase cubic solution with the fluorite structure (Table 9.1) [2, 3, 4–6]. It has also been established that the effect of two jointly present rare-earth ions on the phase composition and properties of solid solutions may be equivalent to the effect of the rare-earth ions with the mean radius equal to $r_m = x, r_1 + (1-x)\cdot r_2$. If the temperatures of the phase

Table 9.1. Effect of the concetration of stabilizing oxide on the eutectic of the phase diagrams of the ZrO_2–R_2O_3 systems [3]

REE	r_{R3+}, Å	C_{eut}, mol%	T_{eut}, °C
La	1.14	5	1850
Ce	1.07	10	1600
Y	0.92	25	300

transformations of the solid solutions are higher than 1600°C, the decomposition of the cubic solid solutions may take place in the growth process and it will affect the dimensions and phase composition of the crystals.

As indicated by Table 9.1, the addition of ions with a small radii together with the 'large' rare-earth ions may reduce the temperature of phase transformations and widen the range of existence of the compositions of the cubic solid solutions of the fluorite type. This circumstance also explains why it is possible to produce a large single phase crystals with the cubic structure containing a high concentration of the oxides of the rare-earth elements situated in the beginning of the series by simultaneous addition to the melt of large quantities of rare-earth element oxides with smaller ion radii.

The maximum concentration of Pr_2O_3 which could be introduced without breakdown of the cubic solid solution was 5 mol% for the ZrO_2–(20–30)mol% Y_2O_3 solid solution and 12 mol% for the ZrO_2–(14–20) mol% Lu_2O_3 solid solution. The x-ray diffraction patterns of the crystals with the ZrO_2–(20–30) mol% Y_2O_3–(3–5) mol% Pr_2O_3 and ZrO_2–(10–30) mol% Yb_2O_3(Lu_2O_3)–(3–12) mol% Pr_2O_3 composition showed the presence of only a fluorite-type cubic solution. In the crystals containing 15 mol% of Pr_2O_3, x-ray diffraction analysis showed the presence of the monoclinic phase.

The nature of distribution of the oxides of the rare-earth elements with large ion radii (cerium, neodymium) is identical with that described in Chapter 7. The effect of Pr_2O_3 on the growth of single crystals and the distribution of praseodymium in relation to its concentration is shown in Fig. 9.2. The boundaries of the crystals are characterised by an increased concentration of praseodymium and by a decrease in the content of yttrium. This indicates the displacement of praseodymium at the solidification front as a result of the low value of the effective distribution coefficient of Pr^{3+}.

In the growth of C-OX crystals, the authors of [1] used large diameter crucibles (400 mm, equipment Kristall-403) and low growth rates (up to 1–2 mm/h). A large diameter cold crucible supports the formation of favorable conditions for the growth of large crystals.

The dependence of several optical and mechanical properties of the C–OX crystals of the ZrO_2–Y_2O_3– Pr_2O_3 system on the chemical composition is presented in Table 9.2.

The comparison of these properties with the properties of fianite crystals shows that the C–OX single crystals are characterised by

Table 9.2. Dependence of the properties of C–OX crystals of the ZrO_2–Y_2O_3–Pr_2O_3 system [1]

Composition, mol%			Refractive index (λ = 582 nm)		Microhardness, kg/mm²	Density. kg/m³
ZrO_2	Y_2O_3	Pr_2O_3	Prior to annealing*	After annealing		
78	19	3	2.04	2.07	1359±51	5626
78	20	2	2.12	–	1359±58	–
77	20	2	2.12	2.10	1272±73	5713
75	20	5	2.10	–	1231±49	5690
65	20	15	–	–	1123±45	5790
70	25	5	2.01	–	1325±56	5694
67	29	4	1.93	–	1499±45	5534
67	30	3	2.03	–	1364±38	5544
65	31	4	2.02	2.03	1155±57	5385
65	30	5	2.05	2.04	1235±49	5422
64	30	5	2.04	2.04	1287±53	5574

*Annealing: 1200–1400°C, vacuum 10^{-3} Pa, 2–5 h

lower density, microhardness and reflective index. This is associated with a high content of the stabilizing oxide. However, these differences are small. The content of oxides with high ion radii (cerium, neodymium) has a strong effect on the microhardness of crystals.

It should be mentioned that after growth a part of the praseodymium ions is situated in the crystals of cubic zirconia in the tetravalent state [7]. Vacuum annealing is followed by reduction to Pr^{3+} and this is accompanied by a large change in the ion radius of praseodymium (from 0.96 A to 1.14 A). In the crystals containing even large quantities of the praseodymium oxide (up to 10–12 mol%), this transition takes place reversibly without destruction of the phase homogeneity and integrity of the specimens. The changes in the absorption spectra of the 80 mol% ZrO_2–14 mol% Y_2O_3–6 mol% Pr_2O_3 after annealing for two hours in a vacuum of 10^{-3} Pa at 1200°C are presented in Fig. 9.3.

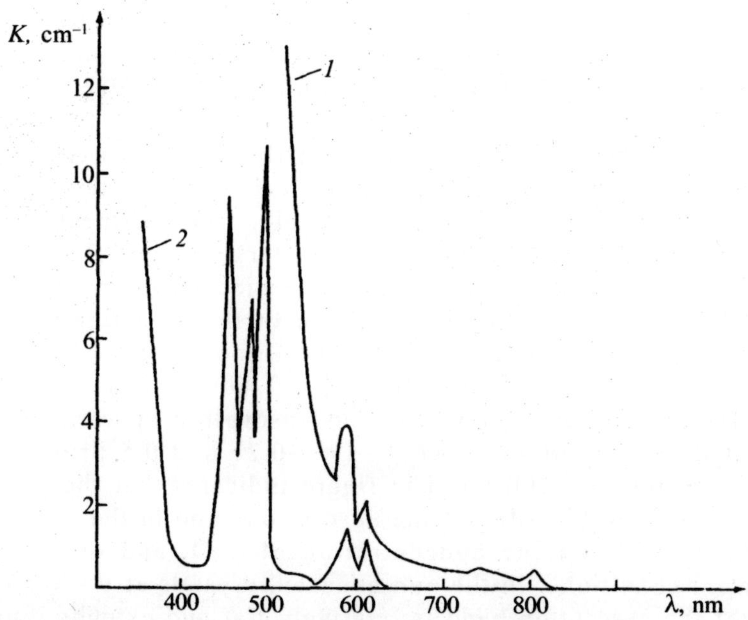

Figure 9.3. Transmission spectra of a 80 mol% ZrO_2–14 mol% Y_2O_3–6 mol% Pr_2O_3 crystal (1 – after growth, 2 – after vacuum annealing in a vacuum of 10^{-3} mm Hg, T = 1200 °C).

9.2. THE SPECTROSCOPIC PROPERTIES OF C–OX CRYSTALS

An increase in the concentration of the stabilising oxide has a strong effect on the optical properties of the crystals of cubic solid solutions based on zirconia. The absorption spectra of the transition elements are sensitive to the local structure of the nearest neighbourhood of the admixture ion. The authors of [1] investigated the effect of composition and heat treatment (in vacuum and in air) on the absorption spectra of 'pure' $Zr_{1-y}Y_yO_{2-y/2}$ crystals and similar crystals activated by transition metals (Co, Mn, V, Ni, Cr) in a wide yttrium concentration range $0.15 \leq y \leq 0.55$, {ZrO_2–(8–38) mol% Y_2O_3}. Investigations were carried out on the absorption spectra of the crystals in relation to the oxidation–reduction conditions of heat treatment. Admixtures were introduced into the melt in the form of oxides and their concentration was varied in the range 0.01–1 wt%. A further increase in the concentration of the admixtures resulted in a large deterioration of the conditions of growth of single crystals and formation of growth defects in these crystals.

The effect of vacuum annealing on the absorption spectra of zirconia and solid solutions based on it depends on temperature, pressure, annealing time and cooling rate. 'Hard' annealing ($T >$ 1400°C and rapid cooling) results in darkening of the crystals caused by the presence of a high intensity wide absorption band with a maximum in the vicinity of 0.5 μm. In the investigations, the authors used 'soft' annealing ($T <$ 3000°C, cooling rate ~10°C/min). The crystals were annealed for 0.5–10 hours in a vacuum of 10^{-4}–10^{-5} Pa. Figure 9.4 shows the absorption spectra of non-activated crystals of $Zr_{1-y}Y_yO_{2-y/2}$, (y = 0.21 and 0.5 prior to and after annealing at 1200°C), The figure indicates that the position and curvature of the edge of the absorption band in the ultraviolet region depends on the concentration of Y_2O_3 and on the heat treatment conditions. In the non-annealed crystals at y = 0.55, the edge of the absorption band is relatively flat and extends into the visible range and, consequently, these crystals are characterised by yellow colour. In crystals with y = 0.21, the edge is greatly displaced into the short wave range. Soft annealing in vacuum reduces the efficiency of absorption in the range λ > 0.270 μm and eliminates the difference in the spectra of the crystals with different concentrations of the yttrium oxide. The 0.5 μm band is not detected

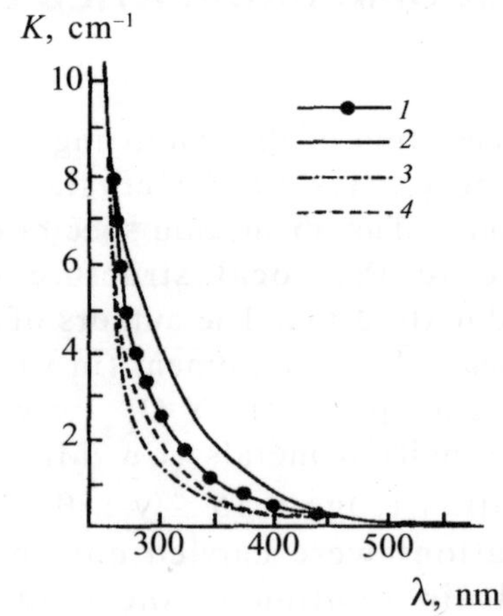

Figure 9.4. Absorption spectra of non-activated $Zr_{1-y}Y_yO_{2-y/2}$ crystals, y = 0.21 (curves 1 and 2) and y = 0.5 (curves 2 and 4) prior to (1, 2) and after vacuum annealing at 1200°C (3, 4).

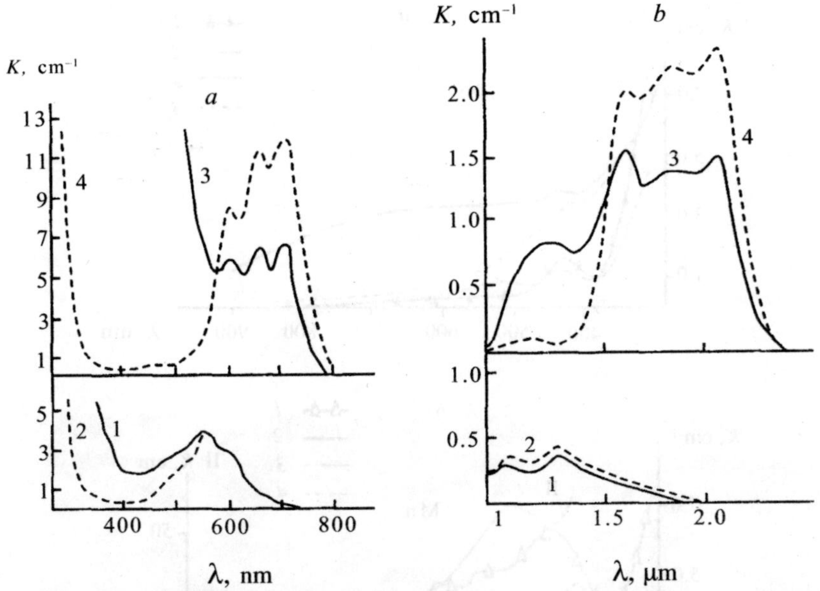

Figure 9.5. Spectra of $Zr_{1-y}Y_yO_{2-y/2}$–Co crystals, y = 0.15 (curves 1, 2) and y = 0.55 (curves 3, 4) in UV and visible (a) and IR ranges (b). Curves 1,3 – prior to, 3, 4 – after annealing of crystals.

in these conditions. The introduction of transition metals results in the formation of an absorption characteristic for these metals which also depends on the concentration of Y_2O_3 and on the heat treatment conditions (Fig. 9.5a, b, Fig. 9.6).

Prior to annealing, the spectra of the activated crystals contained two components: the relatively high intensity structureless absorption in the short-wavelength part of the spectrum detected to various degrees for all activating agents, and absorption bands, characteristic of the introduced activation agent which can be attributed to the *d–d* transitions of the metal ions [8]. An increase in the concentration of the yttrium oxide increases the intensity of absorption in the short-wavelength part of the spectrum and, in addition to this, results in changes in the *d–d* spectra in some cases. An especially strong effect of the concentration of the yttrium oxide on the absorption spectra is detected in the crystals with manganese, where the absorption coefficients for crystals with y = 0.5 in the entire visible range are 10–20 times higher than for crystals with y = 0.15 (Fig. 9.6b). Vacuum annealing in all cases greatly reduces the intensity of the short-wavelength structureless absorption and results in changes in the *d–d* spectra. The spectra of crystals with manganese

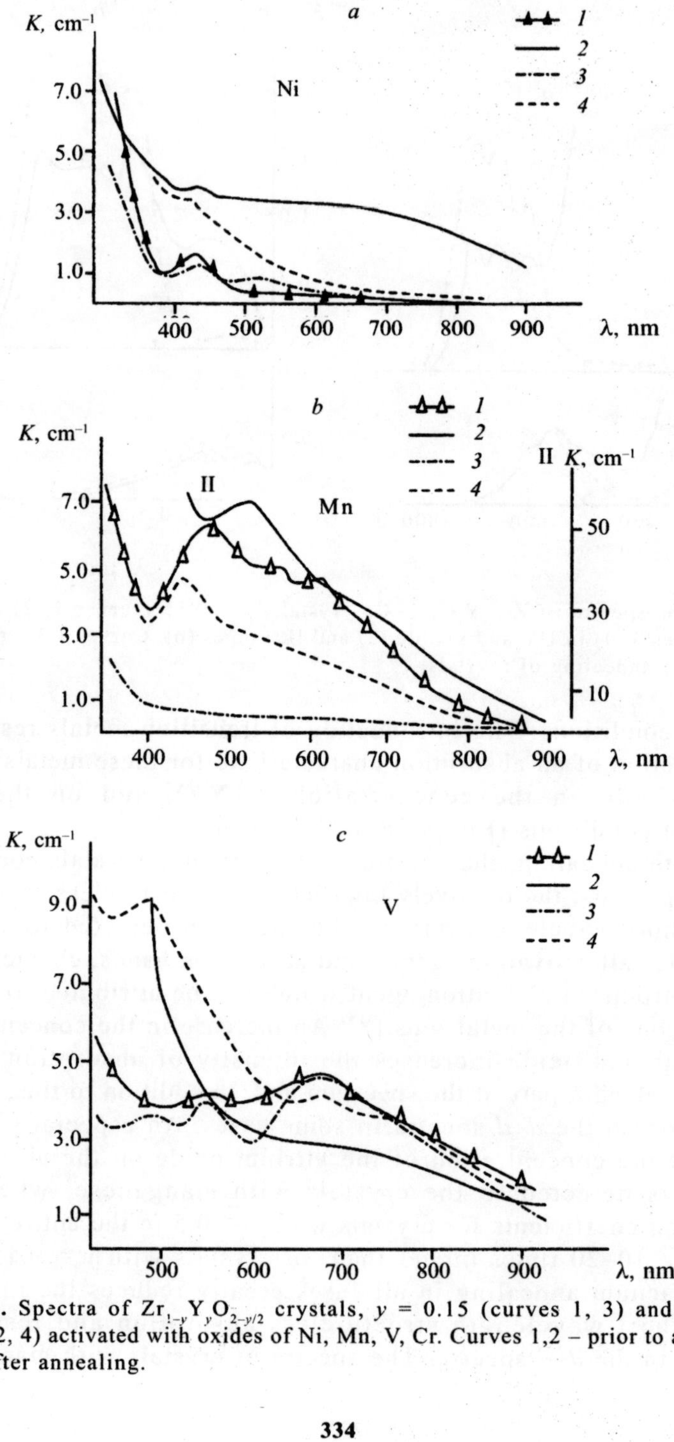

Fig. 9.6. Spectra of $Zr_{1-y}Y_yO_{2-y/2}$ crystals, $y = 0.15$ (curves 1, 3) and $y = 0.55$ (curves 2, 4) activated with oxides of Ni, Mn, V, Cr. Curves 1,2 – prior to annealing, 3, 4 – after annealing.

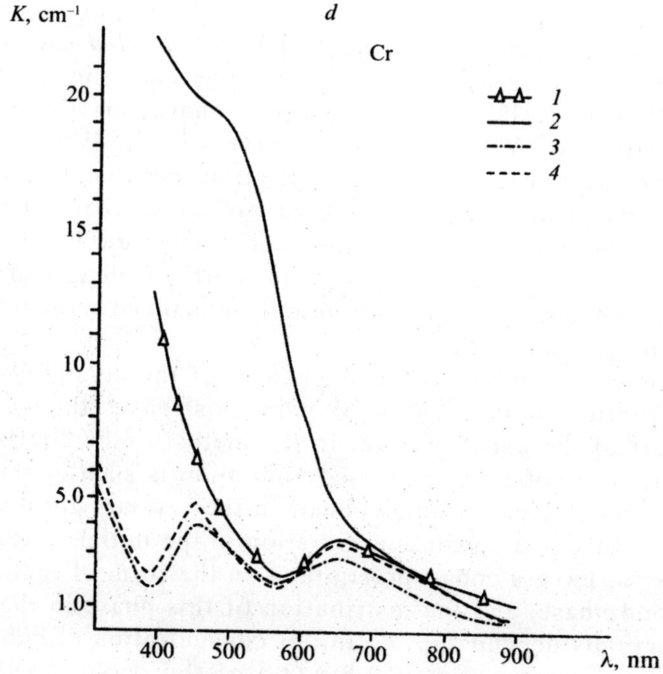

Figure 9.6 (completion).

(y = 0.5) show the largest changes in the *d–d* spectra.

The effect of the composition and heat treatment on the absorption spectra has been studied in detail on crystals with a cobalt admixture. Figure 9.5 shows that at a low Y_2O_3 concentration (curves *1, 2*), the cobalt spectra contain two complicated bands affected only slightly by annealing and causing only a small decrease in the efficiency of absorption in the short-wavelength part of the spectrum. High concentrations of Y_2O_3 (curves *3, 4*), result in the formation of new absorption bands and an increase in the intensity of short-wave structureless absorption in the non-annealed crystals. Vacuum annealing of crystals with a high concentration of Y_2O_3 not only causes 'bleaching' in the short-wavelength part of the spectrum but also results in a large increase of the intensity of the *d–d* bands (with the exception of a band with a maximum at 1.2 μm which weakens after annealing). Isochronous annealing (10 min) at different temperatures in a vacuum of 10^{-5} Pa shows that changes in the spectra take place mainly in the temperature range 600–700°C. Investigations of crystals with different concentration of Y_2O_3 (y = 0.15–0.5) also shows significant changes in the profiles of the *d–d* spectra of Co in $Zr_{1-y}Y_yO_{2-y/2}$ with an increase of the yttrium concentration. At y = 0.28 (16 mol%) the spectra are characteristic of a low concentration of Y_2O_3 (Fig. 9.5, curves *1*, *2*). With a further increase of the yttrium oxide concentration, new bands appear and the intensity of these bands increases with an increase of the Y_2O_3 concentration and at $y > 0.46$ (30 mol%) the new bands become controlling (Fig. 9.5, curves *3*, *4*). The short-wavelength absorption of the unannealed crystals is gradually enhanced with increasing content of yttrium oxide.

The determination of the composition of the crystals by the x-ray diffraction method shows that in most cases they include a small part of the cobalt present in the melt, i.e. the coefficient of distribution of cobalt in stabilised zirconium is smaller than unity. The limiting concentration of cobalt in the crystals did not exceed 0.10–0.15 wt% at a cobalt concentration in the initial mixture of up to 1.0 wt%. Excess cobalt precipitates in the form of inclusions of the second phase, and the distribution of this phase in the volume of the crystals depends on the initial concentration of cobalt. At a small concentration (up to 0.5 wt%) of the cobalt oxide, large inclusions (~10 μm) group at the beginning of the crystal, and the smaller ones form groups in the region of the cellular structure. With an increase of the cobalt concentration in the crystal $y > 0.4$, inclusions form throughout the entire volume, and no inclusions were detected at a concentration of $y < 0.4$.

The effect of heat treatment on the absorption spectra of the $Zr_{1-y}Y_yO_{2-y/2}$ crystals is associated with the changes of the charge state of the defects present in them. With a disruption of oxygen stoichiometry, every oxygen atom moving outside the limits of the crystal, leaves behind an anion vacancy and two electrons because of which the $ZrO_{2-\delta}$ crystal, where δ is the oxygen deficiency, contains 2δ excess electrons distributed between defects depending on temperature, the concentration of defects and the binding energy of the electrons in the defects. The deviation from oxygen stoichiometry increases with increasing temperature and with a decrease in the partial pressure of oxygen in the atmosphere. Thus, vacuum annealing increases the concentration of excess electrons.

Oxygen vacancies form donor levels in the forbidden band of the crystal [9]. The stabilized crystals of zirconia contain large concentrations of the vacancies, comparable in the order of magnitude with the concentration comparable with the concentration of the anion sites of the lattice. This may result in the formation of a 'band' of vacancies below the conduction band. It may be assumed that the probability of electronic transitions from the valence band to the levels of the oxygen vacancies is comparable with the probability of inter-band transitions and in this case, the short-wavelength edge of absorption of the non-activated crystals (Fig. 9.4) should be attributed to transitions to the levels of the vacancies, and not to the conduction band. Consequently, the position of the edge of absorption (approximately 5 eV) does not correspond to the width of the forbidden band but corresponds to the position of the levels of vacancies of oxygen in relation to the valence band of the crystal. The stabilizing oxide introduces, into the crystal lattice of zirconia, vacancies with the effective charge +2 (V_O^{++}) which may, in principle, trap one or two electrons with the formation of a singly ionised V_O^{+} or neutral (V^o) vacancy [9]. Another type of trap is formed by admixtures of transition metals capable of changing valence quite easily, capturing an electron or a hole. The level of the admixture ion maybe regarded as empty or filled, depending on the valence state of the ion in the crystal. Therefore, the presence of, for example, Co^{3+} in the crystal indicates the presence of empty levels of Co^{2+} whose filling corresponds to the $Co^{3+} \rightarrow Co^{2+}$ transition.

In the crystals containing defects, inter-band electronic transitions are added to intra-centre transitions with charge transfer between the defects and the main ion. The charge state of the defect determines the nature of absorption spectra. In particular, transitions from the valence band are possible to the local empty level. In the crystals, grown in air, the deficit of oxygen is smaller, consequently, the levels in the forbidden band are mostly empty. The absorption spectra of these crystals should have the bands corresponding to the transitions from the valence band to the local levels. Vacuum annealing, accompanied by an increase in the oxygen deficit, results in filling of part of the local energy levels and transitions to these levels become impossible. This approach may be used to explain the weakening of absorption in the short-wavelength part of the spectrum after vacuum annealing. The transitions from the valence band to the levels of admixture ions correspond to the bands of charge transfer from a ligand to a metal ion, typical of transition metals in melts and solid solutions [8]. The levels, responsible for similar absorption in non-activated crystals (Fig. 9.4), which are also removed by vacuum annealing, may be associated with non-controlled admixtures of the transition metals or vacancies. Evidently, the band in the vicinity of 0.5 μm, formed after 'hard' vacuum annealing, is determined by these transitions from the level of vacancies to the conduction band; possibly, vacancy levels are situated at a depth of 2.5 eV below the conduction band. The appearance of an electron on the local level changes the nature of intra-centre transitions. For an admixture centre, filling of the level by an electron indicates a decrease by unity of the charge of the ion and the appropriate change of the spectrum.

In the spectra of the $Zr_{1-y}Y_yO_{2-y/2}$–Co crystals (Fig. 9.5), the absorption in the range 0.270–0.500 μm, which disappeared after vacuum annealing, may be attributed to transitions with charge transfer. Cobalt in the crystal lattice of ZrO_2 may exist in theform of Co^{2+} and Co^{3+} ions The presence of the charge transfer band indicates the presence of Co^{3+} ions in the unannealed crystals because only they are capable of accepting electrons from oxygen in the process of charge transfer $Co^{3+}+O^{2-} \rightarrow 5^{2+}.+O^{-}$. Since vacuum annealing changes Co^{+3} to Co^{+2}, the spectra of annealed crystals in Fig. 9.5 should be attributed to bivalent cobalt. As indicated by the figure, Co^{+2} is present in the crystals prior to annealing, therefore unannealed crystals contain simultaneously Co^{+3} and Co^{+2}. Evidently, ions Co^{+3} are related to the band λ_{max} = 1.2 μm which disappears after annealing. In the visible part of the spectra it is not possible to separate Co^{+3} bands from Co^{+2} bands. In Co^{+2} spectra, attention should be given to special features, associated with the Y_2O_3 concentration: 1. existence of two types of Co^{+2} spectra; 2. increase of the fraction of Co^{+3} in the total concentration of cobalt prior to annealing with increasing concentration of yttrium oxide.

The existence of two different spectra of Co^{+3} is naturally associated with the possibility of formation of centres with different coordinations of Co^{+2}. It is well known that an increase in the Y_2O_3 concentration increases the number of oxygen vacancies and, consequently, it may be assumed that the Co^{+2} spectra in the crystals with the Co concentration y = 0.55 belong to centres with a lower coordination number than in the crystals with y = 0.15. Evidently, at y = 0.15 the coordination of Co^{+2} is octahedral and at y = 0.55 it is basically tetrahedral. Consequently, for the sixfold coordination at y = 0.15 transitions in the optical range may be interpreted (Fig.9.7a) as the $^4T_1 \rightarrow {}^4T_2$ transitions (bands with max

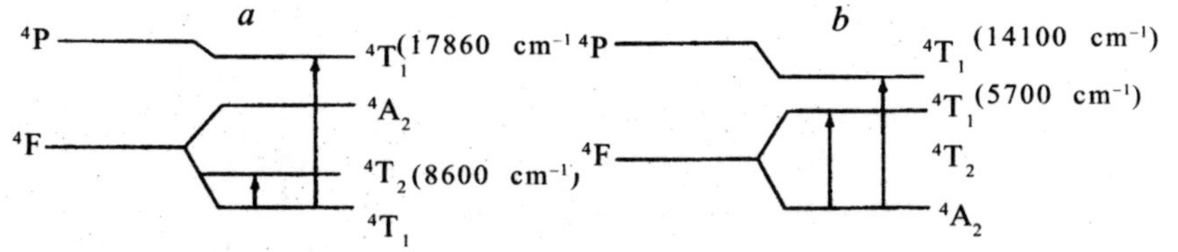

Figure 9.7. Scheme of transitions corresponds to absorption of the Co^{2+} ions in nodes with coordination numbers 6 (a) and 4 (b) in ZrO_2–Y_2O_3 crystals

~1.15 μm) and $^4T_1(F) \rightarrow {}^4T_1(F) \rightarrow {}^4T_1(P)$ (bands with λ_{max}~ 0.56 μm). The $^4T_1(F) \rightarrow {}^4A_2(F)$ transitions are weak and almost completely merged with $^4T_1(F) \rightarrow {}^4T_1(P)$.

The spectrum of bivalent cobalt in quaternary coordination (Fig.9.7b) can be interpreted in the same manner. The band with maximum ~0.68 μm can be attributed to $^4A_2 \rightarrow {}^4T_1(P)$ transitions in the band with max~1.75 μm to $^4A_2 \rightarrow {}^4T_1(F)$ transitions. For Co^{2+} in the tetrahedral environment the $^4A_2 \rightarrow {}^4T_2$ transition usually does not occur [8]. The splitting of the levels of the ion in the tetrahedral coordination is smaller in comparison with octahedral coordination and, correspondingly, the absorption spectra are displaced to longer wavelengths.

The fraction of Co^{3+} in unannealed crystals may be estimated on the basis of the changes in the intensities of Co^{2+} bands after annealing. Fig. 9.5 shows that after annealing the intensity of the 1.15 μm Co^{2+} band in the crystal with y = 0.15 increases by 10%, and the intensity of the 1.85 μm band in the crystal with y = 0.55 increases by 50%. This means that prior to annealing the Co^{3+} content in the crystals with y = 0.15 is less than 10%, and in the crystals with y = 0.55 the fraction of Co^{3+} is 30% of the total cobalt concentration. An increase of the fraction of Co^{3+} with increasing Y_2O_3 concentration is confirmed by an increase in the intensity of the charge transfer band in the crystals with y = 0.55. Consequently, it may be concluded that the oxygen non-stoichiometry in the crystals will also decrease with increasing Y_2O_3 concentration. Evidently, this conclusion also holds for non-activated crystals because in these crystals an increase of Y_2O_3 concentration is accompanied by intensification of absorption in the vicinity of the edge associated with the charge transfer from the valence band to the local level (Fig. 9.5).

The effect of Y_2O_3 concentration on the disruption of oxygen stoichiometry in $Zr_{1-y}Y_yO_{2-y/2}$ crystals is partially explained by the "dilution" of the anion sublattice by the vacancies, i.e. by a decrease of the concentration of O^{2-} in anion sites. However, dilution may result in a change not greater than 20% and not 300% as observed in the experiments. One of the possible explanations of the effect of the Y_2O_3 concentration on the disruption of oxygen stoichiometry is based on an assumption according to which the migration of oxygen from the crystal is affected by the vacancy concentration.

The process of oxygen exchange between the crystal and the atmosphere can be described as a result of three consecutive quasi-chemical reactions [10]:

$$O_O \leftrightarrow V_O + 1/2O_2,\ K_1 = [V_O]P_{O2}^{1/2}, \tag{9.1}$$

$$V_O \leftrightarrow V_O^{+} + e',\ K_2 = [V_O^{+}][e']/[V_O], \tag{9.2}$$

$$V_O \leftrightarrow V_O^{++} + e', \tag{9.3}$$

where O_o is the atom of oxygen in the anion site, V_O, V_O^{+}, V_O^{++} are respectively the neutral (with two electrons), once ionised (with one electron) and twice ionised oxygen vacancy, e' is a free electron.

In a stabilized $Zr_{1-y}Y_yO_{2-y/2}$ crystal with slight disruption of stoichiometry (i.e. $\delta \ll y$) $[V_O^{++}] \approx y/2$, equations (9.1)–(9.3) show that:

$$[V_O] = K_1/P_{O2}^{1/2}, \tag{9.4}$$

$$[V_O^{+}] = (K_1/K_2/K_3)^{1/2}(y/2)^{1/2}P_{O2}^{-1/4}, \tag{9.5}$$

$$[e'] = (K_1/K_2/K_3)^{1/2}(y/2)^{-1/2}P_{O2}^{-1/4}, \tag{9.6}$$

The total concentration of excess electrons is $n = 2\delta = 2[V_O] + [V_O^{+}] + [e']$. Assuming that at the quenching temperature $[e'] \gg [V_O][e'] \gg [V_O^{++}]$, we have:

$$2\delta = [e'] \sim (y/2)^{-1/2} \tag{9.7}$$

In this case, an increase of y from 0.15 to 0.55 halves electron density $[e']$. This is closer to the experimental results. Evidently, in addition to the investigated mechanisms, the stoichiometry is affected by other factors: Y^{3+} may affect the binding energy of O^{2}, and the mobility of vacancies also depends on the composition and quenching temperature, which can differ at the same cooling rate.

Thus, the investigations into the effect of vacuum annealing on the spectrum of absorption of pure and activated (by transition metals) crystals $Zr_{1-y}Y_yO_{2-y/2}$ with $y = 0.15–0.55$, carried out at [1] shows that:

a. Prior to annealing, the crystals have a small oxygen deficit, which decreases with increasing Y_2O_3 concentration. Vacuum annealing increases the oxygen deficit and, consequently, crystals contain excess electrons occupying the local levels in the forbidden band. This results in changes in the absorption spectra;

b. In the non-activated $Zr_{1-y}Y_yO_{2-y/2}$ crystals, the short-wavelength edge of the absorption band in the vicinity of 0.250 μm may be associated with the transition of electrons from the valence band to empty local levels of the oxygen vacancies. "Hard" vacuum annealing results in filling of these levels with electrons and in the appearance of a wide absorption band with a maximum near 0.5 μm;

c. The absorption spectra of activated crystals form as a result of the *d–d* intra-ion transitions and transitions between admixture local levels and the conduction band (transition with charge transfer). Vacuum annealing usually results in the reduction of admixture ions and, consequently, the bands of charge transfer from the oxygen to the admixture ion disappear;

d. Ions Co^{2+} form two types of centres with different coordination in $Zr_{1-y}Y_yO_{2-y/2}$: at a low content of the stabilizing oxides, the coordination number is 6, at a higher content it is 4.

9.3 EFFECT OF ANNEALING ON THE VALENCE STATE OF COBALT IN THE ZrO_2–Y_2O_3 CRYSTALS

The processes of reversible exchange of oxygen between the crystal and the atmosphere, including diffusion in the single crystal and reactions in the volume and on the surface, as shown previously, may be investigated on the basis of changes in the absorption spectra accompanying these processes and associated with valence transitions of admixture elements.

The temperature dependences of equilibrium and rates of solid phase reactions $Co^{2+} \leftrightarrow Co^{3+}$ taking place during annealing of ZrO_2–30 mol%Y_2O_3 crystals with a cobalt admixture, were investigated in [1]. The cobalt content in the crystals was $3\cdot10^{-2}$ wt.%. Plate-shaped specimens with a thickness of 0.5–0.6 mm were subjected to isothermal annealing in a vacuum of $3\cdot10^{-3}$ Pa and in air at 400 and 1400°C and then quenched. The absorption coefficients were measured at the maximum of the absorption band at 715 nm for Co^{2+} and in the charge transfer band for Co^{3+}.

Valence transitions of cobalt in ZrO_2–Y_2O_3 crystals can be represented as a result of two consecutive quasi-chemical reactions [4]:

$$O_O \leftrightarrow V_O + 1/2O_2, \tag{9.8}$$

$$K_1 = K_{10}\exp(-\Delta H_1/kT) = [V_O]P_{O2}^{1/2} \tag{9.9}$$

$$V_O + 2Co^{3+} \leftrightarrow V_O^{++} + 2Co^{2+} \tag{9.10}$$

$$K_1 = K_{10}\exp(-\Delta H_1/kT) = [V_O]P_{O2}^{1/2} \tag{9.11}$$

where O_O is the O^{2-} ion in the anion site, V_O, V_O^{++} are the neutral and twice-ionised oxygen vacancies, $[V_O]$, $[V_O^{++}]$ are the vacancy concentration, c_2 and c_3 are the concentrations of Co^{3+} and Co^{3+}; K_1 and K_2 are equilibrium constants; ΔH_1 and ΔH_2 are the heats of reaction (9.9) and (9.11), p is the partial pressure of oxygen above the crystal. The ratio of the equilibrium concentrations of Co^{3+} and Co^{2+} is determined by the equations:

$$c_3/c_2 = [V_O^{++}]^{1/2}K_0^{-1/2}\exp(-\Delta H/kT)P_{O2}^{1/4};$$

$$K_0 = K_{10}K_{20};\ \Delta H = (\Delta H_1 + \Delta H_2)/2$$

The variation of the enthalpy ΔH was determined by annealing the specimens to equilibrium conditions in air at different temperatures in the range 600–1000°C. After quenching, measurements were taken of absorption of Co^{3+} at 500 nm (k_3) and Co^{2+} at 715 nm (k_2). To establish the equilibrium state, the specimens were held in a furnace for a specific period of time. The dependence $k_3/k_2 = f(1/T)$ is sufficiently approximated by a straight line and by its slope, assuming that $[V_O^{++}] \approx$ const, the variation of enthalpy was calculated to be $\Delta H = 1.9\cdot10^4$ J/mol.

Using the correlation between k_3 and k_2, it is possible to determine the concentration of bivalent and trivalent cobalt in the

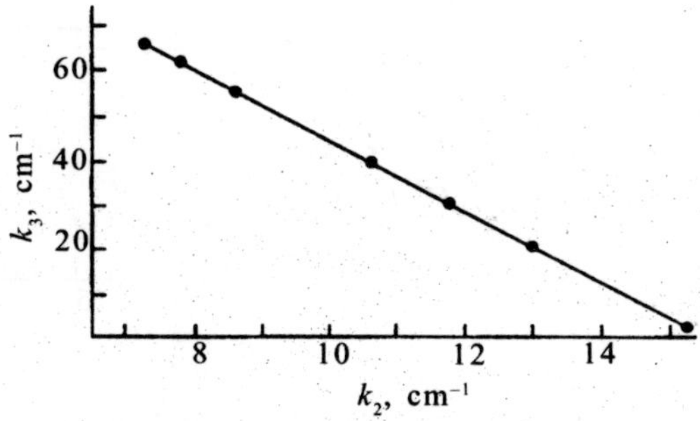

Figure 9.8. Correlation between the absorption coefficient of Co^{3+} at λ = 500 nm (k_3) and Co^{2+} at λ = 715 nm (k_2).

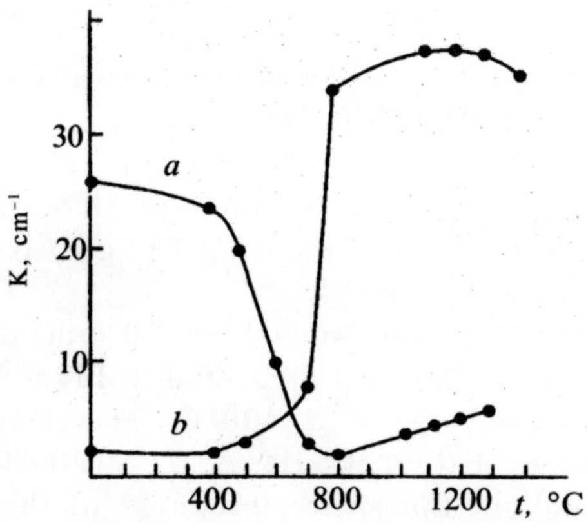

Figure 9.9. Dependence of the absorption coefficient of Co^{3+} at λ = 450 nm on annealing temperature (10 min). a) in vacuum, b) in air after vacuum annealing.

crystals, c_2 and c_3. If the changes of c_2 and c_3 due to heat treatment are caused only by the transitions $Co^{2+} \leftrightarrow Co^{3+}$, then $c_2+c_3 = c_0$=const, and k_3 and k_2 are linked by a linear dependence $k_3 = c_0\varepsilon_3-(\varepsilon_3/\varepsilon_2)\cdot k_2$, where ε_3 and ε_2 are molar coefficients of absorpt-ion. In the case of non-uniform distribution of Co^{2+} and Co^{3+} in the thickness of the crystal at intermediate stages of annealing, this dependence is fulfilled for mean absorption coefficients:

$$k = (\varepsilon/1)\int\sigma(x)dx \qquad (9.12)$$

Figure 9.8 shows a correlation between k_3 (500 nm) and k_2 (715 nm), obtained by measurements in different stages of transition from one equilibrium state (800°C in vacuum) to another (650°C in

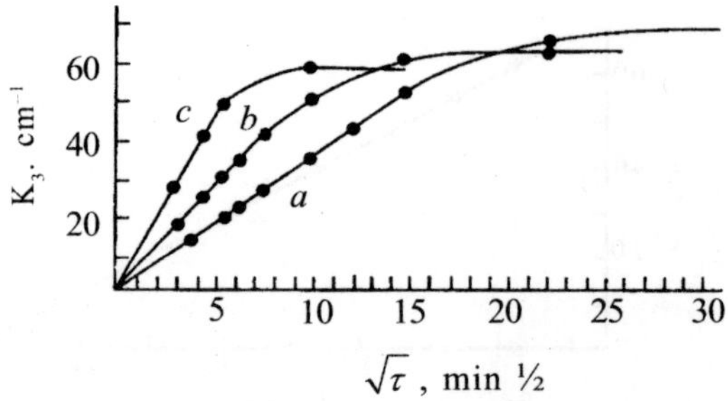

Figure 9.10. Dependence of the absorption coefficient of Co^{3+} at wavelength λ = 500 nm on annealing time at temperatures, °C. a) 650, b) 700, c) 750 in coordinates: k and $\sqrt{\tau}$.

air), which gives $\varepsilon_3/\varepsilon_2$ = 8.5. The resultant ratio $\varepsilon_3/\varepsilon_2$ shows that in transition from the equilibrium state in vacuum at 800°C to equilibrium in air at 650°C, the fraction of Co^{3+} in the total concentration of cobalt changes from 1 to 50% and the transition is accompanied by the absorption of $2\cdot10^{-5}$ g of oxygen by 1 g of the crystal with $c_0=3\cdot10^{-2}$ wt.%.

The curves of isochronous annealing (10 min) of crystal, in the non-equilibrium conditions, are presented in Fig.9.9.

Large changes start at T > 400°C, at relatively extensive diffusion of oxygen in the crystals. At T > 1200°C equilibrium is established throughout the entire thickness of the specimen after annealing for 10 min.

Figure 9.10 shows the kinetics of the $Co^{2+} \rightarrow Co^{3+}$ transition in annealing in air at 650–750°C. At the start of annealing, the thickness of the specimens has no significant effect, and a linear dependence of k_3 on annealing time $\sqrt{\tau}$ typical of diffusion kinetics, is observed.

The diffusion coefficients in initial sections are calculated for the total amount of the substance diffused into the specimens [10]. In the examined case, diffusion takes place through two surfaces and calculations are carried out using the equation:

$$(k - k_0)/(k - k_\infty) = (4/1\sqrt{\pi})\sqrt{D\tau} \tag{13.13}$$

where k_0 is the initial value of the absorption coefficient, k_∞ is the stationary value corresponding to diffusion through the entire thickness. The diffusion coefficients: D = $0.64\cdot10^{-8}$, $2.24\cdot10^{-8}$, $21.5\cdot10^{-8}$ cm^2/s were determined at temperature lower than 650, 700 and 750°C, respectively; activation energy was 1.8 eV. The numerical values of the diffusion coefficients do not characterise the free diffusion of neutral vacancies of oxygen (combined diffusion of ionised vacancies and electrons). Instead, they characterise "reaction" diffusion accompanied by binding of neutral vacancies by Co^{3+} ions by reaction (9.2) and, according to [11], they represent effective coefficients which depend not only on the coefficients of diffusion of free defects but also on the parameters of the chemical reaction and concentration of the defects and admixtures. The activation energy is determined by the potential barrier of diffusion and by the energy of chemical bonding of defects by the reaction (9.2).

9.4 USING C–OX CRYSTALS IN JEWELLERY INDUSTRY

The large differences between C–OX crystals and fianites are associated with the possibility of introducing into them relatively high concentrations of oxides of Ce, Pr, Nd and with changing the optical characteristics by doping with ions of transition metals. A suitable example of this is the behaviour of Co^{2+} ions in the ZrO_2–Y_2O_3 crystals. At a low content of yttrium oxides (fianites) the crystals after vacuum annealing are lilac-coloured and at a high content (>25mol%) they are cornflower-coloured. This is associated with the fact that at a low yttrium oxide content, the coordination of the Co^{2+} ions is octahedral and there is a significant transmission in the red and yellow visible ranges of the spectrum, whereas at high yttrium oxide concentrations, the absorption spectrum of the Co^{2+} ion in the tetrahedral coordination in these spectral regions is greatly increased (Fig. 9.10a,b).

When adding cobalt ions to crystals with a higher Pr (or Ce) oxide content it is possible to produce crystals with a wide range of colours and their shades from blue to green of various intensities [12,13]. Consequently, C–OX crystals may efficiently imitate different specimens of natural crystal sapphire and emerald which could not be achieved previously with fianite crystals. At present, C–OX crystals are used widely in jewellery industry.

10. CONCLUSION

The presented specific examples of synthesis and growth of crystals of different substances clearly demonstrate the individual and specific nature of the conditions dictated by the physicochemical properties of the synthesized compounds. However, it should be mentioned that heating of the melt, its maintenance and control of the temperature are common for the cold crucible technology. The main advantages of the cold crucible as the chemical reactor are an unlimited range of temperature variation, the possibility of using it for substances of different chemical nature in various atmospheres, the complete exclusion of the interaction of the melt with the crucible material, and the simple procedure of extracting the crystalline material from the crucible. All these special features and other potentialities place the cold crucible technology among promising methods of synthesis and growth of crystals of oxide compounds.

References

1. E.E. Lomonova, Technology, properties and application of zirconia-based crystals, Dissertation, General Physics Institute, Moscow, 2001.
2. M. Perez-y-Jorba, *Ann. Chem.*, 7, 479-511, 1962.
3. R. Collong, Nonstoichiometry, Mir, Moscow, 1974.
4. Yoshimura Masahiro, Sata Toshiyuki, *Bull. Tokyo Inst.Technol.*, 108, 25-32, 1972.
5. Alain Rovanet, Diagrammes de Solidification et Diagrammes de Phases Formes par la Zircone Avec les Sesquioxydes des Lanthanides, Colloques Internationaux du Centre National de la Recherche Scientifique, Paris-Grenoble, 1, 180, 221–235, 1970.
6. Note de M.A.Rovanet, presentee par M.G.Chaudron, Etude a haute temperature du systeme Zircone-sesquioxide de neodime, CR. Sci. Paris, 270, C, 802–805, 1970.
7. V.I. Aleksandrov, et al., *Izv. AN SSSR, Neorg. Mater.*, 11, 4, 664–667, 1975.
8. D.T. Sviridov, et al., Optical spectra of ions of transition metals in crystals, Nauka, Moscow, 168–173, 179–180, 1976.
9. F. Kröger, Chemistry of Imperfect Crystals, Mir, Moscow,
10. K. Hauffe, Reactions in Solids and on Their Surface, Vol. 1, IL, Moscow, 1962.
11. E.M. Akul'onok, et al., *J. Sol. St. Chem.*, 26, 1, 17–25, 1978.
12. V.I. Aleksandrov et al., A method of growth of coloured crystals of refractory oxides (its variants), Author's Cert. No. 1086833, 23.12.1983.
13. V.I. Aleksandrov et al., A method of growth of coloured crystals of refractory oxides (its variants), Author's Cert. No. 1112821, 09.02.1983.

Origins and Foundations in the Modernist Age

In memory of Murray, Brian, and Bill

Origins and Foundations in the Modernist Age

Physical Sciences, Social Sciences and Humanities

Stephen Kern

EDINBURGH
University Press

Edinburgh University Press is one of the leading university presses in the UK. Publishing new research in the arts and humanities, EUP connects people and ideas to inspire creative thinking, open new perspectives and shape the world we live in. For more information, visit www.edinburghuniversitypress.com.

Edinburgh University Press Ltd
13 Infirmary Street, Edinburgh EH1 1LT

Typeset in 10.5/13pt Sabon LT Pro
by Cheshire Typesetting Ltd, Cuddington, Cheshire
Printed and bound in the UK using 100% renewable electricity by CPI Group (UK) Ltd

A CIP record for this book is available from the British Library

ISBN 978 1 3995 6588 2 (hardback)
ISBN 978 1 3995 6659 9 (paperback)
ISBN 978 1 3995 6661 2 (webready PDF)
ISBN 978 1 3995 6660 5 (epub)

EU Authorised Representative:
Easy Access System Europe
Mustamäe tee 50, 10621 Tallinn, Estonia
gpsr.requests@easproject.com

MIX
Paper | Supporting responsible forestry
FSC® C013604
www.fsc.org

This book is printed using paper from well-managed forests, recycling and other controlled sources

Contents

Figures

Preface

I had a first inkling of this book's period of focus over sixty years ago when as a Berkeley undergraduate I took a course on modern European cultural history by the distinguished scholar Carl Schorske, who at that time was working on a book on turn-of-the-century Vienna, about a time and place that was exceptionally rich in cultural production.[1] Week after week he introduced one major cultural figure after another from a wide range of fields including philosophy, psychiatry, literature, music, architecture, and art, all from that period that has continued to intrigue me up to the present.

In 1970 I completed a dissertation on a major thinker at the center of that culture, Sigmund Freud.[2] While writing it I became something of a Freudian methodologically, because for five years I had been immersed in his psychoanalytic theory, but at that time I discovered a radically different approach to mental illness and to human experience generally – phenomenology, a method of philosophical inquiry for describing phenomena as they appear to consciousness in order to uncover their essential structures and meaning. Ever since that discovery, phenomenology has grounded my approach to history.

I initially intended my first book to examine Freud's contribution to what was sometimes called "the sexual revolution," but during the research I discovered the French phenomenologist Maurice Merleau-Ponty, who characterized Freud's contribution to sexual theory in a more compelling way as a "discovery of the meaning of the body."[3] Merleau-Ponty's concept of embodiment was more fundamental and more comprehensive than Freud's narrower conception of sexuality in oral, anal, phallic, and Oedipal modes. For Merleau-Ponty the body was not primarily sexual, but more basically a lived foundation of an experiencing subject that shaped perception and action, which ultimately opened a wide range of topics including smell, sound, touch, proximity, taste, temperature, illumination, movement, rhythm, exercise, sports, fatigue, health, disabilities, and death. All humans are universally, essentially, and primordially embodied, and so embodiment became the focus of my first phenomenological approach to history.[4] Embodiment is universal because all humans are embedded in a body, it is essential because it grounds the most fundamental experience of human existence and is necessary for survival, and it is primordial because it is present in earliest infancy and initiates all subsequent developments.

The founder of modern phenomenology Edmund Husserl captured the essence of his method in a famous slogan, "to the things themselves."[5] That pronouncement underscored the necessity of abandoning abstract theorizing and focusing rather on direct experience with an emphasis on perception. Phenomenology also led me to single out for historical analysis universal and essential topics such as embodiment and to describe human experience with language that is as free of presuppositions as possible, meaning language that avoids abstract theorizing and ideally could be understood by any human being in any place or time.

In 1983 I published a general cultural history of the years 1880–1918 that drew on phenomenology to identify focal concepts of time and space.[6] By grounding my analysis on those dimensions that frame every human experience I was able to reduce a wide range of complex activities such as industrial production, diplomatic negotiations, class relations, and military combat to fundamental ideas about time and space. That is, I interpreted industrial production in terms of time management studies, the diplomatic crisis of July 1914 from diplomats' lack of experience with the speedy new electronic communication, class relations as a function of social distance, and combat in World War I as structured by new weaponry that enabled soldiers to kill at unprecedented long distances. Such interpretive reductions allowed me to integrate wide-ranging ideas and experiences and make historical arguments about complex activities in terms of basic experiences of time and space.

In subsequent years I continued to apply phenomenology to the study of history, focusing on other universal and essential concepts such as encounter (love), vision (the gaze), and narrative (story telling).[7] My work was further inspired by learning how phenomenologically oriented psychiatrists correlate three kinds of mental disorder with three kinds of causal understanding: depression with a deterministic causality, mania with a probabilistic causality, and paranoia with an intentional causality. I reasoned that, if individuals can be defined by their distinctive mode of causal understanding, perhaps entire cultures might be defined by their mode of causal understanding, and I elaborated that idea into a cultural history of causality that pivoted around the modernist period.[8] My long-term application of such a "categorical phenomenology"[9] also inspired this new book on changing ideas about origins and foundations from the Victorian to the modernist age.

Notes

1. Carl E. Schorske, *Fin-de-Siècle Vienna: Politics and Culture* (Knopf, 1980).
2. Stephen Kern, "Freud and the Emergence of Child Psychology: 1880–1010" (UMI, 1970).
3. Maurice Merleau-Ponty, *Phenomenology of Perception* [1945] trans. Colin Smith (Routledge, 1962), 157–73.
4. Stephen Kern, *Anatomy and Destiny: A Cultural History of the Human Body* (Bobbs-Merrill, 1975).
5. Edmund Husserl, *Logical Investigations* [1900–1901], trans. J. N. Findlay (Humanity Books, 2000), vol. I, 168. An earlier use of the term "phenomenology" was by Hegel, whose *Phenomenology of Spirit* (1807) traced the development of consciousness or spirit from immediate sense experience up to absolute knowledge as a dialectical, historical unfolding.
6. Stephen Kern, *The Culture of Time and Space: 1880–1918* (Harvard University Press, 1983, 2nd ed. 2003). My early sources on phenomenology were Rollo May ed., *Existence: A New Dimension in Psychiatry and Psychology* (Clarion, 1958); Eugène Minkowski, *Lived Time: Phenomenological and Psychopathological Studies* (Northwestern University Press, 1970); Edmund Husserl, *A Phenomenology of Internal Time Consciousness* (Indiana University Press, 1964); and Calvin Schrag, *Experience and Being: Prolegomena to a Future Ontology* (Northwestern University Press, 1969).
7. Those applications include *The Culture of Love: Victorians to Moderns* (Harvard University Press, 1992); *Eyes of Love: The Gaze in English and French Paintings and Novels 1840–1900* (Reaction Books, 1996); and *The Modernist Novel: A Critical Introduction* (Cambridge University Press, 2011).
8. Stephen Kern, *A Cultural History of Causality: Science, Murder Novels, Systems of Thought* (Princeton University Press, 2004).
9. On "categorical phenomenology" see Henri Ellenberger, "A Clinical Introduction to Psychiatric Phenomenology and Existential Analysis," in *Existence: A New Dimension in Psychiatry and Psychology* (Basic Books, 1958), 101–17. This article surveys the concepts of temporality, spatiality, materiality, and causality, which phenomenologically oriented psychiatrists used in treating patients and which I adapted for historical analysis.

Acknowledgments

I am indebted to several fellow researchers who discussed this project over the years: Sarah Maza, Vincent Sherry, William Everdell, Robert Caserio, and Robert Brenner. I would also like to thank three anonymous readers at Edinburgh University Press, whose incisive comments were most encouraging and valuable. Many others helped with specific chapters: John Beacom, Robert Scherrer, Lane Harper, and Ike Thierry on physics and cosmology; Chris Otter on geology; Gerald Karp on biology; Alice Conklin on anthropology; Richard Metzner and John Burnham on Freud; Robert Holub on Nietzsche; Richard Polt and Richard Capobianco on Heidegger; David Brakke on Christianity; and Linda Dalrymple Henderson, Ara Merjian, and Myroslava Mudrak on art. Peter Hersha was a sounding board for problems and ideas generally.

I am particularly grateful to the members of a modernist study group at Ohio State University who read my chapters along the way. While their area of specialty was primarily literature, they shared a wide range of interests in modernism generally including architecture and physics. They include Katherine Elkins, Todd Gannon, Ellen Carol Jones, Jim Phelan, and Jessica Prinz. The passing of three vital members – Morris (Murray) Beja, Brian McHale, and Bill Palmer – left a gaping hole in our group and in my life, and it is to them that this book is dedicated.

I am ultimately and abidingly grateful to my wife, Sydney, and my children, Justin and Sim, in so many ways.

Introduction

> I have wanted, through drawing and through color, since those were my weapons, to penetrate always more deeply into knowledge of the world and of men.
>
> —Pablo Picasso[1]

Origins and Foundations in the Modernist Age analyzes a period of exceptional achievement in a wide range of fields from physics and biology to literature and art. It surveys historically distinctive ideas about chronological origins as well as theoretical, functional, ontological and formal foundations. Origins are chronological as in cosmologists' theories about the origin of the universe, geologists' theories about the origin of Earth, biologists' theories about the origin of life, and psychologists' theories about the origin of mental life. Foundations are theoretical as in Albert Einstein's theory of relativity, functional as in Bronislaw Malinowski's anthropology of human society, ontological as in Martin Heidegger's philosophy of human existence, and formal as in novelists' and painters' artistic techniques. I link origins and foundations with the unifying adjective *originary* that embraces their meaning and enables me to identify cross-field patterns from the culture of the period.

In an influential talk delivered in 1966, the French philosopher Jacques Derrida linked origins and foundations with the word "originary" and argued that the idea that human existence has an origin or "reassuring foundation" is a tempting objective but is ill conceived and ultimately hopeless. "Henceforth," he proposed, "it was necessary to begin thinking that the center had no natural site, that it was not a fixed locus [or moment in time] but a function, a sort of imprecise non-location in which an infinite number of signs come into play."[2] I concur with Derrida's linking these concepts as originary and with his doubt about finding a single origin, center, or ultimate foundation of language or life, but I disagree that we are left with only floating signifiers and a free play of ideas. Instead, I argue that in the modernist period physical scientists did investigate determinable origins (if not the ultimate singular origin) of objects in their field of study in increasingly accurate ways, that social scientists did analyze foundations of human social experience in increasingly insightful ways, and that artists did develop new techniques to express themselves and recreate their world in ways that they believed constituted a decisive break from earlier techniques.

Origins and foundations are different concepts that enlist different investigatory skills. For example, to understand geometry, one might investigate its historical origin starting with the writings of Euclid or investigate its theoretical foundation starting with its basic axioms and postulates. Those investigations are relatively distinct, but other examples of originary investigations are not so clear-cut. To know how a clock works, one might ask a historian to trace its step-by step construction or ask an engineer to analyze its current structure and functioning, but those inquiries partly overlap. The historian needs some knowledge of how the clock works, the engineer needs some knowledge of how it was assembled and functions, and both experts need some understanding of its purpose. In this book, origins and foundations sometimes overlap as in Franz Boas's study of "primitives," whom I characterize as "living fossils" that reveal much about the remote past as well as the foundations of present society, or in Karl Barth's faith in a "living Christ" who died in the distant past but also inspired contemporary faith. Sometimes the concepts are explicitly differentiated as in Ferdinand de Saussure's distinction between *diachronic* (through time) origins of language and *synchronic* (same time) structures of language, but those features also overlap.

To be clear, this book is not about the origins or foundations *of* modernism but primarily about how those two ideas were conceived by leading cultural figures *in* the modernist period. However, that distinction is not absolute, because changing ideas about origins and foundations did destabilize the sense of specific origins and more secure foundations that appeared to ground life and thought in the Victorian age. Such a destabilization is graphically illustrated by the theory of continental drift proposed by the German geologist Alfred Wegener in 1912, which maintained that continents were moving apart, meaning that the grounds under Berlin and New York were not always in a single place on the Earth's surface and that the coasts of South America and Africa were once contiguous. The broader cultural impact of continental drift was noted in 1922 by the Spanish philosopher José Ortega y Gasset: "The landscape in the neighborhood of the observer seems to break up, vacillate and quake in all directions . . . and the very roads beneath his feet are melting away in serpentine undulations."[3] While continental drift reoriented the Victorian world spatially, the discovery of radioactivity by Henri Becquerel in 1896 reoriented it temporally by vastly extending estimates of the Earth's age.

The historical focus of this book is the modernist period, roughly 1890–1930, against a background of the Victorian period, roughly 1850–1890. *Victorian* does not refer to the full span of Queen Victoria's reign, its moralistic ethos, or even specifically English developments. It is strictly a period label for cultural developments in Europe and America during forty years toward the end of Victoria's reign. Similarly, *modernist* is another period label and does not signify any specific content. That is, for my purposes modernism is not a thing. It does not mean alienated or self-reflective, discontinuous or fragmented, innovative or up to date. As a scholar trained in historical study, I use these two overloaded and confusing terms exclusively as period labels simply to avoid having to endlessly repeat the wordy modifiers "late nineteenth century" for "Victorian" and "early twentieth century" for "modernist" and because they are by now standard terms.[4]

Uses of "modern," "modernism," and "modernist" have a complex history that has contributed to the complexity of their analysis and resistance to any single definition. In a "Syllabus of Errors" promulgated in 1864, Pope Pius IX used "modernism" to characterize a list of eighty propositions that he determined to be anathema for Catholics including liberalism, socialism, naturalism, positivism, religious toleration, and science when it contradicted Catholic dogma. In 1871 the Danish critic Georg Brandes used "modern" as an adjective to identify a "Modern Breakthrough," which was the "modern" revolutionary role of realism and naturalism in challenging traditional culture.[5] By the 1890s, *modern*, *modernism*, and *modernist* were used variously by Germanic scholars to identify developments in literature. The most prominent among them was the Austrian critic Hermann Bahr, who used modernism to identify current literary styles that rejected the previous dominion of realism and naturalism.[6]

In subsequent decades the meaning of "modernist" began to expand from a period label to a wide range of cultural achievements, as it was increasingly replaced by "modernism" and increasingly reified as a cultural historical entity. During that time several scholarly societies were founded for the study of modernist culture, primarily literature, most notably the Modernist Studies Association in the United States in 1998 (although its journal was called *Modernism/Modernity*), the British Association for Modernist Studies in 2011, and the French Société d'Études Modernistes in 2013. The title of those associations retained the adjectival term "modernist" even as the subject of their study became increasingly reified as "modernism."

That proliferation of societies was accompanied by expansions of the scope of modernist studies, which made crafting any comprehensive definition of modernism increasingly difficult. The most consequential set of expansions was proposed by Douglas Mao and Rebecca Walkowitz, who in 2008 urged that modernist studies expand spatially, vertically, and temporally. That is, modernist studies, which they specified as literary studies, ought to expand its area of focus spatially to the entire globe beyond its traditional concentration on Europe and America and expand its source base vertically from "high" canonical literary sources down to a broad range of sources by marginalized authors as well as the production and distribution of those sources. Mao and Walkowitz also called for a temporal expansion "to the late twentieth century or even as far back as the early seventeenth."[7]

Since 2008 scholars have greatly expanded the scope of modernist studies in those three directions to produce a wealth of new findings; however, those expansions, especially the temporal expansion, vitiate any comprehensive definition of the historical specificity of the modernist period. As a historian accustomed to studying historical periods with a thematic coherence in a reasonably limited time span, I find the call for a temporal expansion across four centuries, or even much beyond fifty years, to be unworkable. I view the modernist period as a temporal frame of a culturally transformative series of developments across a wide range of fields from roughly 1890 to 1930.[8]

These three proposed expansions have enormously extended the possible meanings of modern*ism* to be a philosophy of life, set of character traits, political position, social arrangement, economic system, or literary style. My phenomenological approach that focuses on the universals of origins and foundations has enabled me to avoid sorting through such wide-ranging analyses by "bracketing"[9] many scholarly debates about modernism. The historian of psychiatry Henri Ellenberger summarized Husserl's method of bracketing as it was applied by psychiatrists to their patients and which I have applied to historical developments more generally.

> The phenomenologist uses an absolutely unbiased approach; he observes phenomena as they manifest themselves and only as they manifest themselves. This observation is accomplished by means of an operation of the mind which Husserl calls the *epoché*, or "psychological-phenomenological reduction." The observer "puts the world between brackets," i.e., he excludes from his mind not only any judgment of value about the phenomena but also any affirmation whatever concerning their cause and background. . . .

> With this method, observation is greatly enhanced: the less apparent elements of phenomena manifest themselves with increasing richness and variety, with finer gradations of clarity and obscurity, and eventually previously unnoticed structures of phenomena may become apparent.[10]

My use of phenomenology is a method that governs my selection of subjects and interpretation of evidence. While it is impossible to be "absolutely unbiased" or entirely exclude from consideration conceptual frameworks involving values, causes, or backgrounds that help make sense of cultural developments, one can bracket the enormous mass of scholarly studies of modernism to focus on the historically distinctive nature of the concepts of origins and foundations. I bracket this threefold expansion of theories and methods while pursuing a fourth expansion across a wide range of fields that are the chapters of this book on physical sciences (physics, cosmology, geology, and biology), social sciences (anthropology, psychology, and psychiatry), and humanities (philosophy, religion, literature, and art).

The different nature of these fields poses a huge challenge to formulating a general interpretation of the modernist period. Physical scientists' objective is truth about physical phenomena. Their findings need to be correct, but they can also be simply wrong. Social scientists analyze, interpret, and assign values to human experience. They want their evidence to be reliable and their arguments convincing, but one cannot reasonably argue that their analyses or interpretations are flat-out wrong, even in instances of blatantly discriminatory studies such as racist anthropology, sexist psychology, or eugenicist sociology. The field of humanities covers the wisdom-seeking of philosophy, the faith of religious experience, and the creation of beauty in literature and art. With such a wide range of objectives, a single generalization about modernist culture is not feasible. My conclusion nevertheless undertakes to group these diverse undertakings with a survey of cross-field patterns of ideas about origins and foundations across wide stretches of the cultural record.

Modernists' work in these three major areas was conceived in response to the work of their Victorian predecessors.

Modernist physical scientists pushed back against many erroneous Victorian ideas such as that light travels in waves through a luminiferous ether, that the Milky Way Galaxy spans the entire universe, that gravity is a force that acts at a distance instantaneously, that the Earth formed 20 to 400 million years ago, that the continents have always had a permanent location on the Earth's surface, that species evolved according to a goal-directed purpose the way a tree grows from a seed, that parental characteristics are produced in offspring by a mixing of male and female sexual fluids, that a child's temperament is determined by the mental state of its parents at the moment of conception, that acquired characteristics are biologically transmitted from parents to children in the blood, that complex character traits such as an instinct to kill can be transmitted in toto through sexual reproduction, and that mothers' sex organs are "infected" by the character traits of the men who impregnate them and therefore can reproduce in any subsequent child characteristics of any such former sexual partner.

Modernist social scientists rejected Victorian ideas that human beings progressed racially from "savagery" to "civilization," that the "moral faculty" has a specific location in the brain, that people can be born with a "criminal mind," that murderers can be identified by a sloping brow, that women ought to be chaperoned during courting and confined during pregnancy, and that women are properly subject to social sanctions for transgressions of sexual morality according to a "double standard" for women and men, including female prostitutes and their male clients. Victorians also elaborated Darwin's theory of natural selection with crude varieties of social Darwinism to justify racism, imperialism, and social inequalities along with moral "perfection" of families and nations as well as species. The imperialism of European countries in Africa and Asia that ran rife during the Victorian period underscored a morally reprehensible set of attitudes including that imperial powers were justified in using violence against "barbaric" and "semi-civilized" states in contrast to the way they settled disputes among themselves with reason and negotiation.

Modernist novelists and painters did not progress in their knowledge of human existence over their Victorian predecessors as did modernist physical and social scientists, to be sure, but they did develop new techniques to capture their world in ways that they believed earlier techniques were not as well suited to accomplish. Modernist novelists and painters had the benefit of hindsight in that they were able to draw on as well as critically evaluate the work of their predecessors. Many modernists viewed themselves as an avant-garde, by which they meant a group that led the way into formerly tabooed literary topics and new artistic techniques. Some modernists such as the German Expressionists emphasized the moral aspect of their emancipation from bourgeois morality and oppressive family dynamics. Others emphasized their contribution as formal and during the heyday of their formal innovations shared a sense of surpassing what preceded them. While Victorian writers crafted deep foundations of human experience, many pioneering modernist writers believed that they found new ways to craft those foundations at deeper levels of understanding.

A major challenge in generalizing about any culture based on a wide range of sources as this book undertakes to do is "conceptual distance" that works both for and against such a cultural generalization. To explain, the conceptual distance between the thinking of a poet and an architect on any given subject such as the nature of space is likely greater than the conceptual distance between the thinking of two poets or between two architects on that subject. Greater conceptual distance works for a culture-wide generalization by expanding the separation between the sources that support it like a set of guy wires spread out in many directions to hold up a telecommunications tower. However, conceptual distance works against such a generalization precisely because those sources come from different scholarly disciplines and artistic genres that have different procedures and purposes and thereby exert different pulls against any culture-wide generalization. My phenomenological approach addresses that problem by reducing diverse areas of inquiry and creativity to fundamental ideas about origins and foundations in ways that can be more easily grouped into patterns that I analyze historically in the following chapters and generalize in the conclusion.

This book counters modernist scholars who emphasize negative, divisive characteristics of modernism such as *rupture* (Marjorie Perloff), *fragmentation* (Sarah Haslam), *disintegration* (Erich Kahler), *nihilism* (Shane Weller), *crisis* (Jacques Le Rider), *meltdown* (Marshall Berman), *trauma* (Ariela Freedman), and *shock* (Robert Hughes).[11] These studies are partly balanced with considerations of creative and unifying projects, but their emphasis is essentially about varieties of rupture, meltdown, and shock. In an earlier study I also probed fragmenting aspects of modernism,[12] but on balance my emphasis was back then and is now on creative aspects, namely how origins generate positive developments toward some achievement and how foundations ground sciences, philosophies, and arts. My preferred unifying metaphors are *paradigm* (Thomas Kuhn), *matrix* (Sanford Schwartz), *mosaic* (David Kadlec), *cosmic web* (N. Katherine Hayles), *intellectual climate* (Christopher Butler), *cultural dominant* (Brian McHale), and the one that best suits my approach, *patterns* (Virginia Woolf, Ruth Benedict, Susan Hegeman, and Allyson Booth).[13]

As I studied ideas about origins and foundations across many academic fields and artistic genres, a compelling argument repeatedly emerged – that modernists probed these concepts more deeply than their Victorian predecessors did. That claim is manifestly true for the physical sciences, as is particularly evident in my first chapter, which explains how modernist physicists investigated increasingly minute particles and events taking place at the subatomic level and at enormously high relative speeds at the cosmic level and in those respects corrected some specific limitations of Victorian physicists about the nature of time and space as well as matter, light, gravity, and causality. That claim about greater depth is also a plausible argument about the social sciences, if not quite so rigorously, and still relevant to developments in the humanities. At first I resisted that argument because it recalled earlier interpretations of modernist triumphalism, namely that history was a linear progression toward enlightenment, reason, and artistic purity; but evidence for viewing modernists as having probed deeper into the nature of the physical world and human social experience than did their Victorian predecessors became increasingly evident as I examined field after field, especially in the physical and social sciences.

My reliance on the imagery of ever deeper probing may seem annoyingly insistent as I move across the cultural record, but a wide-ranging study such as this one cries out for some unifying argument, and, as I wrote, deeper probing repeatedly emerged from my evidence. Physicists probed ever deeper subatomic particles and events, cosmologists probed ever deeper reaches of outer space, geologists literally explored the composition and movement of ever deeper layers of the Earth, while biologists searched for ever deeper secrets of the origins of life. Social scientists also aligned with the imagery of increasingly deeper analysis – anthropologists into primitive societies and psychologists and psychiatrists in a discipline that took on itself the label of a "depth psychology." I do not argue that modernist novelists probed more deeply into human experience than did Victorian novelists but I do show that the beginnings of their novels dramatize the opacity and mystery of beginnings with their "antibeginnings" and "self-cancelling beginnings,"[14] and I underscore their resistance to the sort of clear beginnings evident at the outset of many typical Victorian novels. Modernist artists were intent on transforming the deepest levels of their craft most notably with their rendering of color, form, space, and "reality." My argument about modernists' deeper probing is, I must concede, in one respect circular, because the history of ideas about origins and foundations would have to go deeper, since depth of analysis is implied by both terms. The search for origins is necessarily to go ever earlier and, in that respect, ever deeper into the past; and the search for theoretical, functional, ontological and aesthetic foundations could hardly go shallower.

My strictest criterion for the historical specificity of any development in the modernist period, or any period, is that it would have been inconceivable, if not impossible, fifty years earlier. Such a claim is valid for developments that rely on technologies that did not exist in an earlier period and enabled deeper investigations. So, for some examples, between 1902 and 1905 Richard Zsigmondy developed the ultramicroscope, which made visible colloidal particles that were invisible under conventional light microscopes. A year after the discovery of X-rays in 1895, physicians adapted the new technology to see into the structure and functioning of the human body. In 1911 Charles Wilson invented the cloud chamber to enable physicists to track subatomic particles by showing their condensation trails. Physicists invented particle accelerators to investigate fundamental particles and probe theoretical foundations of light, matter, and energy; in 1928 Rolf Widerøe, a Norwegian engineer, built the first linear particle accelerator (linac), which accelerated potassium ions to about 50,000 volts using alternating current, while in 1930 at the University of California, Berkeley, Ernest O. Lawrence and M. Stanley Livingston built the first cyclotron. Increasingly precise spectroscopes enabled quantum physicists to understand atomic emission spectra as "fingerprints" of elements in stars and distant galaxies. Astronomers developed increasingly powerful telescopes that made possible increasingly precise observations of increasingly distant celestial objects, which generated increasingly fruitful cosmological theories. Geologists developed radiometric dating to understand the origin of the Earth along with its continents and mountains. Geologists also developed X-ray crystallography in 1912–13, which enabled a clearer determination of the atomic arrangements in crystals; in 1952 this was used by Rosalind Franklin and Raymond Gosling in discovering the structure of DNA. In 1924 the German psychiatrist Hans Berger invented the electroencephalogram (EEG) to probe the brain's electrical activity. In 1928 Hans Geiger invented the Geiger counter, which detected and measured radioactive emissions. In 1931 the German scientists Ernst Ruska and Max Knoll invented the electron microscope and with ever more precise observations researched deeper into the foundations of biological processes. With all that deeper probing, however, it must be noted that all the sciences, including the social sciences, were subject to what I call the *specificity–uncertainty dialectic*, that is, the deeper scientists probed, the more complex, perplexing, and challenging their subject became. Ironically, that delineation of the growing realm of the unknown among modernist scientists was one of their most important achievements.

In addition to new technologies, modernists increasingly improved understanding with more rigorous investigatory methods. Victorian archaeologists undertook numerous projects that made significant findings of ancient remains, but their crude methods damaged artifacts and disrupted sites, leading to deterioration and looting and stripping host countries of their ancient heritage. Modernist archaeologists more carefully handled relics that they interpreted as clues to a larger culture.

Modernist anthropologists also replaced Victorian "armchair anthropology" by immersion studies of remote cultures that required learning new indigenous languages and field techniques. Sigmund Freud's deeply retrospective psychoanalytic technique along with Carl Jung's retrospectively deeper individual analyses put them at the center of an innovative technique that had adherents across Europe and America.[15] In philosophy, Nietzsche's new method of "genealogy" expanded the study of the origins of morality beyond an exclusively philosophical prehistory to a wide range of origins that contributed to understanding like the tributary system of a great river spread over a wide cultural landscape, which for his purposes included the conceptual terrains of philology, psychology, economics, theology, and history as well as philosophy. Heidegger spiraled down with his rigorous phenomenological method ever deeper into the ontological foundations of what it means to be a human being and deeper still into the meaning of "Being" per se.

Modernist novelists distanced themselves from the confident temporal beginnings of conventional Victorian novels by "omniscient narrators" and rather dramatized origins and foundations of human experience in a variety of inventive ways. They not only began stories in medias res but resisted the temptation to flash back to an "actual" beginning as Victorian narrators tended to do by revealing some key originary *éclat* as in Emily Brontë's *Wuthering Heights* (1847), Nathaniel Hawthorne's *The Scarlet Letter* (1850), Charles Dickens's *Great Expectations* (1861), Victor Hugo's *Les Misérables* (1862), Thomas Hardy's *Tess of the d'Urbervilles* (1891), and Theodor Fontane's *Effi Briest* (1895). While modernist novelists continued to use flashbacks to reveal originary moments, they nevertheless were more likely to cut into stories in the middle without seeming to know a precise beginning, which they were convinced they could never know.

An innovative modernist literary technique for probing the foundations of human experience was Woolf's method of tunneling that became increasingly complex the deeper it went.[16] Tunneling generated the interconnected foundational structure of her novels, a deep structure that also applies to James Joyce's Dublin, especially in the interconnected structure of the Wandering Rocks episode of *Ulysses* (1922). Gertrude Stein attempted, ultimately without success, to identify the "bottom nature" of characters in *The Making of Americans* (1925) and dramatized the origins of stories in a new way by repeatedly "beginning again," each time at a slightly different moment, from a slightly different perspective, at a slightly different position in her text. She also explored as she questioned conventional foundational ideas about gender and language as well as roses. In *Tender Buttons* (1914), as Jesse Matz notes metaphorically, she explored "a deeper self, speaking some primal language from the bottom of the human mind."[17]

The modernist dancers Loie Fuller, Isadora Duncan, Mary Wigman, and Martha Graham explored new possibilities of bodily movement as an abstract art parallel to analogous modernist developments in abstract painting and cinema.[18] Modernist artists interrogated the formal foundations of their craft in many ways from their use of color to their rendering of form, perspective, and visual reality. Austrian Secessionists devised new ways to paint ever deeper emotions of their subjects, particularly sexual desire, aggression, and anxiety, as those emotions coursed along muscles, nerves, and veins that the most radical among them, Oskar Kokoschka and Egon Schiele, rendered visible underneath the skin (see Figures 11.8 and 11.9 plus the cover image). While Picasso's main concerns were aesthetic and formal, he was also aware of his work's epistemological significance, as, looking back on his career in 1944, he explained how he "wanted, through drawing and through color, since those were my weapons, to penetrate always more deeply into knowledge of the world and of men."[19] His metaphor of weaponry suggests a combative engagement with the world around him that he inherited from his predecessors, while the action of penetration evokes the main direction of his efforts to know as well as create.

A final deeper probing, which bears on modernist culture across the board, was an assault on the originary myth of creation announced in Genesis. In its place modernists investigated the actual beginning of the universe, Earth, and life based on empirical evidence, using investigative tools of scientific research in cosmology, geology, and biology. The novelists I survey in the chapter on religion – Gide, Joyce, and Lawrence – explored the Christian foundation of morality with passionate attacks on the originary doctrine of original sin and its debilitating effect on the ability of human beings to experience the fullest possibilities of love. Modernists' efforts at deeper probing did not lead to a final vision of rupture or discontinuity, to nihilism or alienation, but to an appreciation of the humanizing and unifying possibilities of scientific knowledge, philosophical insight, human relations, and art.

To fairly assess the deeper probing of modernist culture, it must be noted that Victorians were by no means shallow. They dug as deep as they could with enormous energy and devotion to generate an extraordinarily rich culture of their own. Their top scientists such as Charles Darwin, James Clerk Maxwell, William Thomson, and Ludwig Boltzmann devoted their lives to brilliant research of enduring value on evolutionary biology, electromagnetism, geology, and thermodynamics. Leading philosophers including John Stuart Mill, Herbert Spencer,

Karl Marx, and Søren Kierkegaard produced penetrating studies of the subjection of women, evolution, history, and faith. The germ theory of disease pioneered by Louis Pasteur and refined by Robert Koch revolutionized internal medicine and saved millions of lives. Imaginative writers produced deeply powerful novels and evocative poems, while musicians cultivated expressive techniques of dissonance and chromatism in composing powerfully moving symphonies and operas. Victorian painters refined sfumato techniques to render skin and atmospheric effects, cultivated chiaroscuro and fine hatching to capture subtle folds of fabrics in varying lights, and exploited a vast inventory of symbols and allegories to express richly layered meanings.

Still, on balance, while Victorians probed as deeply as they could, modernists continued to search ever deeper into origins and foundations for many reasons including the obvious one that they had Victorians as sources to inspire and provoke as well as augment and correct. Put simply, Einstein was able to go deeper into the nature of physical reality than were Maxwell and Boltzmann because he knew both his own work and theirs, while they knew their own work but could not possibly know his. While it would be absurd to argue that literature and art progressed from the Victorian to the modernist period as it did in the sciences, or, worse, that Joyce was a better writer than Dickens or that Picasso was a better painter than Courbet, the history of those arts was nevertheless unmistakably a scenario of radical breaks and probing new areas and techniques in contrast to the work of their predecessors.

The division of this study into eleven chapters framed around academic disciplines and artistic genres resists any fully integrated single argument or interpretation. Throughout the text I note occasional interdisciplinary interaction such as the famous one between the artist Wassily Kandinsky and the composer Arnold Schoenberg, which produced their respective efforts at artistic synesthesia,[20] but my specific analyses field by field do not evince any wide-ranging understanding that the modernist age had a single cultural historical significance with respect to ideas about origins and foundations.

I address this huge challenge in my conclusion that offers a model of cultural history as a clustering of ideas presented field by field throughout the chapters into cross-field patterns. For example, one such pattern of modernist ideas is the clustering of unexpectedly significant "trivia" evident across wide stretches of the cultural record, which played analogously innovative originary roles. These not-so-trivial trivia include the Trobriand Islanders' Kula necklaces, Marcel Duchamp's found objects, Freud's slips of the tongue, Joyce's epiphanies, Woolf's "little daily miracles," Marcel Proust's involuntary memories, William James's philosophy of "little words," William Faulkner's Compson brothers' glimpse of their sister Caddy's muddy drawers, and William Carlos Williams's poem about "a red wheel barrow glazed with rain water beside the white chickens," upon which, Williams affirms, "so much depends."

In focusing on "trivia" upon which so much indeed depended, and which had so many innovative ramifications and cultural historical significations, modernist psychologists, novelists, philosophers, anthropologists, and poets reevaluated what is significant in understanding human existence, a major objective of modernist culture generally. My conclusion surveys that and other patterns of cultural artifacts that in the aggregate shaped the culture of the modernist period as a cluster of meanings that are evident in increasingly profound ideas about origins and foundations and increasingly innovative artistic techniques for capturing them.

Notes

1. Interview with Pol Gaillard in *L'Humanité*, Paris, 29–30 October 1944. Quoted by Patricia Leighton in *Re-ordering the Universe: Picasso and Anarchism, 1897–1914* (Princeton University Press, 1989), 97.
2. Jacques Derrida, "Structure, Sign, and Play in the Discourse of the Human Sciences" [1966] in *Writing and Difference* (University of Chicago Press, 1978), 287–88, 292.
3. José Ortega y Gasset, "The Signs of the Times" [1922] in *The Modern Theme* (Harper, 1961), 79.
4. In that same spirit Susan Hegeman argues that "Modernism is for me first and foremost a periodizing concept, characterized by a nexus of related historical, intellectual, technological, and aesthetic developments rather than by a set of formal traits or styles." *Patterns for America: Modernism and the Concept of Culture* (Princeton University Press, 1999), 19.
5. Georg Brandes, *Main Currents in 19th-Century Literature* [1871] (Russell & Russell, 1900).
6. Hermann Bahr, *Zur Kritik der Moderne* (1890) and *Die Überwindung des Naturalismus* (1891).
7. Douglas Mao and Rebecca Walkowitz, "The New Modernist Studies," PMLA 123, no. 3 (May 2008): 737, 738. On other calls for such a temporal expansion see Anne-Florence Gillard-Estralada and Anne Besnault-Levita eds., *Beyond the Victorian/Modernist Divide: Remapping the Turn-of-the-Century Break in Literature, Culture and the Visual Arts* (Routledge, 2018).
8. That periodizing argument is underscored by the critics David James and Urmila Seshagiri, who reaffirm the specificity of the modernist era as a "temporally localizable moment" in contrast to the "cultural diffuseness" implied by excessive temporal expansion. For emphasis they add that "to

characterize modernism as a feature common to centuries of interlocking modernities is to risk depriving works such as *To the Lighthouse* of the formal and temporal inventiveness unique to their era." "Metamodernism: Narratives of Continuity and Revolution," *PMLA* 129, 1 (2014): 93, 88, 92. My periodization is even more restrictive than theirs that ends at "the mid-twentieth century," which I push back to 1930 for greater temporal compression and historical specificity.

9. Fred Kerston's translation of Husserl's *Ideas Pertaining to a Pure Phenomenology and to a Phenomenological Philosophy* (1913) renders the German word *Einklammerung* (literally "putting in brackets") as "parenthesizing." In that work Husserl also refers to this procedure with the Greek "*Epoché*," meaning a "suspension" of judgment, and elsewhere Husserl also refers to it as "excluding." The emphasis is, however, not on entirely excluding, but on setting it aside to make possible a clearer focus on the essential elements of a perception.
10. Henri Ellenberger, "A Clinical Introduction to Psychiatric Phenomenology and Existential Analysis," in Rollo May et al. eds., *Existence: A New Dimension in Psychiatry and Psychology* (Basic Books, 1958), 96. This article also surveys the concepts of temporality, spatiality, and causality, which phenomenologically oriented psychiatrists used in treating patients and which are the focus of two of my previous historical studies.
11. Marjorie Perloff, *The Futurist Moment: Avant-Garde, Avant Guerre, and the Language of Rupture* (University of Chicago Press, 1986); Sara Haslam, *Fragmenting Modernism: Ford Madox Ford, the Novel and the Great War* (Manchester University Press, 2002); Erich Kahler, *The Disintegration of Form in the Arts* (George Braziller, 1968); Shane Weller, *Modernism and Nihilism* (Macmillan, 2011); Jacques Le Rider, *Modernity and Crises of Identity* (Continuum, 1993); Marshall Berman, *All That Is Solid Melts into Air* (Penguin, 1982); Ariela Freedman, *Death, Men, and Modernism: Trauma and Narrative in British Fiction from Hardy to Woolf* (Routledge, 2003); Robert Hughes, *The Shock of the New* (Knopf, 1981).
12. Stephen Kern, *Modernism after the Death of God: Christianity, Fragmentation, and Unification* (Routledge, 2017).
13. Thomas Kuhn, *The Structure of Scientific Revolutions* (University of Chicago Press, 1962); Sanford Schwartz, *The Matrix of Modernism: Pound, Eliot, and Early Twentieth-Century Thought* (Princeton University Press, 1985); David Kadlec, *Mosaic Modernism: Anarchism, Pragmatism, Culture* (Johns Hopkins University Press, 2000); N. Katherine Hayles, *The Cosmic Web: Scientific Field Models and Literary Strategies in the 20th Century* (Cornell University Press, 1984); Christopher Butler, *Early Modernism: Literature, Music and Painting in Europe 1900–1916* (Clarendon Press, 1994); Brian McHale, *Postmodernist Fiction* (Routledge, 1987); Virginia Woolf, "A Sketch of the Past" in Jeanne Schulkind ed., *Moments of Being* (Harvest, 1985), 72–73; Ruth Benedict, *Patterns of Culture* [1934] (Mariner, 2005); Susan Hegeman, *Patterns for America: Modernism and the Concept of Culture* (Princeton University Press, 1999); Allyson Booth, *Postcards from the Trenches* (Oxford University Press, 1996).
14. Melba Cuddy-Keane introduced the concepts of "antibeginnings" and "self-canceling beginnings" in reference to Woolf's *Mrs. Dalloway*, *Jacob's*

Room, *To the Lighthouse*, and *A Room of One's Own*. "Virginia Woolf and Beginning's Ragged Edge," in Brian Richardson ed., *Narrative Beginnings: Theory and Practice* (University of Nebraska Press, 2008), 99–103.

15. Dieter Wyss, *Depth Psychology a Critical History: Developments, Problems, Crises* (Norton, 1966).
16. In an August 1923 diary entry Woolf referred to "my tunneling process" and characterized it as her "prime discovery so far" but did not explain it in any detail. In 1930 she elaborated the method as a way of understanding the paintings of her sister Vanessa Bell, which "yield their full meaning only to those who can tunnel their way behind the canvas into masses and passes and relations and values of which we know nothing" to understand how her paintings "claim us and make us stop [and] give us an emotion." For these sources see Edward A. Hungerford, "'My Tunnelling Process': The Method of 'Mrs. Dalloway'" *Modern Fiction Studies* 3, 2 (Summer 1957): 165.
17. Jesse Matz, *The Modern Novel: A Short Introduction* (Blackwell, 2004), 3.
18. Nell Andrew, *Moving Modernism: The Urge to Abstraction in Painting, Dance, Cinema* (Oxford University Press, 2020).
19. See note 1.
20. On those efforts see Chapter 11, note 43.

Part I

Physical Sciences

Chapter 1

PHYSICS: Foundations of Matter, Energy, Light, Time, Space, Motion, Gravity, and Causality

Physics analyzes the theoretical foundations of the physical world. Victorian physicists explored how mechanical, electromagnetic, thermal, and gravitational forces shaped events in time and space. Modernists were schooled on that physics, most notably James Clerk Maxwell's transformative electromagnetic theory and Ludwig Boltzmann's hugely influential thermodynamics. However, modernists challenged many features of Victorian physics modeled on Isaac Newton's "natural philosophy" with relativity theory and quantum theory that revolutionized ideas about matter, energy, light, time, space, motion, gravity, and causation.[1]

Victorian physicists embraced the Newtonian-styled corpuscular theory of matter that was made of supposedly indivisible particles called atoms.[2] In 1897 J. J. Thomson announced some even more basic "corpuscles" out of which the elements were built, and in 1911 Ernest Rutherford discovered the atomic nucleus and proposed a model of atoms with electrons orbiting a nucleus.[3] In 1898 Marie Curie and her husband, Pierre, discovered the emission of radiation from certain elements, a property of matter that she coined "radioactivity."

Victorians also believed that light was a wave that moved through an invisible medium called an ether, that time and space were unchanging and qualitatively separate frames for tracking events, that gravity was a force that acted at a distance and did so somehow instantaneously, that the shortest distance between two points was always a Euclidean straight line, that causal knowledge was deterministic, and that future physical events could in theory be predicted with certainty. Modernists undercut those ideas with relativity and quantum theory.

Sound waves propagate through air or water but not through a vacuum. A big question was how light can travel through a vacuum, and, to answer it, Victorian physicists posited some sort of luminiferous ether that could not be seen or felt. In 1887, Albert Michelson and Edward Morley used a new technology, the interferometer, to attempt to identify that ether. Following the failure of that experiment to detect any ether, physicists struggled to explain the transmission of light waves through a vacuum without any detectable medium to carry them. The most notable explanation was the Lorentz–FitzGerald hypothesis of 1895, which held that the ether did exist but was undetectable and that Michelson and Morley's interferometer must have contracted physically from motion through the ether just enough to account for a slowing of time and a contraction of space caused by that motion.[4]

That explanation was midway between classical physics and relativity theory. It adhered to the classical concepts of absolute time and space by insisting that the change took place *in the object* as a result of motion through the ether, similar to the way elastic bodies contract in the direction of their motion through a fluid or gas, but it looked forward to relativity theory by suggesting that temporal and spatial measurements are modified by motion itself and that there is a plurality of local times and spaces dependent on the relative motion of observers and the events they observe. While the Lorenz–FitzGerald hypothesis correctly found the length of contraction that would eventually be predicted by relativity theory, it also substituted one ad hoc and erroneous theory for another – the contraction of the interferometer for a luminiferous ether.

Einstein with his special and general theories of relativity and Niels Bohr and Werner Heisenberg with their quantum theory explained the transmission of light without a luminiferous ether and together also reconceived Victorian ideas about matter, energy, time, space, motion, gravity, and causation.

Einstein's Relativity Theory

In 1905, the twenty-six-year-old Albert Einstein, while employed at a Swiss patent office, published five papers in his spare time that transformed physics. One paper, for which he won the Nobel Prize in 1921, argued that light behaved in some ways as a wave but in other ways as a stream of particle-like photons. His most revolutionary paper, "On the Electrodynamics of Moving Bodies," introduced the "principle of relativity" that relativized time and space, necessitated by the fact that the speed of light was constant at 186,000 miles a second even when viewed from a moving reference system.[5] A great irony in this theoretical revolution is that the relativity of time and space was necessitated by the nonrelative or "invariant" speed of light.

When Einstein addressed these matters with his special theory of relativity, Newton's mechanics had reigned supreme for over 200 years as a model of science generally. In *The Principia* (1687) Newton maintained that absolute space is an infinitely extended empty volume where physical phenomena occur and is unaffected by whatever happens in it or how observers measure it. Absolute time flows at a uniform pace throughout eternity and is also unaffected by what happens during it or how observers keep track of it. Einstein would argue that space and time change to account for the invariant speed of light, even when viewed from moving reference systems. He generalized about that theoretical change with his "special theory of relativity" – that in explaining every mechanical event in the physical world including electromagnetism and thermodynamics *the laws of physics are the same for all observers in uniform motion.*

The special theory of relativity was special because it applied only to observers and observed events in uniform motion such as when they are together inside a moving train or on the deck of a moving ship. In 1915 Einstein published his general theory of relativity, which accounted for accelerated motion that he equated with gravity. The general theory applied to all motion anywhere and under any conditions of relative motion, including even accelerating motion, according to the more general theory that "the laws of physics are the same for all observers," period, with no need for the restricting phrase "in uniform motion."

Put together, his two theories of relativity introduced stunning reversals of deep-seated ideas about time, space, motion, and gravity. With the special theory Einstein calculated how time in one reference system moving at a constant velocity slows down when viewed from another reference system at rest relative to it. His general theory explained the time change of accelerated bodies in gravitational fields. Since every bit of matter in the universe generates a gravitational field, and since gravity is equivalent to acceleration, "every reference body has its own particular time," and "unless we are told the reference-body to which the statement of time refers, there is no meaning in a statement of the time of an event."[6] Therefore, time is fundamentally relative. The general theory had the effect, figuratively, of placing a clock in every gravitational field in the universe, each moving at a rate determined by both the intensity of the gravitational field where that clock is located and the relative motion of the object and the person measuring it. In contrast to Victorian physicists who explained mechanics with one correct universal time, Einstein, in theory, filled the universe with an infinite number of clocks, each telling a correct time. By relativizing time, Einstein also pluralized it.

In a manner analogous to the relativization of time, Einstein also relativized space, because, just as relative motion causes time to slow down, it also causes space to contract. These relativized times and spaces are not observational illusions or distortions of "real time" and "real space" that in themselves are stable and absolute everywhere in the universe à la Newton; they are fundamental conditions of knowledge of events everywhere and for all time.

Two more refinements of Victorian physics introduced by relativity theory were (1) that time and space are not different kinds of dimensional frames for events but rather components of a single spacetime continuum and (2) that gravity is not a force acting at a distance along Euclidean straight lines but a field effect generated by matter and energy that curves space and hence the trajectory of anything traveling through it – matter as well as light.

In 1908 Hermann Minkowski, Einstein's former math teacher, undercut the view of space and time as separate dimensions of events. Drawing on Einstein's relativity theory, he announced that "henceforth space by itself, and time by itself, are doomed to fade away into mere shadows, and only a kind of union of the two will preserve an independent reality."[7] He suggested calling that union a "world-line," which he described as the "everlasting career" of a point as it exists across what was formerly regarded as the independent dimension of time. All events should rather be conceived in a four-dimensional spacetime continuum represented by coordinates x, y, z, t, which are to be understood not as three spatial dimensions plus an inherently different sort of temporal dimension but as four components of the same kind of "world-points."

While it had taken just six weeks for Einstein to work out the special theory of relativity, it took him nine years to develop the general theory, because among other difficulties it involved an extremely complex mathematics of four dimensions based on non-Euclidean geometry that Einstein had to learn. In 1952 he summarized the revolutionary nature of his new theory: "It appears therefore more natural to think of physical reality as a four-dimensional existence, instead of, as hitherto, the evolution of a three-dimensional existence."[8]

Einstein also theorized that gravity is not an attractive force generated by matter on a distant object but a local action of a curved spacetime continuum generated by matter and energy, which are related by the equation $e = mc^2$. Those aspects of the theory of general relativity were confirmed during a solar eclipse on 19 May 1919 when Arthur Eddington observed a displacement of the expected position of stars that appeared optically near the Sun as viewed from Earth. That observation confirmed Einstein's theory that accounted for such displacements by the way that centers of matter and energy such as the Sun generate gravitational fields that curve spacetime around them and deflect the observed position of stars, because the path that the stars' light traverses in the shortest possible way to Earth is not a Euclidean straight line but a curved trajectory through a curved space. That finding created an international sensation, as Einstein became the genius of the age.

Newton worked out equations to calculate gravitational forces acting at a distance but admitted that he had no idea what gravity was or how it worked. Victorians followed Newton in believing that gravity was a perplexing force that somehow reached across space instantaneously to attract objects remotely, but such an explanation was impossible to reconcile with the fact that it took 1.3 seconds for light to travel from the Earth to the Moon, so any gravitational force acting remotely, whatever it was, would have to exceed the "cosmic speed limit," the speed of light at 186,000 miles per second. Einstein's theory was that the Moon's orbit is not determined remotely by some mysterious force emanating from Earth and traveling across space at some very rapid speed, possibly the speed of light, but is rather determined locally and instantaneously as the Moon traverses the shortest distance through the curved spacetime field generated by the Earth.

N. Katherine Hayles summarized the revolutionary significance of Einstein's relativity theories in conceiving of the universe as having the field-like structure of a "cosmic web."

> In both the Special and General Theories, then, Einstein arrived at a view of physical reality that transformed the isolated entities of Newtonian mechanics into unified, mutually interacting systems. Instead of seeing time as a series of independent and omnipresent moments, Einstein conceived of it as inextricably linked with space to form the four dimensions of spacetime; instead of thinking of space as a rigid container, Einstein postulated that it took its structure from matter; instead of seeing energy and matter as fundamentally separate and inconvertible, Einstein showed that they are essentially equivalent and potentially interconvertible . . . In seeing fundamental interconnections between entities that had been discrete quantities in classical physics, Einstein helped to prepare the way for a field concept of reality.

Hayles characterized the universe described by that field concept as a "cosmic web."[9]

Quantum Theory

The enormous success of classical physics was its ability to explain the causes of events deterministically and predict with mathematical certainty the continuous future of events such as a billiard ball's trajectory or a planet's orbit. The ultimate predictive power of such an ability was envisioned in 1814 by Pierre Laplace, who proposed that an intellect that knew the precise location and momentum of all the atoms in the universe and all the laws that governed their motion could in theory see infinitely far into the future as well as back into the past with total accuracy. The refutation of the possibility of such certain causal knowledge came from physicists concerned with knowledge of subatomic events.

The classical model dominated thinking about causality until the late Victorian period, when physicists began to encounter phenomena that that model could not account for, such as the spectra and stability of the atom that Ernest Rutherford modeled with electrons orbiting around a nucleus like planets around a star.[10] According to classical electrodynamics, such orbiting electrons should steadily radiate electromagnetic waves, lose energy, and crash into the nucleus in a fraction of a second, and therefore atoms should not exist as stable entities. But electrons do not crash into the nucleus, and atoms do exist and are stable. Efforts to explain such phenomena fed into quantum theory, which, in accounting for subatomic events, limited the universal applicability of the classical model that was grounded in a theory that for over two hundred years had satisfied the physical requirements of determinism and continuity and the epistemological requirements of certainty, predictability, objectivity, visualizability, and causality. In the modernist period those features of the classical model were challenged by quantum theory.

Determinism (Mechanism of Causal Events)

Classical physics in the Victorian period maintained that events were in fact determined according to the formula by which cause *c* is always followed by effect *e*, but, according to quantum theory, for subatomic events cause *c* is not always followed by effect *e*, or, put more strongly, the precise cause of subatomic events cannot be established at all. The most easily visualizable application of the classical model was the mechanistic determinism of objects such as a billiard ball striking and moving another ball in accord with Newton's laws of motion. According to this theory, the position and momentum of an object at one moment determine its position and momentum at the next moment or at any subsequent moment in its trajectory. However, quantum physicists maintained that one cannot measure position and momentum of subatomic particles with sufficient accuracy at the same time to be able to predict a precise future trajectory, so causal events at the quantum level are inherently indeterminate.

Continuity

An example of quantum discontinuity is the behavior of electrons around a nucleus. In the classical model, objects change their position continuously and occupy every position along their trajectory like a baseball along its trajectory to home plate. In describing subatomic objects such as electrons, however, quantum physics replaced continuous pathways by the probability of an electron being in one or another specific region of space and maintained further that electrons cannot be in any position in between those regions or travel continuously through them. To describe that discontinuity with a new model of the atom, in 1913 Niels Bohr introduced the notion of "quantum jumps" between energy levels, which spelled the end of the classical notion of a continuous action between causes and effects for subatomic events. Bohr envisioned atoms as having only specific quantized energy levels and ascribed atomic spectra as the light emitted when there was a "jump" from one quantum state to another. Of course, electrons do not actually "jump," and they do not travel in continuous linear orbits, analogous to the trajectory of the Earth around the Sun, which is a proper continuous orbit. Bohr was acutely aware of the necessity of metaphorical language taken from mechanical and visualizable models, hence his belief that we are "suspended in language."[11] Accordingly, electrons change their position around the nucleus not in continuous "orbits" but only discontinuously in absorbing or emitting discrete amounts of energy while "jumping" from one "allowable" energy level to another.

To understand how physicists came to understand the quantized nature of quantum "jumps," one must understand the nature of Planck's constant and its successive epistemological and ontological functions in quantum theory. In 1900 Max Planck introduced a constant, subsequently known as Planck's constant, a super-minuscule constant approximately 6.626×10^{-34} joule-seconds.[12] Initially, that constant functioned merely as a purely formal assumption; Planck himself referred to it as a "mathematical trick,"[13] used to resolve the ultraviolet catastrophe[14] within the framework of classical physics. At that stage in the history of physics, that constant carried no implication about the actual nature of the physical world; it was merely a mathematical adjustment. However, with the development of quantum theory, Planck's constant came to represent a fundamental limit in the inherent structure of reality. It no longer served simply as a means of organizing experimental data but instead came to define the discreteness of energy exchanges and the granular nature of subatomic processes generally. Thus, a constant that originated as an epistemological necessity was reinterpreted at a deeper level of analysis as a constitutive feature of physical ontology, a quantum of action inherent in nature.[15] In the face of this challenge to classical physics, for a long time Planck himself remained reluctant to entirely recast his ideas about continuity.

In 1905 Einstein took Planck's mathematical trick seriously and proposed that light is quantized as photons and exchanged in discrete packets, thereby underscoring the quantized nature of physical reality at the subatomic level.[16] In 1921 Einstein won the Nobel Prize for that paper on the photoelectric effect, not for his paper on relativity. Finally, Planck himself acceded to the ontological nature of Planck's constant. In 1925 he wrote: "Either the quantum of action was a fictitious quantity . . . or must play a fundamental role in physics and proclaim itself as something quite new and hitherto unheard of, forcing us to recast our physical ideas, which, since the foundation of the infinitesimal calculus by Leibnitz and Newton, were built on the assumption of continuity of all causal relations."[17]

Certainty (Knowledge of Causal Events)

In 1927 Heisenberg announced as a key principle of quantum theory, the "uncertainty principle," which states that there is a fundamental limit, Planck's constant, to how precisely we can know *at the same time* both the position and momentum of subatomic particles such as photons.[18] According to Bohr, quantum uncertainty extends to the inherent nature of objects themselves, or at least to the way nature allows experimenters to interact with objects. Quantum uncertainty does not refer merely to the inability to make precise measurements and predictions. It is not a consequence of experimental error or bad equipment. That is, the behavior of an electron does not just *appear* to be uncertain, but, difficult as it is to grasp this change in thinking, it is *inherently* uncertain.

An example of such quantum uncertainty is radioactive disintegration. Physicists can determine and predict to an extremely precise degree the half-life of a mass of radium atoms based on the average rate of radioactive disintegration of all the atoms in the mass, but they cannot predict precisely when any specific atom will emit an electron and disintegrate. Bohr maintained that quantum theory cannot offer a precise causal description of the physical behavior of any subatomic entity such as an electron but can offer only statistical predictions under certain conditions.

Predictability

A related limit on the classical model was the replacement of a deterministic causality by a probabilistic causality. Classical physics maintained that masses of particles such as molecules of a confined gas behave probabilistically and can be measured only statistically as pressure, but every single individual molecule of that gas behaves in a deterministic fashion and could, in theory, be measured precisely. Quantum theory concurred with those findings for macroscopic events but also held that the behavior of individual subatomic particles is fundamentally probabilistic.[19] Some physicists refused to accept such irreducible probability. Einstein continued to believe that nature was deterministically causal in its ultimate constitution and that physicists would someday discover a theory that accounted for subatomic phenomena by means of "hidden variables" in a classical-like way. He expressed his skepticism about irreducible probability in a remark to Max Born in 1926 that God "does not play dice."[20]

Objectivity

Classical physics also maintained that causal knowledge of physical events is objective, meaning that its experiments were thought to distinguish clearly between the subject (experimenter) and the object (phenomenon being observed) and to reveal an objective reality of causal actions unaltered by the action, technology, or theoretical orientation of the experimenter. But, in view of the quantum uncertainty principle, some experiments make this type of objective knowledge of subatomic events impossible. For example, experiments to determine the nature of subatomic events show that electrons (classically believed to be particles) or photons (classically believed to be waves) exhibit either type of behavior, particle-like or wave-like, depending upon the experimental setup used to detect them. In 1927 Bohr explained: "[T]he quantum postulate implies that any observation of atomic phenomena will involve an interaction with the agency of observation not to be neglected. Accordingly, an independent reality in the ordinary physical sense can neither be ascribed to the phenomena nor to the agencies of observation."[21] As Heisenberg elaborated, "what we observe is not nature in itself but nature exposed to our method of questioning."[22] When an experiment is set up to test for light's wave-like properties, the light behaves like a wave, but when it is set up to test for particle-like properties it behaves like a particle. Quantum physics complicated the distinction between subject and object for subatomic events and in so doing limited the precision of objective knowledge of such events, which necessarily involved an error in measurement equal to some multiple of Planck's constant.

Visualizability

In classical physics mechanical causal action can be easily visualized and conceptualized as one billiard ball striking and moving another billiard ball. The necessity of visualizability in classical physics was underscored by the Victorian physicist William Thomson (Lord Kelvin), who, in 1884, as classical electromagnetic field theory began to dispense with mechanical models, is believed to have remarked, "I am never content until I have constructed a mechanical model of the object I am studying. If I succeed in making one, I understand; otherwise I do not. Hence I cannot grasp the electromagnetic theory of light."[23]

Quantum theorists had to learn to understand certain phenomena without mechanical models or visualizability. By 1923 Bohr had abandoned the visualizable model of the atom as a miniature solar system. Then Heisenberg took the lead by proposing an entirely unvisualizable conception of atoms based on purely mathematical entities known as matrices, which could be used to obtain proper statistical predictions concerning the outcomes of certain experiments but could not enable experimenters to visualize those outcomes. A partly visualizable model was restored in 1926 with Erwin Schrödinger's theory of atomic entities as wave packets with electrons surrounding the atom's nucleus like vibrating strings charged with electricity.[24]

Causality

Confronted with so many upended classical notions, some quantum physicists argued that the concept of causality should be abandoned altogether. Friedrich Waismann tagged 1927 as "the year which saw the demise of the notion of causality as men of science came . . . almost against their will, to recognize the impossibility of giving a coherent causal description of happenings on the atomic scale."[25] Bohr himself argued that quantum physicists "have been forced step by step to forego a causal description of the behavior of individual atoms in space and time."[26] But, while he rejected precise causal understanding at the subatomic level, he retained it at the macrocosmic level.

Quantum theory shattered the goal of explaining all events causally from the microcosm to the macrocosm as the action of simple entities moving continuously according to deterministic forces that could be accounted for by universally valid laws and calculated with unlimited precision. While few physicists seriously believed that such a deterministic reductionism obtained throughout the physical world, let alone for complex human behavior, it still provided a standard of the highest sort of causal knowledge. Quantum theory offered a new standard. The uncertainty and probabilistic causality of quantum theory, even though limited to subatomic phenomena, had considerable impact on some thinkers and imaginative writers concerned with human behavior. They reasoned that, if the most basic events were irreducibly discontinuous and if knowledge of them was probabilistic and uncertain, then causal knowledge of more complex composites of organic events such as human behavior must be even more discontinuous, probabilistic, and uncertain.

Some modernist writers outside of physics were intrigued by aspects of quantum theory, which they believed might hold the key to understanding human experience. The concepts they found most relevant were probability (or randomness), discontinuity, and, especially, the uncertainty principle, although this last concept was frequently misunderstood. Heisenberg did not maintain that the observer always interferes with the observation, making accurate and objective measurements impossible. In fact, for certain kinds of knowledge he argued the opposite, because, for example, experimenters can measure the position or momentum of an electron with unlimited precision, but not both features at the same time. The most important consequence of the uncertainty principle for general cultural history is that it introduced an unavoidable error or uncertainty (albeit of the minuscule size of Planck's constant) that was as absolute and universal as the speed of light. Planck's constant is indispensable for understanding the photoelectric effect, the quantization of energy levels in atoms, and Heisenberg's uncertainty principle, as well as many other phenomena in physics. Whereas classical physics offered the possibility of exhaustive predictive knowledge, quantum theory established that some errors in measurement were unavoidable and that therefore classical-style deterministic causal explanations as well as precise predictive knowledge of subatomic events were impossible.

Put together, relativity theory and quantum theory revolutionized understanding of the theoretical foundations of matter, energy, light, time, space, motion, gravity, and causation.

Notes

1. In addition to sources cited in this chapter, my discussion has drawn on John S. Rigden, *Einstein 1905: The Standard of Greatness* (Harvard University

Press, 2005); John Stachel, *Einstein's Miraculous Year: Five Papers that Changed the Face of Physics* (Princeton University Press, 1998); and Richard Wolfson, *Simply Einstein: Relativity Demystified* (Norton, 2003).

2. A contrasting theory was introduced with Lord Kelvin's theory of the "vortex atom" in 1867. The theory was "widely popular in Great Britain" but around 1890 "ran out of steam" because of its "lack of progress" in explaining other physical phenomena. Helge Kragh, "The Vortex Atom: A Victorian Theory of Everything" *Centaurus* 44, nos. 1–2 (2002): 32–114.
3. J. J. Thomson "Cathode Rays" [1897] in Henry A. Boorse and Lloyd Motz eds., *The World of Atoms* (Basic, 1966), 426.
4. H. A. Lorentz, "Michelson's Interference Experiment" [1881] in H. A. Lorentz et al eds., *The Principle of Relativity* (Dover, 1952), 7.
5. Albert Einstein, "On the Electrodynamics of Moving Bodies" in *The Principle of Relativity*, 38.
6. Albert Einstein, *Relativity: The Special and the General Theory* [1916] (Crown, 1961), 26.
7. Hermann Minkowski, "Space and Time" [1908] in *The Principle of Relativity*, 75, 76.
8. Albert Einstein, "Relativity and the Problem of Space" [1952] in *Relativity: The Special and the General Theory*, 150.
9. N. Katherine Hayles, *The Cosmic Web: Scientific Field Models and Literary Strategies in the Twentieth Century* (Cornell University Press, 1984), 47–48.
10. Friedrich Waismann, "The Decline and Fall of Causality," in A. C. Crombie ed., *Turning Points in Physics* (Palgrave, 1959); Paul Forman, "Weimar Culture, Causality, and Quantum Theory, 1918–1927: Adaptation by German Physicists and Mathematicians to a Hostile Intellectual Environment," *Historical Studies in the Physical Sciences* 3 (1971): 1–115. On these and other problems in the classical model see Morton Tavel, *Contemporary Physics and the Limits of Knowledge* (Rutgers University Press, 2002), 167.
11. Arthur I. Miller, "Visualization Lost and Regained: The Genesis of the Quantum Theory in the Period 1913–1927," in Judith Wechsler ed., *On Aesthetics in Science* (MIT Press, 1978), 74–77.
12. A joule is the energy transferred when one newton of force moves an object one meter in the direction of the force.
13. Helge Kragh, "Max Planck: The Reluctant Revolutionary," *Physics World* (December 2000): 31–35.
14. The ultraviolet catastrophe was an erroneous theory of prequantum physics that hot objects should give off infinite light energy at ultraviolet wavelengths, which is impossible.
15. The image of quantum physics probing deeper into the nature of reality is captured by Max Born, who in 1934 wrote that "the deeper we penetrate, the more restless becomes the universe, and the vaguer and cloudier." *The Restless Universe* (Dover, 1951), 277.
16. Albert Einstein, "On a Heuristic Point of View Concerning the Production and Transformation of Light," *Annalen der Physik* Series 4, Volume 17 (1905): 132–48.
17. Quoted in Alan J. Friedman and Carol C. Donley, *Einstein as Myth and Muse* (Cambridge University Press, 1985), 114.

18. Werner Heisenberg, "On the Perceptual Content of Quantum Theoretical Kinematics and Mechanics," *Zeitschrift für Physik* 43 (1927).
19. Arkady Plotnitsky, *The Knowable and the Unknowable: Modern Science, Nonclassical Thought, and the "Two Cultures"* (University of Michigan Press, 2002), 34.
20. Letter to Max Born, 4 December 1926, in Max Born ed., *The Born–Einstein Letters 1916–1955* (Palgrave Macmillan, 1971).
21. Niels Bohr, "The Quantum Postulate and the Recent Development of Atomic Theory [1927] in *The Philosophical Writings of Niels Bohr* (Ox Bow Press, 1987), vol. I, 54.
22. Werner Heisenberg, *Physics and Philosophy: The Revolution in Modern Science* (Harper, 1958), 58.
23. It is puzzling that Waismann gives no direct source for such an important quotation in "Decline and Fall of Causality," 146–47. Kelvin did make a similar statement, however, in 1884 in his Baltimore Lecture XI: "It seems to me that the test of 'Do we or [do we] not understand a particular subject in physics?' is, 'Can we make a mechanical model of it?'" Robert Kargon and Peter Achinstein eds., *Kelvin's Baltimore Lectures and Modern Theoretical Physics* (Bradford, 1987), 111.
24. "Visualization Lost and Regained," 74–76. See also Arthur I. Miller, *Insights of Genius: Imagery and Creativity in Science and Art* (Copernicus, 1996), 56.
25. "Decline and Fall of Causality," 84–154.
26. Niels Bohr, "Introductory Survey" [1929] in *Atomic Theory and the Description of Nature* (Ox Bow Press, 1987), 4.

Chapter 2

COSMOLOGY: Composition, Center, and Origin of the Universe

Victorian and modernist researchers sought to answer three major originary questions about the material composition, spatial center, and temporal origin of the universe. In centuries past those subjects had been addressed by influential Aristotelian, Judeo-Christian, and Newtonian ideas.

Aristotle's *Metaphysics* (350–330 BCE) located a stationary Earth at the center of a system of concentric transparent spheres that held the Moon, planets, Sun, and stars and rotated at different rates and in different directions. It held also that the natural state of objects was to be at rest as close as possible to the center of the universe, which was the center of the Earth. The origin of the universe Aristotle attributed to a prime mover, a first uncaused cause of all motion. That system was weakened when Galileo looked through a telescope in 1609 and saw mountains on the Moon (not a perfect sphere) and four moons orbiting Jupiter (not embedded in a hollow crystalline sphere). Galileo was concerned primarily with mechanical causes for which he provided mathematical explanations, although under threat of imprisonment he grudgingly acceded to scriptural explanations that God created the universe and caused tidal movements.[1]

Biblical estimates of the origin of the universe begin with a divine creation announced in the first words of Genesis, a chapter that was supposedly inspired by God and written by Moses.[2] But modern scholars place its authorship in the sixth and fifth centuries BCE, 500 to 1,000 years after Moses was supposed to have lived. Over a half dozen Judeo-Christian chroniclers drawing on biblical and supplementary texts estimated the creation of the universe by God during a six-day period between six and nine thousand years ago. The best known among these estimates was Bishop James Ussher's chronicle that pinpointed the day of creation as 22 October 4004 BCE, a date that appeared as a marginal notation in the 1701 edition of the *Cambridge University Press English*

Bible, where it remained until 1900, the beginning of the modernist age. The *Oxford University Press Bible* included the same notation until 1910.[3] In the *New Testament*, *The Gospel According to John* underscores and repeats several times the divine origin of everything that existed at some unspecified time or place: "In the beginning was the Word, and the Word was with God, and the Word was God. The same was in the beginning with God. All things were made by him; and without him was not any thing made that was made." In the thirteenth century Thomas Aquinas's *Summa Theologica* refined that account of creation in arguing that God was not just a one-time creator but a creator and ongoing sustainer of all that is.[4]

Critics challenged biblical chronologies along with Christian scientific authorities over the past several centuries but with special intensity in the modernist period.[5] Those challenges apply to the origins of the universe, Earth, and life, which are subjects of this and the next two chapters.

Beginning in the late seventeenth century and continuing for over two hundred years, Newtonian mechanics held sway among physicists and astronomers. Newton explained the origin of the universe and the laws of motion and gravitation that governed it by an omnipresent God who pervaded the universe as an immaterial ether that kept stars at just the right distances so that their attractions cancelled one another and kept them in balance. In 1905 Einstein's special theory of relativity challenged the applicability of Newtonian absolute time and space to explain all movements in the cosmos, and in 1915 his general theory of relativity challenged the applicability of Newton's law of gravitation to explain macrocosmic as well as microcosmic events.

To the Victorian layperson, dots of light in the night sky were either planets that moved in elliptical orbits around the Sun or stars that rotated across the sky embedded in some sort of dome or solid sphere. Amateur astronomers had a more accurate understanding. They may have used private telescopes, visited observatories, attended public lectures, subscribed to astronomy journals, or read books such as Richard Proctor's popular *Our Place among Infinities* (1875). More highly trained Victorian astronomers discovered an enormous amount of detail about the location of those dots of light by means of increasingly powerful telescopes and two inventions – the spectroscope, invented in 1814 by Joseph von Fraunhofer and used to analyze light from the stars to determine their chemical composition, and photography, invented in 1839 by Louis Daguerre and used to make permanent records of the position of celestial bodies as a basis for measuring their distance from Earth and tracking their movement.

Material Composition of the Cosmos

In 1833 the pioneer historian of science Auguste Comte predicted that "we can never know anything of [the stars'] chemical composition or mineralogical structure."[6] Working together starting in 1859, the German chemist Robert Bunsen and the German physicist Gustav Kirchoff began to prove Comte wrong. Over the next several years, they used spectroscopes to analyze the spectral lines of light coming from the Sun and found that it contained hydrogen, iron, sodium, magnesium, nickel, and calcium. In 1864 the British astronomer William Huggins made the first spectral analysis of a gaseous planetary nebula NGC 6543 (Cat's Eye Nebula), a finding that subsequently led him to distinguish between gaseous nebulae such as Orion and star-filled nebulae (later called galaxies) such as Andromeda. In 1868 the French astronomer Pierre Janssen discovered that the Sun contained helium before it was discovered on Earth. In that period for the first time, as John Farrell explained, the beguiling question of the stars' composition could be answered, and "the field of astrophysics was born."[7] That breakthrough showed that heavenly bodies were made of the same elements as Earth, signifying that there were no Aristotelian special elements in the celestial realm.

An important modification to the notion that the stars were composed of the same elements and in the same proportions as on Earth was made by the British-born Celia Payne, who completed a dissertation in the astronomy department at Harvard University, demonstrating that the Sun and other stars were largely helium and hydrogen, research that transformed understanding of the composition of the stars. Another contribution to understanding the composition of the material universe was made by Annie Jump Cannon, working at the Harvard College Observatory, where she developed a method for classifying stars based on their temperature and the absorption lines in their spectra.

Throughout the modernist period, deeper understanding of the material composition of celestial bodies continued to be generated by spectrographic analysis integrated with investigations into the spatial center and temporal origin of the universe. That research was pursued by modernist astronomers using improved spectrographs and telescopes and theoretical physicists investigating space, time, and energy.

The Spatial Center of the Universe

The second major originary question for cosmologists was about the center of the universe. In attempting to identify that location, astronomers tracked the direction and rate of motion of objects in the universe and reasoned backwards in time and space from those findings to some possible primordial spatial center as well as a temporal beginning. In 1868 William Huggins made spectroscopic analyses of light from the star Sirius, revealing that its color shifted to the red end of the spectrum from the Doppler effect on light, indicating that it was moving away from the Sun at 29.4 miles per second. That first breakthrough in understanding a star's rate of movement relative to the Sun was followed in the modernist period by major astronomical discoveries about stars' movements as well as their distances from Earth.

Confusing the history of astronomy in the modernist period is the use of two terms for the same thing – nebula and galaxy. At the end of the nineteenth century, astronomers generally identified luminous cloudy objects in the sky as nebulae. Stronger telescopes eventually revealed that some of those objects were composed of many stars, and astronomers increasingly identified clusters of stars as galaxies. In an important article of 1922, Edwin Hubble distinguished between galactic nebulae (within the Milky Way) and nongalactic nebulae that included spirals, spindles, and irregulars, but in 1929, when he introduced his revolutionary theory of the expansion of the universe, he also referred to galaxies as nebulae. Throughout the nineteenth century the ancient term galaxy was generally associated with the Milky Way Galaxy with a capital G.

In the beginning of the modernist period, astronomers knew where a great many stars were located but not how distant they were. They erroneously believed that the nebulae (or galaxies) were all contained in the one and only Galaxy, the Milky Way. They calculated distances to the closest stars by using the methods of visual parallax and geometric triangulation that relied on changes in a star's position as viewed from different points in the Earth's orbit at six-month intervals. Unable to use that method to measure stars at greater distances for which the parallax was too small to detect, they instead filled their time making charts of the names and locations of stars and the spectral colors radiating out of them.

In 1895 at the Harvard College Observatory, Henrietta Leavitt, one of the many female "computing assistants" hired to make such charts, began to look through photographic plates for Cepheid variable stars whose brightness varied relative to their size. She discovered hundreds of such stars in two dwarf galaxies, the Large and Small Magellanic Clouds on the outer fringe of the Milky Way. In 1908 she discovered a method for measuring distances to those stars with the "period–luminosity relationship," subsequently known as Leavitt's Law. It allowed her to calculate the distance to Cepheid variables based on their pulsating luminosity. Such stars became "standard candles" by which astronomers could measure the distance to other stars in galaxies up to around 200,000 light-years away. As one researcher characterized this path-breaking discovery, "[n]ow, it appeared, the genuine distance of any Cepheid variable could be determined from knowing only its apparent brightness and the period of its variability. Astronomy had found its third dimension."[8]

In 1913 the Danish astronomer Ejnar Hertzsprung drew on Leavitt's method and first estimated the distances to stars in the Magellanic Clouds, but for stars farther away more refined techniques and stronger telescopes were required. In 1914, using the new sixty-inch telescope installed on Mount Wilson in 1908, Harlow Shapley applied Leavitt's method for Cepheid variables to study constellations of stars called globular clusters. In them he found exceptionally large red giants and super giants that became new standard candles that he used to calculate even more distant stars. Shapley speculated about the location of the Sun with respect to the center of the Milky Way, albeit erroneously. He calculated that the Sun was around 50,000 light-years from the center of the Milky Way, while in fact it is 30,000 light-years from the center. His over-estimation of the size of the Milky Way Galaxy and the presumption that it was the entire universe prompted a widely publicized debate in 1920 over the universe's size and composition.

Opposing Shapley's view was that of Heber Curtis who worked with the thirty-six-inch telescope at the Lick Observatory near San Jose. While Shapley concluded that the observed spiral galaxies were all contained within a huge Milky Way, Curtis thought that they were outside the Milky Way, more remote components of an "island universe." In 1920 the two met in Washington, DC, to debate the matter. Observers concluded that Curtis was correct in his view of an island universe with galaxies outside the Milky Way but wrong in arguing that the Sun was in its center. Astronomers settled on the notion that the space of the universe was expanding in every direction and had no center, not even three decades later in 1948 when astronomers were speculating that the universe might have had a beginning with a Big Bang that did not, however, start in any one spot in the universe but occurred somehow throughout the vast stretches of it.

A jarring implication of Einstein's cosmological theorizing was that astronomers could not definitively determine the center of the universe or even the location of any object in it, because those locations are determined by the point of view of an observer that is endlessly variable and because it must be determined by light that is curved by the curved spacetime manifold or continuum through which it travels. Thus, a major finding of modernist cosmologists was that, although there is a center of the Earth and a center of the Milky Way Galaxy and other galaxies, there is no discoverable center of the universe. In 1933 Sir Arthur Eddington made that point. "We shall therefore no longer regard the phenomenon [of an expanding universe] as a movement away from our galaxy. It is a general scattering apart, having no particular center."[9] While Copernicus delivered a famous first blow to the human ego's notion that the Earth was in the center of the universe, modernist cosmologists shattered that idea altogether with the existentially disorienting notion that there is no center to the universe at all.

The Temporal Origin of the Cosmos

The idea that the space of the universe might be an infinite expansion raised two major questions about its time frame: the possibility of measuring it and the moment of its beginning.

The first serious scientific speculation that the universe might be expanding was made in 1863 by Kelvin, who proposed, in accord with the law of entropy, that the universe was moving toward increasing disorder and so there must have been some time in the past when it was more ordered and compact than at present.[10] He was unable to identify such an ultimate originating moment.

In the modernist period, an argument for the expansion of the universe was first proposed by the American astronomer Vesto Slipher in 1912, working with spectroscopes and the twenty-four-inch telescope at an observatory near Flagstaff, Arizona. His study of spiral galaxies revealed that most of them were redshifted and therefore moving away from Earth.[11] By 1917, he had measured the radial velocities of twenty-five spiral galaxies and found that most were moving away from Earth at substantial speeds, some as much as a thousand miles a second. While some astronomers thought that these findings suggested a general expansion of the entire universe, no such theory could be confirmed without convincing measurements of distances to the stars in those galaxies.

The first source of such measurements was provided by the most celebrated astronomer of the period, Edwin Hubble. The new 100-inch reflecting mirror telescope that he used to make his discoveries was a major accomplishment in itself, evident in the enormous challenges encountered over a fourteen-year period to conceive, finance, build, and install it. In 1903 George Hale, director of the Mount Wilson Observatory, conceived of the project and persuaded the Carnegie Foundation to front $150,000. In 1906 a Los Angeles businessman agreed to purchase a 100-inch-wide and thirteen-inch-thick glass disk from a Parisian factory, and to complete the project the Carnegie Foundation eventually pledged $10 million more (equal to $385 million in 2026). In 1908 the huge disk reached Hoboken aboard a steamer and from there traveled by train over land to Pasadena. When it was discovered to have many tiny air bubbles, it was rejected and put in storage, a disappointment that triggered Hale's nervous breakdown. Two years later a vice president of the Corning Glass Company examined the glass and concluded that the bubbles strengthened rather than weakened it. For nearly five years workers polished away a ton of glass from the fifty-five square feet of surface so that every spot had the same exacting curvature to within one part in about 90,000. Building the observatory dome began in 1913. The construction required 100 tons of reinforced concrete to support the enormous weight of the telescope, its housing, and the machinery to control it. The winding nine-mile footpath leading to the Mount Wilson summit had to be widened and strengthened to support the special truck with solid tires and unyielding springs to transport the telescope. To reduce tension on the massive steel girders of the telescope's supports, two large floats that bore the weight were submerged in tanks of mercury to enable astronomers to move the massive telescope with a touch of the hand. The driving clock that moved the telescope to synchronize its movement with the Earth's rotation and enable it to remain fixed on objects in space for long exposures, sometimes for several days, required more than a ton of brass castings, 1.5 tons of iron castings, and a two-ton driving weight. The movement of the telescope, dome, and shutters were controlled by over thirty electric motors. The 100-inch reflecting mirror was installed on 1 July 1917.[12] This remarkable technology had the bulk of a railroad locomotive and the precision of a Swiss watch.

Hubble's research with this telescope in the 1920s resulted in three major findings regarding the distance to galaxies outside the Milky Way, the increasing speed of objects moving away from Earth, and the expansion of the universe. Hubble failed to identify any center of the universe but did produce findings that formed the basis for theorizing about a possible beginning of the known universe.

Hubble was preceded in analyzing distances to spiral nebulae by the Swedish astronomer Knut Lundmark, who in 1921 identified individual stars in the spiral galaxy M33 in the constellation Triangulum that he estimated at 1 million light-years from Earth (it is in fact 2.73 million light-years away). In 1922 the Estonian astronomer Ernst Öpik argued that the Andromeda galaxy was 1.5 million light-years away. To prove that these galaxies were like the Milky Way but located well beyond it, Hubble had to measure the distance to stars, which he was able to do with measurements to Cepheid variable stars in them based on Leavitt's Law. When Hubble's paper on the Andromeda galaxy was read in 1924 at the meeting of the American Astronomical Society, "the entire Society knew that the debate [between Shapley and Curtis] had come to an end, that the island-universe concept of the distribution of matter in space had been proved, and that an era of enlightenment in cosmology had begun."[13]

Hubble continued researching galactic distances. In 1925 he published findings from two years of research on a clutch of stars named NGC 6822 near what later would become known as the Sagittarius galaxy. There he found fifty Cepheids that enabled him to calculate the galaxy's distance at 700,000 light-years (in fact 1.6 million light-years), clearly outside the Milky Way. After Hubble published his paper on Andromeda in 1928 it was impossible to deny that the cosmos was composed of galaxies far beyond the Milky Way. Such findings increased the distance of the Earth not only from the outer limit of the Milky Way galaxy but from the billions of galaxies beyond that edge, underscoring the Earth's remoteness and isolation from everything else in the universe.

The next challenge was to analyze the galaxies' speed and direction of movement, which Hubble accomplished with telescopic-spectrographic analyses of their light. While a few galaxies were blueshifted, indicating that they were moving toward Earth, the majority were redshifted, indicating that they were receding. Hubble discovered that stars' rate of speed away from Earth increased the farther away they were. From that finding he concluded that the entire universe was expanding. He announced that revolutionary finding in a paper of 1929, modestly titled "A Relation between Distance and Radial Velocity among Extra-Galactic Nebulae" and introduced an equation for calculating distances to the stars that eventually became known as Hubble's Law.

While Victorians believed that the differential rates of expansion of stars were a function of perspective, modernists came to believe that the observed rates were systematic and physical, that galaxies throughout the universe were moving away from one another at different relative speeds because space itself was expanding. From observations on Earth, parts of the universe seemed to be expanding faster than others in proportion to their distance from Earth. Astronomers were unable to locate the Earth at the center of that expansion because there was no privileged point of view on it. The different expansion speeds were a function of the Earth's privileged location at the center of investigation, that is, the place where the telescope happened to be located, but that position could not be determined to be at *the* center of the universe, because there is no discoverable center, and the light coming into those telescopes is curved as it passes by myriad objects throughout the universe. While astronomers could not determine any spatial center to the known universe, their findings did make possible a determination of its beginning, currently set at 13.8 billion years ago.

Modernist Cosmological Theory

Since antiquity astronomers have debated the size of the universe. Those who held that it was finite were unable to explain what was beyond the "edge." Those who held that it was infinite and full of material bodies had to contend with the prospect that the action of gravity generated by those bodies would tear apart stars and planets or recollapse them into a single, extremely dense mass.

Einstein's 1917 paper "Cosmological Considerations on the General Theory of Relativity" rejected the idea that the universe was expanding in arguing rather that it was static but unbounded. To keep it static he introduced into his equations a "cosmological constant" that stabilized his universe mathematically without any observational evidence to justify the theory. While he rejected the idea that the universe was infinite, he nevertheless argued that it was unbounded. That is, no matter how far out into space one traveled, one would never come to the end of the universe but eventually return to where one started owing to the curved spacetime continuum.

In response to theorizing by the Russian physicist Alexander Friedmann and the Belgian mathematician Georges Lemaître during the 1920s, Einstein gradually abandoned his cosmological constant, which he eventually dismissed as his "biggest blunder."[14] In 1922 Friedmann found a mathematical error in Einstein's theory, which Einstein eventually acknowledged. In 1931 Lemaître elaborated a theory of an expanding universe from an extremely dense "primeval atom" that exploded, giving rise to space, time, and all the matter and energy in the universe.[15] Lemaître's theory came to be viewed as a precursor of the theory of the expansion of the universe proposed by George Gamow and others in 1948 and named the following year by Fred Hoyle as "the Big Bang."[16] Thus, modernist theorizing about the nature of the universe pointed to some original, albeit not yet determined, moment when the expansion began.

American observers and European theoreticians remained isolated from one another in the formative years of modernist cosmology. They were brought together by Sir Arthur Eddington in a series of influential articles and books, most notably *The Expanding Universe* of 1933. This surveyed ways in which modernist astronomers had shattered the foundational ideas of Victorian cosmology in arguing that supposedly gaseous nebulae were actually galaxies full of stars, that the known universe extended far beyond the Milky Way, that galaxies were speeding away from one another at velocities that increased in proportion to their distance, that time and space were not separate dimensions but components of a spacetime continuum, that gravity did not act in a straight Euclidean line across a static space but rather that space itself was curved by the action of gravity on it, and, as Einstein argued, that the universe in the aggregate was not infinite but paradoxically finite as well as unbounded. Modernist cosmogonists, those researchers concerned primarily with the origins of the cosmos, were unable to identify any spatial center of the universe but by extrapolating backwards from the expansion of matter in the universe were able to posit some primordial temporal moment at the beginning, later called a singularity, when the universe was infinitely hot and dense, and equations describing that state broke down.

A jarring implication of Einstein's cosmology was that astronomers could not determine the center of the universe because it had no center, and, even if it did, such a center would be determined by the point of view of an observer on an Earth that is orbiting the Sun that is orbiting the Milky Way that is wandering in the universe driven willy-nilly at enormous relative speeds through an enormous number of curved spacetime frames. Cosmogonists did suggest that there might have been a beginning, but under circumstances that could never be fully understood – some sort of enormously powerful explosion unlike any explosion they had ever witnessed, which did not propel matter like pieces of an exploding bomb through a preexisting empty space but, in an altogether different manner difficult to conceive, an expansion of space itself throughout the universe at speeds exceeding the speed of light.

The findings of modernist cosmologists involved an existential paradox. They greatly expanded Victorian estimates of the size of the universe but in so doing greatly contracted the sense of their own stature in it. That paradox conforms to the specificity–uncertainty dialectic. The deeper they probed the nature of the universe, including its lack of a center, the more they realized how much more there was to know and how little they understood what they thought they knew. That deeper modernist probing was emphatically marked historically because it was based on looking through telescopes that had not been built in the Victorian period, on applying laws determining celestial distances that had not been worked out before Leavitt and Hubble, and on understanding space, time, and energy in ways that no one before Einstein had conceived.

Current thinking about the origin of the universe focuses on the Big Bang theory of an event around 13.8 billion years ago that ultimately determined the composition and distribution of matter in the universe along with the origin of space, time, and energy. While scientific knowledge of what happened before the Big Bang is guesswork, knowledge of what happened after it, beginning in the first 10^{-43} seconds (the Planck Epoch) is, if highly speculative, impressively detailed and based on surprisingly abundant evidence. During the Planck Epoch the temperature of the universe "cooled" from incalculably high to 10^{32} Kelvin, while space, time, and energy emerged out of primordial components of those basic features of the universe, clustered in some sort of extraordinarily dense entity subsequently called a "singularity."[17] Within a few seconds there appeared the subatomic particles that made up protons and neutrons to form the nuclei of hydrogen atoms along with the weak and strong forces holding those nuclei together. That deeper knowledge came from extrapolating backwards in time about evidence derived from pioneering observations with spectroscopes and telescopes in the 1910s and theories responding to Einstein's application of his general theory of relativity to cosmology in 1917, two major observational and theoretical products of the modernist period.

Notes

1. Winifred Lovell Wisan, "Galileo and God's Creation" *Isis*, 77, no. 3 (September 1986): 486.
2. "All Scripture is given by inspiration of God": 2 Timothy 3:16, although many scholars believe that the Pentateuch was written by multiple authors over several centuries, sometimes labeled as J (Yahwist), E (Elohist), P (Priestly), and D (Deuteronomist).
3. G. Brent Dalrymple, *The Age of the Earth* (Stanford University Press, 1991), 14–23.
4. Thomas Aquinas, *Summa Theologica*, First Part, Question 104, Article 1. He wrote that text between 1265 and 1273.
5. On criticism of the biblical account of creation, and of Christianity generally, especially Christian sexual morality in the modernist period, see Stephen

Kern, *Modernism after the Death of God: Christianity, Fragmentation, and Unification* (Routledge, 2017).

6. Quoted in Anna Henchman, *The Starry Sky Within: Astronomy and the Reach of the Mind in Victorian Literature* (Oxford University Press, 2014), 141.
7. John Farrell, *The Day Without Yesterday: Lemaître, Einstein, and the Birth of Modern Cosmology* (Thunder's Mouth Press, 2005), 48.
8. Timothy Ferris, *The Red Limit: The Search for the Edge of the Universe*, 2nd ed. (Perennial, 1983), 34–35.
9. Sir Arthur Eddington, *The Expanding Universe* (Cambridge University Press, 1933), 13.
10. Farrell, *The Day Without Yesterday: Lemaître, Einstein, and the Birth of Modern Cosmology*, 50.
11. Ferris, *Red Limit*, 56.
12. These details about construction come from Gale E. Christianson, *Edwin Hubble: Mariner of the Nebulae* (Farrer, Straus, 1995), 114–19.
13. Timothy Ferris reports Adam Sandage's reliance on his colleague Joel Stebbins for this reaction in *Red Limit*, 50.
14. "When I was discussing cosmological problems with Einstein he remarked that the introduction of the cosmological term was the biggest blunder he ever made in his life." George Gamow, *My World-Line: An Informal Autobiography* (Viking, 1970), 149. Quoted in Jim Baggott, *Origins: The Scientific Story of Creation* (Oxford, 2015), n373.
15. George Lemaître, "The Beginning of the World from the Point of View of Quantum Theory," *Nature* 127 (9 May 1931): 706.
16. Ralph Alpher, Hans Bethe, and George Gamow, "The Origin of Chemical Elements," *Physical Review*, 73, 7 (1 April 1948): 803–4.
17. Baggott, *Origins*, 30.

Chapter 3

GEOLOGY: Age of the Earth and Origins of the Continents and Mountains

Geology deals with a broad range of originary inquiries into the Earth's history and structure and the forces that act on it. This chapter focuses on how Victorian and modernist geologists contributed to a fuller understanding of the age of the Earth, the origin of its mountains, and the shape and location of its continents.

Age of the Earth

The underlying theory of many Victorian geologists was a hallmark of Victorian self-confidence – Charles Lyell's uniformitarian theory in his *Principles of Geology* (1830), which held that throughout history the Earth maintained a constant average land area, continental elevation, and balance in the size, distribution, and action of its rivers, seas, and oceans.[1] While many Victorian challenges to uniformitarianism included arguments about catastrophic upheavals modeled on the biblical flood, modernist challenges more typically derived from scientific investigations of naturalistic disruptions from ice ages, volcanic eruptions, earthquakes, strikes from other bodies, and, most disorienting, "continental drift."

While cosmologists could reason backward to calculate when the cosmos began with equations about the direction and movement of galaxies based on specific locations, directions, and speeds of star movement, geologists did not have such data because the Earth's beginning was so amorphous and protracted. The most widely held theory early on was based on the nebular hypothesis about the origin of the solar system proposed by Immanuel Kant in 1755 and elaborated by Pierre Laplace in 1796, that the Earth and other planets formed out of gases and dust circling the Sun. According to current thinking, the Earth was indeed formed out of a solar nebula over a million-year period starting approximately 4.5 billion years ago from electrostatic forces causing an initial clumping of gas and dust, then from gravity-driven aggregations of asteroids, meteoroids, and comets into larger planetesimals, and then into a larger protoplanet that eventually aggregated into the mass of rock that was the Earth.[2] When that mass became large enough, gravity-induced pressure generated heat that melted it as it spun into a sphere, a basic shape it held as it gradually cooled into a solid rock crust encasing underlying molten magma.

According to Genesis, the first thing God did was to create "the heaven and the earth." Genesis does not state that God created water but does state that on that first day "the spirit of God moved upon the face of the waters," and soon God divided heavenly waters or rain from oceanic waters that were "under the firmament." On the second day God separated the waters from dry land, and so the scenario continued with absolute confidence.

The history of geology throughout the Victorian period is a story of increasing secularization. From the 1830s to the 1850s geologists continued to believe that God created the Earth but increasingly rejected the specific periodization of Genesis and viewed geology as a study of methods by which God created Earth through natural causes such as erosion, uplift, and sedimentation, a theory known as theistic naturalism. In the mid-Victorian period, especially after the publication of Darwin's *On the Origin of Species* in 1859 (hereafter *Origin*), geologists increasingly relied on naturalistic explanations but still remained attached to the idea that God played a pivotal role in the Earth's creation.[3] Late Victorian geologists entirely ignored biblical dates in determining the age of the Earth and oceans but did retain the notion that the Earth was a mass of primordial rock that came into existence as a whole at some specific time by some sort of divine action.

While continental geologists in the Victorian period studied crystals, minerals, and strata, British geologists, preoccupied with Darwin's *Origin*, focused on the age of the Earth. They laid down a minimum time frame for natural selection to have produced the great variety of living forms that Darwin analyzed. His one specific estimate of that frame was that the excavation of the Weald in southern England must have required "three hundred million years." From this calculation he hazarded that "in all probability a far longer period than 300 million years has elapsed since the latter part of the Secondary period," indicating that the Earth was far older than that, starting long before the Primary period, to enable life to begin to evolve.[4] Darwin presumed such a long time span, backed by Lyell's calculation that the first animals with shells appeared 240 million years earlier, and so he drew on Lyell in formulating his theory of gradual evolution by the adaptive selection of "slight favorable variations" of species over at least several hundred million years.

In 1862 Darwin's estimate was undercut by the British physicist William Thomson, who dominated Victorian science in the UK as no other researcher of his time aside from Darwin. He was knighted as Sir William Thomson in 1866 and elevated to the peerage as Baron Kelvin in 1882. He published more than 650 papers on electricity, thermodynamics, magnetism, and geology and patented over seventy inventions. In 1848 he devised the temperature scale whose units still bear his name. In 1852 he published "On a Universal Tendency in Nature to the Dissipation of Mechanical Energy,"[5] which drew on the second law of thermodynamics that heat as it dissipates always moves from hotter to colder objects. That theory grounded his entropic theory according to which the energy of the universe was running down, leading to an eventual heat death for the Sun and the Earth. His 1862 paper based on that research, which included one of the most notorious incorrect calculations in the history of science, dominated geological thinking for the next forty years.[6]

In addition to Kelvin's impressive command of thermodynamics and mathematics, his geochronology also drew authority from three popular ideas in his time – the nebular hypothesis, natural law, and the argument from design. The nebular hypothesis held that stars and planets were formed by clumping, driven first by static electricity and then by gravitation. The stature of natural law derived from Newton's iconic elaboration of the "mathematical principles of natural philosophy," which became the model of scientific explanation based on natural laws and calculated by mathematical principles. The argument from design elaborated in William Paley's *Natural Theology* (1802) had enormous authority with nineteenth-century scientists including Darwin, who added a line to the conclusion of *Origin* about "laws impressed on matter by the Creator."[7]

Kelvin's 1862 paper on the age of the Earth began with the assumption that the Earth was originally a molten sphere. The source of the heat he assumed to be gravitation-induced pressure that increased as meteorites impacted the Earth's original aggregation of gases and dust. The scientific challenge was to determine its age by calculating its rate of cooling back to a molten state. His calculation was based on three factors for which he had unreliable evidence – the initial temperature of a molten Earth, the thermal conductivity of its rocks, and its outer part's temperature gradient (change of temperature as a function of depth). Without adequate experimental evidence Kelvin estimated that the initial temperature at which the rocks would be molten was 7000°F, which was three to four times the actual temperature. He did tests to determine the thermal conductivity of three major kinds of rock that he then averaged and projected onto the entire Earth. He based the temperature gradient on the currently accepted value of an increase of 1°F for every fifty feet of depth. From these factors he calculated that the cooling Earth began to consolidate a solid outer crust between 20 and 400 million years earlier, most likely around 100 million years, not the many hundreds of millions of years that Darwinian evolution required.[8]

Kelvin based these calculations on the assumption that heat transferred to the surface by conduction through static rocks, but he did not fully consider that there are variations with different kinds of rocks all over the planet, that heat was generated by the increased gravitational force from meteorite collection, and that heat transfer through solid rocks by conduction was slower than heat transfer through liquids by convection. He was, of course, also unaware of the subsequent discovery in 1903 of heat generated by radioactively decaying elements. Nevertheless, as a leading expert in thermodynamics, his findings had enormous authority.

While Kelvin's most influential theories about the age of the Earth and opposition to Darwin's theory were based on laboratory tests and calculated with mathematics, they were also motivated by religious assumptions of his Presbyterian faith. As a child he heard his father's weekly Sunday readings from Scripture. At age ten he attended lectures on moral philosophy at the University of Glasgow, which emphasized how the natural world was created and directed by intelligent design. At twelve he attended lectures by an astronomy professor at Glasgow who proposed an intelligent design to the solar system. At Cambridge University he was influenced by Paley's *Natural Theology* and in 1851 wrote that no destruction of energy could occur "without an act of power possessed only by the supreme ruler."[9] Fifty years later Kelvin still held to the same religious presupposition of a divinely created universe.

Other Victorian geologists made different estimates of the age of the Earth far less than its actual age of 4.5 billion years as they struggled to make sense of the challenging problem of understanding what had been going on layer by layer in myriad areas beneath the Earth's surface. In 1876, T. Mellard Reade based his estimate of the Earth's age on chemical denudation, the removal of elements from continents into the oceans. Using figures for concentrations of sulfates of calcium and magnesium in the oceans' waters he calculated that age to be 25 million years, and for concentrations of chlorines about 200 million years. In 1893 the American geologist Clarence King ruled out an initial molten state and recalculated the age of the Earth to be around 24 million years.[10] Kelvin returned to the subject in 1897 and underscored King's estimate that was based on his own calculations that originated from the starting state of a molten Earth. In 1899 the Irish geologist John Joly estimated the age of the Earth to be around 90 million years based on oceanic sodium concentrations. These calculations overestimated the significance of human existence on Earth by underestimating the length of its overall history. As Leonard Wilson concluded, "Kelvin developed his argument for the limited age of the Earth with the intent to discredit uniformitarianism upon which Darwin's theory of evolution by natural selection rested."[11]

In 1899 the American geologist Thomas Chamberlin critiqued Kelvin's 1897 article on the age of the Earth. He credited Kelvin for challenging "ultra-uniformitarians" who made "reckless drafts on the bank of time" but criticized him for erring in the opposite direction by proposing an age of the Earth that was far too short. "The fascinating impressiveness of rigorous mathematical analysis," Chamberlin charged, "with its atmosphere of precision and elegance, should not blind us to the defects of the premises that condition the whole process. There is perhaps no beguilement more insidious and dangerous than an elaborate and elegant mathematical process built upon unfortified premises."[12] The reason Kelvin foundered over the age of the Earth was his determination to undermine Darwin's theory, which he believed undermined divine creation. Kelvin did not endorse "Mosaic" estimates of a divine creation 6,000 years earlier but did propose a specific time span far short of the many hundreds of millions of years that Darwin required. Kelvin also neglected the effect of meteorite aggregation increasing the Earth's mass, causing an increase in the gravitational pressure on its interior, in turn increasing heat that would slow the cooling and lengthen calculations of the Earth's age. Chamberlin also proposed presciently that "the internal constitution of the atoms may be . . . the seats of enormous energies."[13]

In 1903 Ernest Rutherford and Frederick Soddy transformed understanding of the Earth's age with research on the release of heat in radioactive decay. Properly appreciating that radioactivity generated heat throughout the Earth's interior enormously extended estimates of its age based on cooling rates. In 1905 Bertram Boltwood determined that lead might be a decay product of uranium, suggesting a way to use radioactivity to calculate geological time by measuring the relative proportions of uranium and lead in rocks.

For the next twenty years geologists used quantitative calculations based on laboratory experiments to augment qualitative interpretations based on field work. They studied radioactive decay in uranium, thorium, and radium to determine rates of radioactive decay in numerous regions all over the Earth. Most reliable for calculating radioactive decay were ratios of uranium to its decay product lead. Arthur Holmes's important study from 1911 on "The Association of Lead with Uranium in Rock-Minerals, and Its Application to the Measurement of Geological Time" affirmed that researchers can use lead as an index of age.[14] G. Brent Dalrymple concludes that these findings "were the first quantitative indication, based on physical principles rather than scientific intuition that Earth might be billions, rather than a few tens or hundreds of millions, of years old. The discovery of radioactivity and the heat that it produced made Kelvin's cooling calculations obsolete."[15]

These early calculations were used to determine the ages of rocks, which made it possible to calculate the age of the Earth as a whole. The first such calculation was made in 1921 by Henry Russell, professor of astronomy at Princeton University. Drawing on estimates of the total amount of uranium and lead in the Earth's crust, he calculated "4×10^9 [4 billion] years as a rough approximation to the age of the Earth's crust."[16] Six years later Arthur Holmes suggested an upper limit of just over 3 billion years.[17] Holmes deemphasized the influence of sodium accumulation and sediment thickness and did not even mention cooling calculations, methods key to Kelvin and other Victorian geologists, which were by then sidelined as imprecise and obsolete.

Radioactive decay rates had another advantage. They were not limited to calculating the age of the Earth when fully formed but also included the age of materials added during the Earth's formation out of the original solar nebula as well as subsequent planetesimal and protoplanet stages. The Earth's constitutive elements were extremely diverse and originated in many widely separate places, and its beginning was a protracted development over at least a million years. These discoveries conformed to the specificity–uncertainty dialectic, in that the more researchers probed into the Earth's age, the more they realized how much more there was to learn and how approximate their findings were and how even the entity to which the singular concept of "the Earth" referred was an approximation, the product of a long and complex generative process involving many components over many geological stages.

Origin of Continents and Mountains

Victorian and modernist geologists alike analyzed many changes that occurred in geological formations from the long-term action of heating and cooling, rain and wind, glaciers and volcanoes, gravity and tidal movements. One difference between the two approaches is that modernists did without divine creation and continuing divine oversight. In an essay "On the Sculpture of Mountains" (1856), the quintessential Victorian aesthetician John Ruskin combined scientific observation and natural laws with religious interpretation of the creation and evolution of mountains. In a distinctly Victorian religious and teleological style, he analyzed geological changes that transformed the "original contours" created by "the Great builder" in "working out the one consistent plan" that was "beneficently adapted to the welfare of the human race."[18] But, for all his detailed observations of geological changes, neither Ruskin nor other Victorians ever seriously questioned whether mountains had resulted from land movements across great distances or whether continents had moved across vast oceans. That questioning was carried on by modernist *mobilists* who believed that the continents had been moved by natural forces, in opposition to *fixists*, who believed that continents had been stationary since time immemorial.

A notable moment in mobility thinking occurred in 1788, when James Hutton discovered the Siccar Point "unconformity" in southeastern Scotland. That striking landform suggested that mountains were formed from both upward and lateral movements. While Hutton did not have technology to date the rocks or fossils, he knew he was viewing evidence of massive movements of rocks over vast stretches of time.[19] The lower level of gray sedimentary rock formed 425 million years earlier. Over the next 65 million years, the ocean closed and the horizontally sedimented rocks rotated vertically to point upward. During that period some red desert sandstone, also sedimented horizontally, moved laterally over the older rocks. The "unconformity" refers to the gap between these layers when there was no sedimentation or active erosion. Although Hutton's discovery unmistakably evinced lateral movement in mountain formation, such an idea was neglected in face of the commonsense notion that massively heavy rocks could not possibly move horizontally over great distances. For the next hundred years, thinking about mountain formation focused not on horizontal movement but on vertical thrusting caused by the buoyancy of rocks on a denser substrate.

Arguments in support of mobilism centered on three kinds of moving forces: the fundamental force of gravity; mechanical forces of blowing wind and moving water across land or in tides; and thermal forces of conduction, convection, and radiation of heat moving through rock and causing it to move. Modernists mixed these forces in elaborating theories about what came to be called "continental drift."

The most widely held theory in the Victorian period was that mountains had formed from primarily thermodynamic forces – cooling-induced contraction that generated an upward thrust in the Earth's crust.[20] A leading proponent of that view was the Austrian geologist Eduard Suess, who spent his life studying the Swiss Alps. In four volumes over 2,000 pages published between 1883 and 1904 (translated into English between 1904 and 1909 as *The Face of the Earth*), he analogized the effect of territorial contraction of the Earth to a drying apple, a contraction that resulted from long-term cooling from a molten state. As the Earth cooled it shrank, causing portions of the outer crust to collapse under gravitational force and form the ocean floor, dividing the Earth into oceans and continents with protruding mountains. With further cooling, some uplifted parts collapsed to form a second generation of oceans, while portions of former oceans rose to become dry land. Those developments explained the presence of marine fossils on dry land adjacent to "platforms" or "bridges" across oceans over which both flora and fauna migrated. Suess also theorized that oceans and continents were formed by alternate heat-induced rising and gravitational-induced sinking of portions of a single supercontinent.

The popular compression theory was easy to visualize: the cooling Earth shrank, causing its crust to wrinkle and make mountains push up, while the other surfaces became continents and oceans. But the theory faced numerous problems. First, the wrinkling of an apple would be distributed randomly over its surface, but the world's mountains appeared in patterns along the edges of continents such as the chain from the Rockies to the Andes. Second, if the Earth's radius had shrunk, the length of the day would have shortened, but it had not. Third, a former enthusiastic supporter of contraction theory, the English geophysicist Osmond Fisher, calculated that the amount of compression of the Earth's radius from secular cooling could be no more than six miles and produce an elevation above sea level of only twenty feet.[21] In support of that argument and against cooling-contraction, Fisher quoted Clarence Dutton of the United States Geological Survey, who wrote that Fisher's calculations had "utterly destroyed" the contraction theory, adding that "this hypothesis is nothing but a delusion and a snare, and that the quicker it is thrown aside and abandoned the better it will be for geological science."[22] Finally, the discovery of the heat generated by radioactivity undercut the energy source for compression theory based on the Earth's cooling.

While some mid-Victorian geographers noticed that the shape of separate continents looked like pieces of a plate that had broken apart, particularly the fit between the coastlines of South America and Africa, no major geologist seriously believed that these continents had once been together and then actually moved apart, dragging their enormous weight across the vast expanse of the Atlantic Ocean. However, evidence of earlier continental contiguity was compelling. In addition to visual inspection of matching coastlines on maps, there were fossils of the same flora and fauna that matched up on both sides of oceans, suggesting that the areas where the fossils had been found must have once been contiguous. To explain these facts contractionists speculated that between these areas there may have once been "land bridges" that allowed movement of fauna and flora across them or that the Atlantic basin had formed out of a collapsed ancient paleocontinent that allowed passage of living forms across its expanse before sinking and filling with water. But even those wild theories could not explain the presence of gold or diamonds in similar rock structures at the same latitudes near South American and African coasts.

In 1855, George Airy, Astronomer Royal of the UK, challenged contraction theory by suggesting that mountains had "roots" and were floating on a liquid substratum like a log floating in water.[23] Under the long-standing dominance of contraction theory, that floating-across theory was largely ignored until taken up by Fisher in the 1880s and reconceptualized by Dutton in 1889, who named that process hydrostatic equilibrium or "isostasy," meaning equal standing.[24] He explained the upthrust of mountains from horizontal pressures according to the following scenario of movement across, down, back across, and then up. Elevated land areas became denuded by the movement of wind and water producing runoff that flowed toward and then sedimented from gravitational force into low lying coastal areas. The weight of that sedimentation caused subsistence that in turn caused lateral displacement from mechanical force underground made possible by the plastic nature of the magma. That lateral movement caused uplift in neighboring areas, generating mountains.

Explanation of the origin of mountains and the position of continents was based on such lateral movement. Mountains were created not from the buoyancy of a lighter continental crust *on* a denser ocean substrate but from horizontal pressures generated by land masses moving laterally *through* a highly viscous substrate. The most significant version of that theory was by the German geologist Alfred Wegener. In two oral presentations of 1910 and then in an article of 1912, he formulated the classic theory of "continental displacement," which he eventually called "continental drift."[25] He reformulated the theory in a by-now-classic ninety-four-page book of 1915, substantially revised in 1920, 1922, 1924, and 1925, and then in a 231-page edition in 1929. The 1922 edition was translated into English in 1924 as *The Origin of Continents and Oceans*. Wegener rejected Suess's contraction theory and the idea that the continents and oceans were permanent in shape and that the identity of floral and faunal fossils on widely separated coasts could be explained by land bridges or sunken continents. Wegener countered:

> The basic 'obvious' supposition common to both the land-bridge and permanence theory – that the relative position of the continents has never altered – must be wrong. The continents must have shifted. South America must have lain alongside Africa and formed a unified block which was split in two in the Cretaceous [145–66 million years ago]; the two parts must then have become increasingly separated over a period of millions of years like pieces of cracked ice floe in water . . . If drift theory is taken as [our] basis, we can satisfy all the legitimate requirements of the land-bridge theory and of permanence theory. This now amounts to saying that there were land connections, but formed by contact between blocks now separated, not by intermediate continents which later sank.[26]

Wegener presented five kinds of evidence in support of continental drift. The evidence that struck him first in 1910 was viewing a globe and noticing "the congruence of the coastlines on either side of the Atlantic."[27] He also cited paleontological evidence of similar faunal and floral fossils between Africa and Brazil as well as between Australia and India, Madagascar and India, and Europe and North America. More significant was a "conformity of rocks" along the separate coasts. As he put it metaphorically, "It is just as if we were to refit the torn pieces of a newspaper by matching their edges and then check whether the lines of print run smoothly across," which they did.[28] A fourth source of evidence was from climate change, most notably that parts of South America, Africa, India, and Australia had all experienced glaciation indicated with special clarity by rock striations in the Permian era 300–250 million years earlier. That coincidence, which would be impossible to explain based on their current locations in warm regions, did make sense if these areas had once concentrated around the South Pole before "drifting" north. The fifth source of evidence was that the theory was testable, based on land surveys of Greenland in 1823, 1870, and 1907 that revealed a westward drift. His dedication to drift theory was tragically memorialized by his death in 1930 during his fourth expedition to Greenland to confirm longitudinal drift.

For all the evidence in favor of Wegener's drift theory, there was still considerable resistance for several reasons. Wegener's book-length statement was published in German in 1915 during World War I, alienating his ideas from non-German readers and out of strong political divisions. Mountains long symbolized the permanent and stable foundation of all things, memorably captured in Shelly's poem *Mont Blanc*, which begins by celebrating the mountain as a symbol of "[t]he everlasting universe of things" and continues with images of the mountain as "[s]till, snowy, silent, and serene." Wegener's book destabilized ideas about the "stable" ground under the feet of Victorians and violated their commonsense notion that mountains or continents could not move laterally any appreciable distance. The study of drift theory decentered geology itself within science because evidence for it came from several other sciences including chemistry, astronomy, physics, meteorology, biology, oceanography, and geography, all of which had different sources, methods, and criteria for proof.[29] Moreover, the term "drift" was misleading because it suggested random, undirected movement, but geological formations were driven by directed forces. But the greatest weakness in Wegener's theory was the lack of a clear mechanism or force driving the drift. He made one tentative suggestion based on a presupposition that the continents that were made of a lighter sial drifted like rafts over the underlying layer of the Earth's denser sima, but he had no idea what pushed the continents with enough force to overcome enormous frictional resistance. He speculated that a westward tidal force was caused by the gravitational pull of the Sun and Moon, and a centrifugal force caused by the Earth's equatorial bulge that may have moved continents away from the poles and toward the equator, but he had no direct evidence for these theories, and they raised widespread resistance.

Still, in the early twentieth century several new originary scenarios supported lateral movement or some version of drift. Most prominent were Reginald Daly's theory of gravity sliding, John Joly's theory of periodic fusion, and Arthur Holmes's theory of subcrustal convection currents, all battling tradition and the commonsense notion that massive continents with their huge mountains could not possibly move across vast distances.

While Wegener patterned drift after icebergs, Daly called his theory gravity slidings and patterned them after lava flows. In "The Earth's Crust and Its Stability" (1923) he proposed that crustal plates can be dragged horizontally as they are dragged down into the Earth's interior, and he kept open the possibility of large-scale "continental displacements."[30] The conspicuously placed epigraph to his *Our Mobile Earth* (1926) was the phrase "*E pur si muove*" ("and yet it does move"), words that Galileo supposedly muttered to himself after he was forced to recant publicly in 1633 before the inquisition that condemned his argument that the Earth orbited the Sun. Continental drift violated everyday commonsense as radically as Galileo's utterance violated commonsense geocentrism. Daly concluded with a sweeping affirmation of the likelihood of drift. "Our solid earth, apparently so stable, inert, and finished, is changing, mobile and still evolving . . . The lava floods and intriguing volcanoes tell us of the plasticity, mobility, of the deep interior of the globe."[31]

John Joly, whose work on radioactivity expanded the age of the Earth well beyond Kelvin's estimate of 20–40 million years, turned to tectonics of the Earth's crust and eventually aligned with the central idea of continental drift. The title of his book of 1909 suggested the framework of his argument: *Radioactivity and Geology: An Account of the Influence of Radioactive Energy on Terrestrial History*. Erosion leads to massive sedimentation and pressure that generates heat. The additional heat from radioactive decay forces a convective lateral movement and eventual uplift of deep-lying, liquefied molten rock. He explained: "In the deposition of sediments—their uplifting into mountain chains—their subsequent removal and their re-deposition elsewhere—there is a continual convective movement of radioactive materials on the surface of the earth . . . A mountain is not a mass of piled-up igneous rocks but (save in the case of volcanic cones) is composed of a complex of sedimentary rocks ridged up, folded, and often horizontally over-thrust, by intense compressive forces."[32]

Joly's earlier emphasis on lateral movement creating mountains applied equally to the movement of land masses, which he elaborated in *The Surface-History of the Earth*, published in 1925 when Wegener's drift theory was at the center of geological discussion. Whereas Daly assumed continuous solidifying of a molten Earth, Joly maintained thermal cycles – periodic fusion and solidification of the subcrust from the buildup and dissipation of radiogenic heat. Joly explained the lateral movement of continents with his theory of periodic thermal cycles.[33] The formation of ocean crusts was the mechanical force for the movement of continents. In a section on "The Building of the Mountains" Joly underscored the dynamic function of magma erupting through seafloor rifts. "The sea floor has been the passive instrument of orogenesis [mountain building]. It has forced inwards the continental margin and deformed and rent the continental crust to its depths. The lava wells out, caught in gaping fissures beneath and forced upwards. And as the stresses increase, the sediments themselves, folded and crushed, emerge above the surface of the sea."[34]

In 1926 the highly honored British geologist Arthur Holmes proposed an explanation of continental drift that accorded with Joly's theory of alternating periods of warming and cooling. "Periodically repeated cycles of fusion and consolidation provide fully for a varied geological history of just the rhythmic and discontinuous type that the earth has actually experienced."[35] Two years later he explained the energy driving those periodic convection currents as radioactive: "[T]here seems no escape from the deduction that slow but overwhelmingly powerful currents must have been generated in the underworld at various times . . . as a result of differential heating by radioactivity."[36] The following year he specified the texture of the magma that would be able to latch on to the underside of continents to move them laterally. "Only currents in a highly viscous glass could get sufficient grip on the continental under-surfaces to exert the required drag on the overlying material."[37] While Daly, Joly, and numerous others contributed to drift theory, Holmes's theory was the most significant, as Naomi Oreskes noted: "it was Holmes who most explicitly, unambiguously, and consistently advocated convection as the *cause* of continental drift."[38] She concluded that "the same driving force that is generally accepted today for plate tectonics was proposed in the 1920s."[39]

Modernist geologists calculated the age of the Earth as far older than the time span proposed by the Victorians, while their theory of continental drift destabilized thinking about the Earth's territorial foundation. The impact of these modernist theories was analogous to the ego hit from modernist cosmological theories about the universe's enormous age and accelerating expansion. One difference between these originary theories is that Wegener reasoned from the current state of continents back to when they were together in a single supercontinent that he called Pangea, whereas the cosmologist George Lemaître reasoned from the current state of matter in the universe back to a "primeval atom." Another notable difference between these originary inquiries is that the cosmological beginning of the universe was the start of an expansion of matter *to* widely different places, while the geological beginning of the Earth was the culmination of a contraction of matter *from* widely different places. Either way these originary theories generated similar existential blows to the human ego as they reduced the significance of human existence in the larger scheme of things. In *Ulysses* (1922), James Joyce vividly expressed that reduction in the tenure of human existence on Earth as "a parenthesis of infinitesimal brevity" (*U* 573).

Fifty years after Wegener elaborated his theory of continental drift, contributions from a half-dozen sciences drawing on data from the key technological inventions of sonar and the magnetometer elaborated the theory of plate tectonics, one of the most heralded scientific achievements of the twentieth century.

Notes

1. Joe D. Burchfield, *Lord Kelvin and the Age of the Earth* (Science History Publications, 1975), 213.
2. Estimates and historical reconstructions throughout this chapter are drawn from G. Brent Dalrymple, *The Age of the Earth* (Stanford University Press, 1991); G. Brent Dalrymple, *Ancient Earth, Ancient Skies* (Stanford University Press, 2004); David Christian, *Maps of Time: An Introduction to Big History* (University of California Press 2011); and Kieran D. O'Hara, *A Brief History of Geology* (Cambridge University Press, 2018).
3. Charles Darwin, *On the Origin of Species* [Facsimile of the first edition of 1859] (Harvard University Press, 1964). Subsequent text references are to this edition.
4. Ibid., 287. He also hazarded that there may have been "an almost infinite number of generations" (481). He was immediately attacked by several geologists and removed that specific calculation by the third edition of *Origins* in 1861. Burchfield, *Lord Kelvin and the Age of the Earth*, 72.
5. William Thomson, "On a Universal Tendency in Nature to the Dissipation of Mechanical Energy," *Philosophical Magazine* 4, 4 (1852): 304–6.
6. William Thomson, "On the Secular Cooling of the Earth," *Transactions of the Royal Society of Edinburgh* 23 (1862): 157–69.
7. Darwin, *Origin of Species*, 488.
8. Calculations from Dalrymple, *The Age of the Earth*, 31–40.
9. On this argument see Leonard G. Wilson, "Religious Assumptions in Lord Kelvin's Estimates of the Earth's Age," *Earth Sciences History* 29, no. 2 (2010): 188, 193, 208.
10. Clarence King, "The Age of the Earth," *American Journal of Science* 3rd series, 45 (1893): 1–20.
11. Wilson, "Religious Assumptions," 208.
12. T. C. Chamberlin, "Lord Kelvin's Address on the Age of the Earth as an Abode Fitted for Life," *Science*, Part I (30 June 1899): 889–90.
13. Ibid., Part II (7 July 1899): 12.
14. Arthur Holmes, "The Association of Lead with Uranium in Rock-Minerals, and Its Application to the Measurement of Geological Time," *Proceedings of the Royal Society of London* series A, vol. 85 (1911): 255. Dalrymple, *Ancient Earth, Ancient Skies*, 50.
15. Ibid., 51.
16. H. N. Russell, "A Superior Limit to the Age of the Earth's Crust," *Proceedings of the Royal Society of London* series A, vol. 99 (1911): 86.
17. Arthur Holmes, *The Age of the Earth: An Introduction to Geological Ideas* (Benn's Sixpenny Library, 1927), 72.
18. John Ruskin, "On the Sculpture of Mountains," in *Modern Painters* (Smith, Elder & Co., 1856), vol. 4, 130–35, 152.
19. The significance of Hutton's discovery was recorded by his companion John Playfair. "The mind seemed to grow giddy by looking so far back into the abyss of time; and whilst we listened with earnestness and admiration to the philosopher who was now unfolding to us the order and series of these wonderful events, we became sensible how much further reason

may sometimes go than imagination may venture to follow." John Playfair, *Transactions of the Royal Society of Edinburgh* (1805) vol. V, pt. III. The giddiness that Hutten felt may well have come from the sudden vertiginous expanse of time he was experiencing far beyond the 6,000-year age of the Earth recorded in Genesis, which he had originally been taught to believe.

20. Naomi Oreskes, *The Rejection of Continental Drift: Theory and Method in American Earth Sciences* (Oxford, 1999), 14.
21. Osmond Fisher, "On the Mean Height of the Surface Elevations of a Solid Globe through Cooling," *Philosophical Magazine* 25 (1888): 7–20, cited in O'Hara, *Brief History of Geology*, 96–98.
22. Osmond Fisher, "Physics of the Earth's Crust," *Nature* 27 (23 November 1882): 76–77.
23. G. B. Airy, "On the Computation of the Effect of the Attraction of Mountain-Masses, as Disturbing the Apparent Astronomical Latitude of Stations in Geodetic Surveys," *Philosophical Transactions of Royal Society of London* 145 (1855): 101–4.
24. C. E. Dutton, "On Some of the Physical Problems of Geology," *Bulletin of Philosophical Society of Washington* 11 (1889): 51–64.
25. Wegener, "Die Entstehung der Kontinente," *Petermanns Geographische Mitteilung* 58 (1912): 185–95, 253–56, 305–9. The third edition of *The Origin of Continents and Oceans* [1915] (Dutton, 1924) begins with a section on "The Essentials of the Displacement Theory." The theory came to known as continental drift theory, a term that Wegener himself and most everyone else eventually used.
26. Wegener, *The Origin of Continents and Oceans* [1922] trans. 1924 (Dover, 1966), 17.
27. Ibid., 1.
28. Ibid., 77.
29. Peter J. Bowler, *The Environmental Sciences* (Norton, 1993), 407–8.
30. Reginald Daly, "The Earth's Crust and Its Stability," *American Journal of Science* 205 (1923): 371.
31. Reginald Aldworth Daly, *Our Mobile Earth* (Scribner's, 1926), 320.
32. John Joly, *Radioactivity and Geology: An Account of the Influence of Radioactive Energy on Terrestrial History* (Archibald Constable, 1909), 94, 95.
33. John Joly, *The Surface-History of the Earth* [1925] 2nd ed. (Clarendon Press, 1930), 125.
34. Ibid., 125.
35. Arthur Holmes, "Contributions to the Theory of Magmatic Cycles," *Geological Magazine* 63 (1926): 307. Quoted in Oreskes, *Rejection of Continental Drift*, 115.
36. Arthur Holmes, "Radioactivity and Continental Drift" 65 *Geological Magazine* (1928): 237–38.
37. Quoted in Oreskes, *Rejection of Continental Drift*, 117.
38. Ibid., 119.
39. Ibid., 120.

Chapter 4

BIOLOGY: Origins of Life, Species, and Human Beings

The most intriguing originary questions in biology were about the origin of life, the origin of species, and the origin of human beings. Answers about the beginning of life included when, where, and how the first forms of life came into being. Answers on the origin of species centered on the causal mechanisms that generated the evolution of species throughout the eons. Answers about the origin of individual human beings concerned the mechanics of hereditary transmission in sexual reproduction.

The theories of the two most transformative Victorian researchers, Gregor Mendel and Charles Darwin, were more fully appreciated in the twentieth century. Mendel published findings of his experiments on the particulate nature of hereditary transmission in 1865, but his key articles were largely ignored until rediscovered by three researchers in 1900. In 1859 Darwin published his theory of natural selection, which was one of the most significant revolutionary breakthroughs in the entire history of science, but until his death in 1882 he continued to retain progressivist and non-Darwinian notions of evolution with a Lamarckian-style theory of the hereditary transmission of acquired characteristics to explain variations of parental traits in offspring.[1] His theory of evolution rose to the high stature it ultimately did after his death as researchers merged his theory of natural selection with a reworked Mendelian genetics combined with continuous variation and statistical analysis in the "modern synthesis" from the 1920s to the 1940s.

The Origin of Life

Victorians supported three main theories about the beginning of life: from divine creation, extraterrestrial sources, and inorganic matter, none of which was based on verifiable empirical evidence.

For those who affirmed biblical inerrancy, God created life by uttering a single sentence: "Let the earth bring forth grass, the herb yielding seed, and the fruit tree yielding fruit" (Genesis 1:11). In the late Victorian period especially, divine creation increasingly lost traction among scientists. The theory of extraterrestrial sources was first suggested by Anaxagoras in the fifth century BC. In the Victorian period it was endorsed by Kelvin in 1871, Hermann von Helmholtz in 1879, and Svante Arrhenius, who in 1903 coined the term *radiopanspermia* for the way microscopic forms of life might have propagated through space by radiation energy from stars. However, most biologists rejected that theory as they noted the vast distances of extraterrestrial travel and the ultraviolet rays that would kill anything traveling across them. Among biologists the most popular theory was abiogenesis – life from nonlife. From the Victorian through the modernist period, research into that subject generated an increasing appreciation of the complexity of life, numerous gradations between nonlife and life, and the enormous time likely necessary for the evolutionary emergence of life out of inorganic matter.

Darwin found massive evidence for the origin of species but nothing probative for the origin of life. He sailed around the world gathering samples of diverse species from 1831 to 1836, worked out his theory of evolution by natural selection in 1838, and spent the next twenty years gathering more evidence to support that theory before publishing *Origin*. For another twelve years he gathered more evidence yet for applying his theory to the origin of human life, which he published in *The Descent of Man* (1871), but throughout that forty-year period he offered only a hunch about how the beginning of life might have taken place – "in some warm little pond with all sorts of ammonia and phosphoric salts, light, heat, electricity etc. present."[2] His "etc." covered myriad biological processes that neither he nor anyone else ever since has fathomed.

Victorian biologists could not explain what might have happened in that pond or anywhere else let alone recreate it in a lab. The most popular theory rested on "spontaneous generation," for which scientists found no verifiable evidence. The term is confusing because "spontaneous" could have a temporal or causal meaning, namely the temporal span of an instant or the causal mechanism of chance. Among those who focused on the suddenness of generation, there was no agreement about how sudden "sudden" was, and those who emphasized the causal mechanism of chance could not explain what causal efficacy chance involved. Further vitiating research, Victorians did not have an accurate understanding of *when* life might have begun because they considerably underestimated the age of the Earth. And no experiment could possibly succeed because no one could know let alone reproduce conditions when life might have appeared, which scientists now think was between 3.5 and 4.0 billion years ago.

In the last two decades of the Victorian period biologists increasingly appreciated the huge gap between life and nonlife and the extraordinary complexity of living organisms. "After 1880," concluded John Farley, "the belief that complex cellular organisms could arise spontaneously never again appeared, and in that very restricted sense the issue of spontaneous generation was abandoned."[3] Into the modernist period researchers directed increasing attention to ways that subcellular living entities might have arisen by long term evolution. Practitioners of the new specialty of biochemistry, named by Carl Neuberg in 1904, began to probe borderline stages between life and nonlife. In 1908 Archibald Macallum, biochemistry professor at the University of Toronto, speculated how combinations of amino acids and proteins could have generated "ultramicroscopic particles endowed with the chemical properties of ultramicroscopic organisms."[4] In his theory and many other theories "could have" was the most frequently invoked causal mechanism. In 1914 the Harvard biochemist Leonard Troland conjectured how a protein enzyme could have become surrounded by an oily liquid creating a droplet that could have been "the origin of the first and simplest life substance."[5] In 1917 the French microbiologist Félix d'Hérelle began studying bacteriophages (viruses that kill bacteria), which he eventually proposed as primordial living entities. His studies suggested that life could have resulted from "a special physico-chemical state of matter, the protein micella" that is the smallest particle of matter in a colloidal state.[6] In 1921, Ben Moore, the first professor of biochemistry at Liverpool University, proposed that "the territory of this spontaneous production of life lies not at the level of bacteria or animalculae" but rather "at a level of life lying deeper [and smaller] than anything the microscope can reveal and possessing a lower unit than the living cell."[7]

In 1924 the Russian biochemist Alexander Oparin traced a history of theories about the beginning of life from the ancient Greek philosopher Thales to his contemporaries with a highly speculative theory about how evolution from nonlife to life could have occurred in stages over long stretches of time. After the Earth cooled to 100° C, water droplets could have formed and fallen as rain that combined with hydrocarbons and ammonium oxide. These droplets eventually could have aggregated into organic substances such as proteins and amino acids, "the organic foundations of life." He continued to speculate that these substances could have increased in size and complexity to form colloidal precipitates or gels. These first bits of organic slime were the first organisms with "characteristics of life." As they grew in complexity and acquired greater internal coherence, they transmitted their ability to absorb and metabolize nutrients to generate energy in a "fierce struggle for existence" in accord with the processes of "natural selection" leading to new forms of life. Oparin concluded with a candid statement of scientific reserve and promise: "We still know very, very little about the structure of colloidal gels and even less about the physico-chemical structure of protoplasm. But this ignorance of ours is certainly only temporary. What we do not know today we shall know tomorrow. A whole army of biologists is studying the structure and organization of living matter, while a no less number of physicists and chemists are daily revealing to us new properties of dead things."[8] That breakthrough to the secret of life did not come "tomorrow" or ever since. Nevertheless, in 1936 Oparin expanded on this speculative essay with a full-length book, *Origin of Life*, that aspired to set the stage for experimental efforts to recreate those life-generating processes in research labs.

Efforts from the new field of biochemistry that began in the modernist period produced a wealth of experimental findings and speculation about the origin of life. Thirty years after Oparin elaborated his summary of those developments, Stanley Miller and Harold Urey attempted to replicate experimentally Darwin's warm little pond with a tabletop glass apparatus to simulate conditions for the generation of life, which produced some amino acids and other bio-building blocks.[9] In the 1980s other researchers studied the possibility that life originated in deep sea black-smokers near jets of scalding water from the Earth's interior as it mixed with mineral-rich, frigid sea water. Still others studied rudimentary life forms in streams flowing out of mine waste, inside frozen Antarctic rocks, and in atmospheric dust miles above the Earth.[10] While modernist biochemists and their successors made considerable headway, none of their findings produced a credible explanation of the origins of a living organism capable of eating, metabolizing, excreting, moving, growing, reproducing, responding to external stimuli, and evolving by means of natural selection, a tall order for a set of processes necessary to qualify as life. To date, all the proposed theories remain speculative scenarios of "could haves." With the progress of biological knowledge throughout the modernist period and after, whatever unknown biological processes Darwin's "etc." might have entailed has seemed to grow ever larger and more puzzling. The deeper biologists probed into the origins of life, the more they realized how much more there was to know, how much deeper they would have to probe, and how little they understood what they thought they knew.

The Origin of Species

In most of the sciences I cover, modernists surpassed the Victorians. However, with the exception of Einstein in physics, the historical stature of Darwin in biology is greater than that of any other single modernist scientist including the cosmologist Hubble and the geologist Wegener. Darwin's theory of natural selection ultimately transformed biology along with embryology, physiology, paleontology, psychology, anthropology, medicine, and history as well as religious cosmogony, and by the end of the nineteenth century *Origin* had been translated into twenty-four foreign languages. He was buried in Westminster Abbey next to Newton.

The nineteenth century was an age of progressivist historical thinking, as major systematic thinkers saw history as having a progressive direction. For Hegel that progress was toward the realization of the idea of freedom. For Auguste Comte science progressed from theological to metaphysical to positivist knowledge grounded in scientific method. For Herbert Spencer, history, sciences, social formations, and the evolution of species followed a trajectory of increasing complexity. In *Creative Evolution* (1907), Henri Bergson criticized Darwin's passive and mechanical theory of evolution with a theory of evolution driven by a dynamic life force, a progressive *élan vital*, toward higher forms. The Victorians viewed their own age as the pinnacle of history, an outlook elaborated by self-confident "whiggish" historians.[11] Darwin repeatedly fell back on progressivist notions. In the conclusion to *Origin* he proclaimed: "As natural selection works solely by and for the good of each being, all corporeal and mental endowments will tend to progress towards perfection."[12]

During five years sailing around the world as a naturalist, Darwin observed species in many different environments. At the outset of that trip he knew, of course, that every living form has a progenitor or pair of sexed progenitors but did not know how their characteristics were transmitted to offspring. He pondered that subject for twenty-three years before publishing *Origin*. That work led to a revolution in thinking about how a multitude of species evolved by means of natural selection that included three key elements – reproduction with variation, a struggle for existence, and adaptation for survival and reproduction in the immediate local environment.

The religiously orthodox Victorian idea of the origin of species was that an omniscient, omnipotent, and omnibenevolent God created all species in a hierarchy of slightly varied forms from the lowest and simplest to the highest and most complex with human beings at the top. Some believed that that hierarchy was fixed at the moment of creation by a single divine utterance, while others such as William Paley in *Natural Theology* (1802) argued that God created simple living forms with an inherent impulse to evolve toward increasingly perfected forms. Paley also believed that individual life was modeled on biological growth as a seed grows toward its goal of becoming a tree. That powerful notion of purposive change, evinced all over the Earth trillions and trillions of times in every living thing, was hard to omit from the force driving evolution. Technically, Darwin's theory of natural selection made that distinction, but he was unable to rigorously accept its fuller consequences.

Darwin's key "dangerous idea," as Daniel C. Dennett noted, was that human beings originated not in a linear manner from any progressive development according to a goal-directed logic but rather from a mechanical process governed by an indifferent algorithm that proceeded as a series of mindless steps initiated by blind chance. Darwin remained unable to accept how mindless and mechanical, how nonteleological and nonprogressive his theory was. But, Dennett asked rhetorically, "Why couldn't the most important thing of all be something that arose from unimportant things [like a mindless algorithm]? Why should the importance or excellence of *anything* have to rain down on it from on high, from something more important, from God?" Darwin's potential inversion, Dennett continued, "suggests that we abandon that presumption and look for sorts of excellence, of worth and purpose, that can emerge, bubbling up out of 'mindless, purposeless forces.'"[13]

Natural selection worked perfectly well without God's direction. Darwin added a line in *Origin* about "laws impressed on matter by the Creator," possibly after feedback from his devoutly religious wife or from Charles Lyell, who advised him to exercise caution and protect himself from religious reproval. The enormous variety of species and the immense complexity of life's beginning, which no laboratory scientists could explain or reproduce, further diminished the likelihood that life began by a single divine act or utterance. *Origin* offered an exceptionally productive naturalistic theory about how species evolved over long periods of time according to natural selection. The three essential elements of that theory disturbed Victorians' comfort and confidence along with their knowledge of the world, including most notably knowledge of their origins. Ends are only ends if they have a beginning, and, since Darwin's theory described a process with no clear end, it implied that his theory might also do quite well without a clear beginning, even though common sense suggests that it must have had a beginning. Modernist biologists were increasingly better equipped to deal with such uncertainties about ends as well as beginnings. That willingness of modernist biologists to deal without an understanding of the biological beginning of life creates a cross-field pattern with modernist writers' willingness to tell stories without determinative beginnings in their novels, which I survey in Chapter 10.

Descent with Variation

In *Origin*, Darwin explained how "[v]ariability is governed by many complex laws—by correlation of growth, by use and disuse, and by the direct action of the physical conditions of life" (466). He repeated that Lamarckian notion in his final paragraph that explained evolution "from the indirect and direct action of the external conditions of life, and from use and disuse."

Variability was a problem that he never solved. After much theoretical struggle he eventually proposed a highly speculative "provisional hypothesis" of pangenesis by which an organism generates tiny gemmules in every part of its body.[14] By some unexplained mechanism these gemmules register changes from use and disuse in every part of the body and then somehow circulate through bodily fluids to the gonads, where they somehow become included in sexual substances that somehow mix during sexual reproduction and somehow blend to produce their acquired changes in offspring. Darwin devised pangenesis to align his theory of natural selection with the theory of the hereditary transmission of acquired characteristics, which made evolution a purposeful development and avoided the disturbing idea that parents did not pass along all their hard-earned acquired characteristics because variations were random. But Darwin had no evidence to support his theory of pangenesis, which, as Jim Endersby estimates, he "probably first formulated in 1841 but only published in 1868."[15] The fact that Darwin mulled over pangenesis for almost thirty years before publishing it suggests uncertainty about its plausibility in explaining the phenomenon that he introduced it to explain – hereditary transmission. In a letter to Charles Lyell in 1868, Darwin confessed that pangenesis was "a mere rag of an hypothesis," a speculation that "has yet been a considerable relief to my mind" by offering "some sort of idea how the inherited characters are developed and transmitted."[16] His theory of evolution rose to the high stature it ultimately did only after researchers merged it with a reworked Mendelian genetics in the "modern synthesis" later in the modernist period and entirely dropped any Lamarckian theoretical residue.

Struggle for Existence

The significance of Darwin's use of struggle as a key mechanism of natural selection is best seen in a broad historical perspective. Eighteenth-century philosophers viewed the history of knowledge, ethics, and history generally as a story of additive progress governed by natural law. Their positivistic epistemology was a story of increasingly accurate knowledge acquired by increasingly reliable scientific methods of observation and experiment supported by increasingly versatile mathematics and rigorous theory. Their positivistic ethics of utilitarianism was based not on avoiding sin but on maximizing the greatest pleasure for the greatest number of people. Their positivist view of history peaked with Condorcet, who viewed history as a story of progress based on Newtonian physics, the scientific method, sensationist epistemology, and the advance of political and economic liberty.

In 1798 Thomas Malthus slammed the door on that optimistic world view in his *Essay on Population*, a resoundingly pessimistic view of history centered on a life-and-death struggle to survive. Darwin was struck when he learned about Malthus's theory that history is a scenario of inevitable struggle for food and viable offspring that leads to major "checks" to the population from famine, disease, and war unless people can exercise "moral checks" to population growth by postponing sexual initiation, reducing the number of sexual relations, and having fewer children. Darwin explained that, "as more individuals are produced than can possibly survive, there must in every case be a struggle for existence . . . It is the doctrine of Malthus" (63). Malthus's focus on struggle as the mechanism of change influenced other major thinkers who elaborated other theories of progress through conflict, starting with Hegel's dialectical method applied to metaphysics, epistemology, and history. Marx announced the dialectical nature of all history in the opening line of *The Communist Manifesto* as "the history of class struggles." Darwin's theory was grounded on a struggle for existence among populations under pressure from geographic upheavals and environmental changes as well as rivals, especially in overpopulated regions.

Adaptation to the Local Environment (Survival of the Fittest)

The third key element of his theory of natural selection was adaptation and survival, a thinly veiled term for the centrality of a life-and-death struggle for existence, or, as he put it bluntly in the last paragraph of *Origin*, "from the war of nature, from famine and death." The idea that the highest species, the supposed goal of all life created by an all-wise and omnibenevolent God or even by natural processes, could come from such immoral and animalistic impulses was deeply troubling to Victorians.

After around 1880, Darwin's theory of evolution fell into "eclipse" for many reasons.[17] (1) The fossil record provided material evidence of ancient life forms and extinct species, but the enormous gaps in that record undermined his theory of a natural selection by slight, successive, favorable variations, and the collection of fossils that the theory inspired encouraged imperialist ventures around the world, which themselves incurred moral disapproval. (2) Darwin's theory of adaptation for survival of the fittest introduced dubious ethical justifications for imperialism, racism, war, class oppression, power politics, and a kill-or-be-killed mentality that further turned off many potential adherents. (3) Darwin could not explain the crucial element of variability and tentatively offered the wild theory of pangenesis, for which he had no evidence. (4) For sexually reserved Victorians, Darwin's theory of sexual selection demeaned humans by emphasizing their animal origins, their "descent" by means of crude sexual selection, and their lowly bodily foundations generally. (5) The mindless and mechanistic nature of natural selection and the increasing roles attributed to capricious geographical changes, population surges, and environmental circumstances reduced personal human autonomy. (6) The hugely expanded time scale that the theory demanded and managed to secure following the discovery of radioactivity accented the insignificance of the brief temporal span of life during the Earth's several-billion-year history. (7) Natural selection's reliance on extinctions evoked the depressing thought that God's creation was not, as Genesis repeatedly claimed, so "good" after all. (8) Most germane to my focus on the history of originary thinking, Darwin's theory undermined the account of creation in Genesis and the conviction that from the beginning of time God bestowed on humans a special place of honor in the hierarchy of living beings. While most biologists did not rely on God's creation to explain the origin of species, it remained part of the cultural record and influenced their thinking, including that of Darwin himself, who could not resist retaining at least a progressivist view of the origin of species with some hints of design as "progress toward perfection" (489).

The Origin of Human Beings

While the preceding chapters emphasize modernist scientific advances over Victorian science, this chapter emphasizes Victorian ideas that degraded social relations, especially erroneous ideas about human sexual reproduction. The major alarm fired up by *Origin* stemmed from its implications for the origin of humans, even though only one sentence addressed the subject – "Light will be thrown on the origins of man and his history" (488). This critically important subject was the focus of many Victorian erroneous theories of origins from the most distant past to the present in *animal progenitors*, *remote human ancestors*, *recent human ancestors*, *the moment of conception*, and *pregnancy*.

Animal Progenitors

The idea that human beings inherit traits in a seemingly degrading manner from animal progenitors was supported by recapitulation theory and Darwin's theory of evolution.

Recapitulation theory emerged in Germany in the early nineteenth century among "teleo-mechanists" who traced the differentiation of structures in the developing embryo to show how the individual life recapitulated the history of the species. These researchers included Johann Meckel and Johannes Müller, who explained phenomena with teleological models based on biomechanical principles rather than divine or occult forces.[18] In 1821 Meckel argued that embryonic development of complex animals recapitulates the hierarchy of life from lower animals to fish, to reptiles, to mammals: "The development of the individual organism obeys the same laws as the development of the whole animal series: that is to say, the higher animal in its gradual evolution essentially passes through the same permanent organic states which lie below it."[19] In 1857 the American biologist Louis Agassiz concluded that "the phases of development of all animals correspond to the order of succession of their extinct representatives in past geological times."[20] Some viewed recapitulation as a function of divine purpose or a life force inherent in organic matter; for others it came from the repetition of chemical or biological processes in ever more complex patterns; but all investigators agreed that human instinct and anatomy derived from animal ancestors, which are recapitulated somehow during embryological development.

Recapitulation's most famous German proponent was Ernst Haeckel, whose *General Morphology* (1866) was based on a "biogenetic law" that "ontogeny recapitulates phylogeny."[21] His *Natural History of Man* (1868) depicted an evolutionary tree covered with illustrations of tiny organisms at the roots up to humans at the top branches. That image expressed an abiding error in evolutionary thinking of the time – that evolution has a preordained pattern and grows toward a goal as a seed grows into a tree. The English read an erroneous account of mental recapitulation in Henry Maudsley's *Body and Mind* (1870): "[E]very human brain does in the course of its development pass through the same stages as the brains of other vertebrate animals, [and] the stages of its development in the womb may be considered the abstract and brief chronicle of a series of developments that have gone on through countless ages in nature."[22]

The idea of animal progenitors became particularly threatening after the appearance of *Origin*, which shocked Victorians with the implication that humans originated from the germ plasm of an ape. In place of a comforting belief in the divine creation of each species "in the image of God," Darwin proposed a troubling thought about common descent from some primate ancestor or even before that from some microorganism drifting through primeval ooze. All species may have "one progenitor," Darwin hypothesized, and may "have descended from one ancient but unseen parent" (412). In an early application of Darwin's theory, Thomas Huxley linked the human brain to a rat's by arguing in *Man's Place in Nature* (1863) that "Nature has provided us . . . with an almost complete series of gradations from brains little higher than that of a rodent, to brains little lower than that of man."[23]

While evolution by natural selection was phrased in terms of favorable adaptations and chances for survival and reproduction, its central mechanism included killing off the weak, the sick, and the old. Darwin offered a striking example of killing in nature: "It may be difficult, but we ought to admire the savage instinctive hatred of the queen-bee, which urges her instantly to destroy the young queens her daughters as soon as born . . . for undoubtedly this is for the good of the community" (202). Elaborations of this inherited instinct to kill also influenced imaginative writers. Alfred Tennyson's elegy *In Memorium* (1850) evoked the killing ground of nature with a disturbing image of "Nature, red in tooth and claw." Atavistic murder impulses surviving in humans wreak havoc in Émile Zola's novels about the Rougon-Macquart family. The title of one of them, *The Human Beast* (1890), refers to humans' beastly nature. The novel's protagonist Jacques inherited a taint from his animal forbears: in killing "he had been carried away by inherited violence, the instinct for murder that in the primeval forests had hurled one beast onto another."[24]

Remote Human Ancestors

The critic George Levine notes that the word *origin* "spoke directly to an apparently insatiable Victorian urge to determine origins—of language, of the Nile, of the human species, of the cosmos, or (in novels from *Oliver Twist* to *Daniel Deronda*), of parenthood."[25] Victorians' ignorance about the mechanism of hereditary transmission fueled exaggerated fears about how children inherit their parents' as well as their more distant ancestors' birth defects, diseases, and vices along with their ideas, property, and social standing. Victorian ideas about remote human origins drew on the theories of Lamarck, Darwin, and organic memory.

Lamarck's theory of the hereditary transmission of acquired characteristics includes a checklist of Victorian ideas and values – that evolution follows from human intentions and efforts and from the residue of utilitarian use and disuse to give a meaningful direction and overall purpose to the progressive achievements of habits and traditions from distant ancestry to the present. As the critic Gillian Beer notes, Lamarck's theory "gave primacy to mind—to intention, habit, memory, a reasoned inheritance from generation to generation in which need engendered solution, and solutions could be genetically preserved by means of an act of will."[26] This "soft" theory of heredity was challenged by Darwin's "hard" theory of natural selection that technically did not allow hereditary transmission of any acquired characteristics, although Darwin repeatedly contradicted his own hard theory in arguing that parents' acquired characteristics can somehow be transmitted to offspring, although he had no clear idea how.

The theory of organic memory relied on recapitulation theory and a Lamarckian mechanism of hereditary transmission to explain how instincts and ideas originate in memories and then in some compressed form recapitulate the past experiences of ancestors.[27] Organic memory was popularized in Germany in 1870 in a lecture by Ewald Hering, *On Memory as a General Function of Organized Matter*, which defined memory as a basic reproductive capability of living matter and argued that the entire history of the race is evident in an individual's memory that retains and in a condensed form recapitulates the life experiences of ancestors. In 1878 the British novelist Samuel Butler incorporated that theory in arguing that the small ovum from which humans have sprung "has a potential recollection of all that has happened to each one of its ancestors."[28] The theory grounded human origins in a dynamic embodied memory that united mind and body in a sequential development through species, ancestors, races, and nations. Like so many tradition-minded Victorians, Butler was unable to believe that the prodigious efforts of a lifetime let alone countless generations in the past entirely disappeared with an individual's death – poof.

Recent Human Ancestors

The increasing influence of recapitulation theory meant that earlier tendencies were becoming increasingly potent as species became more complex. Among those tendencies, deficiencies and vices increasingly overshadowed strengths and virtues, leading to degeneration, a process supported by findings in medicine, sociology, psychiatry, criminal anthropology, and reproductive biology.

In medicine, discoveries about contagion from the 1860s to the 1880s revealed how germs were supposedly pooling in ever greater concentrations and virulence from generation to generation, causing hereditary degeneration. The apparent rise in tuberculosis and syphilis, spiked by an overworked and oversexed population, seemed to demonstrate the spread of killer bugs everywhere from congested cities to the individual bloodstream and from brothels to marriage beds. In a letter of 1881, the French historian Hippolyte Taine warned of "the morbid germs which entered the blood of a diseased society [and] caused fever, delirium, and revolutionary convulsions."[29] Social thinkers believed that environmental poisons and social pressures made their way into germ cells, further intensifying the transmission of degeneration from parent to child. From 1885 to World War I, concluded Robert Nye, degeneration theory "asserted practically an imperialistic sway over all public and professional discourse concerning individual and social pathologies," and during the war degeneration theory ran wild.[30]

Psychiatrists in Europe and America saw degeneration corrupting ancestral lines with physical, mental, and moral deterioration. In 1857 the French psychiatrist Benedict Morel introduced the term *dégénérescence* in a treatise that outlined how a family line could die out over four generations beginning with nervousness and moral depravity in the first generation, neurosis and alcoholism in the second, defective intellect in the third, and congenital deformities and sterility in the fourth.[31] This sequence was triggered by biological and social factors and could supposedly cause a spectrum of medical diseases, personality disorders, and social problems including alcoholism, impotence, suicide, moral corruption, insanity, and crime. This frightening mix of pathological scenarios was funneled through a theory that centered on the concept of degeneration, which became a catch-all for many antithetical ills: overwork and laziness, hyperrefinement and primitivism, hypernervousness and nervous exhaustion, boredom and mania, and premature senility and infantilism. As Daniel Pick concluded, *dégénérescence* became "the ultimate signifier of pathology." "It explained everything and nothing as it moved back and forth between the clinic, the novel, the newspaper, and the government investigation."[32]

Moment of Conception

While many Victorians managed their worries about ancestral origins by displacing them to shadowy stretches of family history, remote antiquity, or even caveman times, some directed them toward the state of parents at the moment of conception. Victorians believed that during impregnation a lifelong accumulation of the physical and psychic experiences of both parents is channeled into offspring through every fiber of their being. Charles Rosenberg found that in America "both popular and scientific treatises warned throughout the century of the need to conceive children only when both parents were well-rested and affectionate. Hostility, even exhaustion during intercourse could result in weak and unhealthy offspring."[33] The father was thought to contribute body stature and intellect, while the mother contributed inner organs and temperament. In *Transmission; or Variation of Character through the Mother* (1877) the American essayist Georgiana Kirby urged that both parents' condition during sex was crucial: "Never run the risk of conception when you are sick or over-tired or unhappy; or when your husband is sick, or recovering from sickness, exhausted, or depressed," she warned, "for the bodily condition of the child, its vigor and magnetic qualities, are much affected by conditions ruling this great moment."[34] In 1881, the popular moralist J. H. Kellogg cautioned that "unhappy or immoral thoughts . . . at the critical moment when life is imparted may fix for eternity a foul blot upon a character yet unformed."[35] Others argued that if a mother is intoxicated at the moment of conception the child will be an alcoholic, or if she is overly aroused it will be demented.

Misunderstanding of impregnation leading to exaggerated parental responsibility is evident in the overblown rhetoric of O. S. Fowler in a popular marriage manual:

> It is this *combined and concentrated*, as well as *high-wrought*, inter-commun-ion of *every physical, every intellectual, every moral element and function of humanity* in generation as it is *by constitution*, which renders the pleasure attendant on this *double* repast so *indescribably* exalted and beatific to those who *spiritually love* each other or in proportion thereto; besides being *the* ONLY means of augmenting and perfecting the intellectuality and morality of its product – redoubling more and more as its handmaid love becomes more and more perfect, and thereby enhances, and also unites, in this holy alliance, faculty after faculty, till finally, when both love and generation have their perfect, and of course *united*, work, they embrace within the wide range

> of their sanctified enjoyment, every animal, every intellectual, every moral, organ and function, and element of man's entire constitution![36]

Fowler's frenetic prose replicates the sex act itself: laced with italics and prolonged as though he could not bear to bring his orgasmic sentence to an end, it strains to convey the exclusivity and totality of sexual as well as spiritual love yielding a child that originates in the parents' thoughts, morals, and physical traits in proportion to the intensity and spirituality of their love, indelibly united under the witness and sanction of God. Victorians were suffocated by such a heightened sense of parental responsibility stemming from confusion about what is and is not transmitted to children at conception. Modernist theories of the independence of sex cells and the complex nature of sexual reproduction generally liberated parents from the burden of erroneous ideas about the hereditary transmission of acquired characteristics and the excessive parental responsibility that that theory engendered.

Pregnancy

Even though motherhood was the ultimate purpose of a Victorian woman's life, pregnancy was concealed by a woman's "confinement" and by a conspiracy of silence about sex generally. Without a clear understanding of blood types, the placental barrier, or sexual reproduction, Victorians speculated wildly about how during pregnancy the fetus can be affected by its mother's blood and sex organs, her food and drink, what she sees and feels, and even her experiences prior to pregnancy, especially if she has had a child.

Victorians did not know about different blood types, and the year of their discovery marks a sharp transition to the modernist age. In 1900 Karl Landsteiner discovered why some transfusions were successful and others lethal as he worked out the ABC blood group system. He proposed that the death of a newborn following a transfusion was not from illness or disease but from mixing incompatible blood types. In 1907 Ludvig Hektoen proposed that the danger of blood transfusions can be avoided by selecting a donor whose corpuscles do not agglutinate from the serum of the recipient. That same year Ruben Ottenberg performed the first blood transfusion using blood typing and cross-matching, making the transfusion safer.

In addition to misunderstanding the independence of blood types from the life experience of their parents, Victorians also believed erroneously that mothers share fetal blood through the placenta, and, since they also believed that character was transmitted through the blood, the moment of conception as well as the entire course of pregnancy were of paramount importance.

Some Victorians believed that a child could even be shaped by its mother's physiological experiences prior to pregnancy, especially by a former pregnancy. This worrisome influence on the origins of character fascinated breeders and researchers especially after 1820, when Lord Morton reported to the British Royal Society that his mare had bred two generations of foals with stripes that came not from her or from the black Arabian that sired them but from the sire of her first foal, a striped horse called a quagga.[37] Darwin risked undermining his theory of natural selection by explaining such a scenario of hereditary transmission by means of gemmules from the sire of a mother's earlier pregnancy collecting in her germ cells and then being passed to her new fetus. In 1868 he maintained that Lord Morton's studhorse affected the character of the offspring subsequently delivered by his mare and concluded that "it is certain that [the mother's] ovaria are sometimes affected by a previous impregnation, so that the ovules subsequently fertilized by a distinct male are plainly influenced in character."[38]

Such erroneous thinking about hereditary transmission prevailed throughout the Victorian period. In 1892 August Weismann named that phenomenon *telegony*, meaning literally offspring at a distance or, more precisely, the direct influence of male gametes on a female's reproductive system and consequently on her subsequent offspring. "The concept of telegony," writes Harriet Ritvo, "was almost universally believed by nineteenth-century breeders and fanciers and widely accepted within the zoological community."[39]

Lacking a clear understanding of how the placental barrier blocks the exchange of blood cells between mother and fetus, Victorians suspected that mothers share some fetal blood, if not all of it. They offered two scenarios for telegony, both with scary names: (1) *maternal infection* – the impact of a pregnancy on a female's sex organs, which subsequently impacts her reproductive organs and any subsequent fetus, or (2) *maternal inoculation* – the impact of the sire's hereditary characteristics on a fetus that pass to the mother during pregnancy from exchanges of blood through the placenta and then are transmitted back through a subsequent placenta to any subsequent fetus.

Such confused theorizing about what is and is not exchanged between mother and fetus fueled a host of anxieties about virginity, sexual fidelity, and family pedigree among Victorian physicians and imaginative writers. Some even envisioned a mother's psychological experiences during pregnancy affecting her child. In 1868, William Hammond, a New York physician, cited over a dozen writers who gave examples of pregnant women seeing something shocking and then imprinting a mark of that shock on their baby. One woman saw a leech on someone's foot and then bore a baby with the mark of a leech on its foot. Although Hammond conceded that there is no general acceptance that the mother's mind can directly affect the fetus, he nevertheless reported that "many of the most eminent physiologists hold to it, and among the people at large throughout the civilized world . . . it is received with unquestioning faith." He suggested an explanation based on misunderstanding of blood types and the placenta: "The blood which circulates through [the fetus's] body is separated from her blood by an exceedingly thin membrane and is thus impressed with the mental characteristics . . . which arise from her brain and nervous system." He therefore urged pregnant women to avoid anger, envy, and even sexual desire because those emotions might adversely affect their child. "Within the womb a new connection is set up between the ovum and the mother – that of the blood; and it is through this medium that all impressions from her mind to her offspring must pass."[40] The consequences of this exchange were ominous: "it is not uncommon to see the children of a widow by her second husband resemble in mind and body her first husband, provided she has had children by him."[41] Another physician, Alexander Harvey, offered further evidence of maternal inoculation in a medical tract with a frightening title for pregnant women: *On the Fetus in Utero as Inoculating the Maternal with the Peculiarities of the Paternal Organism* (1886). Harvey explained how "the fetus, partaking, as it must, of the character or peculiarities of its father *inoculates* therewith the blood, and, generally, the system, of its mother."[42]

The respected German philosopher Eduard von Hartmann (1895) believed that during pregnancy the mother experiences "an interchange of blood with a second body, whose composition is only half-conditioned by qualities inherited from the mother, the other half being contributed by paternally inherited characteristics." Some of her former lover's traits therefore "lie dormant within her but can manifest themselves all the more strikingly in the children of a later marriage." Von Hartmann elaborated: "The husband of a widow does not therefore find a clean page, but one written over by his predecessor, with whose hereditary tendencies his own must enter into conflict."[43] For Victorians with such a view of hereditary origins, virginity had especially high significance.

While some of Zola's hereditary killers are tainted from remote ancestry or even animal forebears, others are tainted by their mother's prior pregnancy. He was influenced in this thinking by the French physician Prosper Lucas and the eminent historian Jules Michelet, both of whom argued that pregnancy imprints a woman forever and leaves traces that may appear in children sired by another man.

While Lucas accepted such instances as atypical,[44] Michelet believed them to be typical, arguing that "woman, once impregnated, carries her husband everywhere in herself . . . A widow with a second husband often has children who resemble the first."[45] In Zola's analysis of the Rougon-Macquart family's heredity in the last novel in the series, *Dr. Pascal* (1893), there is "one example of transmission by influence – Anna, the daughter of Gervaise and Coupeau, who bore a striking resemblance especially in her childhood to Lantier, her mother's first lover."[46] The Swedish playwright August Strindberg was tormented by fear of both maternal infection and inoculation, which convinced him that his children had been contaminated by the infidelity of his wife, who drove him mad with uncertainty about his paternal role. In *The Father* (1887), an anxious father, questioning whether his daughter is really his own, asks a doctor, "Is it true that if you cross a mare with a zebra, you get striped foals?" Panicked about a phenomenon like Lord Morton's mare and obviously aware of that specific example that provoked Darwin, the father asks further: "[I]f the breeding is then continued with a stallion, the foals may still be striped?" The doctor assures him that such influence from a previous pregnancy is possible. Strindberg subsequently dramatized those anxieties in *The Creditors* (1888) by having a woman's first husband, the biological "creditor," return to haunt and eventually destroy her second marriage. Her second husband is fearful because "when the child was three years old it began to look like him, and in that sense belong to him, her former husband."[47] In Strindberg's *To Damascus*, a former husband challenges his male successor directly: "Your child shall be mine, and I shall speak through its mouth."[48]

Telegony was doubtless on the mind of Thomas Hardy when he created *Tess of the d'Urbervilles* while also reading Weismann. In the novel the contaminating effects of a previous pregnancy surface in the explanation for murder because Tess not only inherited her remote ancestor's destiny to kill ("a murder, committed by one of the family, centuries ago") but presumably also absorbed some traces of her rapist, Alec, who left her pregnant with a child who died in infancy. On Tess's wedding night her husband, Angel, is shocked to learn of her former pregnancy and rejects her in disgust, charging that she is not the woman he thought she was. He meant that charge characterologically, but also literally and biologically because he suspected that she had absorbed Alec's essence from the pregnancy and was therefore forever tainted, so Angel tells her that any child they might have together would be tainted, adding: "You were one person; now you are another," and he no doubt meant that charge biologically. Angel returns to this obsession with the bitter question: "How can we live together while that man [Alec] lives? He being your husband in Nature, and not I." If he married her, he fears, their children would become "wretches of our flesh and blood growing up under a taunt."[49] Angel's use of *taunt* also suggests *taint*, because any children he might have with Tess would be exposed to taunts because of their taint. Angel's despair is a culmination of the anxiety about impure blood and hereditary taints that plagued Victorians from their misunderstanding of hereditary transmission.

Victorians were unable to distinguish clearly between the processes of hereditary, infectious, and cultural transmission from parent to child and hence confused how a child might get from a parent hereditary syphilis, tuberculosis, or sexual perversion, respectively. Moralists warned that the entire sexual prehistory of parents could shape the sexual constitution of children. A French educator observed: "There has been some time – a day, an hour – in which the fate of an entire family has been cast, so that mutual moral responsibility binds parents to children."[50] Victorian researchers frightened parents by heaping on them enormous responsibility for problems facing children.

This survey of the many fears caused by misunderstanding pregnancy among the Victorians highlights the historical significance of the triumphant pregnancy of Anna Brangwen in the chapter titled "Anna Victrix" from D. H. Lawrence's novel *The Rainbow* (1915). Anna flaunts her pregnant belly as she dances naked in front of a raging fire while her sexually repressed husband shrinks back while looking at her, a scene that enacts a woman's *victory* over Victorian ignorance about pregnancy and man's anxiety from sexual shame. "Her fine limbs lifted and lifted, her hair was sticking out all fierce, and her belly, big, strange, terrifying, uplifted to the Lord. Her face was rapt and beautiful, she danced exulting before her Lord and knew no man."[51] This scene is a modernist celebration of Anna Victrix's pregnancy as a victory over a sexually debilitated male, a flagrant repudiation of the Victorian convention that required a woman to hide her pregnancy especially during her confinement.

Modernist Genetics

Modernist genetics was prepared by a series of crucial inventions and discoveries by Victorian predecessors including the oil immersion lens in 1840, aniline dyes in 1856, and microtome slicers in 1865, which facilitated more detailed observations and precise explanations of reproductive activity at the cellular level. In 1875 the German embryologist Oscar

Hertwig observed that a single sperm fertilized a sea urchin egg, not a swarming of male sexual fluid that caused hereditary blending. In 1879 the Swiss founder of modern cytology Hermann Fol observed a sperm penetrating an egg and the fusion of the two nuclei, demolishing ideas about preformation of the embryo in either the egg or the sperm and other theories that fertilization is caused by some mechanical or chemical force exciting the egg to develop. In the 1880s researchers focused on the behavior of material in the nucleus, which during cell division formed threadlike structures that were believed to carry out the reproductive function, which in 1888 the German zoologist Theodore Boveri named *chromosomes*. Back in 1871 Friedrich Miescher had discovered that the material agent of that hereditary function was deoxyribonucleic acid (DNA), which in 1953 James Watson and Francis Crick elaborated in their theory of heredity based on the structure of a two-stranded helix of DNA. Victorian researchers probed as deep as they could into the origins of sexual reproduction, however much their efforts were limited by false biological presuppositions.

The key Victorian geneticist was Mendel, who in 1865 published results of experiments crossing pea plants with different characteristics, which, when ultimately accepted in the modernist period, demolished four key features of Victorian hereditarianism. First, Mendel's findings showed that characteristics of parents are transmitted in discrete units of inheritance (later called genes) that are present as pairs (later called alleles) throughout life. During sexual reproduction, members of each pair of genes segregate into different reproductive cells. As a result, Victorian notions of a blending of parents' traits from a blending of their reproductive fluids became untenable. Second, a trait such as blood type or hair color derives from the influence of both male and female alleles, although one allele is often dominant and the other recessive. This finding that both parents contribute equally refuted Victorian theories about specific traits coming exclusively from one parent. Third, according to Mendel's law of independent assortment, hereditary influence on one trait is independent of influence on another trait. This independence undercut the notion that complex behaviors such as an instinct to kill, which is based on many genetically determined traits modified by many environmental circumstances, could be inherited from one parent in toto as a single flaw in character. Finally, when first-generation hybrids self-fertilize, individual traits in the second generation occur in fixed ratios, roughly 3:1, further undercutting any gradational multiple-particle theory such as Darwin's theory of pangenesis.

Widespread prejudices in favor of blended parental contributions transmitted according to Lamarckian principles prevented Mendel's work from being properly appreciated in his time. His contemporaries were more concerned with breeding new species by hybridization than with understanding the mechanism of heredity, and they could not detach evolution from individual growth, a crucial distinction for the emergence of scientific genetics. Seeds do grow into trees according to a goal-oriented, biologically inherent purpose, but species do not evolve according to any similarly inherent telos (orthogenesis). His contemporaries were also committed to blended parental contributions transmitted according to Lamarckian principles. For these reasons, scholars have ventured that Mendel's work was roundly "ignored."[52] An accurate interpretation of his results undercut Victorian notions of the conflation of evolution with growth, a blending of sexual fluids, male or female primacy, and the inheritance of complex behaviors in toto.

Modernist genetics effectively began in 1885 with Weismann's announcement of the continuity of the germ plasm from generation to generation and its independence from influence by body cells (soma), thereby severing hereditary transmission from influence by acquired characteristics.[53] He argued for a continuity of mindless germ plasm instead of a continuity of mindful soma. Five years later the Canadian biologist George Romanes summarized Weismann's theory and challenged the intractable Lamarckism that had for so long justified Victorians' confidence in the hereditary continuity of their acquired characteristics. Romanes was triply blunt with three genealogical roadblocks: all acquired variations are "ruled out" as regards the species, "blocked off in the first generation," and "entirely lost to progeny."[54] Weismann's theory had the effect of shutting off traditional Victorian notions of a continuous flow of parental experiences and characteristics to offspring as though in the legal and economic sphere they had blocked children from inheriting their parents' property and money.

In 1900, also a turning point in the discovery of blood types, Mendel's findings came into broad scientific notice when Mendel was rediscovered by the independently working Dutch botanist Hugo De Vries, the German geneticist Carl Correns, and the Austrian agronomist Erich von Tschermak, suggesting a widespread shift in thinking about sexual reproduction and hereditary transmission. In 1901 De Vries introduced the term *mutation* for variations but erroneously thought that mutations involved a sudden change in appearance rather than a change at the genetic level, and he had no idea how mutations occurred. The English biologist William Bateson coined *genetics* in 1905. Another group of geneticists who founded the "biometric school" began to study heredity using continuous variation and statistical analysis. This approach, which directly countered the Mendelian view of heredity caused by discrete, inherited units (genes), included Francis Galton and Karl Pearson along with the particularly innovative, if not immediately influential, Walter Weldon, whose multifactorial approach included internal and external (environmental) factors that influenced the hereditary function of genes but whose work was cut short by his sudden death in 1906.[55]

In 1909 the Danish geneticist Wilhelm Johannsen coined *gene* for the material basis of heredity. Two years after that, Thomas Morgan introduced the term *crossing over* for the recombination of traits in offspring coming from a breakdown in linkage during sexual reproduction. Since the extent of recombination was determined in part by the distance that separates genes, he and his student Alfred Sturtevant reasoned that recombination frequencies offered a way to map genes on chromosomes. Subsequent gene mapping vividly illustrates modernists' more precise understanding of the hereditary origins of human characteristics. Morgan and his successors' work also increased the complexity of genetics by showing that a trait can be influenced by several different genes, that a single gene may affect more than one trait, that any gene's influence could be modified by neighboring genes, and that genes affect traits at different stages of development.

The increasing specificity of genetic explanations in the modernist period sharpened understanding of the origins of human life in biology, biometrics, neurobiology, endocrinology, and other medical sciences concerned with human origins. Modernists' genetics inspired many innovative ideas with broad cultural implications. For them, evolution is sparked by random variations that are not caused by remote ancestral or immediate parental experiences, and the resulting evolution is therefore discontinuous emergence, not continuous growth. These variations are in a sense "selected" by the Darwinian mechanism of natural selection involving a struggle for existence and "survival of the fittest," but the metaphor of selection erroneously implied a choice, which it is not, and the term "fittest" had misleading moral connotations. Most importantly, the overall itinerary of evolution was not conceived by any designing mind or driven by any teleological force and therefore did not proceed toward any goal or have any progressively emerging meaning. In fact, referring to its history as a story was additionally misleading because the pattern of its form was rather multiple branching and divergent without any coherent itinerary or predetermined goal. Where it was "headed" was unpredictable and open-ended. For Victorians the notion of ancient animal origins degraded human existence, and more recent origins made existence seem accidental and capricious if not also meaningless or absurd. Modernist scientists accepted those features of life, while existentialist philosophers and novelists from the 1930s and 1940s, starting with Jean-Paul Sartre and Albert Camus and later Samuel Beckett, accepted the ultimate absurdity of existence and viewed the creation of life's meaning as an abiding existential challenge.

The Modern Synthesis

The thirty-five-year gap between the publication of Mendel's findings and the scientific community's rediscovery of their significance is a prime example of the force of cultural values shaping scientific discovery. Cultures hang together through material and traceable as well as immaterial and invisible circuits that link their sources and set them off from previous as well as subsequent cultures. These circuits precluded biologists in Mendel's time from fully appreciating the significance of his experiments, because they contradicted the notion that parents' acquired characteristics, including their habits and values as well as physical traits, are directly transmitted to their offspring. Developments in the modernist period led to a fuller understanding of Mendel as well as Darwin as the findings of those two major Victorian biologists were integrated in the "modern synthesis."

While Darwin viewed natural selection as an "agent" of evolutionary change and therefore a metaphysical concept, modernist biologists increasingly saw natural selection as a "mechanism" that was measurable and testable and patterned after the physical sciences. Such an approach was taken in *Genetics and the Origin of Species* (1937) by Theodosius Dobzhansky, who based genetics on the gene that functioned as the basic unit of analysis analogous to the way the particle functioned in Newtonian physics. In a summary of this math-based modeling of the evolutionary synthesis, the historian of biology V. B. Smocovitis argued that Dobzhansky "converted the variables of natural selection, mutation, population structure, random genetic drift, migration, and systems of mating into causal *explanations* for evolutionary change."[56] Dobzhansky integrated the material basis of evolution (the gene) with the mechanism of evolution (natural selection) to make a mechanistic and material science of evolution, relying on the scientific methods of observation and experiment.

Five years later, in *Evolution: The Modern Synthesis* Julian Huxley extended these features of the evolutionary synthesis to include the origins of humans, the mind, and culture. The modern synthesis of evolutionary theory, he argued, has unified the biology of Darwinism and Mendelism "to rival the unity of older sciences like physics, in which advance in any one branch leads almost at once to advance in all other fields, and theory and experiment march hand-in-hand."[57] While Huxley rejected progressive forces driving natural selection from parent to offspring in the origin of species, he nevertheless drew on a broad view of the progress of the biological sciences to predict future progress of human culture. That view was well received, especially as he published during the most discouraging year, 1942, when racist thinking was derived from perversions of hereditarian theories that had progressed so much during the previous century.

Modernists' biology of origins challenged and surpassed Victorians' biology in a variety of ways. Modernists made considerable progress over the Victorians in understanding possible origins of life but remained ultimately baffled by that subject. Modernists' huge advances in understanding the origin of species built on the breakthrough theories of Darwin and Mendel, although those Victorian scientists were held back by their contemporaries' prejudices about blending of parental characteristics in offspring and relying on Lamarckism to explain variations. Modernists' theory of sexual reproduction based on the independence of reproductive cells from body cells liberated people from the sort of worries that plagued Victorians who believed that parental experiences including thoughts, values, or sexual experiences (especially a previous pregnancy) could infect their offspring. The modernist evolutionary synthesis unified formerly disparate sciences and liberated understanding of sexual reproduction from many frightening erroneous notions.

Notes

1. In the last paragraph of *Origin*, Darwin concluded with the strongly Lamarckian statement: "Variability from the indirect and direct action of the external conditions of life, and from use and disuse; a Ratio of Increase so high as to lead to a Struggle for Life, and as a consequence to Natural Selection, entailing Divergence of Character and the Extinction of less-improved forms" (Harvard University Press, 1964), 489–90.
2. Letter no. 7471 from Charles Robert Darwin to Joseph Dalton Hooker, 1 February 1871. Darwin Correspondence Project.
3. John Farley, *The Spontaneous Generation Controversy from Descartes to Oparin* (Johns Hopkins University Press, 1974), 7.
4. A. B. Macallum, "On the Origin of Life on the Globe," *Transactions of the Canadian Institute* 8 (1908): 437.
5. L. T. Troland, "The Chemical Origin and Regulation of Life," *The Monist* 24 (1914): 103, quoted in Farley, *Spontaneous Generation Controversy*, 159.
6. Félix d'Hérelle, *The Bacteriophage and Its Behavior* (Williams and Wilkins, 1926), 359. Quoted in Farley, *Spontaneous Generation Controversy*, 160.
7. Quoted in Farley, *Spontaneous Generation Controversy*, 155.

8. A. I. Oparin, "The Origin of Life" [1924] trans. Anne Synge in J. D. Bernal ed., *The Origin of Life* (World Publishing, 1967), 229, 230, 234.
9. Stanley L. Miller, "Production of Amino Acids under Possible Primitive Earth Conditions," *Science* 117, no. 3046 (1953): 528–29.
10. Robert M. Hazen, *The Story of Earth: The First 4.5 Billion Years, from Stardust to Living Planet* (Viking, 2012), 132–35.
11. Whiggish history was originally applied to English historians who viewed all of history as a story of progress leading to a highpoint in contemporary England, but it is applied more generally to "any subjection of history to what is essentially a teleological view of the historical process." John Burrow, *A History of Histories* (Allen Lane, 2007), 134.
12. Charles Darwin, *On the Origin of Species* (Harvard University Press, 1964), 489. Subsequent text references are to this edition.
13. Daniel C. Dennett, *Darwin's Dangerous Idea: Evolution and the Meaning of Life* (Touchstone, 1995), 58–59, 66.
14. Charles Darwin, "Provisional Hypothesis of Pangenesis" in *The Variation of Animals and Plants Under Domestication*, vol. II (1868), 367.
15. Jim Endersby, "Darwin on Generation, Pangenesis and Sexual Selection," in Jonathan Hodge and Gregory Radick eds., *The Cambridge Companion to Darwin*, 2nd ed. (Cambridge University Press, 2009), 73. Peter J. Bowler notes that the pangenesis hypothesis was "seriously in error" but "reflected typical mid-nineteenth-century attitudes toward growth and reproduction." *Evolution: The History of an Idea* (University of California Press, 1989), 210.
16. Francis Darwin ed., *The Life and Letters of Charles Darwin* (John Murray, 1887), vol. 3, Chapter XV, 66–67.
17. These reasons are surveyed by Peter Bowler in *The Eclipse of Darwinism: Anti-Darwinian Evolution Theories in the Decades around 1900* (Johns Hopkins University Press, 1983); *The Non-Darwinian Revolution: Reinterpreting a Historical Myth* (Johns Hopkins University Press, 1988); and *The Mendelian Revolution: The Emergence of Hereditarian Concepts in Modern Science and Society* (Johns Hopkins University Press, 1989). Bowler also relates these developments to modernism in "Darwinism and Modernism: Genetics, Paleontology, and the Challenge to Progressivism, 1880–1930" in Dorothy Ross ed., *Modernist Impulses in the Human Sciences 1870–1930* (Johns Hopkins University Press, 1994), 236–54.
18. Timothy Lenoir, *The Strategy of Life: Teleology and Mechanics in Nineteenth-Century Biology* (University of Chicago Press, 1982).
19. Johann Meckel, *System der vergleichenden Anatomie* [1821], 1:345; quoted in Ernst Mayr, *The Growth of Biological Thought: Diversity, Evolution, and Inheritance* (Harvard University Press, 1982), 471.
20. Quoted in Mayr, *Growth of Biological Thought*, 474.
21. "The rapid and brief ontogeny [of each organism] is a condensed synopsis of the long and slow history of the stem," and therefore "phylogeny is the mechanical cause of ontogeny." Ernst Haeckel, *The Evolution of Man* [1874] quoted in David J. Depew and Bruce H. Weber, *Darwin Evolving* (Harvard University Press, 1955), 179.
22. Henry Maudsley, *Body and Mind: An Inquiry into Their Connection and Mutual Influence* [1870], quoted in William Greenslade, *Degeneration,*

Culture, and the Novel, 1880–1940 (Harvard University Press, 1994), 69.

23. Thomas H. Huxley, *Man's Place in Nature* [1863] (University of Michigan Press, 1959), 115.
24. Émile Zola, *The Human Beast*, trans. Leonard Tancock (Penguin, 1977), 332.
25. George Levine, *Darwin and the Novelists: Patterns of Science in Victorian Fiction* (University of Chicago Press, 1988), 95.
26. Gillian Beer, *Darwin's Plots: Evolutionary Narratives in Darwin, George Eliot and Nineteenth-Century Fiction* (Arc, 1983), 25.
27. Laura Otis analyzes organic memory and its development in *Organic Memory: History and the Body in the Late Nineteenth and Early Twentieth Centuries* (University of Nebraska Press, 1994).
28. Samuel Butler *Life and Habit* (Trübner, 1878), 297.
29. Taine quoted in Daniel Pick, *Faces of Degeneration: A European Disorder, c. 1848–1918* (Harvard University Press, 1989), 72.
30. Robert Nye, "Heredity, Pathology, and Psychoneurosis in Durkheim's Early Work," *Knowledge and Society* 4 (1982): 104ff.
31. B. A. Morel, *Traité des dégérescences physiques, intellectuelles et morales de l'espèce humaine* (Paris, 1857).
32. Pick, *Faces of Degeneration*, 8.
33. Charles Rosenberg, "The Bitter Fruit: Heredity, Disease, and Social Thought," in *No Other Gods: On Science and American Social Thought* (Johns Hopkins University Press, 1976), 26.
34. Georgiana Kirby, *Transmission; or, Variation of Character through the Mother* (S. R. Wells & Company, 1877), 11.
35. J. H. Kellogg, *Plain Facts for Old and Young* (Segner & Condit, 1881), 109, 112.
36. O. S. Fowler, *Love and Parentage Applied to the Improvement of Offspring* (Fowler Wells, 1846), 74.
37. Richard W. Burkhardt, Jr., "Closing the Door on Lord Morton's Mare," in William Coleman and Camille Limoges eds., *Studies in the History of Biology* 3 (1979): 1–21.
38. Charles Darwin, *The Variation of Animals and Plants under Domestication* (John Murray, 1868), vol. 2, 388. Darwin accepted the possibility of telegony and cited ten sources for it including Lord Morton's mare. He also cited twenty more sources on parallel phenomena in plants. Burkhardt, Jr., "Closing the Door," 18n14.
39. Harriet Ritvo, *The Platypus and the Mermaid* (Harvard University Press, 1997), 108–9.
40. William A. Hammond, "On the Influence of the Maternal Mind over the Offspring during Pregnancy and Lactation," *Quarterly Journal of Psychological Medicine* 2 (1868): 7, 16, 20.
41. Ibid., 21–22.
42. Alexander Harvey, *On the Foetus in Utero as Inoculating the Maternal with the Peculiarities of the Paternal Organism* (Sutherland & Knox, 1886), 6.
43. Eduard von Hartmann, *The Sexes Compared* (Swan Sonnenschein & Co., 1895), 12.

44. Prosper Lucas, *Traité philosophique et physiologique de l'hérédité naturelle* (J.-B. Baillière, 1847–50), 1, 58ff.
45. Jules Michelet, *L'Amour* (Louis Hachette, 1895), 325–26. On Michelet's influence see Marcel Cressot, "Essai sur la genèse de deux romans de jeunesse: 'La Confession du Claude', 'Madeleine Férat,'" *Revue d'Histoire littéraire de la France* 35 (1928): 382–89; and Hilde Olrik, "La théorie de l'imprégnation," *Nineteenth-Century French Studies* (Fall–Winter 1986–87): 128–40.
46. Émile Zola, *Dr. Pascal* [1893] trans. Mary Serrano (Cassel, 1901), 108.
47. August Strindberg, *Five Plays by Strindberg* (Anchor, 1960), 16.
48. Quoted in Marvin Carlson, "Ibsen, Strindberg, and Telegony," *PMLA* 100 (October 1985): 777, 781.
49. Thomas Hardy, *Tess of the d'Urbervilles* [1891] (Norton, 1979), 191, 203–4.
50. Gabriel Compayré, *The Intellectual and Moral Development of the Child* (Appleton, 1896), 258.
51. D. H. Lawrence, *The Rainbow* [1915] (Cambridge University Press, 1989), 171.
52. Elizabeth Gaskin, "Why Was Mendel's Work Ignored?" *Journal of the History of Ideas*, 20 (June 1959): 60–84. See also, even more provocatively, Robert Olby, "Mendel No Mendelian?" *History of Science* 8 (1979): 53–72.
53. August Weismann, *Die Continuität des Keimplasmas als Grundlage einer Theorie der Vererbung* (Gustav Fischer, 1885).
54. George J. Romanes, "Weismann's Theory of Heredity," *Contemporary Review* 57 (1890): 695.
55. These developments are the subject of Gregory Radick's *Disputed Inheritance: The Battle over Mendel and the Future of Biology* (University of Chicago Press, 2023).
56. V. B. Smocovitis, "Unifying Biology: The Evolutionary Synthesis and Evolutionary Biology," *Journal of the History of Biology* 25, 1 (Spring 1992): 23.
57. Thomas Huxley, *Evolution: The Modern Synthesis* (Harper, 1942), 26.

Part II

Social Sciences

Chapter 5

ANTHROPOLOGY: Societal Origins and Foundations in Malinowski and Boas

A universal, essential, and primordial distinction in human relations is between self and others. That distinction is universal because everyone makes it, it is essential because it concerns the most fundamental originary experience of the self in the presence of others, and it is primordial because it begins in earliest infancy and shapes subsequent developments. In making that distinction the self is also instinctively judged as more important than any other self. My hand is more important than any other person's hand, especially if both are in pain. Altruism occurs but only on a foundation of instinctual egocentrism.

Encounters with others necessitate a way of classifying them, and the simplest way is by visible characteristics. Among these, skin color is immediately obvious along with the nose, hair, and eyes. More meaningful are the other's intellect, emotions, and morals, but those characteristics take more time and are more difficult to know. The primacy of visual perception makes superficial physical features of primary importance. They were the initial focus of the study of human relations in the mid-nineteenth century when researchers attempted to characterize such types "scientifically" based on visible features and tagged those types as "races." Racial judgments were back then and remain blatantly superficial, although the inner perception of one's own racial identity runs deep.[1]

Anthropology is the social science most directly concerned with the chronological origins and social foundations of human society. Its broad range of subjects include archaeology, linguistics, physical characteristics, and culture. Anthropologists from both the Victorian and modernist periods probed origins in ruins and historical sources and analyzed foundations in current languages, social practices, and culture. In the Victorian period, as Alice Conklin notes, "the principal reason for anthropologists to study 'primitives' was to reveal new information about the supposedly similar Stone Age predecessors of their own societies."[2] Anthropologists bridged origins and foundations by studying contemporary "primitives"[3] as living fossils, an oxymoron that I use to capture the origins of human society in remote past times but also as the foundations of human society in contemporary times.

Probing origins and foundations of "primitive" races in the Victorian period was an abiding temptation but also a source of anxiety. In *Heart of Darkness* (1899), Joseph Conrad used a double negative to dramatize the emotional double flip of a white[4] man suddenly grasping the humanity of a Black man: "that was the worst of it—this suspicion of their not being inhuman." Conrad's narrator Marlow elaborates the fascination of these dark-skinned human beings:

> [W]hat thrilled you was just the thought of their humanity—like yours—the thought of your remote kinship with this wild and passionate uproar. Ugly. Yes, it was ugly enough; but if you were man enough you would admit to yourself that there was in you just the faintest trace of a response to the terrible frankness of that noise, a dim suspicion of there being a meaning in it which you—you so remote from the night of the first ages—could comprehend.

This pivotal novel on the dawning modernist consciousness of the threatening nearness as well as remoteness of these living fossils captures the shock of recognition of seemingly remote people being all-too-human, as Conrad's dying colonialist Kurtz acknowledges with his famous outburst, "The horror! The horror!"[5]

In the late Victorian period, the increasing network of railroads, steamships, and telegraph lines that facilitated colonial expansion of empires increased the involvement of "civilized" Europeans and Americans with racially different "primitives." The horror of racial mixing sparked increasing racial analysis and eventually a new science of eugenics that was conceived to prevent such mixing and protect if not improve racial purity. Victorian anthropologists ranked races hierarchically in ascending order from "lower" colored races to the "highest" white race and ordered races historically in a progressive evolution from lower to higher levels. The temptation to peer into the remote origins of the white race by studying "primitives" in the present was a driving motive for Victorian anthropology. Some anthropologists thought that with improved living circumstances and education the "lower races" might catch up with the white race, but all concurred with a fundamental discriminatory racial ranking of the white race over inferior red, yellow, brown, and black races.[6]

This chapter focuses on studies of the origins and foundations of race that became increasingly popular around the turn of the century.

My previous chapters center on knowledge of nonmoral physical phenomena such as the nature of time and space, the size of the universe, and the age of the Earth, which had no moral status and were therefore neither good nor bad. My chapter on biology touches on a moral bias in discussing Darwin's penultimate paragraph of *Origin* where he wrote: "As natural selection works solely by and for the good of each being, all corporeal and mental endowments will tend to progress towards perfection." That sentence implied that the evolution of species was historically progressive and, in that sense, good. Darwin underscored those judgments with his later adoption of Spencer's term "the survival of the fittest" for a key element of natural selection. Darwin first used the phrase in *The Variation of Animals and Plants under Domestication* (1868) where it meant simply "better designed for an immediate, local environment," without any moral implications. He added the phrase to the fifth edition of *Origin* in 1869. But "fit" was widely interpreted with positive moral implications that Darwin did not intend but that many others drew. Those implications became more pronounced in the later Victorian period with the emergence of social Darwinism, the application of natural selection to the evolution of increasingly "perfected" human beings who accordingly created increasingly "perfected" societies and nations.

The self-satisfaction of Victorians' sense of superiority was typified by the Earl of Devon's Inaugural Address at the Royal Archaeological Institute of Great Britain and Ireland in 1873: "Few of us can observe such indications of the habits and physical condition of the earliest inhabitants of this island as are afforded by the remains of their rude dwellings, and by the rude implements occasionally found, without a sense of thankfulness that our lot has been mercifully cast in times of improved knowledge, of advanced civilization, and more refined habits."[7] Such sentiments informed early anthropological studies that could not avoid condescending judgments of current "primitive" populations that were routinely described as "savage" or "barbarian." In accord with such judgments, the teleological implications of social Darwinism centered on the evolutionary progress from "lower" to "higher" races. As the historian of anthropology Marvin Harris noted: "the greatest figures of late nineteenth-century anthropology fell under the spell of racial determinism. Even when the racial factor simply hovered over a discussion without actually coming to roost, it crippled the attempt to apply scientific methods to sociocultural phenomena."[8] According to late Victorian anthropologists, a superior white civilization was either created with their pristine greatness in the image of God from which lower races degenerated or emerged from the lowest colored races that evolved from even lower primate animal ancestors according to God's plan.

Victorian anthropologists made moralistic racial judgments to rationalize killing, slavery, and colonialism. European settlers killed millions of indigenous people in the Americas intentionally by murder or unintentionally by spreading disease and social disruption. The Western nations that were engaged in the slave trade banned the practice during the nineteenth century. England banned the transatlantic slave trade in 1807, and the United States banned it in 1808, while France banned it in 1815, although the French ban was not effectively enforced until 1823. Slavery continued in colonies of those countries for some years thereafter. The United States abolished slavery in 1865, but it remained a persistent foundation for racist thinking. Colonialism in Africa and Asia had a similar impact. In Africa, for example, between 1870 and 1890 colonial acquisitions by European nations rose from less than 10 per cent of the continent to 90 per cent, a figure that includes people killed and cultures destroyed as well as land overtaken and resources confiscated.

Under the umbrella of anthropological study, Victorians engaged in numerous archaeological projects, which made important discoveries in Iraq (Ninevah), Turkey (Troy), and Egypt (Ebers Papyrus), but they also frequently involved looting artifacts from colonial regions while posing as scientific explorations. Their crude methods also damaged archaeological sites and artifacts. Private European collectors and institutions, particularly from Britain, sponsored projects in Egypt and Mesopotamia where they generally disregarded local laws and failed to get local consent. Artifacts were shipped to Western museums, stripping local regions of their cultural heritage. Victorian archaeology prioritized imperial prestige over ethical stewardship, leading to its reputation as cultural theft. To rationalize such abuse, Europeans demeaned the humanity of their victims by identifying them as savages or barbarians, a task willingly shouldered by anthropologists at the very least in theory.

Motivated by economic, geopolitical, military, and Christian missionary projects, Victorians persuaded themselves that they were not abusing "primitive" populations but rather distributing the blessings of England's "White Man's Burden," Germany's *Deutsche Kultur*, and France's *mission civilisatrice*. Since 1845, Americans justified their overspreading the North American continent as a divinely sanctioned "manifest destiny" and between 1898 and 1903 claimed territory or exerted influence over Cuba, Puerto Rico, Guam, Hawaii, Samoa, the Philippines, and the Panama Canal Zone. Those slogans imperfectly screened the historical foundations of carnage, slavery, and colonialism that underlay early anthropological theory.

German researchers explored these matters in detail and played a major role in collecting archaeological remnants of early cultures and displaying them in museums, but in the modernist period German *Anthropologie* was overwhelmed by discriminatory and increasingly anti-Semitic *Rassenkunde* (racial science), which beginning in the 1920s fed into Nazi theorizing. As Robert Proctor explains, in the post–World War I period, after the Treaty of Versailles dismantled the German overseas empire, anthropologists focused on the "internal other," namely "Gypsies" (Roma), Jews, and the indigenous races of Europe. Anthropology became governed by "a logic directed towards the rescue of Germanic races from a host of perceived threats—enemies from 'within' and enemies from 'without.'"[9] In the Weimar period, as Peter Gay notes, Germany was beset by a "hunger for wholeness" and preoccupied with racial purity.[10] For these reasons morally impartial, "scientific" anthropology in Germany trailed advances in Britain and America.

In the mid-nineteenth century, French physical anthropology emerged out of Paul Broca's pioneering research in local brain anatomy with studies of the skull and brain and corresponding mental functions. He accurately located the third left frontal convolution as the seat of speech, but his overall method and analytical frame distributed people according to biologically determined and hierarchically arranged racial types. That orientation continued to shape French research until in the 1920s the ethnologists Marcel Mauss and Paul Rivet took charge and organized the first university training in *ethnology*, their term for cultural aspects of the discipline in contrast to the racial focus of physical anthropology.[11]

This chapter focuses on British and American anthropology for several reasons. The two countries also played a leading role in collecting and preserving artifacts of earlier cultures and displaying them in museums. They were both in the forefront of new immersion studies, learning indigenous languages, and pushing back against biological racism. They were also home to a prominent single anthropologist whose impressive career helps to focus this wide-ranging subject – Bronislaw Malinowski in England and Franz Boas in America. For those reasons, I track a major shift in thinking by comparing the achievements of these two modernist anthropologists with the achievements of their most prominent Victorian predecessors – Edward B. Tylor in England and Lewis H. Morgan in America.

British Anthropology

Two major literary sources of Victorian anthropological thinking were Darwin's *Origin*, followed by his application of natural selection to humans in *The Descent of Man.* The massive evidence of natural selection that he presented in those books plus countless bits of evidence of evolution from several hundred million years of fossils ingrained in Victorians the ideas that human social existence also evolved from lower to higher forms "for the *good* of each being" and that "all corporeal and mental endowments" tended to "progress towards perfection." That powerful teleological scenario was further magnified by ubiquitous examples of natural growth from seeds into mature plants and from embryos into adults.

An analogous sense of inevitable progression in human societies grounds Tylor's pioneering *Primitive Culture: Researches into the Development of Mythology, Philosophy, Religion, Language, Art, and Custom* (1871),[12] a 943-page treatise that in his lifetime was reprinted ten times and translated into Russian, German, French, and Polish.[13] Tylor defines culture as "that complex whole which includes knowledge, belief, art, morals, law, custom, and other capabilities and habits acquired by man as a member of society." Those multiple components of culture derive from diffusion across space and from evolution across time. Evidence from these cultures reveals a hierarchy of levels of achievement from "lower" to "higher" or, as Tylor puts it with the dominant three-stage rhetoric that was prevalent in his time, from "savage" to "barbarian" to "civilized," a system of types first defined by Montesquieu in the eighteenth century.

Tylor elaborates those levels as progressing in three stages evident in subsistence activities from hunting to herding to irrigated farming and in family structure from sexual promiscuity to consanguineous marriage to monogamous marriage. Further evidence of those three stages are material artifacts in textiles that progressed from matting to netting to weaving, in weapons that progressed from the wheel lock to the flintlock to the breech-loading rifle, and in astronomical instruments from the astrolabe to the quadrant to the sextant.

This pattern of achievements was enacted on a grand scale by "savage," "barbarian," and "civilized" cultures generally. "So thoroughly is this notion of development at home in our minds," Tylor argues, "that by means of it we reconstruct lost history without scruple, trusting to general knowledge of the principles of human thought and action as a guide in putting the facts in their proper order" (15).

That confident habit of thought made it possible to trace "the connection which runs through religion from its rudest forms up to the status of an enlightened Christianity" (23). Tylor's racial thinking was so "at home" in his mind that he could generalize "without scruple" the "proper order" of his analysis with one race initially mired in "inveterate ignorance."

Tylor's analysis was structured according to a fundamental parameter: "The educated world of Europe and America practically settles a standard by simply placing its own nations at one end of the social series and savage tribes at the other, arranging the rest of mankind between these limits" (26). While Tylor's research was prodigious, his judgments about high and low culture were based on "simple" classifications of excellence levels in the industrial arts, agriculture, architecture, science, morals, and political organization. These classifications enabled ethnographers to create a "scale of civilization" about which "few would disagree" (27). He insists that his overall progression of stages is not one from inherent and intractable evil to unqualified good. Given more time and opportunity, savages can catch up to civilized people, but the basic framework of his analysis and the abundance of evidence that European and American white people were more civilized made that prospect far in the future and of questionable likelihood. The one historical mechanism he proposed to account for an unequal advance toward civilization, "the continual interference of degeneration," is another indication of racial bias implying that savages are more morally degenerate as well as historically delayed (21).

While Tylor did travel in Mexico to gather information for his first book in person, his primary sources for the later important studies were those Victorians who relied on missionary accounts, explorers' journals, and historical studies and were sometimes tagged by modernists as "armchair anthropologists."[14] For example, in *Primitive Culture* Tylor introduces a discussion of the "Development of Culture" with a two-page quotation from Edward Gibbon's *Decline and Fall of the Roman Empire* (1776–89). Tylor contrasts that quotation with a long quotation from the French conservative philosopher Joseph de Maistre's degeneration theory from the early nineteenth century to document civilization's decline (33–35).

In contrast, modernists studied "primitives" with an immersion approach or "participant observation" by trained researchers who studied their subjects' language on site for long periods to increase understanding and empathy. They sought to preserve a record of the societies they studied as evidence of their historical origins and social foundations before those societies disappeared. The switch to immersion was not immediate, however, as George W. Stocking, Jr., notes in challenging the conventional notion that Malinowski personally transformed British anthropology, because many researchers used modernist fieldwork techniques in the decades separating his work from that of Victorians such as Tylor. The most significant of these transitional studies was the Cambridge exploration to the Torres Strait in 1898, a turning point in the history of British anthropology.[15]

Modernists were aware of and empirically tested for superficial racial indicators such as skin color, facial features, and, most importantly, the size and shape of the skull to determine brain sizes and brain weights that Victorians believed indicated mental characteristics including the "moral faculty." While a major purpose of modernist anthropology was to reject biology-based theories of white racial superiority and instead appreciate the complexity and richness of "primitive" intelligence and "primitive" society generally, traces of white *cultural* racism continued to ground their studies and surface in their texts.

Bronislaw Malinowski

Malinowski's classic ethnographic case study of the Trobriand Islanders in *Argonauts of the Western Pacific* (1922) is far from the condescending characterization of savages in Tylor's *Primitive Culture*. Whereas Tylor looks down condescendingly on savages, Malinowski views the Trobrianders as sources of wonder over the many possibilities of human existence evident in their thought and customs. *Argonauts* is an exhaustively detailed study over 500 pages long plus sixty photos of the society of the Trobriand Islands off the coast of New Guinea. In contrast to Victorians' reliance on second-hand accounts, Malinowski immersed himself in Trobriand society among people he generally calls "natives," sometimes calls "primitives," and occasionally slips and calls "savages," although mostly when quoting the racially prejudiced terminology of his predecessors. He trained in anthropology at the London School of Economics under C. G. Seligman, who was a member of the Torres Strait expedition and passed on many new field techniques.

World War I contributed indirectly to the emergence of modernist anthropological skills for the Polish-born Malinowski. When the war started, he was in Australia and as a Polish national was considered an Austrian subject and therefore an enemy alien who could not return to Britain, so he remained in the Trobriand Islands about which he ultimately wrote seven monographs between 1922 and 1935. During protracted stays there between 1914 and 1918, he lived among the indigenous population where he learned their Kiriwinian language and experienced intimate features of their daily life.

Malinowski clarifies the historical achievement of his work by contrasting it with "the often fragmentary, incoherent nature of much of the present ethnological material" collected by administrators, traders, and missionaries. Their idolatry of the "cult of 'pure fact,'"[16] as Malinowski describes it, harkens back seventy years to the narrow-minded educator Thomas Gradgrind in Charles Dickens's *Hard Times* (1854), who orders the schoolteacher M'Choakumchild to fill (choke?) his students with "Facts . . . Nothing but Facts," because "Facts alone are wanted in life." The emphasis on facts per se without attention to the broader contextual forces that conditioned those facts limited understanding in anthropology as it did among "positivist historians."[17] Malinowski details the complex interconnections of Trobrianders' character, language, technology, society, and art. He also distinguishes his work from that of ethnologists who were full of "biased and pre-judged opinions" that were "repulsive to a mind [such as his own] struggling after the objective, scientific view of things."[18]

His work developed two primary objectives of anthropological research – to interpret "primitive" experience in response to basic human needs and integrate a wide range of seemingly trivial facts as part of the larger experiential whole. Aspects of culture fulfill human needs however trivial they may seem in isolation, and to be properly understood they must be interpreted in the larger social context of their function, hence the characterization of his theoretical approach as functionalism.

Argonauts describes the "innumerable threads which keep together the family, the clan, the village community, the tribe" (18). These threads include private activities such as cooking, eating, and toilet practices as well as "quarrels, jokes, family scenes, and events usually trivial, sometimes dramatic, but always significant" (7). These revaluations of so-called trivia are components of a broader rehierarchization of what was valued and devalued across society as evinced in the broader cultural record, which I return to in my conclusion.

To illustrate the richness and complexity of Trobriand society Malinowski focuses on "the Kula," an elaborate, ritualized exchange of jewelry – armbands made of shells and necklaces made of beads and ribbons. He views Trobrianders as modern-day heroes modeled after the original argonauts – fifty heroes from the ancient Greek poem *Argonautika* who sailed in the *Argo*, named after its builder, Argus, to bring back the Golden Fleece that symbolized authority and kingship. Associating Trobrianders with heroic characters from that poem lends them added prestige. Trobrianders gave the Kula jewelry to others across the eighteen islands in the Massim Archipelago by sailing in dugout canoes across hundreds of miles over dangerous waters in threatening weather also on behalf of authority and chieftainship.

Malinowski describes at great length the essentials of the Kula, which include the creation of the armbands and necklaces, their ownership and value, and the many rules and rituals that govern their exchanges during long canoe journeys. While no one pays for the specific gifts exchanged or owns them personally, they are publicly owned and highly valued like the British Crown Jewels. Kula exchanges are regulated by strict codes of behavior that include a complex system of social conventions, technical procedures, artistic efforts, and magical rites.

For example, for over seventy pages Malinowski meticulously records the numerous details involved in constructing the canoes used in the Kula exchanges. Their core is a hollowed-out log. An elaborate ceremony accompanies finding a suitable tree, and then a special magical ceremony is performed before cutting it down. The chief canoe builder inserts a bit of food into a cut in the tree as an offering to evil wood sprites, while a magician performs an incantation to the sprites to go away so that the tree can become a canoe. To make the stripped log lighter before pulling it to the village, a piece of dry banana leaf is put on it, and the builder beats the log with a bunch of dry grass and utters another spell to drive out causes of heaviness from rot and fungus (128–29). After several days the master builder performs a lengthy magical rite as he wields his adze to hollow out the log while repeatedly uttering words to make the canoe eventually skip over waves. Another rite is performed as the tree is put on logs. For several months the master canoe builder and specially trained assistants prepare planks, boards, poles, sticks, and decorative prow-boards. Then comes the lashing of these parts together with vines accompanied by more rites that convey "flying canoe magic" (126). While many help with the preliminary scooping out, as the walls get thin only the master carpenter is permitted to finish that task. During the final hollowing out other parts of the canoe are constructed – broad planks for the gunwale, L-shaped boards for the ribs, long poles for longitudinal support of the ribs and platform rafters, main supports of the outrigging, and the outer float.

Hundreds of people witness painting the canoe black, white, red, and green. Two subsequent chapters detail the magical rites performed during the launching of Kula voyages – the speed magic rite, exorcisms against evil magic, and exorcisms of the wood sprites, all governed by an elaborate mythology. The rest of the construction involves waterproofing the wood, caulking the seams, and building the sail, accompanied by more rituals, spells, community singing, avoiding taboos, and community input for each of these processes carried out by a wide range of specialists.

Malinowski describes the Kula as a vital social function that gave meaning to the Trobriander society. He explicitly rejects nationalist or racist chauvinism in judging Trobrianders as inferior to white Europeans. "We Europeans," he writes, "accustomed to our extraordinarily developed means of water transport, are apt to look down on a native canoe and see it in a false perspective—regarding it almost as a child's plaything, an abortive, imperfect attempt to tackle the problem of sailing, which we ourselves have satisfactorily solved" (105–6). But he does not look down on the Trobrianders or their canoe building. Quite the contrary. He lavishly praises Trobrianders, for whom the canoe is "a marvelous, almost miraculous achievement, and a thing of beauty" (106). It is a powerful contrivance for the mastery of nature, an embodiment of living hopes celebrated in songs and stories.

To appreciate the boldness of Malinowski's analysis of the Trobrianders, one might recall that at the same time when he celebrated canoe building, the British were building steamships with strong steel hulls and driven by triple-expansion steam engines. The easy judgment for Malinowski to make would be that the British seacraft were superior because they made long-distance travel faster, safer, more comfortable, and more reliable than in canoes. Furthermore, Malinowski initially traveled to the Trobriands on a steamship, and, while steamships replaced canoes in the Trobriands and canoes did not replace steamships in Britain, Malinowski nevertheless insists boldly that those rudimentary canoes were in no way inferior to steamships when considered *in the context of their own culture*. Malinowski's celebration of the Kula is a monument of modernist cultural relativism, although it does seem a bit forced, because it is difficult to set aside the obvious advantages of travel being faster, safter, more comfortable, and more reliable.

Malinowski's work helped to counter the biological racism of earlier anthropologists who harbored the idea that "savages" and "barbarians" were born inherently inferior to "civilized" white people. His investigatory energy went into documenting and valorizing Trobriander cultural achievements and rejecting biological racist analyses of them, and, insofar as racist thinking monopolized earlier anthropological thinking about the origins and foundations of society, his work marked a significant historical change in the direction of greater understanding and multicultural appreciation generally. A similar response to biological racism is evident in the history of anthropology in America.

American Anthropology

A defining feature of Victorian anthropology in both England and America is its centering on progressive stages of culture from "savage" to "barbarian" to "civilized." A founding American text for those stages is Lewis H. Morgan's *Ancient Society*, published in 1877. Its subtitle highlights that three-stage "progression" – *Researches in the Lines of Human Progress from Savagery through Barbarism to Civilization.* Morgan introduces that scenario with the self-congratulatory attitude that pervades it: "The latest investigations respecting the early conditions of the human race are tending to the conclusion that mankind commenced their career at the bottom of the scale and worked their way up from savagery to civilization through the slow accumulations of experimental knowledge." Thus, human society originated at a morally deficient "bottom" that is currently still evinced in modern "primitive" societies. He continues: "These three distinct conditions are connected with each other in a natural as well as necessary sequence of progress."[19] The Victorians could never get beyond the "natural" and "necessary" components of their racism and so viewed the lower races as the primitive chronological origin and lowly existential foundation of a divinely inspired progress toward the higher civilized status of the white race. Although there were intervals of reversal or degeneration, the overall historical trajectory confirmed the Victorian way of thinking that the languages, social relations, and cultures of "savages" were less developed than moderns and that white people were at the highest level of civilization – intellectually and artistically as well as morally.

The mass of evidence in Morgan's treatise documents the three-stage progress in seven major areas – subsistence, architecture, government, language, family, property, and religion – that progressed through the savage and barbarian to the current civilized stage based largely on innovations in technology and technique. For example, subsistence progressed from savage collecting of fruits and nuts to barbarian fishing and use of fire, to civilized domestication of animals and irrigation of farms. Architecture progressed from savage huts to barbarian communal houses to civilized single-family homes (5). Morgan affirms a fundamental unity among humans in proposing that "the experience of mankind has run in nearly uniform channels" (8). His view mixes national, religious, and racial chauvinism: "There is something grandly impressive in a principle which has wrought our civilization by assiduous application from small beginnings; from the arrowhead, which expresses the thought in the brain of a savage, to the smelting of iron ore, which represents the higher intelligence of the barbarian, and finally, to the railway train in motion, which may be called the triumph of civilization," a scenario that was foreseen, overseen, and sanctioned by God (563).

In 1886, Morgan's theory was authoritatively summarized in a landmark talk "From Barbarism to Civilization" by the director of the United States Bureau of Ethnology, John Wesley Powell, the prestigious geologist and explorer who, among many other accomplishments, explored the Colorado River and the Grand Canyon in 1869 and in 1881 was appointed director of the US Geological Survey, a man whose career embodied manifest destiny. The talk, published in 1888, probed the chronological origins of human society back to racial primitivism. "The age of savagery," Powell argued, "is the age of stone; the age of barbarism, the age of clay; the age of civilization, the age of iron." Savage music is only rhythm; in barbarism it is rhythm and melody; in civilization it is rhythm, melody, and harmony. Savages group socially in clans, barbarians group in tribes, and civilized people group in nations. Writing progresses from picture-writing, to hieroglyphics, to alphabetic languages. Religion progresses from "beast polytheism," to "nature polytheism," to monotheism. Powell concluded this triumphant survey of the progress of these stages with a celebration of the pinnacle of history in contemporary research such as his own: "The greatest intellectual achievement of civilization was the discovery of the physical explanation of the powers and wonders of the universe, and the intellectual superiority of man, by which he becomes the master of those powers and the worker of wonders."[20] Powell's talk became an influential text that summarized the confidence of Victorian anthropology and set a target against which the modernist Franz Boas would direct his ideas.

Franz Boas

Victorian thinking about the origins and foundations of human society focused on a hierarchical world where people were divided into races based primarily on visible physical features such as skin color, and those differences were believed to reflect underlying intellectual, moral, and characterological differences, with the white race superior to other "colored" races. For Europeans, the first contact of the Spanish with the "Indians" of the Americas in the fifteenth century magnified racial prejudice in the modern Western world against "primitive" populations easily identified by their darker skin.[21]

The American researcher particularly well prepared to challenge that theory of the racial focus of human origins and foundations was Franz Boas. His internationalist credentials stemmed from his German-Jewish background; his philosophical credentials came from studies at universities in Heidelberg, Bonn, and Kiel; his scientific aptitude was honed by early training in physics, statistics, and geography; and his ethnographic sensitivity was cultivated by immersive experiences with the Inuit on Baffin Island, where he learned the Inuktitut language, and with the indigenous Kwakiutl population in the Pacific Northwest. His curatorial skills developed during stints as director of the anthropometry display at the Columbian Exposition in Chicago in 1893, curator of anthropology at the Field Museum of Natural History in Chicago in 1894, and curator of ethnology at the American Museum of Natural History in New York in 1896, followed by a professorship and founder of the Department of Anthropology at Columbia University from 1899 to his retirement in 1936. There he spearheaded a transformation of American anthropology and trained students who set up anthropology departments across America, most notably at Berkeley (Alfred Kroeber and Robert Lowie), Yale (Edward Sapir), and Northwestern (Melville Herskovits). He also trained pioneering female anthropologists who expanded interest in anthropology beyond academia with some best-selling and highly influential publications: Margaret Mead's *Coming of Age in Samoa* (1928), Ruth Benedict's *Patterns of Culture* (1934), and Ella Cara Deloria's *The Sun Dance of the Oglala Sioux* (1929) and *Dakota Texts* (1932). In the face of brutal prejudice against Blacks at that time, Boas also sponsored the work of his Black female student Zora Neale Hurston, whose *Mules and Men* (1935) collected African American folktales, followed by *Tell My Horse* (1938) on voodoo in Haiti and Jamaica. Between those publications Hurston managed to publish her famous autobiographical novel, *Their Eyes Were Watching God* (1937).

While from the outset Boas rejected the distinction between higher and lower races based on skin color, between 1891 and 1910 he nevertheless did extensive research in the anthropological subspecialty of anthropometry. In that effort he measured bodies of different racial types but most importantly skulls and cranial capacities that were thought to indicate brain weights that supposedly, although incorrectly, indicated levels of intelligence. He assiduously measured the bodies and skulls of "Indians" and "half-breeds," children and adults, and Black and white people and calculated his findings with complex statistical methods to sort out the respective influences of heredity and environment in the origin of types, which he occasionally, albeit with increasing reluctance, called "races." He devoted his early research to showing that visible physical differences did not indicate levels of intelligence, language, character, or civilization. While he did find the brain size of the most "advanced whites" to be slightly larger than that of the most "advanced blacks" at the "high end" of the population range, he could not determine that the majority of white brains in the middle range were larger than those of the majority of Black brains in the middle range.[22] Most importantly, he could not determine that human brain size at any level was an accurate predictor of intelligence or character. His larger achievement was, as Lee D. Baker concluded, to orchestrate a "shift in racial theory within anthropology by distinguishing race from culture and language and by proving that biologically based racial hierarchies were scientifically untenable."[23]

That historical shift was a long time coming. Boas spent nearly twenty years measuring skulls to determine brain sizes and weights among different age groups, nationalities, and races. His research into fundamental types started in 1891, when he began to measure school children in Worcester, Massachusetts to study growth and mental development. During the 1893 World's Fair in Chicago, he continued measuring the skulls and other body dimensions of fairgoers, using calipers to measure the length of the skull, and using goniometers to measure facial angles.[24]

In the opening paragraphs of an article of 1894, Boas summarized the pervasive racial prejudice against which his research was directed.

> Proud of his wonderful achievements, civilized man looks down upon the humbler members of mankind. He has conquered the forces of nature and compelled them to serve him. He has transformed inhospitable forests into fertile fields. The mountain fastnesses are yielding their treasures to his demands . . . His genius has molded inert matter into powerful machines which wait a touch of his hand to serve his manifold demands . . . Such is the contrast that presents itself to the observer. What wonder if civilized man considers himself a being of higher order as compared to primitive man if it is claimed that the white race represents a higher type than all others.[25]

Boas spent his entire career combating such racial prejudice that drew energy from two issues that became particularly prominent in the United States around that time: racism and immigration.

The first decade of the twentieth century witnessed increasing racial tension exacerbated by the US Supreme Court ruling in *Plessy v. Ferguson* (1896), which held that racial segregation laws did not violate the US Constitution if the facilities for each race were equal in quality, a policy known as "separate but equal." That ruling sanctioned segregation in education and public facilities and led to widespread racism that was elaborated in a system of laws and practices known as Jim Crow,[26] which included a high level of lynchings across the Southern states in the turn-of-the-century period. Boas addressed the racial problem directly with a pathbreaking course at Columbia University on "The Negro Problem" in the academic year 1907–8.[27]

Around that time, American leaders were becoming increasingly concerned that the racial purity of Americans of Anglo-Saxon and Germanic descent would be degraded by racial mixing from a wave of new immigrants who were streaming into the country from Eastern and Southern Europe and other countries around the world. In 1908 the US Congress established a commission to study immigration, chaired by Senator William Dillingham, who asked Boas to prepare a report on the racial impact of immigrants who had recently arrived in the United States. Boas proposed to measure heads, take family histories, and amass sufficient data to respond to these concerns. Across New York City he and his assistants measured 17,821 subjects with calipers and measuring tapes, even specially prepared eye color meters to help analyze the racial makeup of those immigrants. In 1911 he published his report, *Changes in Bodily Form of Descendants of Immigrants*, which argued that immigrants were adapting to social forces more than he had anticipated but were not polluting American racial stock. He determined that human beings were so responsive to environmental pressures that the underlying notion of hereditarily fixed racial types was untenable. Put simply, Boas showed that there was no such thing as a biologically determined purely physical type of a "Jew," a "Pole" or a "Slovak."[28]

That same year Boas published his pioneering work in anthropology, *The Mind of Primitive Man*. It began with the same words about the way "civilized man" looked down on the other races, which he had used in his 1894 essay. The fact that he repeated those exact words in that book and repeated the sense of them again in his revised edition of it in 1938 suggests how powerful and abiding that racially discriminatory outlook was. His misnamed classic on the mind of "primitive man" was indeed a study of the mental capacity of a group of people who in his time were called "primitive," but, as he insisted, they were not primitive as in rudimentary or inferior, just different in ways that his Victorian predecessors were unable to appreciate and routinely denigrated. Boas showed how the *human* mind ought to be analyzed, and he discredited the word "primitive" as a meaningful adjective altogether for the human mind, character, or language. This work shifted away methodologically from Victorian practices of deduction from fixed racial types to induction from a variety of human examples.

In a concluding discussion of the "race problem in the United States" Boas offered an impassioned defense of the largely equal biological and psychological capacities of "Negroes" and white people as far as his research was able to determine. He affirmed, with unintended but unmistakable white cultural bias, that "when given facility and opportunity" the Negro "will be perfectly able to fulfil the duties of citizenship as well as his white neighbor." In a passage cut from the 1938 edition he added in that vein: "It may well be that [the Negro] will not produce as many great men as the white race, and that his average achievement will not quite reach the level of the average achievement of the white race; but there will be endless numbers who will be able to outrun their white competitors."[29] In spite of Boas's condescending speculation that Black people might never equal white people, on balance he boldly confronted and produced masses of evidence to refute the Victorian notion of heritable biological racial inferiority.

Modernist anthropologists challenged Victorian ideas and practices in several ways. Their ideas about the origin of different "races" shifted from polygenism (many origins) to monogenism (a single origin), although the modernists' single origin theory was based not on a *single moment* of divine creation of a primordial pair in the Garden of Eden but on the evolution of different races through the *single mechanism* of natural selection. Victorians harbored their own skepticism about a single racial type. As early as 1871 Darwin registered his impatience with wild speculations about the number of races by earlier thinkers in noting the reigning controversy whether human beings "should be classed as a single species or race, or as two (Virey), as three (Jacquinot), as four (Kant), five (Blumenbach), six (Buffon), seven (Hunter), eight (Agassiz), eleven (Pickering), fifteen (Borey St. Vincent), sixteen (Desmoulins), twenty-two (Morton), sixty (Crawfurd), or as sixty-three, according to Burke."[30] In place of the Victorian jumble of multiple races, modernists identified multiple cultures, and they shifted their analyses of them from a condescending judgmentalism to relativism. They expanded the range of "world-historical nations" that were deemed worthy of anthropological study from Europe and America to the entire world. They questioned the hierarchy of races from high to low and the possibility of defining "race" as a distinct physiological or psychological type. Modernists grouped artifacts in museums inductively according to their social function, geographical distribution, and cultural significance and not deductively, as did most Victorians, according to a priori classification theories based on stages of racial development. In their museums modernists worked to preserve remnants of dying "primitive" cultures with projects subsequently called "salvage anthropology," although in more recent times that practice came to be identified as imperialist and the focus of numerous repatriation trials over remains and artifacts.

While many other anthropologists contributed to the modernist transformation of thinking about the origins and foundations of society, the works of Malinowski and Boas stand like pillars in the larger edifice of that corpus. Their writings occasionally reveal traces of white cultural racism, but the bulk of them were devoted to refining studies of origins and foundations, which led them to challenge the crude racist thinking that pervaded Victorian anthropology. They both flatly rejected Victorian notions of a hereditary, biologically based racial hierarchy of higher and lower races and the presumption that the white race was fundamentally and inherently superior to other races. They redirected thinking about what constituted a higher status among human beings from superficial visible features to deeper lying qualities of intelligence, moral character, and artistry. Their practice of participant observation and learning indigenous languages increased their knowledge of the societies they studied, deepened their empathy with them, and enhanced understanding of the origins and foundations of those societies. Perhaps most important, their commitment to evaluating people based on universal humanistic principles instead of provincial chauvinism demonstrated their commitment to a more cosmopolitan and humanistic outlook of cultural relativism. In comparison with the Victorians, modernists' understanding of origins and foundations was geographically broader, analytically deeper, emotionally more empathetic, morally more defensible, and less marred by racial or nationalist bias.

The transformative ideas of anthropologists of the modernist period, however, was no panacea for the crude division of human beings along racial lines. The 1930s witnessed an eruption of race-based thinking and behavior, which continued in various forms to the present day, an unfortunate situation highlighted by a global movement founded in 2013 that was obliged to remind Americans of a basic humanistic idea, namely that "Black Lives Matter." While the modernists shifted markedly away from biological racism, cultural racism proved to be deeply ingrained and seemingly intractable. Modernists did not solve this racial problem, to be sure. Before one might go about solving the problem, one had to see it clearly, and that is something that even the most conscientious modernists anthropologists whom I have surveyed in this chapter failed to do.

Notes

1. The deep sense of racial identity based on skin color was dramatized by John Howard Griffin, a white man who darkened his skin to pass as a Black man in the deep South and recorded the profound transformation of his inner sense of himself in *Black Like Me*, published in 1961. The high point of that transformation is the shame he feels talking to his own white wife back home from a phone booth, dramatizing the extent to which a superficial alteration in his skin color had internalized social discrimination toward Blacks.

2. Alice L. Conklin, *In the Museum of Man: Race, Anthropology, and Empire in France, 1850–1950* (Cornell University Press, 2013), 2.
3. I put scare quotes around "primitive" to distance myself from its racist implications but retain the word to indicate the people it referenced in both the Victorian and modernist periods.
4. I lowercase "white" and capitalize "Black" in accord with the Associated Press's policy. "AP style will continue to lowercase the term white in racial, ethnic and cultural senses. This decision follows our move last month to capitalize Black in such cases. We consulted with a wide group of people internally and externally around the globe and considered a variety of commentary in making these decisions." Reported by Nicole Meir in "Why We Will Lowercase White," *The Associated Press* (20 July 1920).
5. Joseph Conrad, *Heart of Darkness* (Dover, 1990), 32, 64.
6. Johann Blumenbach's *On the Natural Variety of Humankind* (1775) introduced that five-part typology of color-coded and geographically located racial types as African "Ethiopians," American and Arctic "Americans," Asian "Mongolians," Pacific Ocean "Malayans," and European "Caucasians."
7. Quoted in Bruce G. Trigger, *A History of Archaeological Thought* (Cambridge University Press, 1989), 110.
8. Marvin Harris, *The Rise of Anthropological Theory: A History of Theories of Culture* [1968] (Altamira, 2001), 141.
9. Robert Proctor, "From *Anthropologie* to *Rassenkunde* in the German Anthropological Tradition," in George W. Stocking, Jr., ed., *Bones, Bodies, Behavior: Essays in Biological Anthropology* (University of Wisconsin Press, 1988), 139; Benoît Massin, "From Virchow to Fischer: Physical Anthropology and 'Modern Race Theories' in Wilhelmine Germany (1890–1914)," in George W. Stocking, Jr., ed., *Volksgeist as Method and Ethic: Essays on Boasian Ethnography and the German Anthropological Tradition* (University of Wisconsin Press, 1996), 79–154.
10. Peter Gay, "The Hunger for Wholeness," in *Weimar Culture: The Outsider as Insider* (Harper, 1968).
11. Conklin, *In the Museum of Man*.
12. Edward B. Tylor, *Primitive Culture: Researches into the Development of Mythology, Philosophy, Religion, Language, Art, and Custom* (Brentano's, 1871), 1. Further text references are to this edition.
13. Jerry D. Moore, *Visions of Culture: An Introduction to Anthropological Theories and Theorists* (Altamira, 1997), 26.
14. Recent scholars now generally put "armchair anthropologists" in quotation marks to indicate its dubious accuracy. Efram Sera-Shriar rejects the "so-called armchair period of the discipline" and notes that term is overly dismissive of the efforts of many nineteenth-century practitioners "who were highly attuned to the problems associated with their research techniques and continually sought to transform their methodologies." "What Is Armchair Anthropology? Observational Practices in 19th-Century British Human Sciences" *History of the Human Sciences* 27, no. 2 (2013): 28, 27. Other researchers who make similar objections to the term include Henrika Kuklick in "The British Tradition" in Henrika Kuklick ed., *A New History of Anthropology* (Blackwell, 2008), 57 and H. Glenn Penny, "Traditions in the German Language" in Kuklick, *New History of Anthropology*, 84.

15. George W. Stocking, Jr., "The Ethnographer's Magic: Fieldwork in British Anthropology from Tylor to Malinowski" in *The Ethnographer's Magic and Other Essays in the History of Anthropology* (University of Wisconsin Press, 1992), 12–59.
16. Bronislaw Malinowski, "Baloma: Spirits of the Dead in the Trobriand Islands," *Journal of the Royal Anthropological Institute* 46 (1916): 419, quoted in James Urry, *Before Social Anthropology: Essays on the History of British Anthropology* (Harwood, 1993), 32.
17. "The refusal to judge the facts came to mean that history could only be the history of external events, not the history of the thought out of which these events grew." R. G. Collingwood, *The Idea of History* [1946] (Galaxy, 1956), 132.
18. Bronislaw Malinowski, *Argonauts of the Western Pacific: An Account of Native Enterprise and Adventure in the Archipelago of Melanesian New Guinea* (Benediction Classics, 2016), 5. Further text references are to this edition.
19. Lewis H. Morgan, *Ancient Society: Researches in the Lines of Human Progress from Savagery through Barbarism to Civilization* (Charles H. Kerr, 1871), 3. Further text references are to this edition.
20. John Wesley Powell, "From Barbarism to Civilization," *The American Anthropologist* 1, no. 3 (April 1888): 122–23.
21. Thomas F. Gossett, *Race: The History of an Idea in America* (Southern Methodist University Press, 1963), 1–16.
22. "No proof of an inferiority of the negro type could be given, except that it seemed possible that perhaps the race would not produce quite so many men of highest genius as other races, while there was nothing at all that could be interpreted as suggesting any material difference in the mental capacity of the bulk of the negro population as compared with the bulk of the white population." Boas, *The Mind of Primitive Man* (Macmillan, 1911), 268.
23. Lee D. Baker, *From Savage to Negro: Anthropology and the Construction of Race, 1896–1954* (University of California Press, 1998), 100.
24. Charles King, *Gods of the Upper Air: How a Circle of Renegade Anthropologists Reinvented Race, Sex, and Gender in the Twentieth Century* (Doubleday, 2019), 69.
25. Franz Boas, "Human Faculty as Determined by Race," *Proceedings of the American Association for the Advancement of Science* 43 (August 1894): 301–27; Boas, *Mind of Primitive Man*, 1–2.
26. The term "Jim Crow" has been attributed to "Jump Jim Crow," a song and dance that caricatured Black people, first performed in 1828 with that term that eventually became a derogatory term for "Negro." Laws passed by southern legislatures were known as Jim Crow Laws.
27. King, *Gods of the Upper Air*, 96.
28. Ibid., 96–98.
29. Boas, *The Mind of Primitive Man*, 273.
30. Charles Darwin, *The Descent of Man, and Selection in Relation to Sex* (Princeton University Press, 1981), 226.

Chapter 6

PSYCHOLOGY: Mental Origins and Foundations in Freud and Jung

The discipline of child psychology began in the late nineteenth century in the authoritative theoretical framework of Darwin's theory of evolution and Ernst Haeckel's theory of recapitulation, which together held that a child's mental development recapitulated the mental development of the species. Freud's selection from their work, reflections on his own childhood, and clinical findings fed into the synthesis of psychoanalytic theory that impacted how modernists understood childhood and the origins of mental life more broadly. In *Three Essays on the Theory of Sexuality* (1905), he elaborated the childhood origins of adult mental life as it emerged in successive oral, anal, phallic, and Oedipal stages; in *Totem and Taboo* (1912–13) he speculated about the origins of the Oedipus complex back to a mythic primordial family; and in *Beyond the Pleasure Principle* (1920) he speculated about a repetition compulsion to restore an even earlier, tension-free state of inorganic homeostasis.

The first part of this chapter focuses on how Freud's ideas probed ever further back in time to the origins of mental life and ever deeper psychologically into its unconscious foundations. The second part focuses on Carl Jung's probing even further back temporally into ancestral origins and even deeper into the archetypes of the collective unconscious.

Freud on the Origins of Mental Life

Freud's earliest significant publication on the origins of mental illness was *Studies on Hysteria* (1895), coauthored with his fellow Viennese physician Joseph Breuer. The now outmoded term "hysteria" referred to psychogenic disturbances of sensory, vasomotor, and visceral functions such as blindness, contractures, tics, pain, anesthesia, and difficulties in seeing or hearing.

In 1880 Breuer took the lead in treating hysteria with his young patient Bertha Pappenheim, referred to in their book as Anna O. She had numerous hysterical symptoms including headaches, loss of appetite, semiparalyses, and partial losses of vision and hearing. During periods of self-hypnosis, she began telling stories that temporarily relieved her symptoms, a procedure that she called her "talking cure" or "chimney sweeping," playful ways of naming her procedure that included the essentials of future psychoanalytic treatment that was eventually used on unhypnotized patients. A particularly vivid instance of talking cure relieved her of a fear of drinking water when, while under self-hypnosis, she recalled a traumatic experience of disgust from seeing a dog drinking out of a glass, reexperienced the emotions that emanated from that trauma, and put that experience into words. Upon emerging from hypnosis, her hydrophobia disappeared. The successful treatment of that symptom became a pivotal moment in the history of psychoanalytic therapy and in understanding mind–body influences generally.[1]

Freud and Breuer found that the causes of hysteria were not brain lesions, as Victorian psychologists thought, but psychological traumas that were structured and eventually treated chronologically, concentrically, and dynamically.[2]

Their treatment aimed to help patients abreact (express or release) emotions that have been repressed or kept unconscious. That process proceeded chronologically but in reverse order to their occurrence. That is, while under hypnosis patients were led to abreact the most recent traumas first followed by an abreaction of earlier traumas. To introduce an account of their method Breuer wrote that in treating Anna O: "It was our regular experience that the patient did not hear when she was spoken to" (*SE* II, 36). The vast scale and analytical rigor of their clinical method is illustrated by Breuer's notes on Anna O's 438 instances of hysterical deafness: 108 instances of deafness when someone entered a room, 27 instances of not understanding when several people talked, 50 instances of not hearing when directly addressed, 15 instances of deafness brought on by being shaken, 37 instances of deafness from a frightening noise, 12 instances of deafness during a distraction, and 54 instances of deafness from listening intently for a long time. Breuer could not have totaled and classified such specific numbers had not he and Freud recorded each incident with exceptional thoroughness.

A second figurative structuring of traumas was "stratified concentrically round the pathogenic nucleus" (*SE* II, 289). Their therapy was conceived in a spatial and temporal framework starting at the conceptual periphery of the psyche, overcoming resistances in probing origins, following threads to find multiple starting points and interconnecting pathways zigzagging through the mind, looking for ever "deeper" causal layers. *Studies on Hysteria* introduced a method of inquiry into the temporally remote, deep lying, and complex origins of mental life that contrasted sharply with comparatively superficial psychiatry of the time that was preoccupied with naming and classifying symptoms and treating them superficially with diet, rest, massage, drugs, hypnosis, and electrotherapy.

Freud's dynamic method also required penetrating forces of intrapsychic repression to find the earliest origins of his patients' problems that were buried in the conceptually deepest levels of the psyche. To achieve those investigatory goals, he practiced intensive psychoanalysis, which in 1913 he recommended should be an hour a day, six days a week, for several years, and he joked about "Monday crust" after the Sunday break when resistant forces reassembled to protect against further analysis.[3] Freud typically conducted six to eight sessions a day, unable to take notes during sessions for fear of distracting his patients' concentration or threatening their privacy. For its time, psychoanalysis was a uniquely long and demanding deep probing of the chronological origins and psychodynamic foundations of mental life.

Although Freud was dedicated to probing the temporal origins of mental life in his theory and practice of psychoanalysis, he proposed that we can never accurately remember the most important traumatic incidents from earliest childhood because they are overlaid by "screen memories"[4] that disguise the original repressed memory and only allow it into enter consciousness disguised by a host of unconscious mental processes, which his daughter Anna Freud later called "defense mechanisms."[5]

Around the turn of the century modernist psychologists began to publish long case histories.[6] Freud published five of his own, averaging around 100 pages. A sixth long case history was the self-analysis that began in 1897 with letters to his friend Wilhelm Fliess, then continued in the fifty personal dreams that he interpreted in *The Interpretation of Dreams* (1899), and after that in letters and publications to the end of his life. Not all commentators celebrated Freud's self-analysis as appreciatively as his biographer Ernest Jones, but Jones was a contemporary and sensitive to pressures in the emotionally restrictive atmosphere of Vienna, so his assessment is worth recalling. "In the summer of 1897," Jones wrote, "Freud undertook his most heroic feat—psychoanalysis of his own unconscious. It is hard for us nowadays [c. 1953] to imagine how momentous this achievement was . . . Once done it is done forever. For no one again can be the first to exploit those depths."[7] The singularity of that achievement is more plausible if one adds to Jones's claim – exploring those depths unassisted by anyone else. Freud did share this self-analysis in letters to Fliess, but for the difficult parts he pushed through obstacles alone to discover the origins of his own neurotic symptoms that included obsessional ideas, intellectual paralysis, fear of traveling, feelings of impotence, tabooed desire for his mother, guilt about anger toward his father, and psychosomatic heart and gastrointestinal problems.[8] He specifically referred to "my own neurosis" and "my little hysteria."[9] Those symptoms centered on difficulties in loving that he became aware of around age forty as he discovered the childhood origins of his problems.

Freud's troubled early love life was shaped before age three by an intense relationship with his Czech Catholic nanny Theresa ("Resi") Wittek. He viewed her (not his mother) as the originator of his love life, although the two women were intertwined in his mind.[10] He began to theorize about those emotions on 3 October 1897, when he wrote to Fliess that "in my case the 'prime originator' was an ugly, elderly, but clever woman [Resi], who told me a great deal about God Almighty and hell and who instilled in me a high opinion of my own capacities."[11] On 15 October Freud wrote about a scene "which in the course of twenty-five years has occasionally emerged in my conscious memory . . . My mother was nowhere to be found; I was crying in despair. My brother Philipp (twenty years older than I) unlocked the wardrobe for me, and when I did not find my mother inside it either, I cried even more . . . Why did my brother unlock the wardrobe for me, knowing that my mother was not in it and that thereby he could not calm me down? . . . I had asked him to do it. When I missed my mother, I was afraid she had vanished from me, just as the old woman [Resi] had a short time before. So, I must have heard that the old woman had been locked up and therefore must have believed that my mother had been locked up too . . . The fact that I turned to [Philipp] in particular proves that I was well aware of his share in the disappearance of the nurse."[12] This recurrent memory (not a dream) reveals that he had conflicted feelings about Resi and his mother, because she and not his mother had vanished and because his tears were for her.

At a deeper level was the possibility that the wardrobe symbolized the uterus and that Philipp was responsible for what was inside it. The evidence is compelling that Philipp was having an affair with his stepmother Amalie (Freud's mother) and may have even sired one of her children.[13] In 1924 Freud wrote about himself:

> The child of not yet three had understood that the little sister who had recently arrived had grown inside his mother . . . The wardrobe or cupboard was a

> symbol for him of his mother's inside. So, he insisted on looking into this cupboard and turned for this to *his big brother, who (as is clear from other material) had taken his father's place as the child's rival*. Besides the well-founded suspicion that this half-brother had had the lost nurse 'boxed up,' there was a further suspicion against him—namely, that he had in some way introduced the recently born baby [Anna Freud] into his mother's inside.[14]

Amalie and Philipp were less than a year apart in age and lived across the street from one another, while Freud's father, Jacob, was twenty years older than both of them and frequently out of town for work. These circumstances might explain the sudden disappearance of Resi, who must have learned about the affair, prompting Philipp to take the lead in getting her arrested and eventually out of the way. Thus did Freud traumatically lose the person whom he loved most and who cared for him more than anyone else during his first thirty-two months because of the actions of his stepbrother who had likely become his mother's lover.

The enormity of Freud's loss of Resi's love was magnified by the impoverished relationship he had with his mother. Amalie did not have much time for baby Freud. Five months after he was born on 6 May 1856, she became pregnant with Julius, who died less than a year later. Her daughter Anna was born on 31 December 1858, around the time when Resi disappeared. While later Amalie repeatedly said that she favored Sigmund, she did not write that she loved him, and I have not found any letter in which he wrote to her that he loved her. While no explicit aggression against Jacob appears in Freud's dreams or memories, aggression against fathers became increasingly evident in his theories about the origins of family life, morality, and religion. No doubt the intensity of Freud's inquiry into the childhood origins of his patients' neuroses derived partly from discoveries about his own neuroses, but those personal determinants also likely motivated him to help patients come to terms with their problems in a give and take exchange with him similar to the way he had explored them in letters exchanged with Fliess.

Beginnings of Case Histories

Victorian case histories were rarely more than a few pages, starting with the patient's age, occupation, major complaint, and family history.[15] At a time when the new germ theory of disease revealed single identifiable causes of medical illnesses, these histories often ended with lab tests for pathogenic germs causing mental illness, and if the illness was fatal they may have included autopsy reports on brain lesions. As Mark Micale concluded, "[b]efore the final quarter of the nineteenth century, psychiatric cases published in medical textbooks and monographs tended to be short and mechanical recitations of hereditary background, symptom profile, diagnosis, and prognosis."[16] The prevailing Victorian theories about organic causation of mental illnesses skirted the sort of complex psychodynamic origins on which Freud based psychoanalytic theory.

Freud wrote his first long case history about eighteen-year-old Dora in 1901 during a two-week period after she terminated her three-month therapy, and, even though it totaled 115 pages, he referred to it as "Fragment of an Analysis of a Case of Hysteria," which he published in 1905. He prefaced the case history with a forty-eight-page "clinical picture" of many obstacles that impede understanding the origins of mental disorder. He compared patients' first accounts of their illness to "an unnavigable river whose stream is at one moment choked by masses of rock and at another divided and lost among shallows and sandbanks." That imagery sets up a critical "wondering" about conventional case histories: "I cannot help wondering how it is that the authorities can produce such smooth and precise histories in cases of hysteria" (*SE* VII, 16). His navigations of the rivers of neurosis typically disclosed many obstacles. Patients consciously withheld information, forgot other things, repressed still more, and created false substitute memories because they in fact did not want to be relieved of the neuroses that were their default way of managing problems, as is alcohol for alcoholics.

Dora's symptoms included migraines, nervous coughing, voice loss, suicidal impulses, depression, and fainting. Unlike Victorian psychiatrists who routinely explained such symptoms erroneously as inherited through the blood, Freud was concerned with family psychodynamics. Not wanting to entirely dismiss the importance of heredity, he conceded that Dora inherited a "sufficiently heavy taint," but, aside from a mention of her father's earlier bout of syphilis, Freud moved on to the core foundations of his theory of neurosis – "a psychical trauma" and "a disturbance in the sphere of sexuality" (*SE* VII, 24). The core family problem and the source of Dora's sexual trauma was her father's affair with the married woman Frau K., and as a cover her father turned a blind eye to Herr K.'s sexual advances toward Dora, including kissing her on the lips at age fourteen. Freud's interpretation of this event includes complex unconscious mental processes. The pleasure from the kiss was transformed into disgust, a *reversal of affect*, and the locus of the sensation in her genitals was *displaced* upwards to her throat, causing the nervous cough. During the kiss she felt this erection pressing against her, *repressed* that disgusting sensation, and replaced it with a sense of pressure further displaced onto her thorax. The disgust was a symptom of *repression* of the erotogenic oral zone that had been over-indulged in Dora's infancy with sensual sucking. To explain the wide range of causes at various levels of consciousness, Freud introduced the concept of *overdetermination*, meaning that each symptom had more than enough causes to make it occur. Other mental processes included *somatic compliance* between psychic causes and bodily manifestations, *sublimation* of forbidden sexual impulses into socially acceptable forms, *fixation* of emotions resulting from overwhelming psychic traumas, and a cluster of *Oedipal dynamics* that were the heart of her childhood sexual trauma.

In contrast with the formulaic simple beginnings of typical "smooth" Victorian case histories, Freud's history of Dora began with a tangle of analytical challenges. His treatment evinces the specificity–uncertainty dialectic in that the more he learned about the case, the more he realized how much more there was to know and how little he understood what he thought he knew in that, as it turned out, the case history was incomplete (a fragment), the analysis was inconclusive, and the outcome was unsuccessful.

Developmental Psychology, Childhood Origins

The overall temporal arc of Freud's thinking was to look for ever earlier origins of mental life and ever deeper analysis of its psychic foundations. In an autobiographical study of 1925, he recalled how the age of traumatic origins gradually receded from the "age of understanding," back to puberty, to the second dentition (eight to ten years), to the age of the Oedipus complex (five to six years); and subsequently before the age of three. In this search, he reported, "I was carried further and further back into the patient's life and ended by reaching the first years of his childhood . . . [T]he impressions of that early period of life, though they were for the most part buried in amnesia, left ineradicable traces" (*SE* XX, 33).

His findings on childhood origins came out of analyses of neuroses and dreams along with his self-analysis. In *Studies on Hysteria*, he theorized that hysteria can be caused by "impressions from the pre-sexual period" that produce "no effect on the child" but gain traumatic power later when a child acquires a moral sense (*SE* II, 133). Soon Freud scratched the concept of a presexual period because he came to believe that all children are normally sexualized from birth in rudimentary ways. His use of "sexual" was misleading, and it distorted evaluations of psychoanalysis throughout Freud's career to the present day. It did not refer merely to genital excitation but to a variety of effects that he also called libidinal, erotic, sensuous, somatic, instinctive, bodily, energetic, sublimated, or simply interested. Ironically, although the purpose of his libido theory was to demoralize sexuality for his highly moralistic contemporaries, it had the opposite effect of hypermoralizing sex. Freud's sex instinct was far more foundational in human affairs than his contemporaries were accustomed to believing, and, while he viewed that pervasive originary function as normal and healthy, they saw it as morally suspect.

Freud's first theory of neurosis, "the seduction theory of neurosis," which he worked out in May and June of 1896, undercut Victorian mythologizing about childhood innocence, particularly about sexual desire. In "The Aetiology of Hysteria" (1896) he introduced that theory that identified a sexual assault on a supposedly sexually inexperienced child as the key causal factor, but he also wondered whether "even the age of childhood is not wanting in slight sexual excitations" (*SE* III, 202). Two years later, as he abandoned the seduction theory, he no longer wondered. In place of that theory he substituted the psychoanalytic theory of neurosis that was based on the universal and normal existence of sexual feelings from earliest infancy: "[I]n my experience," he wrote, "children are capable of every psychical sexual activity, and many somatic ones as well."[17] He did not deny that children were subject to assaults, but the historically significant part of his argument was that children are capable of being traumatized during such assaults precisely because they experience powerful, albeit rudimentary components of sexuality in oral, anal, and phallic zones during specific stages of development, further erotic stimulation from seeing and being seen, and erotogenic sensations all over their body, a condition that he characterized misleadingly as "polymorphous perversity."

While to traditionally minded readers these views seemed outrageous and degrading to children, for Freud they revealed rich psychological possibilities. To take one example, Freud valorized the generally degraded significance of the anal function as it is managed during toilet training as a foundation for important adult character traits that emerge along a pairing of contrasts – clean–dirty, cooperative–stubborn, and generous–stingy. Unless fixated at the anal stage, the maturing child moves through this stage normally, acquiring a productive balance between these contrasting traits. Freud further valorized childhood experience in "Creative Writers and Day Dreaming" (1908), where he argued that children at play cultivate creative capabilities that are frequently glossed over by adults. "Every child at play behaves like a creative writer, in that he creates a world of his own, or, rather, re-arranges the things of his world in a new way which pleases him" (*SE* IX, 143). Child's play cultivates creative powers, forges the ability to handle disappointments, fosters emotional independence, dramatizes the vital distinction between phantasy and reality, teaches basic skills necessary to become an adult, conjures up a treasure-house of myths and fairy tales, and provides a normal and healthy training school for their adult sexuality, beginning with digital self-stimulation.

Mythic and Primal Origins

Freud's commitment to a historical originary approach to human experience led him to search ever further back in time, even before his patients' birth. The metaphor he used in Dora's case history to characterize psychoanalytic inquiry, "like a conscientious archeologist [investigating] mutilated relics of antiquity," portended his later inquiries into a mythic prehistory and before that to primordial instinctual origins (*SE* VII, 9, 12). To supplement those inquiries, he had a personal collection of over 2,000 statuaries from antiquity. His investigation into the origin of morality and religion in *Totem and Taboo* (1913) uncovered what he believed were the origins of human society generally, evinced in multicultural evidence of what he posited as three universal taboos against murder, cannibalism, and incest. The unifying emotion associated with those acts was ambivalence, a simultaneous love and hate toward both parents that in 1910 he named the Oedipus complex.

The origin of ambivalence in ancient history, he theorized, was in a mythic "primal horde" in which a primal father monopolized all the women sexually and was hated by the jealous sons. One day the sons killed the primal father out of jealousy and ate him to partake of his potency, but they soon experienced remorse because they also revered the primal father, and so they projected their collective ambivalent feelings about him onto sacred totem animals. Thereafter everyone in the totem clan was forbidden to kill or eat those animals or have sex with anyone from the same clan who worshipped them, hence the prohibitions against the tabooed crimes of murder and cannibalism that the sons had committed and against incest that they had contemplated. But during festive times the clan was permitted to engage in all three prohibitions collectively – they could kill and eat those sacred animals during a "totem meal" and engage in forbidden sexual activity. Modern children inherited this ambivalence, and so, Freud concluded, "The beginnings of religion, morals, society, and art converge in the Oedipus complex" (*SE* XIII, 156). Thus did he run the causal scenario back in time as far as he could in 1913 to emphasize its primordial historic origins as well as its deep foundational function.

Deepest Foundations of Mental Life

World War I impacted psychoanalytic theory by reducing Freud's clinical practice, giving him time to reexamine the foundations of his theory in key "metapsychological papers." The war also presented cases of "shell shock" that forced him to revise his notion of the psychosexual foundations of mental life. Investigating shell shock led back in time to the personal beginnings of mental illness, and the metapsychological papers led deep into the unique metaphorical spatiality of Freud's conceptualization of mental life.

One of those metapsychological papers was on *repression* (*Verdrängung*), which in 1914 Freud maintained was "the foundation stone on which the structure of psychoanalysis rests" (*SE* XIV, 16). Different uses of "repression" contrast Victorian and modernist thinking. "Repression" in Victorian discourse meant rather conscious *suppression* that Freud referred to as *Unterdrückung*. Examples of such misnamed "repression" occur in George Eliot's *Middlemarch* (1872), where Mary Garth was "unable to repress her sarcasm" or in Charles Dickens's *Oliver Twist* (1838) where "Mr. Sikes appeared to repress a rising tendency to jealousy." But, for Freud, repression did not mean conscious deference to social pressures but intrapsychic, unconscious efforts to forcefully keep something from consciousness.

One example of confusion between Victorian and Freudian ideas about repression is in John Kucich's *Repression in Victorian Fiction*. Even though the book prominently accents "repression" in its title and repeatedly uses that word to refer to Victorian strategies for exalting interiority and intensifying emotions, the book explicitly rejects "the Freudian psychic mechanism of repression."[18] But that rejection misfires, because Kucich provides abundant evidence not for unconscious *repression* but for conscious *suppression*. Michel Foucault also failed to take note of Freud's crucial emphasis on the unconscious nature of repression in offering as examples of "the repressive hypothesis" Victorian moralist tracts that advised men to consciously suppress their impulse to masturbate and warned couples to consciously regulate their sexual relations. In characterizing such texts as "incitements to discourse" that promoted "intensification, reorientation, and modification of desire" Foucault erred, because such texts were not examples of unconscious repression, and the practices they described did not incite desire but thwarted it.[19]

Repression is not always pathogenic. To get on with life, normal individuals repress all sorts of memories and impulses. Repression is pathological when it results from traumatic fixation and is excessive and drains the psyche of energy that could be used for productive purposes. Repression underlies many mental processes such as regression, displacement, denial, undoing, and reaction formation that occur along a diagnostic continuum from normal to abnormal and range along a psychic (metaphorical and spatial) continuum from a conscious to an unconscious level. That latter level is the subject of one metapsychological paper on "The Unconscious" (1915). Its title led to confusion, because the definite article and the noun *unconscious* suggested that *the* unconscious was a region in the mind. Freud used the term most effectively as an adjective to characterize a theoretical level of mental processing or, as he put it, a "graphic illustration" of mental life (*SE* XIV, 175). Together with the other "levels" of the *conscious* and *preconscious*, the three levels created a "topographical" model of the mind.

A full metapsychological account of mental life included also a "dynamic" model that tracked the direction and intensity of forces that drive the psyche. As Freud explained, "[w]e seek not merely to describe and to classify phenomena [as did typical Victorian psychologists], but to understand them as an interplay of forces in the mind, as a manifestation of purposeful intentions working concurrently or in mutual opposition. We are concerned with a *dynamic view* of mental phenomena" (*SE* XV, 82). In 1923 Freud added a *structural* model of the mind that graphically illustrated an ordering of psychic depths starting at the "top" with the moral censorship of the superego, moving "down through" the practical ego, and "further down" to the source of primal instinctual urges, the narcissistic id (*SE* XIX). Freud's metapsychological papers probed "spatial depths" of mental life along a variety of ways "down," conceptually ever deeper.

Freud also probed the foundations of mental life further back in time in an essay of 1920 that led to his controversial theory of a "death instinct." Its title, *Beyond the Pleasure Principle* (*Jenseits des Lustprinzips*), was misleading, because *Jenseits* suggested spatial remove (literally "on the other side of"), while his intended meaning was temporally before or, as he added, "more primitive than" the action of the pleasure principle (*SE* XVIII, 17). In this essay Freud abandoned the primacy of libido theory (sexual causation) that he had fought so tenaciously to maintain in the face of enormous public resistance. One source of evidence for his journey "beyond" the pleasure principle was drawn from cases of shell shock. These threatened the theory of the pleasure principle, the notion that people are driven by a fundamental instinct to seek pleasure and avoid pain (which dominated Victorian utilitarian moralizing), because the shell-shocked soldiers' dreams were not of pleasurable experiences but of pain and terror, and they did not repress painful memories but obsessively recalled them. Such obsessional recollections derived from an instinct more fundamental than the pleasure principle, namely the repetition compulsion, an instinct to repeat earlier experiences regardless of how painful they had been.

Freud probed ever deeper for foundational evidence of a repetition compulsion that undercut the pleasure principle in animals, primitive life forms, and even inorganic matter. A premise for that compulsion was that the mind strives to keep excitation as low as possible or at least constant. That striving leads ultimately to restore an earlier state of things throughout organic life, because life itself disturbs the inorganic static equilibrium of, say, rocks. The challenges of sexual reproduction subordinate pleasure to pain. Fish make stressful migrations to deposit their spawn far from their normal haunts. Migrating birds struggle to fly great distances to reproduce, more evidence of a compulsion to repeat that is enacted during sexual reproduction. In a footnote added in 1925 Freud warned his readers that "what follows is the development of an extreme line of thought" (*SE* XVIII, 37n2). Indeed. In that highly speculative way of thinking, he posited further that "the aim of all life is death" and that, even more fundamental than living forms, "inanimate things existed before living ones" (*SE* XVIII, 38). There is an instinct in living things to return to the inanimate and tension-free state from which life emerged (*Todestrieb*), and another name for that state is death. Although not quite, because prelife is not the same as postlife. Still, Freud was led by bold, and some thought reckless, theorizing to the theory of the death instinct. He summarized that life is dominated most fundamentally by the death instinct (Thanatos) in a perpetual conflict with the life instinct (Eros) and acknowledged that these arguments are so speculative that "it may be that I have over-stated their significance" (*SE*, XVIII, 59). Freud's search for originary foundations continued a lifelong search for ever earlier origins and more fundamental processes of mental life even if those searches led to overstatements and imperiled the success of psychoanalytic theory. The theory of a death instinct was criticized by many of his most loyal followers including Alfred Adler, Karl Abraham, Otto Rank, and his good friend and future biographer Ernest Jones, who together unloaded a cluster of criticisms including that the theory lacked observable evidence, contradicted current biological science, and had no clinical use.

After a painful self-analysis and psychoanalyzing many patients over thousands of clinical hours and writing hundreds of dream interpretations, five long case histories, and a pathbreaking developmental psychology along with numerous books and articles on jokes, art, literature, religion, and history, Freud risked the reputation of psychoanalysis with highly speculative theorizing about primordial processes of life and even inorganic matter for which he had no empirical evidence. The lasting value of this body of work has come under criticism from many quarters, but there can be no denying that Freud's depth psychology marked a radical break from the comparatively superficial Victorian psychiatry and its "smooth" case histories.

Carl Jung and Depth Psychology

In 1910 the Swiss psychiatrist Eugen Bleuler introduced the term "depth psychology" (*Tiefenpsychologie*) in an article about Freud's psychoanalysis.[20] Four years later Freud used the term in his own *History of the Psychoanalytic Movement*. Depth psychology refers to the way he probed ever deeper into his patients' repressed unconscious, a technique that was particularly evident in dream interpretation. For example, in Freud's own breakthrough dream of 1895, known as "Irma's Injection," he analyzed ever deeper layers of referents behind the image of his patient Irma, who represented herself. However, as his analysis revealed, the position Irma assumed in a chair as he examined her came from another woman he would have preferred to treat. The image of Irma's diphtheritic membrane recalled deep lying anxieties about his oldest daughter, and behind her was someone with his daughter's name who had been poisoned. Irma also associated with a child he had examined in his neurological department, while Irma's reluctance to open her mouth suggested yet another patient as well as Freud's wife. The dream persona of Irma was thus a condensation of numerous individuals and circumstances, and in dreaming about her he attempted to resolve anxieties or fulfill wishes concerning all of them. What is true of this pivotal dream, which he interpreted over sixteen pages, is true of all mental life, and he went on to theorize that every present moment is similarly lengthened with the residue of ever deeper-lying past experiences that ultimately go back to childhood memories, wishes, and impulses.[21]

A variant of Freudian depth psychology was developed by Freud's Swiss disciple Carl Jung, who recalled that when he began his career in psychiatry in 1900 at the Burghölzli Psychiatric Hospital in Zürich, students were taught merely how to describe superficial symptoms, make diagnoses, and compile statistics, while "the psychology of the mental patient played no role whatsoever."[22] Jung soon adopted Freud's psychoanalytic version of depth psychology.[23] In 1907 he visited Freud in Vienna and published the first in-depth psychological study of a psychotic. In 1909 the two men traveled to America to speak at Clark

University on their psychoanalytic method and conversed freely while interpreting one another's dreams. In 1910 Jung became the first president of the International Psychoanalytic Association and played a major role in the psychoanalytic movement from that year to 1913, when he broke with Freud over Freud's insistence on an unqualified acceptance of his sexual theory and the primacy of the Oedipus complex, disagreements that Jung publicized in his study of a schizophrenic patient in *Symbols of Transformation* (1912).

Jung made four major contributions to the new "depth psychology" of the early twentieth century. In contrast to Freud, he probed deeper back in time before the birth of the patient to the primordial ancestral origins of the psyche, elaborated on the deeper realm of those origins that he called the "collective unconscious," investigated new sources of myths, symbols, and archetypes for the contents of the collective unconscious, and interpreted those deeper sources more positively and holistically than primarily as sources of mental illness as he charged Freud with doing.

Freud did not treat psychotics because, he believed, they were too detached from normal human relating to engage in the crucial transference relationship between psychiatrist and patient necessary for psychoanalytic therapy, and their minds were too disorganized to assist in interpreting their dreams and symptoms or reconstructing their past in a clinically useful way. However, Jung did treat psychotics, as he explained in his autobiography. To that end in 1909 he began to study mythology as a key to universal images and thought processes to understand "what really motivated the patient" (*MDR*, 131).

A breakthrough to previously unexplored depths of the psyche and to his rejection of Freud was a dream during their trip to America. In the dream Jung was in a house that represented "a kind of structural diagram of the human psyche" (*MDR*, 161). The upper story furnished in rococo style was the conscious mind. The ground floor, with medieval furnishings, was the personal unconscious with sources from his own past. The cellar furnished from Roman times represented the collective unconscious from the prehistory of his ancestors, and below that level was an even more primitive cave in which he found two skulls. Reluctant to tell Freud what the two skulls symbolized, he lied and said that they represented his wife and sister-in-law, but he in fact believed that they symbolized his impending break with Freud over two theories regarding sexuality and the Oedipus complex. Other findings in the cave he interpreted as "remains of a primitive culture, that is, the world of the primitive man within myself" (*MDR*, 160). That dream, he recalled, "was my first inkling of a collective a priori beneath the personal psyche," which he subsequently identified as a set of primordial archetypes (*MDR*, 161).

Freud viewed dreams as a "royal road to the unconscious" and practiced dream interpretation as a crucial technique in his psychoanalytic therapy. In contrast, Jung interpreted dreams as a path to the collective unconscious, a key to understanding his self and working out his therapeutic method of "individuation." That effort went ever deeper psychologically into the collective unconscious and further back temporally through studies of myths, symbols, archetypes, mandala drawings, and Gnosticism.

In reflecting on myths that hold cultures together, Jung realized that he could not accept the current Christian myth, with its strongly patriarchal focus on Jesus, the heavenly Father, the Pope, and the male church hierarchy and its neglect of any female figures from Christian history such as the Holy Mother, "the Mother church," or the "Church's womb."[24] He found in Gnosticism figures of female authority and a greater sense of "the other face of God," the dark side of the God-image that lay deep in the collective unconscious and helped account for the powerful attraction of images of a threatening as well as beneficent God in modern Christianity.

The major contents of the collective unconscious are the archetypes, sources of recurring images throughout all cultures and all ages of such figures as the mother, the wise old man, the hero, the shadow man, and the demon, but, most importantly and powerfully, "two crepuscular figures of the dark hinterland of the psyche," the anima and the animus.[25]

These paired concepts are among the most significant in Jung's corpus. He believed that their power to shape individuals derives from the antiquity of their origin, the multiplicity of their sources, and their depth in the human psyche. The anima is a composite of the experience of all male and female human beings vis-à-vis all variety of women in their many attributes back to the most distant human origins as sisters, lovers, mothers, grandmothers, great grandmothers, and so on. While Jung rejected any specific Lamarckian mechanism for the hereditary transmission of memory traces from ancestral predecessors, he nevertheless held to the existence of such a transmission so that every contemporary male is somehow born programmed with an anima in his collective unconscious that accounts for his many "female traits." Such an inheritance underlays his personal unconscious of experiences with women. His anima accounts for why his system "is tuned into woman from the start, just as it is prepared for a quite definite world where there is water, light, air, salt, carbohydrates [and] why a man, in his love-choice, is strongly tempted to win the woman who best corresponds to his own unconscious femininity—a woman who, in short, can unhesitatingly receive the projection of his soul."[26] Love relations initiated in such a way are based not on a true understanding of his beloved as an independent other self but on a projection of his own psyche. A more fulfilling love emerges after many years only as the man learns to distinguish his narcissistic projections from genuine caring for the other in a process of "individuation," which can be aided by Jungian therapy. Analogously, the woman is born with an animus, a composite of the experience of her female ancestors vis-à-vis all men, which functions analogously to the anima.

In addition to venturing back in time to the personal, ancestral, and prehistorical beginnings of himself in time, Jung found a way to explore the symbolic center of his existence in space. In 1918 he discovered the meaning and therapeutic value of drawing mandalas. These are Buddhist devotional images elaborated with geometric and organic forms that have a center point that symbolizes the temporal origin and spatial center of the self. To grasp that doubly originary beginning and center is the primary goal of Jungian therapy. When he began drawing mandalas, he recalled, he saw that all the paths he had been taking "were leading back to a single point—namely to the mid-point." The mandala is "the path to the center, to individuation." Drawing mandalas helped him literally center himself. Life does not follow a linear development; rather, it circulates from a temporal beginning and around a foundational center. "There is only a circumambulation of the self," he wrote. "Uniform development exists, at most, only at this beginning; later, everything points toward the center" (*MDR*, 196–97). Drawing mandalas focuses the psyche on the process of individuation that is the goal of a fuller life, enhanced, if possible, under the direction of a Jungian therapist.

While Jung's break with Freud liberated him to explore temporally deeper sources of the psyche, both men spearheaded innovative experiments in depth psychology distinctive to the modernist period that went on to include a wide spectrum of depth psychologists including Karl Abraham, Alfred Adler, Otto Rank, and Anna Freud. In contrast to Victorian psychiatrists who as Jung reported were taught primarily to describe superficial symptoms, make diagnoses, and compile statistics, Freud and Jung rivaled one another in probing ever deeper into the recesses of the psyche. While Freud sampled theorizing with organic memory of ancestral experience, he ultimately held back and remained concerned with personal memory and the personal unconscious alone. Jung on the other hand threw himself into such theorizing to understand his own and his patients' powerful responses to recurring archetypical images in myths and symbols. While Jung's break with Freud was at times bitter, he retained an abiding respect for his teacher. His autobiography credits Freud with vitally important destructive and creative probes of psychological depths. "Like an Old Testament prophet, [Freud] undertook to overthrow false gods, to rip the veils away from a mass of dishonesties and hypocrisies, mercilessly exposing the rottenness of the contemporary psyche," but he also "reached a deeper understanding of mental illness than had hitherto been possible" (*MDR*, 168–69).

For all the deep probing of psychoanalytic inquiry, Freud remained fundamentally modest in his conclusions, and his work clearly aligns with the specificity–uncertainty dialectic. While he was at times personally domineering among colleagues, tyrannically demanding theoretical loyalty to psychoanalysis, he viewed his own studies, even the long case histories, as "preliminary communications," always incomplete and subject to revision. He held that all mental events in dreams as well as waking life generally are "overdetermined," meaning that there are always far more sufficient causes in play than are necessary for any dream image to occur, so all dream interpretations, even those most detailed, are incomplete, as is every psychoanalysis. The French phenomenologist Paul Ricoeur called Freud's thought a "hermeneutics of suspicion" that interrogated surface beliefs and sought to demystify them to disclose deeply hidden, powerfully repressed memories, wishes, and impulses that govern mental life.[27] As Freud dug ever deeper into his patients' unconscious, ever deeper gaping holes became increasingly evident.

Notes

1. On this episode see Peter Gay, *Freud: A Life for Our Time* (Norton, 1988), 63–69.
2. Sigmund Freud, *Standard Edition of the Complete Psychological Works of Sigmund Freud* (Hogarth, 1957), vol. II, 288. Subsequent references to this edition are hereafter referred to as *SE* followed by volume and page number.
3. Sigmund Freud, "On Beginning the Treatment," *SE* XII, 127.
4. Sigmund Freud, "Screen Memories" [1899] *SE* III, 301.
5. Anna Freud, *The Ego and the Mechanisms of Defense* (Hogarth, 1936).
6. These case histories included Carl Jung's "On the Psychology and Psychopathology of So-Called Occult Phenomena" (1902); Théodore Flournoy's *From India the Planet Mars* (1904), a 400-page case history of multiple personality; and Pierre Janet's "*L'amnésie et la dissociation des souvenirs par l'émotion*," *Journal de psychologie* 1, 5 (1904): 417–453. See Anne Sealey, "The Strange Case of the Freudian Case History: The Role of

the Long Case History in the Development of Psychoanalysis," *History of the Human Sciences* 24 (2011): 36–50.

7. Ernest Jones, *The Life and Work of Sigmund Freud* (Basic Books, 1953), vol. 1, 319.
8. "Under the influence of analysis my cardiac problems are now very frequently replaced by gastrointestinal problems." Letter to Wilhelm Fliess, 31 October 1897. In Jeffrey Moussaieff Masson ed., *The Complete Letters of Sigmund Freud to Wilhelm Fliess 1887–1904* (Harvard University Press, 1985), 276.
9. Didier Anzieu, *Freud's Self-Analysis*, trans. Peter Graham (International Universities Press, 1986), 231, 232.
10. Samuel Slipp, "Freud's Two Mothers" in *The Freudian Mystique: Freud, Women, and Feminism* (New York University Press, 1993), 66–70, 78–79.
11. Masson, *Complete Letters*, 268–69.
12. Ibid., 271–72.
13. For this interpretation see Marianne Krüll, *Freud and His Father* (Norton, 1986), 120.
14. Footnote added in 1924 to Freud, *Psychopathology of Everyday Life* (*SE* VI, 70–73).
15. Sealey, "Strange Case of the Freudian Case History," 36–50.
16. Mark S. Micale, *The Mind of Modernism: Medicine, Psychology, and the Cultural Arts in Europe and America, 1880–1940* (Stanford University Press, 2004), 6.
17. Sigmund Freud, "Sexuality in the Aetiology of Neurosis" [1898] (*SE* III, 280).
18. John Kucich, *Repression in Victorian Fiction: Charlotte Brontë, George Eliot, Charles Dickens* (University of California Press 1987), 1, 2.
19. Michel Foucault, *The History of Sexuality, Volume I: An Introduction* [1976] trans. Robert Hurley (Pantheon, 1978), 23.
20. Eugen Bleuler introduced the term *Tiefenpsychologie* in an article of 1910 on Freud's psychoanalysis that appeared in their jointly edited journal, *Jahrbuch für psychoanalytische und psychopathologische Forschung*. Henri Ellenberger, *The Discovery of the Unconscious: The History and Evolution of Dynamic Psychiatry* (Basic, 1970), 562, n308.
21. Sigmund Freud, "The Interpretation of Dreams" (*SE* VI, 106–21).
22. C. G. Jung, *Memories, Dreams, Reflections* [1961] trans. Richard and Clara Winston (Vintage, 1989), 114. Subsequent text references are to this edition.
23. On the history of depth psychology and Jung's role in it see Ira Progoff, *Depth Psychology and Modern Man* (McGraw-Hill, 1959); Dieter Wyss, *Depth Psychology: A Critical History* (Norton, 1966).
24. C. G. Jung, "The Archetypes of the Collective Unconscious" [3 ed., 1926] in *Two Essays on Analytical Psychology* (Meridian, 1956), 115.
25. C. G. Jung, "Anima and Animus" in Jung, *Two Essays on Analytical Psychology*, 221.
26. Ibid., 199.
27. Paul Ricoeur, *Freud and Philosophy: An Essay on Interpretation* (Yale University Press, 1970), 32–36.

Chapter 7

PSYCHIATRY: Origins of Mental Illness and Crime

While the previous chapter dealt with the origins and foundations of normal mental life, this one deals with abnormal mental life, personal mental illnesses, or public disorders and crime. My first four chapters on the physical sciences dealt with cultural developments that are devoid of moral judgments, but in the social sciences that distinction becomes less clear. As causes of behavioral disorders, Victorians included moral failings such as drunkenness, debauchery, dissipation, vicious habits, lustful excesses, indolence, and sloth.[1] English-speaking professionals who treated the mentally ill were referred to as alienists who proposed a wide range of moral causes of mental illness that included luxurious habits, gluttony, and "terrifying tales in early life."[2] In 1838 William Mosley listed thirty-one "moral causes" of crime including gambling and the sight of a public execution. In 1854 the French alienist J.-P. Falret argued that "madness is more often engendered by moral causes than by physical causes," although his moral causes were not actual causes but rather outward signs of socially "deviant" behavior.[3]

American jurists judged murderers who exhibited no clear motives such as greed or revenge as acting as out of "moral insanity," which for some killers was renamed "moral imbecility" or "moral monstrosity."[4] Moral insanity was popularized in France by Jean-Étienne Esquirol and in Germany by Johann Heinroth, who gave it a religious connotation in arguing that its cause was sin. In popular culture the morally insane were thought to be possessed by the devil.[5] Possible effects of this disorder included any antisocial behavior, which made moral insanity a catch-all diagnosis for baffled experts trying to explain, circularly, why the "morally alienated" committed immoral acts.

A shift in the analysis of crime took off in the late nineteenth century as experts rejected moral diagnoses. In England, moral judgments gave way to naturalistic explanations in terms of cause and effect, evidence of what Martin Wiener tags "the de-moralizing of criminality."

The interpretation of behavior "as good or evil was replaced by analyses of constitutions and dispositions that differed from the normal only in degree."[6] After the turn of the century, psychiatrists began to substitute for sweeping moralistic social judgments more individualized causal analyses such as those pioneered by Freud.

Victorian Explanations of Murder

Victorian explanations of the mental life of murderers derived from investigatory practices such as craniology (Franz Gall, 1807) and phrenology (Johann Spurzheim, 1819) and diagnostic concepts including the born criminal (Cesare Lombroso, 1876), the criminal brain (Moritz Benedikt, 1879), and moral epilepsy (Edward Mann, 1883), terms that disappeared from standard use in modernist psychiatry.[7]

In the early nineteenth century, cerebral localization was advanced by the Austrian physiologist Franz Gall, who identified regions of the brain that he believed were the seat of mental faculties. He paired local brain anatomy with faculty psychology to develop a theory of the mind that in 1807 he called *craniology*. It held that the mind was divided into twenty-seven faculties whose function depended on the organization of the cerebral cortex in specific "organs." The amount of the faculty that each organ contained varied with its size, which was also reflected in the shape of the skull, making it possible to infer an individual's faculties by viewing bumps on the skull. He proposed diagrams of these organs as diagrams of the mind. Whereas Victorian anthropologists attempted to analyze inner character from observing superficial skin color, Victorian craniologists or, as they were subsequently called, phrenologists attempted to analyze inner character from observing superficial head bumps. Those paired developments illustrate the extent to which Victorian studies of the mind were superficial in comparison with deeper probing modernist studies.

Gall's craniology mixed "instincts, propensities, and talents" with "moral and intellectual dispositions" including such vague mental capacities as "firmness of purpose" and "veneration of duty." His four-volume treatise *On the Function of the Brain* included seventy pages on the supposed faculty of "Destructiveness—Carnivorous Instinct; Disposition to Murder," located on the surface of the cerebral cortex next to the ears.[8] To document this location he examined the heads of around 500 inmates from prisons and asylums. The theory was based on faulty analogical theorizing that the organ of destructiveness was inside the skull next to the ears because this area was bigger in carnivores than in herbivores and, even more dubious, was observed in a sadistic tyrant who delighted in watching executions and an apothecary who became a public executioner.[9]

Gall performed skillful dissections of the brain to identify medullary structures, the optic tracts, and the nuclei of cranial nerves. He attempted to connect local cerebral anatomy with associationist psychology theoretically but never managed to do so in practice. He admitted being unable "to circumscribe exactly the extent of each organ," none of which conformed to the convolutions of the brain.[10] His location of the faculty of destructiveness was a lucky guess in that later neuroscientists did have moderate success in reducing aggressive behavior by destroying areas of the amygdala and the hippocampus which are within the region of the brain that Gall designated as the organ of destructiveness.

Gall's craniology became increasingly popular after Johann Spurzheim renamed the discipline *phrenology* and took the practice on the road.[11] People were willing to believe most anything however imprecise or downright fabricated in order to fathom the mind's mysteries. Phrenology was popularized by the Scottish philosopher George Combe, who emphasized its ability to reveal complex inner secrets of the mind with easy-to-understand visual inspections. By 1832 there were twenty-nine phrenological societies in Great Britain, with Combe drawing over a thousand listeners to some lectures. During its heyday there were hundreds of public lecturers on phrenology and thousands of publications on the subject, while phrenological heads adorned physicians' offices. Their precise lines on ceramic models of the skull, showing the exact location of supposedly distinct faculties such as mirthfulness and musicality, are graphic reminders of how nineteenth-century explanations of human behavior relied on superficial evidence such as bumps on the head.

By 1840 phrenology was discredited in scientific circles, although along with physiognomy, an analogous system for reading character in the face, it continued to influence popular thinking. By the 1870s popularizers were illustrating phrenological charts of mental faculties with pictures for each organ such as the image of a tiger attacking a lamb for the faculty of destructiveness. As Roger Cooter recorded in a study of 1884, "phrenology in the second half of the nineteenth century became in many ways more deeply entrenched than ever in everyday thought and expression" even as it fell out of favor among more serious researchers.[12]

That historical claim is supported by further evidence from the pioneering Italian criminologist Cesare Lombroso, whose theory of "the born criminal" with a "criminal brain" revived the reputation of phrenology and physiognomy in the new specialty of forensic psychiatry. Lombroso was inspired by Comte's positivism and used empirical observation and measurement, which led to his work and that of his followers being tagged the "Positivist School of Criminology." He was also influenced by evolutionary theory, from which he added the idea that criminality was inherited as an atavism from savage ancestors. As an Italian army physician in the early 1860s he measured the bodies and heads of 3,000 soldiers to find patterns of anatomical differences. That research directed his attention to soldiers' obscene tattoos, which he correlated with crime, and to mental patients who provided more evidence for his work in forensic psychiatry.[13]

In 1866 Lombroso made an apparent breakthrough while studying the brain of the executed murderer Vilella. "At the sight of that skull," he recalled, "I seemed to see all of a sudden . . . the problem of the nature of the criminal – an atavistic being who reproduces in his person the ferocious instincts of primitive humanity and the inferior animals." This insight was followed by a huge leap to an explanation: "[T]hus were explained anatomically the [born criminal's] enormous jaws, high cheek-bones, prominent superciliary arches . . . excessive idleness, love of orgies, and irresistible craving for evil for its own sake, the desire not only to extinguish life in the victim, but to mutilate the corpse, tear its flesh, and drink its blood." Vilella's brain exhibited two anomalies that Lombroso generalized into the brain of the criminal type – a depression in the brain that he named the *median occipital fossa* and a vermis so enlarged that it resembled the middle cerebellum in birds.[14] He followed that insight with extensive research on the brains of criminals, leading to his publication in 1876 of *Criminal Man*, the grounding text for the new discipline of "criminal anthropology."

The idea of a criminal brain was "confirmed" by a flood of studies following Lombroso.[15] In 1879 a French physician reported that the brains of thirty-six guillotined assassins shared pathological cerebral lesions.[16] That same year the Austrian physician Moritz Benedikt reported that beheaded criminals revealed a deficiency of brain matter and a confluence of three major fissures (Rolando's, third frontal, and parietal), which tend to unite with the Sylvian fissure.[17] Numerous other Victorian anatomists supported Benedikt's specificity about the "confluent-fissure type," as it came to be known.[18]

Modernist forensic psychiatrists rejected Lombroso and his positivist followers. The scholarly stake in the heart of Lombroso's criminal anthropology was Charles Goring's *The English Convict: A Statistical Study* (1913).[19] This massive analysis of 3,000 English convicts, begun in 1902 and lasting six years, was based on statistical calculations by a team of researchers, and it took another five years to write. In the opening sentence Goring indicts Lombroso's imprecision: "The recent application of exact and standardized methods to the study of anthropology has revealed the extent to which this science has been dominated and confused by conventional prejudices and unfounded beliefs." Lombroso's science was based on the notion that inward dispositions are revealed by outward bodily configurations. That invalid assumption is the foundation for "the misnamed 'sciences' of phrenology, chiromancy, and physiognomy." Goring challenged Lombroso's sweeping generalizations based on the Vilella epiphany, his erratic definition of concepts such as criminality, his confusion of legal and biological terms, his inadequate controls (no comparing criminals with the general population of the same economic and social class), his impressionistic method of gathering data, his confusion of correlation with causation such as tattoos with criminality, and his deficient statistical skills.[20]

The development of more sophisticated statistical techniques in the modernist period revolutionized social science. Statisticians sharpened understanding of origins and foundations by identifying margins of error and discriminating more precisely effective causal variables. In so doing they helped sort out genuine causality as a statistical probability from fanciful absolute causality based on anecdotal data and random correlations. Sophisticated statistical theorizing is testable by other researchers and can be used as a basis for prediction, neither of which could be done successfully with Lombroso's theory of the born criminal.

Monomania and Split Personality

In the nineteenth century, the mental disorders most frequently invoked to explain murder were monomania and split personality. Monomania was named by Esquirol around 1808, and by 1830 it was one of the most frequent diagnoses at French mental hospitals.[21] Its popularity coincided with the authority of physical causes, which for monomania Esquirol theorized to be something that no anatomist could localize – "lesions of the will." In 1819 Esquirol classified several monomanias by their ideational obsession as in *theomania* (religious) and *erotomania* (sexual).[22] A pamphlet of 1825 by his student Etienne-Jean Georget named another variant, *homicidal monomania*, to explain cases in which someone suddenly killed in a determined way but for no apparent reason.[23] Like other monomanias, homicidal monomania could be caused by a lesion of the intellect that was responsible for the obsessional idea, but it could also be caused by a mysterious "lesion of the will" that made it impossible to resist the impulse. In the following few years psychiatrists debated its nature and function, although their theories were based on case histories that did not specify what caused a lesion of the intellect or will.

Alienists were also unable to explain how homicidal monomania led to a murder beyond reference to the murderer's delusion that the victim was a threat or that some voice ordered the murder. The critical focus of such speculation was the degree of criminal intent involved in murders. Such speculation was the focus of the new professional specialty of forensic psychiatry. Georget's 1825 pamphlet defended the new profession, which he proposed should be staffed by new *médecins des aliénés*, who would play a major role in murder trials. The following year an alienist used the diagnosis of homicidal monomania to defend a murderess who had suddenly and inexplicably killed a neighbor's child. His testimony was the first time this diagnosis had been used in a criminal case.[24]

Homicidal monomania was a confusing concept. It had no clear cause, it referred to a great many symptoms and disease patterns, it caused many possible effects, and accounts of them centered on a metaphor – a lesion of the will. This confusion was magnified by alternate terms for this disorder: melancholic homicide, homicidal mania, homicidal insanity, instinctive monomania, moral epilepsy, and, beginning in the late 1880s, paranoia.[25] The term nevertheless was widely used by forensic psychiatrists because the scientific term gave a gloss of professional expertise to their subject and because it seemed to make murders more easily comprehensible with several features: a unified personality, a single explanation, and a simple label.

The increasing specificity of psychiatric and jurisprudential analyses of the causal role of mental illness in murder drew from increasingly probing analyses of ideas about character, identity, free will, the guilty mind, and criminal responsibility. Victorian thinkers believed that the seat of the self or soul was in the brain or mind of human beings and operated like the pilot in a ship as it guided a person through life. Character was made up of a set of specific faculties, which, if not anatomically localizable, were present in all human beings as variable strengths or weaknesses. Personal identity evolved additively in time as experiences were registered in consciousness, recorded in memory, and integrated into the personality. While subject to external pressures, individuals were nevertheless internally coherent agents with the capacity to choose courses of action for which they were fully responsible. As such they were moral agents. If they killed someone with "malice aforethought" and at the time of the killing were aware that what they are doing was wrong, they had committed murder, for which they were responsible and had to be punished, no matter how emotionally distraught their mind might have been.

This view of a unified and autonomous self was the subject of broad questioning in the modernist period. Modernist thinkers concurred that the mind is centered in the brain, but they became increasingly aware of new causal factors: interactions between the brain, hormones, and neurotransmitters as well as early childhood traumas, sexual drives, unconscious mental processes, social circumstances, and historical forces. Personal identity evolved in time not additively as a sum of memories but psychodynamically as the outcome of traumas and crises involving challenges to the autonomy of a self that appeared to be increasingly fragmented and uncertain. Individuals were not simply free to choose between good and evil but buffeted about in a whirlwind of existentialist freedom, struggling to find a normative basis for action as the moral-religious basis lost its authority. The Victorian image of a moral agent struggling before social pressures and the witness of God to resist evil and embrace goodness was for the moderns a quaint psychological antique that involved self-deception and screened the multifarious choices that modernists faced alone and without any reliable moral code or divine witness. Evidence of this cultural historical shift was vividly dramatized in novels.

Murder Novels

In melodramatic Victorian novels, moral and intelligent men had high brows, while criminals had low ones, although more sophisticated Victorian writers, acutely aware of the complexity of mental life, were wary of such simplistic symbolism and invoked phrenology and physiognomy as decorative flourishes more than serious explanations. Balzac referenced phrenology to make his novels current with popular thinking but did not rely on the shape of skulls to identify or explain character. In *Old Goriot* (1835) a medical student jokes about a "paternity bump" to explain why Goriot is so devoted to his daughters, and after Vautrin's wig is knocked off "everyone present instantly understood the manner of man Vautrin was." Phrenology influenced many other writers including Poe, Melville, Baudelaire, Whitman, Emerson, Hawthorne, Twain, and Charlotte Brontë.

Victorian novelists also freely drew on physiognomy, albeit with measured credulity. In Elizabeth Gaskell's *Mary Barton* (1848) the public prosecutor explains to the court "I am no physiognomist," but then he reads a murderous character into the face of the defendant Jem Wilson with his "low, resolute brow, his downcast eye, and his white compressed lips," concluding, "I have seen a good number of murderers in my day, but I have seldom seen one with such marks of Cain on his countenance as the man at the bar."[26] Graeme Tytler, who documents dozens of other European novelists influenced by physiognomy, concludes that the theory was most often invoked to engage popular thinking rather than explain a character, and it was typically used ironically. These writers include Dickens, Eliot, Flaubert, Fontane, Hugo, Maupassant, Meredith, Radcliffe, Scott, Stifter, and Tolstoy.[27]

Serious Victorian novelists produced richly layered dramatizations implying a medley of factors that shape behavior including economic and social pressures, moral and religious ideas, the ambient setting, and divine providence. Still, in their novels criminals are more likely than they are in modernist novels to have oddly shaped skulls, low foreheads, protruding jaws, and shifty eyes, and those phrenological and physiognomic characteristics are believed to at least hint at something about the mind within.

Late-Victorian crime novelists still found Lombroso intriguing. Zola drew support for his theory of the hereditary origins of the Rougon-Macquart family's criminal propensities in Lombroso's hereditarian theory of the born criminal with tell-tale atavisms in the form of cranial bumps and facial stigmata. In Zola's *The Human Beast* (1890), signs of degeneration are evident in brutal or insane types. Roubaud is a walking checklist of Lombrosian criminal stigmata, with a flat head, low forehead, and eyebrows that meet over his nose. The maniacal killer Cabuche has a low brow and prominent jaw.[28]

The icon of a Lombrosian-styled born criminal is the protagonist in Bram Stoker's *Dracula* (1897), whose criminal propensities are very much in the blood. As Minna Harker tells Professor Van Helsing, the Count is a criminal type. "Nordau and Lombroso would so classify him, and *qua* criminal he is of imperfectly formed mind." Lombroso's characterization of the criminal included an aquiline nose like a bird of prey, bushy eyebrows that meet over the nose, and pointed ears. As Jonathan Harker notes, Dracula's face is "aquiline, with the high bridge of a thin nose and peculiarly arched nostrils." His eyebrows are "massive, almost meeting over the nose," and the tops of his ears are "extremely pointed."[29] The Count's main behavioral characteristic comes straight out of Lombroso's epiphany over Vilella's brain, which, Lombroso believed, evinced the murderer's destiny "to mutilate the corpse . . . and drink its blood." The broad impact of this theory in the late nineteenth century, as Daniel Pick argues, came from an effort to "represent, externalize and kill off a distinct constellation of contemporary fears," namely that the moral fiber and spiritual vigor of the Western world was degenerating.[30] Stoker adapted Lombroso to externalize and make manageable those fears in the form of a fictional monster who in the end, as he reassured his readers, was thoroughly finished off by being stabbed through the heart as well as beheaded.

In *The Secret Agent* (1907), Conrad used Lombroso to craft the appearance of criminalistic characters and suggest a reason for their behavior. One of those characters is the mentally disabled Stevie, who has "a deep hollow at the base of his skull" and "degenerate" ear lobes – Lombrosian signs of criminality. But this novel also marks the waning of Lombroso's influence, because by the end of the story Conrad impugns Lombroso's credibility. The anarchist Ossipan is a fake doctor who espouses Lombroso with "nervous, jerky phrases" to dupe audiences. Another anarchist character, Karl Yundt, thinks that Lombroso was an "ass" who "made his way in this world of gorged fools by looking at the ears and teeth of a lot of poor, luckless devils."[31]

Monomania in Novels

In Dickens's *Our Mutual Friend* (1865), Eugene Wrayburn dismisses Bradley Headstone as "a curious monomaniac."[32] In Mary Elizabeth Braddon's *Lady Audley's Secret* (1862), Audley diagnoses her pursuer with monomania in a thinly disguised confession of her own homicidal madness. In this condition, she explains to her husband, "the mind becomes stationary; the brain stagnates . . . the thinking power of the brain resolves itself into a monotone . . . and perpetual reflection upon one subject resolves itself into monomania."[33] Audley's explanation concentrates on the obsessive idea and offers no account of the deeper characterological structures and developmental experiences that make someone vulnerable to obsession. In *Crime and Punishment* (1866), Dostoevsky tags Raskolnikov as a monomaniac, as does a physician in the story, Dr Zosimov, whose diagnosis includes "a certain fixed idea, some indication of monomania."[34]

The classic Victorian monomaniac is Captain Ahab from *Moby-Dick* (1851). A dozen times Melville refers to him as such, for example by "his narrow-flowing monomania," "his monomaniac revenge," "his monomaniac mind."[35] The ultimate source of Ahab's monomania is not the loss of his leg but rather "the direct issue of some former woe." Yet, like alienists of that time, Melville offers no details about that woe, and of course in his time a determinative childhood trauma would have been absurdly out of scale with the immensity of Ahab's tragic character and his gargantuan nemesis. A specific etiology is, for Melville, unknowable because, as he explains, "to trail the genealogies of these high mortal and different feelings are attached to every one of them. The human brain has happily split things apart, [and] what everyone of us needs today more than anything else is simplicity" (I, 65–66).

Musil adapts the language of phenomenology, which he learned as a student of Husserl, to describe a gruesome murder of a prostitute through the mind of the murderer Moosbrugger. Panicked because his victim would not go away and confused about the border between himself and her, he stabs her over thirty times as if to cut her out of himself in a desperate act of self-definition, which Musil reconstructs with a detailed description of the murderer's consciousness from the moment he tries to escape her clutches.

When the court attempts to make sense of the killing, Musil's phenomenology of the murder becomes evident. The judge's questions such as "Why did you wipe the blood off your hands?" and "Why did you throw the knife away?" are intended to establish knowledge of right and wrong as a basis for conviction, but, while the judge sees those actions as a coherent whole, Moosbrugger experienced them as "a series of completely separate incidents having nothing to do with one another, each of which had a different cause that lay outside him." In the judge's eyes Moosbrugger was the source of those acts for which he is responsible, while "in Moosbrugger's eyes they had perched on him like birds that had flown in from somewhere or other." Confused and incapable of expressing "the strange, shadowy reasonings of his mind," he was found guilty and sentenced to death, although by the end of the novel he is still in prison (I, 75).

In the following weeks Ulrich becomes obsessed with this case. The clarity and finality of the guilty judgment and death sentence contrast with Ulrich's sense of the arbitrariness of all value judgments. To him everything seems scattered and blurred. "Nothing in him was moved by cause, purpose, or physical desire, but everything went rippling out in circle after ever-renewed circle" (I, 131). The ready-made distinctions of innocence and guilt become incomprehensible and unreliable as Ulrich contemplates Moosbrugger's fate, and it flashes on him that "one could just as easily devour people as build monuments" (I, 136). Without belief in God, there is no sufficient reason why things turn out as they do. Musil narrates: "This uncertainty gave Ulrich's personal problem a broader context. In earlier times one had an easier conscience about being a person than one does today." Nowadays personal autonomy and private experiences are swamped by impersonal forces and public symbols.

This confusion of self and other is exaggerated in Moosbrugger, who does not know whether his frightening voices and visions come from within or outside of himself. Judgments by the prosecutor, court, press, and gossips are worlds away from what he experiences from within, and Musil uses that contrast to indict the Austrian judicial system and reigning ideas about personal identity, free will, morality, and criminal responsibility. These criticisms stem from Ulrich's philosophy of "living hypothetically," which he does to avoid committing to any given order, because for him "no thing, no self, no form, no principle is safe, everything is undergoing an invisible but ceaseless transformation" (I, 269).

Musil tags this philosophy "essayism," which means living experimentally by exploring life from many sides. Like James's pragmatism and Nietzsche's perspectivism, this philosophy rejects absolute truths and single points of view and embraces partial, perspectival truths. It views causal understanding as probabilistic and uncertain. "The course of history was therefore not that of a billiard ball – which, once it is hit, takes a definite line – but resembles the movement of clouds, or the path of a man sauntering through the streets, turned aside by a shadow here, a crowd there, an unusual architectural outcrop, until at last he arrives at a place he never knew or meant to go to" (I, 396). Human actions must be experienced in many ways with understanding depending on shifting circumstances. "All moral events take place in a field of energy whose constellation charges them with meaning" (I, 270). Morality is not grounded absolutely in Judeo-Christian commandments but is a provisional and unstable system of values that shift with social and historical developments.

The originary qualities that Ulrich are without are prefabricated conventions that have nothing to do with him. They pressure individuals analogous to the way Heidegger's Dasein is shaped by "what one does" ("the 'They'").[45] Ulrich is free of them but thereby condemned to fashion a personal identity that is more fully his own. In that sense Ulrich lines up with a host of modernist heroes who achieve the higher insight that a psychoanalyzed person might achieve, or a level of individuation that a patient of Jung might reach, or a degree of authenticity that Heidegger would sanction. As a friend notes about Ulrich, "[h]e is gifted, strong-willed, open-minded, fearless, tenacious, dashing, circumspect—why quibble, suppose we grant him all those qualities—yet he has none of them . . . What he thinks of anything will always depend on some possible context—nothing is, to him, what it is: everything is subject to change" (I, 63–64). As a man without conventional qualities, Ulrich finds that the conventional criminal justice system, based on knowledge of Christian qualities of good and evil, cannot do justice to the slipperiness and complexity of the mind in the modern world, even in its most obvious criminal deviations.

Across the years of this study understanding of the origins and foundations of human behavior, especially when sharply focused regarding murder, shifted from Victorians' reliance on narrowly conceived concepts such as monomania, the split personality, and the born criminal in the direction of an appreciation of traumatic childhood origins, complexities of psychosexual development, and unconscious mental processes. Musil adds to these concepts the idea that individuals do not develop according to some identifiable teleological order as a tree grows from a seed but from historically unique circumstances "From Which, Remarkably Enough, Nothing Develops."

In the opening paragraphs Musil elaborates on this cryptic chapter title with a description of the contending rhythms of life in a modern city that summarizes modern life from trivial everyday happenings to a brutal murder.

> Like all big cities it was made up of irregularity, change, forward spurts, failures to keep step, collisions of objects and interests, punctuated by unfathomable silences made up of pathways and untrodden ways, of one great rhythmic beat as well as the chronic discord and mutual displacement of all its contending rhythms. All in all, it was like a boiling bubble inside a pot made of the durable stuff of buildings, laws, regulations, and historical traditions. (I, 4)

Musil strains to craft a hero who is without conventional qualities and is yoked to an insane misfit who kills without any conventional motive and is in turn judged by a system that is narrowly oriented and unjust. The two men live in a city that is without a single rhythm in a country that is on the verge of a great war that will destroy the monarchy along with its religious justification by virtue of the doctrine of the divine right of kings that had held it together.

Notes

1. These causes were culled from Robley Dunglison ed., *Cyclopedia of Practical Medicine*, American Edition, 4 vols (Philadelphia, 1845) by K. Codell Carter in *The Rise of Causal Concepts of Disease* (Ashgate, 2003), 20.
2. These causes are from British psychiatric treatises from 1824 and 1853, excerpted in Vieda Skultans, *Madness and Morals: Ideas on Insanity in the Nineteenth Century* (Routledge, 1975), 33, 55.
3. Quoted in Robert Castel, "Moral Treatment: Mental Therapy and Social Control in the Nineteenth Century," in Stanley Cohen and Andrew Skull eds., *Social Control and the State: Historical and Comparative Essays* (Oxford University Press, 1983), 265, n12.

4. This pair of diagnoses emerged at the 1881 trial of the man who assassinated President James Garfield when one expert in support of his insanity defense explained that "some authors call that moral insanity which I term moral imbecility or moral monstrosity." Edward D. Spitzka, "Review of the Trial of Charles J. Guiteau," *American Journal of Insanity* 32 (January 1881): 303–448.
5. Henri Ellenberger, *The Discovery of the Unconscious* (Basic, 1970), 212. Robert Nye documents the mix of medical and moral concepts in nineteenth-century French medicine and psychiatry. *Crime, Madness and Politics in Modern France: The Medical Concept of National Decline* (Princeton University Press, 1984), 63, 229.
6. Martin J. Wiener, "The De-moralizing of Criminality," in *Reconstructing the Criminal: Culture, Law, and Policy in England, 1830–1914* (Cambridge University Press, 1990), 215–56, 253.
7. Henry Maudsley, *Responsibility in Mental Disease* (H. S. King & Co., 1874), 413; Edward C. Mann, *A Manual of Psychological Medicine and Allied Nervous Diseases* (J. & A. Churchill, 1883), 120.
8. F. J. Gall and J. Spurzheim, *Anatomie et physiologie du système nerveux en general et du cerveau en particulier* (chez F. Schoell, 1810–19), vol 4: 50–119.
9. Robert M. Young, *Mind, Brain, and Adaptation in the Nineteenth Century: Cerebral Localization and Its Biological Context from Gall to Ferrier* (Oxford University Press, 1970), 28, 24.
10. Ibid.
11. *Phrenology* was first used by Benjamin Rush in 1805, and then in 1815 in an article by Thomas Forster on Gall and Spurzheim. Spurzheim claimed the term as his own, and Gall refused to use it. Edwin Clarke and L. S. Jacyna, *Nineteenth-Century Origins of Neuroscientific Concepts* (University of California Press, 1987), 222–23.
12. Roger Cooter, *The Cultural Meaning of Popular Science: Phrenology and the Organization of Consent in Nineteenth-Century Britain* (Cambridge University Press, 1984), 258. Cooter reproduces such a head from 1878 and lists Victorian thinkers influenced by phrenology: Auguste Comte, Henri de Saint-Simon, John Stuart Mill, Alfred Russell Wallace, Herbert Spencer, and Paul Broca (268, 7).
13. Cesare Lombroso, *La medicina legale nelle alienazioni mentali studiata col metodo sperimentale* (Stabilimento P. Prosperini, 1865).
14. Cesare Lombroso, Introduction by Gina Lombroso-Ferrero, *Criminal Man According to the Classification of Cesare Lombroso* [1911] (G. P. Putnam's Sons, 1972), xxiv–xxv.
15. In the 1860s the British researchers James Bruce Thompson, George Wilson, and David Nicholson published on distinctive mental characteristics and cerebral anomalies of criminals. See C. H. S. Jaywardene, "The English Precursors of Lombroso," *British Journal of Criminology* 4 (1963): 164–70.
16. A. Bordier, Étude anthropologique sur une série de cranes d'assassins," *Revue d'anthropologie* (1879): 264–300.
17. Moritz Benedikt, *Anatomical Studies upon Brains of Criminals: A Contribution to Anthropology, Medicine, Jurisprudence, and Psychology* (Wm. Wood & Company, 1881), 157.

18. E. P. Fowler, "Are Brains of Criminals Anatomical Perversions," *The Medical-Chirurgical Quarterly* 1 (October 1880): 1–32; William Osler, "On the Brains of Criminals," *Medical & Surgical Journal* 10 (1882): 385–98; James Weir, "Criminal Anthropology," *Medical Record* 45 (January 1894): 42–45.
19. Charles Goring, *The English Convict: A Statistical Study* [1913] (Patterson Smith, 1972).
20. Lorenz Krüger, Lorraine J. Daston, and Michael Heidelberger eds., *The Probabilistic Revolution*, 2 vols. (Harvard University Press, 1987). Theodore Porter surveys the discovery of a number of these techniques of modern mathematical statistics beginning in 1893, when Pearson first contacted Galton and took up the study of statistics. "The correlation coefficient was defined mathematically in 1896, contingency analysis and the chi-square test in 1900, the t-test and its distribution defined by W. S. Gosset (Student) in 1908, [and] analysis of variance derives from a paper by Fisher in 1918." *The Rise of Statistical Thinking 1820–1900* (Princeton University Press, 1986), 315. These techniques enabled social scientists to assign more precise magnitudes to the relative causal efficacy of an increasing number of factors, thereby refining but also limiting their explanations.
21. Jan Goldstein, *Console and Classify: The French Psychiatric Profession in the Nineteenth Century* (Cambridge University Press, 1987), 153–55.
22. Esquirol, "Monomania," *Dictionaire des sciences médicale*, 34 (1819): 125.
23. Etienne-Jean Georget, *Examen médicale des procès criminels* (Migneret, 1825).
24. Goldstein, *Console and Classify*, 166–67.
25. Kraepelin and Krafft-Ebing introduced paranoia in the 1880s. Arthur E. Fink, *Causes of Crime: Biological Theories in the United States, 1800–1915* (University of Pennsylvania Press, 1938), 29–30, 54.
26. Elizabeth Gaskell, *Mary Barton* (Penguin, 1985), Chapter 32, 385.
27. Graeme Tytler, *Physiognomy in the European Novel: Faces and Fortunes* (Princeton University Press, 1982).
28. Emile Zola, *La bête humaine*, trans. Leonard Tancock (Penguin, 1977), 22, 123–24.
29. Bram Stoker, *Dracula* (Penguin, 1993), 28.
30. Daniel Pick, *Faces of Degeneration: A European Disorder, c. 1848–1918* (Cambridge University Press, 1989), 167–75.
31. Joseph Conrad, *The Secret Agent* [1907] (Penguin, 1996), 259, 77–78.
32. Charles Dickens, *Our Mutual Friend* (Oxford University Press, 1998), 294.
33. Mary Elizabeth Braddon, *Lady Audley's Secret* (Oxford University Press, 1987), 287.
34. Feodor Dostoevsky, *Crime and Punishment* [1866] trans. Jessie Coulson (Norton Critical Edition, 3rd ed., 1964), 23, 275.
35. Herman Melville, *Moby-Dick* [1851] (Norton, 1867), 161–62, 385.
36. Ibid., 386.
37. Ann Harrington, *Medicine, Mind, and the Double Brain: A Study in Nineteenth-Century Thought* (Princeton University Press, 1987), 248, 100, 105–9.
38. Robert Louis Stevenson, *The Strange Case of Dr. Jekyll and Mr. Hyde* (Dover, 1991), 42.

39. Ibid., 43.
40. André Gide, *Lafcadio's Adventures* (*Les caves du Vatican*, 1914) trans. Dorothy Bussy (Vintage, 1953), 82.
41. *The Journals of André Gide* (Knopf, 1948), vol. 2, 238.
42. Theodore Dreiser, *An American Tragedy* (New American Library, 1981), 442. Further text references are to this edition.
43. Robert Musil, *The Man without Qualities* [1930, 1943] trans. Sophie Wilkins (Knopf, 1995), 65. Further text references are to this two-volume edition. A similar idea was expressed in 1907 by Henry Adams, who surveyed the "sudden eruption of forces totally new" around the turn of the twentieth century including electrical, magnetic, thermal, chemical, osmotic, cohesive, elastic, vibratory, capillary, and sexual forces, an eruption that left him "with his historical neck broken." *The Education of Henry Adams* (Houghton Mifflin, 1973), 382.
44. Robert Musil, "The Polemic against the Concept of Causality; Its Replacement by the Concept of Function," in *On Mach's Theories* [1908]; trans. Kevin Mulligan (The Catholic University of America Press, 1982), 44–56.
45. On this terminology see Chapter 8.

Part III

Humanities

Chapter 8

PHILOSOPHY: Genealogy of Morals in Nietzsche and Foundation of Being in Heidegger

The two key modernist philosophers Friedrich Nietzsche and Martin Heidegger challenged major originary foundations of Victorian philosophy – most prominently Nietzsche's essay on the historical origins of morality, which he reconceived and renamed as a *genealogy* of morality, and Heidegger's deep phenomenological probing into the ontological foundations of human existence and deeper yet into the foundations of Being per se. They both came from strong Christian traditions and in their youth intended religious vocations, but their mature philosophy included strongly negative views of Christian church dogma and its institutions. Their apostasy and skepticism led not to nihilism but to a fuller appreciation of the challenges and possibilities of human existence, hence their roles as founders of modern existential philosophy.

Friedrich Nietzsche on the Genealogy of Morality

Nietzsche's reputation was built on a powerful positive philosophy that centered on inspiring his disciples to make their lives more fulfilling. Equally influential was his negative philosophy. He worked like a surgeon lancing boils and removing sick tissue to foster healthy existential growth. He specifically targeted Victorian ideas about the origins of morality in psychology, education, justice, government, and selfhood and about foundations of thinking about God, causality, human relations, and history.

His iconoclasm struck across the board of Victorian ideas, values, and institutions. He criticized the long-standing idea of a fundamental pleasure principle motivating human behavior, and instead posited a fundamental motivating "will-to-power," a will to make or create meaning and inspire his followers to love their life so much so that they could will it to recur eternally. He hated newspapers that "vomit their gall" and mocked scholars who "knit the socks of the spirit."[1] He derided celebrations of loving one's neighbors who are conveniently nearby and easy to access and proposed rather loving who or what is far away and difficult to access, the proper stimulus for great human achievements. Although he admired Richard Wagner's family and especially Wagner's wife, Cosima, he dismissed typical marriages as a "poverty of the soul in pair!" (Z, 70). He ridiculed the ideal of equality as a sanction for mass mediocrity and condemned conventional justice as a rationale for revenge. He scorned the "New Idol" of the state that "tells lies in all the languages of good and evil" (Z, 49). No proto-Nazi was he. In place of democracy and socialism that gave authority to the masses, he promoted an existential elitism (not social snobbery) dedicated to cultivation of "the highest specimens," especially in philosophy and the arts. He dramatized this individualistic philosophy in *Thus Spoke Zarathustra* (1883–5), where his spokesperson motivates select followers to avoid the values of "the herd" and the easy comforts of "the last man" and rather achieve the individual greatness of an "overman" (*Übermensch*).

His announcement of the death of God in *The Gay Science* (1882) spearheaded an attack on Judeo-Christianity's theology centered on belief in God and a cosmology starting with God's creation of the universe in six days 6,000 years ago. The death of God for the masses who desperately sought to believe in Him brought the potential of a wide-ranging nihilism that swept away a trio of foundations: the ability to know what is true (epistemological nihilism), what is good (ethical nihilism), and the ultimate meaning of life (existential nihilism). The scenario for a meaningful life on the way to becoming an overman is not charted on a fixed timetable or designed for everyone but is a dynamic process that each person must work out individually with only provisional goals.

On the Genealogy of Morality (1877) was Nietzsche's major contribution to a new method of originary moral analysis.[2] Subtitled *A Polemic*, this essay questioned Victorian ideas about first principles of morality – Christianity's Divine Word, William Paley's natural theology, and Lamarck's theory of the hereditary transmission of acquired characteristics. In place of a single, continuous, and purposeful chronological process, Nietzsche sketched a "genealogy" of multiple historical processes beginning at different and uncertain moments, proceeding at different paces through accidental interruptions and reversals like the tributaries of a river that begins all over an irregular landscape. That pluralistic philosophy of history accorded with his pluralistic epistemology of perspectivism, as he explained: "The *more* affects we allow to speak about a matter, the *more* eyes, different eyes, we know how to bring to bear on one and the same matter, that much more complete will be our 'concept' of this matter" (*GM*, 85). His many-sourced genealogy challenged four versions of Victorian morality: utilitarian morality based on maximizing the greatest pleasure for the greatest number of people, Kantian morality based on realizing an inner sense of duty about what is right (a categorical imperative), ethical Darwinism based on adaptation to the environment (later "survival of the fittest"), and Christian morality based on carrying out God's commandments.

He also rejected popular histories of morality as overly metaphysical, singular, continuous, utilitarian, and teleological. The preface announced that he "no longer sought the origin of evil *behind* the world" in some metaphysical foundation (*GM*, 2). Rather he substituted a naturalistic foundation in human nature mediated by human senses and the brain. While he agreed with English utilitarian psychologists' deriving morality from human nature, he accused them of arguing circularly and from the wrong point of view. They summarized current moral values based on usefulness and projected them back to an origin of morality that they then tracked forward to the current time. That circular reasoning obscured their seeing multiple origins of current morality. Victorians frequently viewed the logic of history as analogous to the purposeful growth of plants; he countered that "the cause of the genesis of a thing and its final usefulness, its actual employment and integration into a system of purposes, lie apart *toto caelo* [by the whole extent of the heavens]" (*GM*, 50). Utilitarians defined a single origin of the good as usefulness from the point of view of those who found a "good" action of use to them, while he sought origins of the good as defined by those empowered to make their lives authentic, meaningful, and great.

He found positive evidence for a genealogy of morality in philology, history, psychology, economics, and theology.

From philology and history he found synonyms for *good* among ancient Iranians, Slavs, Greeks, and Romans, whom he fused into a mythic-historical "master race." Their code of behavior centered on a concept of good that indicated nothing morally right, merely their good luck from being powerful, rich, and happy. Subordinate to these masters were mythic-historical slaves who were powerless, poor, and unhappy. Nietzsche's models for these slaves were the Jews of the Babylonian exile, which began around 600 BCE, not the Jews of the United Israelite Kingdom around 1000 BCE.[3] Those ancient masters identified values based on their own privileges as good and defined slaves as bad, not morally deficient but simply unlucky, unable to be as powerful, rich, and happy as their masters. In time the slaves resented their subordination and desired revenge. Led by priests (Nietzsche uses this term for Jewish and Christian religious leaders alike), the Jews managed a "slave revolt in morality" that did not liberate them physically but did so morally as they transvalued the masters' good-versus-bad moral code into a good-versus-evil code with themselves as good and the masters as evil. Their new morality was motivated not by love but by hatred or, more precisely, by *ressentiment*.[4] They carried out that revolt and "held it fast with teeth of the most unfathomable hate" (*GM*, 16). The priests seeking power said that the masters were immoral because they were free to choose how they wielded power and were therefore responsible for their actions. They could choose not to mistreat slaves, so what they did was morally wrong. In this new moral code, the priests called the masters "evil" and called slaves – still powerless, poor, and unhappy – "good." Weakness and poverty came to be celebrated in the Judeo-Christian world: the Old Testament taught that the "meek shall inherit the earth" (Psalms 37:11), and in the *New Testament* Jesus said "It is easier for a camel to go through the eye of a needle, than for a rich man to enter into the kingdom of God" (Matt. 19:24). Put together, both testaments reassured the downtrodden that their own meekness and poverty would triumph over the masters' power and wealth and ultimately be rewarded by God, while the evil masters would "eternally be the wretched, accursed, and damned!" (*GM*, 17).

Drawing on psychology, Nietzsche argued that instinctual aggression was healthily discharged against others. Christianity blocked such discharge with its urging to turn the other cheek (Matt. 5:39), and so man "tore apart, persecuted, gnawed at, stirred up, maltreated himself." Man "became the inventor of 'bad conscience.'" In him "the greatest and most uncanny of sicknesses was introduced, one from which man has not recovered to this day, the suffering of man *from man*, from *himself*" (*GM*, 57).

From economics Nietzsche found theorizing to link the origin of moral guilt with indebtedness. He compared the authority of parents and more remote ancestors to the authority of a creditor over a debtor who agreed to suffer pain, perhaps "a pound of flesh," for failure to pay a debt. As the debt increased, ancestors took on increasing authority and eventually changed into a frightening, punitive God. "The rise of the Christian God as the maximum god that has been attained thus far therefore also brought a maximum of feelings of guilt into appearance on earth" (*GM*, 62). The words for debt and guilt in German are the same – *Schuld*. That polysemy bolstered Nietzsche's argument that increasing indebtedness and guilt became so great that only a God could pay it off with innocent Jesus's suffering and self-sacrifice on the cross. That sacred moment that was supposed to give Christians hope for salvation ironically plunged them into an indebtedness that they had no hope of ever repaying and laid on them a load of guilt that they could never remove because they were born in sin and incarnated with desires that perpetually tempted them to further sins.

The last part of *Genealogy* addresses the question "What Do Ascetic Ideals Mean?" Victorians celebrated positivism and prosperity, Whiggish self-confidence and self-satisfaction, military might and imperial greatness, and a Christian morality based on the seemingly primordial distinction between good and evil. But underlying that bubbling confidence Nietzsche discloses how arbitrary and historically inflected the foundations of that morality were. He also identifies darker sources of meaning, ultimately no meaning at all – the persistent threat of nihilism. To avoid that existential dead end, "ascetic priests" gave meaning to life by encouraging asceticism. They made their followers sick to justify their role as caregivers and in so doing reversed the direction of the original resentment that led to the slave revolt in morals. In place of a resentment against the masters, ascetic priests directed self-hatred toward the slaves themselves, saying in effect, that "you alone are to blame for yourself" (*GM*, 92). Churches became havens for the weak and helpless, and priests became purveyors of asceticism – confessions, fasting, and self-recrimination. "This abhorrence of the senses, of reason itself, this fear of happiness and of beauty, this longing away from all appearance, change, becoming, death, wish, longing itself—all this means—let us dare to grasp this—*a will to nothingness*, an aversion to life" (*GM*, 118).

Yet, he qualifies, "with some fairness one could also add that it was on the soil of this essentially dangerous form of human existence, the priestly form, that man first became an interesting animal, that only here did the human soul acquire depth in a higher sense and become evil" (*GM*, 15–16). He also notes that asceticism derives from the Greek *ascesis*, meaning training or discipline central to the cultivation of great things in sports, in morals, and in spirit as well as in literature and the arts. For all their dehumanizing function, priests can bear a great load, discipline themselves for a long time in a single direction, and make their followers deeper as well as eviler. While Nietzsche ripped up the foundations of Victorian morality, he also made his readers more reflective about those originary foundations. His pushing back through all the twists and turns of his genealogy – drawing on sources from philology, history, psychology, economics, and theology – made his readers conscious of the complex origins of their worldview and increasingly self-aware.

Nietzsche's last book *The Anti-Christ* (1888) offered a nuanced estimate of Jesus, who had an admirable childlike innocence that led him to say *yes* to everything except whatever denied his message. He was admirably willing to die for his ideas, but that devotion came from blind determination as well as faith. Jesus's message was not to impart knowledge but to display his life. His "glad tidings" were not that blessedness is promised in an afterlife but that it is here and now as "the *only* reality." He did not so much preach a religion as live in a way that inspired one, but the Christian Church that came out of his life would not have accorded with his fundamental beliefs. The Judaism that Jesus abolished as the Evangel was "the Judaism of the concepts 'sin,' 'forgiveness of sin,' 'faith,' 'redemption by faith'—the whole Jewish *church doctrine*."[5] Jesus was a free spirit, but not the sort that Nietzsche ultimately revered. Jesus was free of sin; however, that freedom was not earned by reflection and struggle but was endowed on him by God and by being God. He was moral by divine fiat, by default; he could not possibly choose evil or good because he was all good and therefore could not model a way to make difficult moral choices. Nietzsche did provide such a foundation, worked out most fully through a collection of aphorisms poetically expressed in *Thus Spoke Zarathustra* as a journey toward self-fulfillment with no clear end, just some intriguing clues about a process that everyone must work out for themselves.

In assailing the originary foundations of Victorian institutions and values, Nietzsche offered newer foundations – always in the plural and always provisional – for his positive existential philosophy for people to make their lives as fulfilling as possible. In place of routine repetitions at the beginning of prayers that bless God for his greatness and glory, Nietzsche told his followers that they themselves could become greater and more glorious if they tried. The goal of life for Nietzsche was not prosperity and happiness in Victorian society or eternal felicity in the Kingdom of Heaven but an ongoing struggle to make life authentic, meaningful, and great in the here and now wherever they happened to be, a challenging personal journey about which one should be able to affirm with joy a regret-free motto derived from his beloved music – *da capo*, play it all again from the beginning.

Martin Heidegger on the Ontology of Human Existence

While Nietzsche analyzed Victorian philosophy to its rudimentary genealogical origins, Heidegger analyzed human existence to its deep ontological foundations.

Heidegger was born in 1889 and grew up in a home of staunch Catholic believers in the conservative rural atmosphere of Messkirch, Germany. Early on Catholicism shaped his morality with the routine of confession and penitence, remission of sins by a priest, and an abiding conviction of his inherently sinful nature reinforced by images of the crucifixion with Jesus's hands and feet pierced by nails. He studied at a Catholic seminary from 1903 to 1906. Between 1910 and 1912 he wrote reviews for a journal that supported Catholic conservatism and denounced modernism. In a review of May 1910, he proposed that "the Church will, if it is to remain true to its eternal treasure of truth, justifiably counteract the destructive influences of modernism."[6] His faith was deflected in 1917 by marriage to a Protestant, when he attempted to embrace Protestantism. From the Protestant theologian Friedrich Schleiermacher, he took a view of religion as a humanizing experience grounded on personal emotions, in contrast to Catholic faith grounded on revealed truth. Religion increasingly became a matter of emotions over beliefs, and the emotion most characteristic of the divine and symbolized by the crucifixion was suffering.

He remained outwardly committed to Catholicism until 1919, when he renounced the Church. He wrote: "Epistemological insights that extend to the theory of historical knowledge have made the *system* of Catholicism problematic and unacceptable to me—but not Christianity and metaphysics, which, however, [I now understand] in a new sense."[7] His emerging philosophy wrested agency from the Christian God and located it in human beings surrounded by circumstances they cannot change, contemplating an absurd death at any moment that they must face alone, and obliged to rely on themselves for salvation. He could not accept Catholicism's closed system of the world governed by God throughout eternity or Protestantism's preoccupation with sin, suffering, and death. His philosophy became atheistic, and his personal faith followed suit. Thereafter he occasionally wrote about "other gods" or "the last god" but ceased to believe that the Christian God was incarnated in Jesus Christ and resurrected after his crucifixion as the Messiah.[8]

Heidegger turned against Christianity because of doctrinal concerns, substantive philosophy, and his phenomenological method.

Doctrinal problems surfaced as his philosophy clashed with Papal authority. In 1907 Pope Pius X issued a decree that condemned sixty-five "modernist" propositions. In 1910 he ordered teachers at Catholic institutions to take an "oath against modernism." Four years later he made Aquinas the sole source of doctrinal authority and forbade Catholic scholars from analyzing theology with historical and scientific methods without acknowledging "the supernatural origin of Catholic tradition" and promising "to preserve all revealed truth forever."[9]

Heidegger's independent philosophy rebelled against such restrictions. His philosophy, based on ever more difficult questions, clashed with Catholic memorization of absolute truths in catechisms and recitations of prescribed articles of faith in prayers. He clashed with the static institution of the Church as defined by Roman canon law based on the authority of Church dogma and the doctrine of "papal infallibility." He defined the opposition between Christianity and phenomenology in a lecture of 1927, arguing that "faith, as a specific possibility of existence, is in its innermost core the mortal enemy of the form of existence that is an essential part of philosophy." He concluded vividly that "there is no such thing as a Christian philosophy; that is an absolute 'square circle.'"[10] In *Introduction to Metaphysics* (1935) he characterized Christian faith as inimical to doing philosophy and leading an authentic life. "Anyone for whom the Bible is divine revelation and truth already has the answer to the fundamental question, 'Why are there beings at all instead of nothing?' before it is even asked." Christians who believe that "In the beginning God created heaven and earth" cannot answer that fundamental question because they cannot come into a genuinely questioning attitude if they cannot think for themselves and if all difficult questions have already been asked and answered.

Despite the vehemence of his apostacy, he remained strongly influenced by a Catholic upbringing and the study of Christian thinkers, especially Paul, Augustine, Aquinas, Luther, and Kierkegaard along with his contemporaries Rudolf Bultmann and Karl Barth. The unknowable God of his earlier Christian belief evolved into an unknowable fundamental Being that he spent the rest of his life trying to understand. His analysis of the meaning of Being per se was the ultimate goal of an ever-deeper analysis of the meaning of human being. My use of "deeper" here and throughout follows Heidegger's method of probing ontological foundations rather than chronological origins.

Heidegger's Spiraling Down into the Foundations of Human Being

In *Discourse on Method* (1637) René Descartes published the first statement of his famous argument for his existence, "I think, therefore I am." In *Meditations on First Philosophy* (1641) he inaugurated a daring modern approach to philosophy by doubting everything he previously believed to be true, including the evidence of his senses. His famous "universal doubt" included even belief in God who might be an "evil genius" and a deceiver, and so he resolved to admit as true only what he knew indubitably, by which he meant clearly and distinctly. But his "universal doubt" did not last long. By the "Second Meditation" he reassured himself that he knew indubitably that he existed as "a thing which thinks." By the Third Meditation he knew indubitably that God exists, and, because this God was by definition perfect and therefore not a deceiver, Descartes was confident that he could admit into the realm of indubitable knowledge experience that was submitted to rigorous rational review. That proof of his existence and subsequent proof of his knowledge of the external world presupposed the self-knowing "I" in the premise of his famous argument, namely "I think." In sharp contrast, Heidegger would spend a lifetime trying to fathom the nature of his existential self in pursuit of a kind of understanding that was anything but indubitable but rather fundamentally and increasingly dubitable.

In a classic defense of liberty that ought to be the right of every individual, John Stuart Mill's *On Liberty* (1859) also begins like Descartes *Meditations* with the presupposition of a self-knowing individual. His opening sentence announces the purpose of his essay to define "the nature and limits of the power which can be legitimately exercised by society over the individual." Mill questions many threats to individual liberty but does not question his own starting point as a self-aware and autonomous *individual.* "Over himself, over his own body and mind," Mill presupposes, "the individual is sovereign."[11]

In opposition to the philosophies of Descartes and Mill, modern phenomenologists rejected philosophies that start with a confidently self-aware and sovereign subject. In 1913 Max Scheler published the first phenomenology of love (or encounter), which rejects previous theories that begin with a primary self that builds up a sense of others from external experience. For Scheler what occurs is rather "an immediate flow of experience, *undifferentiated as between mine and thine*, which actually contains both our own and others' experience intermingled."[12] Out of this flow begins a gradual sorting out of the self as separate from the other. Put simply, his philosophy begins with "we" instead of "I."

Scheler bolsters that argument with speculation about the child's earliest absorption of what is external to it as the origin of self-definition. In contrast, Heidegger does not base his analysis of human existence on a chronological tracing of progress from a child to an adult, à la Freud, but on an ontological reconstruction of *existentials*, which are modes of existence in adult human beings that upon analysis reveal fundamentally what it means *to be*. His existential inquiry replaces the traditional assumption of an initial sovereign individual by an inquiry that begins with a human being immersed in a world of others. Without targeting Mill explicitly, Heidegger shows how the social factors that Mill viewed as threats to individual liberty are rather inextricable features of human existence as it is enriched in a world of intending and caring as well as threatening others.

Heidegger's most important book, *Being and Time* (hereafter *BT*), which he began writing in 1923 and published in 1927, addresses the meaning of human existence as well as the most fundamental originary questions about the foundational meaning of Being per se. To proceed some definitions are necessary. Heidegger's uncapitalized "being" or "beings" (translation of *Seiende*) refers to things that exist and to which one can point such as chairs, dogs, and triangles as well as human beings. The standard translation of Heidegger's German title *Sein und Zeit* is *Being and Time.*[13] *Zeit* is German for time. The German verb *sein* means "to be," while the noun *Sein*, which is translated as "Being," means "the to-be," so a literal translation of his title would be *The To-Be and Time*. The two German meanings of *Sein* and *sein* allow Heidegger to interlace the differences between the thingness of the noun "Being" and the action of the verb "to be," an interlacing that suggests that Being is an originary, foundational aspect of all that is as well as a way of actively be-ing.

The first word in that translated title, "*Being*," refers to pure existentiality that somehow underlies every existing thing. After reading *BT* several times and consulting numerous commentaries, including particularly helpful ones by Richard Polt and Thomas Sheehan, I still have no clear understanding of what Heidegger's Being is, although I do understand that Heidegger's primary inquiry is into Being or what it means to be.[14] In standard translations of Heidegger, "Being" is capitalized, while "being" is not. Ontology comes from the Greek words "*on*," which means "being," and "*logia*" that means "study," so, while social sciences such as anthropology and psychology study facts about human beings, ontology studies the Being that underlies those human beings or, as Sheehan puts it, the "source or origin of meaningful presence." Before human beings can philosophize about what is good (ethics), beautiful (aesthetics), or true (epistemology), they must exist, and for that reason ontology is the most fundamental philosophical inquiry.

Heidegger's lifelong ultimate project was to understand the fundamental nature, ground, and meaning of Being itself, which he never succeeded in doing. That task was complicated by the rigor of his analysis and a relentless questioning method that made such understanding increasingly unattainable. His student Hans-Georg Gadamer recalled being in thrall to this interrogative method: "Who among those who then followed him can forget the breathtaking swirl of questions that he developed in the introductory hours of the semester only to entangle himself in the second or third of these questions and then, in the final hours of the semester, to roll up deep-dark clouds of sentences from which the lightning flashed to leave us half stunned?" Karl Löwith remembered how "his lecturing method consisted in constructing an edifice of ideas, which he himself then dismantled again so as to baffle fascinated listeners, only to leave them up in the air."[15] However, for all his students' puzzlement, they ultimately became increasingly illuminated. His followers included a pantheon of leading philosophers and theologians including Rudolf Bultmann, Karl Barth, Hans-Georg Gadamer, Paul Tillich, Ludwig Binswanger, Viktor Frankl, Jürgen Habermas, Jean-Paul Sartre, Maurice Merleau-Ponty, Louis Althusser, and Michel Foucault. Heidegger had a brief and naïve embracing of National Socialism from 1930 to 1934, which marked a complete deviation from the central line of his philosophy, but he nevertheless profoundly influenced the Jewish philosophers Herbert Marcuse, Emmanuel Levinas, Jacques Derrida, and Hannah Arendt, with whom he had an adulterous affair that began in the early 1920s. The affair broke off in 1926 when he was already working on *Being and Time*, but their communication resumed after World War II.[16]

Descartes's moment of epistemological skepticism, his universal doubt that earned him the reputation of being "the first modern philosopher," pales compared to Heidegger's protracted existential skepticism in addressing three formidable questions: Why is there something rather than nothing? What does it mean to be? And what is it about human beings that lets Being have a meaning for them? Heidegger was never able to fully answer these questions to his satisfaction, but he made a valiant effort to answer the third, which is the main subject of *Being and Time* and the basis of his stature in the history of philosophies of human existence. He complicated that effort by addressing numerous variants of *sein* in *BT* and other works, which include being itself (*Sein selbst*), that which is (*Seiende*), is more fully (*seiender*), beingness (*Seiendheit*), forgottenness of being (*Seinsvergessenheit*), nonbeing (*Nichtsein*), potentiality-for-being (*Seinkönnen*), the question of Being (*Seinsfrage*), and so forth. Sheehan lists sixty-one variants of *sein* or Sein from *BT* and other works.[17]

Early in *BT* Heidegger introduces the focus of his study – the nature of human beings who might be capable of understanding the meaning of human existence. He refers to human beings in the singular or plural as "Dasein," which translators leave capitalized but unitalicized. Dasein captures the spatial, situated nature of human existence immersed in its world because the locative prefix "Da" means "there" or "here," so Dasein is/are always located in the world as a "to-be there" or a "to-be here." On occasion Heidegger separates the "Da" from "sein," as in "Da-sein," to accent the spatial nature of human existence, but Dasein are also temporal beings in the three ecstasies of past, present, and future and for that reason particularly well constituted to grasp the meaning of their temporality. "Ecstasy," from the Greek *ekstasis*, means to stand outside or transcend oneself. Heidegger uses that term because it suggests how the capabilities of experiencing past, present, and future extend an individual beyond itself in three interrelated temporal orientations.

In contrast to a tradition of beginning inquiry into human existence with some version of Descartes's self-assured and isolated "I," Heidegger begins with a fundamental "we," mindful of Dasein's immersion in a temporal stream and spatial context, always already in a world of others. Repeatedly inquiring ever deeper into the nature of Being, he admits that "one is constantly compelled to face the possibility of disclosing an even more primordial and more universal horizon from which we may draw the answer to the question, 'what is Being'?" (*BT*, 49). He never does disclose that ultimate horizon, and so his philosophy remains inconclusive. Polt describes Heidegger's argument as a "*spiral* structure: [in which] each turn around the 'circle' reaches a deeper level."[18] I follow that approach. By the end of *BT*, which stops before Heidegger completes his down-spiraling search for a secure foundation for his or anyone else's Being, the nature of Being is partly disclosed but still partly unknown.

Heidegger identifies two levels of existential analysis. *Ontical* analysis concerns facts about entities. My chapters on cosmology, geology, and biology concern such ontical facts as the age of the universe, the age of the Earth, and the origin of species that can be explained by rational argument based on empirical observations. At a deeper level is *ontological* analysis of the Being of galaxies, continents, and living forms that the scientists I survey appear not to have even wondered about. Heidegger's analysis of Being attempts to probe that deeper level. Polt concludes his summary of Heidegger's distinction between ontical and ontological analyses as the "first time around the circle [that] will be deepened with each new turn," without ever reaching the ultimate ontological foundation.[19]

While Freud begins his depth psychology by exploring chronologically back to the temporal beginning of his patients' existence as a nursing infant, Heidegger begins his existential analysis by tunneling "down," so to speak, ontologically toward the foundation of Dasein's Being. To pursue that inquiry Heidegger shifts from investigating the nature or meaning of Being to the nature of those beings who can inquire into the nature or meaning of Being, namely Dasein.

Heidegger's analysis of Dasein initially focuses on existentials. He begins by distinguishing between the essence of entities that have a specific essence (such as the essence of a lamp that is a single thing, a "what" that illuminates) and the essence of Dasein, a "who," whose essence "lies in its existence," which includes many possible ways of choosing who or what it is. Those ways are characterized first by the existential of "mineness" (*Jemeinigkeit*) (*BT*, 68), a first effort to grasp the meaning of Dasein.

Dasein begins its existence in the midst of its world as a "Being-in-the-world" (*In-der-Welt-sein*) (*BT*, 78). Heidegger hyphenates those words to emphasize the inextricable unity of Dasein and its world. Dasein does not begin as an isolated Cartesian subject that opens the vizor of perception and – lo – beholds objects in an objective world "out there" but rather is inextricably and inherently in, with, and amid its world that is a holistic field of meanings and purposes.[20] As Heidegger explains, "When Dasein directs itself towards something and grasps it, it does not somehow first get out of an inner sphere in which it has been proximally encapsulated, but its primary kind of Being is such that it is always 'outside' alongside entities which it encounters and which belong to a world already discovered." He vividly illustrates this important aspect of Dasein: "the perceiving of what is known is not a process of returning with one's booty to the 'cabinet' of consciousness after one has gone out and grasped it" (*BT*, 89). Here Heidegger rejects the long history of empirical psychology that viewed the mind as "in" the body, perceiving an external world like a ship's pilot who peers out to see and thereby comes to know and then navigates through the external world. Dasein is always already in, of, and with its world.

Heidegger then probes the ontological constitution of another existential, "inhood" (*BT*, 79). Dasein is in the world not the way water is in a glass but rather the way it dwells in and is reciprocally absorbed by its world. Thus does Heidegger strain to invent new language to probe deeper into the meaning of "in" plus the supremely important and elusive meaning of "is." His radical reinventions of language to express new ideas are analogous to Gertrude Stein's reinventions of language to restore the ability of words to capture what things fundamentally are, most famously what a rose is, after centuries of being buried in "wornout literary words" and clichés.[21] Polt makes a similar argument about Heidegger: "We have to wean ourselves away from the habit of reducing experience to common-sense [Victorian] platitudes and train ourselves to be surprised by the ordinary."[22] I added "Victorian" to Polt's proposal to align it with my argument about Heidegger's probing the deeper foundations of originary experiences as a challenge to the ideas of his Victorian predecessors. While he rarely mentions Victorians explicitly, their ideas and methods of inquiry are the target of numerous arguments in *Being and Time*.

Further analysis of how Dasein experiences the world or, more precisely, its interactive worldhood, includes many more possibilities of "Being-in," namely: "having to do with something, producing something, attending to something and looking after it, making use of something, giving something up and letting it go, undertaking, accomplishing, evincing, interrogating, considering, discussing, determining" (*BT*, 83). Dasein is also "fascinated with" and "absorbed in" its worldhood. Heidegger finds added meanings in deficient modes of *un*concern about its world such as "leaving undone, neglecting, renouncing." By elaborating positive as well as deficient possibilities of Being, Heidegger probes ever deeper into frequently overlooked features of Dasein such as the in-ness of its inhood, the spatiality of its worldhood, and the withness of its Being-with.

In discussing how the mind is a streaming of consciousness, William James drew attention to those neglected little words and the in-between experiences they denote that have rich significance. He wrote: "We ought to say a feeling of *and*, and a feeling of *if*, a feeling of *but*, and a feeling of *by*, quite as readily as we say a feeling of *blue* or a feeling of *cold*. Yet we do not: so inveterate has our habit become of recognizing the existence of the substantive parts alone, that language almost refuses to lend itself to any other use."[23] Heidegger develops a language for just such uses to probe the meaning of in, with, for, and alongside in rendering Dasein's worldhood.

Heidegger also probes deeper into Dasein's spatiality, not as a static being at some measurable distance from something or someone but rather as an active engaging in a dialectical "de-severance" (*Entfernung*) from other entities. De-severance is another existential of Dasein, suggesting its dynamic capability of being attracted and attracting, its active capability of "making the farness vanish" (*BT*, 138–39). In bringing someone or something closer, Dasein dynamically unmakes separation. Heidegger writes, "In Dasein there lies an essential tendency towards closeness" (*BT*, 140). Dasein also experiences the existential of "belongingness" (*Gehörigkeit*) (*BT*, 145). Although elsewhere Heidegger argues that Dasein is "thrown"[24] into existence without any ultimate reason for its being where it is, it is also paradoxically able to, indeed, destined to learn to belong wherever it happens to be. Another spatial existential of Dasein is a capacity for "making room" for other things and especially other Dasein. Sheehan views Dasein as an attractive force field: "We are like a hermeneutical field of force, like a magnet that draws things together into unities of sense insofar as these things are connected with a possibility of ourselves as the final reference point."[25]

In contrast to Mill's view of others as impinging on individual liberty, Heidegger stresses the inextricable connection between Dasein and others as a function of the fundamental existential of Being-with (*Mitdasein*). Heidegger probes deeper than Mill into the possibilities of "withness" in Dasein's relations with others. "By 'Others' we do not mean everyone else but me—those over against whom the 'I' stands out [à la Mill]. They are rather those from whom, for the most part, one does *not* distinguish oneself—those among whom one is too" (*BT*, 154).

Mill argues against the "tyranny of the majority" over the rights of the minority.[26] Heidegger is concerned about an even more tyrannically intrusive circumstance – the "dictatorship of the 'they'" (*das Man*). "Dasein as everyday Being-with-one-another, stands in subjection to Others" (*BT*, 164). However, these Others are not identifiable others such as Mill's majority, but an unidentifiable, anonymous, and ultimately internalized "one" or "they" who coerce Dasein as the source of conventions about what "one does" or what "they do." These pervasive and indefinite Others are the source of the particularly potent "dictatorship of 'the 'they'" but not as a historical political development but as a fundamental element of Dasein's inherent nature. He does not demonize "the 'they'" but sees it, or them, as an essential feature of Dasein's everyday condition as a Being-with-others.

All the existentials of Dasein can be experienced in authentic and inauthentic modes. "Authentic" is a translation of the German *eigentlich*, which is based on "*eigen*" that means one's own. Inauthentic denotes what is not one's own. While Heidegger explicitly rejects the judgment that authenticity is ethically superior to inauthenticity, his language occasionally suggests otherwise. He speculates that if Dasein "discloses to itself its own authentic Being, then this discovery of the 'world' and this disclosure of Dasein are always accomplished as a clearing-away of concealments and obscurities, as a breaking up of the disguises with which Dasein bars its own way" (*BT*, 167). Many of Heidegger's students could not accept Heidegger's claim that he did not intend his pointed analysis of "the dictatorship of 'the 'they'" in a negative ethical sense, but surely clearing away Dasein's self-concealments and obscurities and breaking up its disguises *are* ethically commendable.

With the introduction of the authentic-inauthentic distinction Heidegger spirals down to a deeper level of analysis that applies to all the existentials of Dasein that he has previously analyzed and will continue to analyze such as discourse, guilt, and death. In each way that we exist we can either more inauthentically conceal from ourselves what it means to be in that way or more authentically reveal to ourselves what it means to be in that way.[27] For example, inauthentic discourse is gossip, while authentic discourse such as poetry genuinely reveals matters at stake; inauthentic guilt is self-recrimination, while authentic guilt is the realization that one is not all one can be; and inauthentic death is a misfortune that Dasein undergoes passively and helplessly, while authentic "Being-towards-death" is an active state of anticipatory resoluteness about life's brevity.

In common usage, authentic seems to be a better way of being than inauthentic, but Heidegger insists that when using these terms in a "strict sense" he makes no such judgment. He generally begins with the way the existentials are lived inauthentically in the "everyday" mode, without reflecting on their meaning. We normally do that when we are interested in something (studying) or excited (at a sports event) or ready for pleasure (eating a meal) and are not at such times thinking about what it means to be a human being studying, cheering, or eating. Heidegger insists that those inauthentic modes are not ethically objectionable. When we begin to think about the meaning of existing, we become more authentic. The "dictatorship of 'the 'they'" thrusts Dasein into an inauthentic way of relating to others, but it is possible to act authentically in accord with the dictates of "the 'they'" so long as that action is a reflective choice among one's own possibilities, not an unreflective reflex, and so one can choose to follow a law or go through a traditional marriage ceremony authentically if one does so reflectively in a self-aware manner.

After many pages analyzing Dasein into its many existentials, Heidegger proposes a synthetic existential of *care* (*Sorge*) that embraces them all. His care does not mean being concerned about some specific worrisome matter such as a sick child but rather being aware of and involved in one's existence as well as everything in one's ambient world including religion and politics, animals and trees, doorknobs and lamps, beauty and goodness. After all the jarring terminology that Heidegger uses in straining to characterize the many existentials of Dasein such as worldhood, inhood, and belongingness, his using such an ordinary word as "care" to pull together the many existentials of Dasein is refreshingly simple and seemingly clear. But the simplicity of the term veils the breadth and depth of its application to every element of Dasein as becomes evident when Heidegger goes on to define care more fully with its temporal and spatial aspects as "ahead-of-itself-Being-already-in-(the world) as Being-alongside (entities encountered within-the-world)" (*BT*, 237). And no, that is not a joke. That super-complex definition with a generous use of hyphens and parentheses illustrates the super-complexity of a seemingly simple experience such as care. If nothing else, Heidegger reveals highly profound complexities in the most fundamental originary (trivial?) aspects of existence. You thought you knew what care is? Think again.[28]

Heidegger's analysis of Dasein contrasts strikingly with Descartes' epistemology and Mill's sociology. Dasein is not a Cartesian thinking thing that comes to *know* things by looking at them objectively and locating them in the surrounding world but rather is a network of ways of doing and being as it comes to *understand* its world more fully and deeply by dwelling in it. Neither is Dasein a self-sufficient Millian individual threatened by others impinging on its liberty but rather a center of caring concerns about its entire world.

Early on Heidegger anticipates his ever-deeper penetration into the nature of Being, when he explains how "one is constantly compelled to face the possibility of disclosing an even more primordial and more universal horizon from which we may draw the answer to the question 'What is *Being*'?" (*BT*, 49). Down, down into Being Heidegger leads his readers ontologically. Indeed. Two hundred pages later he prepares the reader for that deeper penetration into the Being of Dasein through an ontological analysis of care, as he explains: "we must pursue the ontological question even further until we can exhibit a *still more primordial* phenomenon which provides the ontological support for the unity and the totality of the structural manifestation of care" (*BT*, 241). What could that more primordial phenomenon be? Heidegger is no mystery writer, but with his engaging interrogative method he does present a dramatic disclosure of ever-deeper-lying existential aspects of Dasein.

He first discloses that "the primordial ontological basis for Dasein's existentiality is *temporality*" (*BT*, 277). Following that disclosure he adds: "We must go back [to what he explored in painstaking detail starting 200 pages earlier] and lay bare in their temporal meaning the ontological structure of Dasein which we have previously obtained." And "by thus recapitulating our preparatory fundamental analysis of Dasein we will at the same time make the phenomenon of temporality itself more transparent" (*BT*, 277–78). As Michael Gelven explains this move, "the major existentials must be more deeply examined to reveal the underlying ontological structure of time."[29]

First, Heidegger explains what he means by the temporality of Dasein in contrast to conventional notions of scientific time as a line of identical instants from past to future. Rather, Dasein experiences the three ecstasies of temporality simultaneously and interactively. Authentic Dasein does not just wait passively for things from the future to happen to it but anticipates the future actively with resolution in choosing its possibilities and assuming responsibility for those choices. The ultimate choice is authentic Being-towards death (*Sein-zum-Tode*).

The awareness that every instant of life that brims with hope and future expectations encapsulates an end state of impending death. Authentic Dasein is not suicidal or morbidly preoccupied with dying but actively affirms life with a heightened commitment to live as fully as possible while resolutely anticipating and accepting life's inevitable and absurd end in death.

The most significant ecstasy for Dasein is not the present but the future (*Zu-kunft*, "coming to"), which authentic Dasein approaches with "anticipatory resoluteness." The future is not just some moment along the road of time that is not yet and that Dasein will eventually come upon but is rather an interactive projection from Dasein ahead in time as well as back toward the past as it draws on futurity as well as pastness in every present moment. For example, in authentically experiencing my age, I encompass simultaneously and interactively several key temporal aspects of Being – an awareness of when I *was* born, numerous memorable experiences I *have been having* since then, the fact that I *am* currently alive, and what my future life-expectancy *might be* as I approach that inescapable absurd end in death. I am also simultaneously aware that time is always passing and running out and of how my age differs from that of others whom I immediately put into three temporal categories – younger than me, older than me, or about the same age as me.

The deeper Heidegger penetrates the temporality of Dasein, the denser his language becomes. That challenging feature of his philosophy is particularly evident with his explication of the major temporal elements of *care* – two of which are understanding and mood. Put simply, the primary temporal ecstasy of *understanding* is the future, and the primal temporal ecstasy of *mood* is the past, although these aspects of care involve all three temporal ecstasies.

Understanding is a projection of possibilities, so its primary temporal ecstasy is the future. Authentic futurity is active *anticipation*, while inauthentic futurity is passive *expectation*. In authentic futurity Dasein "lets itself come toward itself as its ownmost potentiality-for-being" (386). That is, in authentic futurity Dasein sees the future as its *own* possibilities, while in inauthentic futurity Dasein sees the future as possibilities emanating from specific others or the impersonal action of "the 'they'" coming toward it. Authentic *present* of understanding, called "moment-of-vision," is Dasein's active freedom, while inauthentic present of understanding, called "making-present," accepts situations without resolve and responds rather the way "they" do as part of a routine. Authentic *past* of understanding involves authentic mindful repetition of the way Dasein understands itself in history, while inauthentic past of understanding involves Dasein's forgetting its historical context. Dasein's complex experience of understanding thus combines the three interacting ecstasies of time in both authentic and inauthentic modes. The deeper Dasein probes existence, the deeper its depth seems, just as the more questions it answers, the more questions emerge.

One historically noteworthy aspect of Heidegger's philosophy is his attention to *moods*, which traditional philosophers regarded as somewhat trivial secondary experiences. Victorian philosophers investigated lofty human projects such as doing good, creating beauty, and seeking truth, but not moods. Victorians studied the higher "faculties" of perception, judgment, imagination, and will. The "lower" emotions or moods began to attract attention largely at the margins of Victorian philosophy. In the 1850s the nonacademic Søren Kierkegaard probed sickness, fear, and anxiety in his complex philosophy of faith. In the 1880s Nietzsche probed the lowball emotions of guilt, self-pity, and *ressentiment* in Christianity.

To analyze moods Heidegger begins by contrasting fear and anxiety. While he lists both conditions as moods, the pointed orientation of fear suggests that it is more a specific emotion than a nonspecific mood. The German word for mood is *Stimmung*, sometimes translated as *disposition*. Unlike the specific emotional reaction of fear of some specific object such as a charging bear, anxiety is a generalized mood or disposition that impacts Dasein's entire world like smoke pervading a home.

Heidegger's contrasting of fear and anxiety reveals their complex temporal as well as spatial foundations. Although frightening objects come at Dasein in time seemingly out of the future, the dominant temporal ecstasis of fear is the past. In fear Dasein backs away in "depression or bewilderment" from what seems threating (*BT*, 342). The depression "forces Dasein back to its thrownness" in space and time that accents the arbitrariness of its spatial-temporal location and consequent inability to act effectively in the present in a frightening world. Dasein's bewilderment is based on being cut off from the past, which leads Dasein to leap from one thing to the next, unable to mount any effective action. For example, if on a hike I come across a bear, I may become paralyzed by fear, unable to move in the present to ward off a threatening future and wishing that I was somewhere else, safe back at home and in the past before I began the hike. Although fear disables me from acting with anticipatory resoluteness in the present because I am cut off from my past in forgetting all my defensive skills, the past is still the dominant ecstasy of the fear, albeit in a deficient mode, that is, the thought that I am not back home and in the past where and when

I was safe. The thought that "if only I was back there and back then" paralyzes me from dealing authentically with danger in the here and now of the present.

In contrast to fear, in anxiety the entire world is threatening as I feel the fundamental nature of my entire existence as thrown into the world and not at home (uncanniness).[30] There is no safe space or time back home or anywhere else. The entire world is threatening *because the threat comes from myself*. But in anxiety, unlike fear, my posture is not inauthentic paralysis but authentic anticipatory resolve, even though there may be no clear plan of action. The present is a moment of vision even though there is nothing specific to see as threatening, another source of anxiety. Put graphically, there is no bear on my path as in fear, because the bear is myself. The past is not forgotten but is repeated in preparing for effective action even if no clear course of action is evident. While fear closes Dasein off from its Being-in-the-world, anxiety reveals Dasein as more fully aware of the uncanniness and thrownness of its condition. This contrast between the more inauthentic emotion of fear and the more authentic mood of anxiety does not mean that anxiety is more humanizing or meaningful than fear but it does show how complex is the existence of Dasein in time as well as space and how such a phenomenological analysis probes that existence to historically new depths.

At the beginning of the final part of *BT* Heidegger announces that the "primordial phenomenon of temporality" is a key to "the possible totality, unity, and development of those fundamental structures of Dasein which we have hitherto exhibited" (*BT*, 352). He goes on to explore temporality for the next 150 pages, and the book ends with a final unanswered question: "Does *time* itself manifest itself as the horizon of *Being*?" After almost 500 pages of *Being and Time*, he still has not fathomed what time is or how it is the ultimate horizon of Being. His method is indeed, as Gadamer and Löwith noted, a breathtaking swirl of questions that come one after the other as their professor plunged ever deeper into the nature of Dasein on his way to answering, and then again posing, a cascade of questions about Dasein's foundation in Being. In 1928 Heidegger estimated the ultimate appeal of his philosophy – that it conceals and reveals "a richness that again and again demands a renewed awakening."[31] In 1929 he again underscored the depth of his interrogative approach: "the idea of 'logic' itself disintegrates in the turbulence of a more original questioning."[32]

Heidegger's exploration of the question of Being, especially in *BT*, does not provide a simple or ultimate answer to the question of Being. Instead, he argues that the question of Being itself has been neglected in Western philosophy and so must be approached in a new way. In *BT* he argues that we must approach the question of Being through an analysis of Dasein. His main insight is that Being cannot be fully grasped through abstract definitions or theoretical concepts but must rather be *experienced* as inquirers reflect upon it, as they philosophize in the context of their lived experience. Being is fundamentally tied to time and temporality. Our awareness of time's passing, our sense of mortality, and our ability to project ourselves into the future are central to how we understand our own existence.

Heidegger never offers a conclusive or definitive answer to the question of Being but rather calls for a new way of thinking and engaging the world – one attuned to the mystery of Being and that resists the temptation to reduce it to concepts or objects. His ultimate goal is to establish a more authentic ongoing relationship with Being, which is grounded in our emerging but never complete awareness of it in an ever-deeper probing.

Notes

1. Friedrich Nietzsche, *Thus Spoke Zarathustra* [1883–5] trans. Walter Kaufmann (Viking Penguin, 1954), 125. Subsequent text references are to this edition.
2. Friedrich Nietzsche, *On the Genealogy of Morality: A Polemic* [1887], trans. Maudemarie Clark and Alan J. Swensen (Hackett, 1998), 9. Subsequent text references are to this edition.
3. On Nietzsche's complex views of Jews see Robert C. Holub, *Nietzsche's Jewish Problem: Between Anti-Semitism and Anti-Judaism* (Princeton University Press, 2015).
4. Nietzsche used the French word for the English "resentment," because there was no direct equivalent in German that captured the bitterness mixed with a desire for revenge that he sought to express.
5. Friedrich Nietzsche, *The Anti-Christ: A Curse on Christianity* [1888], trans. Judith Norman (Cambridge University Press, 2005), 30, 31.
6. Martin Heidegger, review of F. W. Foerster's *Authority and Freedom: Thoughts on the Problem of Culture in the Church* in *Der Akademiker* 2, no. 7 (May 1910): 109–10, quoted in Theodore Kisiel and Thomas Sheehan eds., *Becoming Heidegger: On the Trail of his Early Occasional Writings, 1910–1927* (Northwestern University Press, 2007), 13.
7. Quoted in Kisiel and Sheehan, *Becoming Heidegger*, 96.
8. On his relation to God to the end of his life see George Kovacs, *The Question of God in Heidegger's Phenomenology* (Northwestern University Press, 1990).
9. Quoted in Judith Wolfe, *Heidegger and Theology* (Bloomsbury, 2014), 15.
10. Martin Heidegger, *Pathmarks* (Cambridge University Press, 1998), 53.
11. John Stuart Mill, *On Liberty* [1859] (Norton, 1975), 11.

12. Max Scheler, *Zur Phänomenologie und Theorie der Sympathiegefühlen und von Liebe und Haas* [1913], trans. Peter Heath as *The Nature of Sympathy* (Archon, 1970), 246. Martin Buber argues similarly in *I and Thou* (1923).
13. Martin Heidegger, *Being and Time*, trans. John Macquarrie and Edward Robinson (Harper, 1962). All text references are to this edition.
14. Richard Polt proposes that Being refers to "what marks beings out as beings rather than nonbeings," and he subsequently defines Being as the reason "why things count as beings in the first place" in *Heidegger: An Introduction* (Cornell University Press, 1999), 3, 28. Thomas Sheehan argues that "the 'essence' of Being [is] that which accounts for and is the source or origin of meaningful presence at all" in *Making Sense of Heidegger: A Paradigm Shift* (Rowman & Littlefield, 2015), xv.
15. Gadamer and Löwith quoted in Polt, *Heidegger: An Introduction*, 19–20.
16. Heidegger's engagement with Nazism is one of the great conundrums of intellectual history. How *could* a founder of existential philosophy, a man who had such a profound impact on philosophies of the meaning of human existence, become a Nazi? This episode is the subject of scores of books and articles. I choose to sidestep it here because it is not the reason why he is a major figure in continental philosophy and had an enormous influence on thinking about what it means to be a human being. I offer a fuller account of his Nazism and his affair with Arendt in *Modernism after the Death of God*, 140–48.
17. Sheehan, *Making Sense of Heidegger*, 5–8.
18. Polt, *Heidegger: An Introduction*, 31.
19. Ibid., 35.
20. In 1919 Heidegger introduced an active quality to the world in reappropriating the noun "world" into the active verb "to world." Richard Capobianco makes this point: "As he puts it, *es weltet'*—that is, the surrounding world 'worlds; things 'world' 'everywhere and always.'" *Heidegger's Way of Being* (University of Toronto Press, 2014), 9.
21. Quoted in Thornton Wilder, "Introduction" to Gertrude Stein, *Four in America* (Yale University Press, 1947), v–vi.
22. Polt, *Heidegger: An Introduction*, 98.
23. William James, *The Principles of Psychology* [1890] (Dover, 1950), vol II, 246–47.
24. Heidegger discusses the "thrownness" (*Geworfenheit*) of Dasein in *BT* on page 223.
25. Thomas Sheehan, "What, After All, Was Heidegger About," *Continental Philosophical Review* (published online 29 October 2014): 9.
26. Mill, *On Liberty*, 6.
27. Here I use the qualified terms "more inauthentically" and "more authentically" to emphasize that these conditions are not absolute disjunctive states but positions or orientations along a continuous gradation. However, in the following discussion I use the terms without qualification.
28. For an analysis of seemingly trivial experiences across the entire cultural record, which prove on analysis to have nontrivial meanings, see my conclusion, pages 260–63.
29. Michael Gelven, *A Commentary on Heidegger's* Being and Time (Harper, 1970), 190.

30. On the broader modernist realization that human beings are not at home anywhere, see my conclusion, pages 255–59.
31. Martin Heidegger, *The Metaphysical Foundations of Logic* (Indiana University Press, 1984), 76.
32. Martin Heidegger, "What Is Metaphysics?" quoted in Polt, *Heidegger: An Introduction*, 11.

Chapter 9

RELIGION: Original Sin and Love in Gide, Joyce, and Lawrence

The religious origins and foundations of Western society center on Christianity as it emerged out of Judaism. Stretching back to at least the conversion of Emperor Constantine and the Edict of Milan in 313, Christianity grounded philosophy and science, inspired music and art, was the foundation for personal morality and political sovereignty, ordained priests and ministers who legitimated births, marriages, and deaths, and proposed God's creation as the origin and foundation of everything that was, is, or will be.

Challenges to these extensive originary functions intensified with the sixteenth-century scientific revolution that displaced the Earth from the center of the universe, questioned the unchanging perfection of the heavenly spheres, and replaced the authority of truth from revelation and tradition by truth from observation and experiment. Eighteenth-century deists replaced a Christian providential God by a Newtonian watchmaker God. Nineteenth-century historians analyzed new sources to document a historical-critical approach to Christianity, as when in 1835 the German theologian David Friedrich Strauss applied that approach in his influential *Life of Jesus, Critically Examined*, a historical-critical analysis of the life of Jesus, while in 1863 Ernst Renan did something similar in a biography describing Jesus as "just a man."[1] Auguste Comte replaced metaphysical religious dogma by positivist science, Marx tagged religion as an opiate, Darwin undercut the biblical account of creation, and Nietzsche proclaimed not only that "God is dead" but, more importantly, that "we killed him."[2] Early in the Victorian period, a powerful dramatization of the tension between love of God inspired by God and love of other human beings inspired by a natural "Mighty Spirit" is the tortuous interlude in Charlotte Brontë's *Jane Eyre* (1847). Over 100 pages the ascetic Evangelical missionary St John Rivers uses every resource of religious exhortation at his disposal to coerce Jane into going to India with him as a missionary's wife, while she resists with every fiber of her being. Rivers threatens that "if you reject [his offer], it is not me you deny, but God . . . Refuse to be my wife, and you limit yourself forever to a track of selfish ease and barren obscurity, [and] you should be numbered with those who have denied the faith and are worse than infidels!"[3] Increasingly desperate to secure her acceptance, Rivers threatens most unlovingly that "the fearful, the unbelieving . . . shall have their part in the lake which burneth with fire and brimstone."[4] Jane narrates how she "entreated of Heaven" to tell her what to do, and she was ironically saved by the very God whom Rivers had invoked to coerce her into accepting his proposal. God foiled his seductive ploy by transmitting the words that Rochester uttered while thirty miles away – "Jane! Jane! Jane!" At that moment she realizes that Rochester is still alive, loves her, and wants her to come to him. Through this long interlude Rivers draws on "the word of God" to entice Jane to give up sensuous love from the heart and succumb to obedient duty from the head in accord with doctrinal demands, which that very God contravenes by bringing the true lovers together. This moment of Victorian melodrama nevertheless strains the authority of orthodox faith as it raises an intriguing question about which side God is on – dogma or spirit? In the end Brontë leaves no doubt that the concluding happy marriage of Jane and Rochester is, as it must be, sanctioned by God, but Rivers's coercive evangelism as well as God's telecommunications miracle raise doubts about the value and proper function of the word of God. Brontë's religiosity was ethically serious and antidogmatic; she believed fiercely in God but not in religious authority that denied human desires or crushed the spirit.

In the modernist period social and political developments challenged Victorian ideas about Christian origins and foundations. Around the turn of the twentieth century, in American universities new departments in anthropology, sociology, and history began to study religion scientifically, free of Christian apologetics.[5] The concentration of increasingly large and diverse populations in cosmopolitan cities weakened traditional Christianity that flourished in more conservative rural areas. Educational reforms, increasing literacy rates, and mass circulation newspapers engendered a broader critical thinking readership. In 1905, during the latter years of the Dreyfus Affair, when the French Catholic Church sided with the army's blatant miscarriage of justice against a Jewish officer, the French government separated church and state and established secularity throughout the country. Evidence of the modernist threat to the foundational ideas of Catholicism is the 1907 encyclical of Pope Pius X, *Lamentabli Sane Exitu* (a Lamentable Departure Indeed), which condemned a lamentable "modernism" by which he meant rationalism, liberalism, positivism, materialism, religious freedom, and anticlericalism. World War I raised troubling questions about how a supposedly benevolent and all-powerful God could allow such senseless carnage, although it also drove many back into churches, seeking consolation for the death of their loved ones and hope for a reunion with the dead in the hereafter. In the war's wake the stature of Christianity was diminished by the collapse of the Romanov, Habsburg, and Hohenzollern monarchies that had ruled by the authority of a Christian God according to the principle of the divine right of kings. After its establishment in 1922, the Soviet Union adopted a policy of state atheism that sought to suppress Christianity and other religions, marginalize their social influence, and prevent religious revival, though these efforts varied in intensity over time and never fully eliminated religious belief.

In the wake of these developments, many modernist intellectuals and artists could no longer believe in the existence of an omniscient and omnipotent God or in the incarnation, the virgin birth, the resurrection, Jesus's miracles, Christ's ability to save all of humankind for eternity, or, for Catholics, the transformation of the Eucharist bread and wine into the body and blood of Christ "really, truly, and substantially." This loss of faith entailed that there was no God, that no message was divinely revealed, and that no one could know the ultimate origin or essential foundation of human existence.

When prominent Victorians such as Feuerbach or Darwin lost their faith, they generally continued to respect religious traditions and muted their most troubling doubts. In contrast, leading modernists did not just become tactfully agnostic but interpreted Christianity as harmful and turned against it forcefully and publicly. Opposition to that argument, sometimes called "the secularization thesis," is based on the number of people who attend church or believe in God, on "rampant religiosity in the United States," or on the intensification of the Muslim faith around the world, but church attendance, numbers of believers, and word religions are not my subject. I am concerned with highly accomplished and influential intellectuals and artists of the modernist period, among whom I count over a hundred agnostics and atheists[6] who assailed Christianity as they pursued their cultural efforts.[7]

In *The Anti-Christ* (1888) Nietzsche characterized Christianity as a "crime against life." Christianity weakens the will to power (that is, the will to make or create) and the will to life, leading to decadence. It hypocritically proclaims itself as a religion of love but is grounded on hate, resentment, and punishment signified by the Day of Judgment and the wrath of God. Its promise of an eternal afterlife is a myth that degrades the frightening brevity of actual life. The "hatred-inspired counterfeiter" Paul nailed Jesus to his own cross, dressed up his own hallucinations as proof that the redeemer still lived, and invented his own history of Christianity. Christianity's "sly, secretive, invisible, anemic vampires" such as St Augustine defiled sex and art. But the most dangerous feature of contemporary Christianity was that the faithful could no longer believe and, because Christianity dictated morality, they faced moral nihilism. Nietzsche concludes with a paroxysm of indignation: "I call Christianity the one great curse, the one great innermost corruption, the one great instinct of revenge . . . the one immortal blot on humanity."[8]

In *The Protestant Ethic and the Spirit of Capitalism* (1905), a pioneer of modernist sociology, Max Weber, argued that the Protestant ethic is "puritanical tyranny." It engendered capitalism by valorizing hard work, self-denial, and the accumulation and reinvestment of money to make even more money as a crude sign of salvation. Weber pilloried the Calvinist version of Protestantism as a religion that views all humans including newborn infants as tainted by sin and destined from the beginning of time to an eternity of punishment in hell, which they deserve, although a merciful god, for reasons that are incomprehensible to humans, has chosen to save a select few. Calvinism preaches "the absolute worthlessness and remoteness from God of all mere creatures" and constitutes "a fundamental rejection of every kind of culture of the senses." Furthermore "the Calvinist idea of the depravity of the creature . . . 'feeling like a worm,'—can lead to a deadening of the energies in working life."[9]

For Freud, faith in God is a fantasized compensation for a disappointed childhood belief in an all-powerful father, religious rituals are caricatures of an obsessive-compulsive neurosis, the religious "oceanic feeling" is a regressive return to primary narcissism, and Christian morality derives primordially from crimes of murder, cannibalism, and incest. Heidegger rejected Catholicism as a body of fixed dogma that was incompatible with his cherished method of relentless philosophical questioning and therefore "a tangled, non-organic, dogmatic hedgerow of propositions and proofs that are theoretically wholly unclarified."[10]

In 1912 Bertrand Russell laid the foundation for his later argument against the idea that God was a first cause: "The word 'cause' is so inextricably bound up with misleading associations as to make its complete extrusion from the philosophical vocabulary desirable . . . The reason why physics has ceased to look for causes is that, in fact, there are no such things. The law of causality, I believe, like much that passes muster among philosophers, is a relic of a bygone age, surviving, like the monarchy, only because it is erroneously supposed to do no harm."[11] Fifteen years later he elaborated on that harm in a lecture that surveyed a number of answers to his titular question, "Why I Am Not a Christian," including "the first cause argument" that erroneously attributes that first cause to God. He concludes flatly, "There is no reason why the world could not have come into being without a cause . . . There is no reason to suppose that the world had a beginning at all."[12] My chapter on literature surveys modernists who crafted antibeginnings in novels about characters who themselves have no clear beginning but emerge into existence as they enter the story.

Woolf railed against Christianity throughout her life. A letter of 1902 refers to "the Church—which seems utterly gone over to the bad."[13] In 1911 she charged that one congregation "sings without understanding, and as for the psalms . . . I never heard anything so senseless in my life."[14] In 1928 she denounced Anglo-Catholicism with a comment on "poor dear Tom [T. S.] Eliot, who may be called dead to us all from this day forward. He has become an Anglo-Catholic, believes in God and immortality, and goes to church . . . A corpse would seem to me more credible than he is. I mean, there's something obscene in a living person sitting by the fire and believing in God."[15]

Woolf further maligned Christianity in two novels. *The Voyage Out* (1915) dramatized her disgust over the smugness and misunderstanding of Christianity through the thoughts of her protagonist Rachel Vinrace listening to a minister. Rachel is scandalized by the "forced solemnity" of churchgoers who pretended "to feel what they did not feel." Then Rachel thinks of other churches where "this blundering effort and misunderstanding were perpetually going on."[16] With the character of Doris Kilman in *Mrs. Dalloway* (1925), Woolf's fictional assault on Christianity peaks. Outcast and unloved as well as "bitter and burning," Kilman had turned to a church. Out of this unhappy mix, her Christianity became a tortuous brew of arrogance and revenge inspired rage. "Whenever the hot and painful feelings [generated by church preaching] boiled within her, this hatred of Mrs. Dalloway, this grudge against the world, she thought of God."[17]

Modernist artists assailed Christianity with irreverent imagery. The German expressionist Emile Nolde's *Crucifixion* of 1912 depicts Christ with blood streaming from the nail punctures in his hands down his emaciated forearms to his elbows and from the nail puncture in his feet into a stream flowing off his feet and down to the ground. His despondent facial expression with downward-looking deep-set eyes is designed to inspire despair more than faith as the Virgin Mary and Mary Magdalene to the left of the cross appear entirely defeated, while in blatant contrast a Roman soldier in the right foreground irreverently rolls dice for Jesus's tunic. In the aftermath of a senseless and bloody war fought for "God, King and Country," Lovis Corinth's 1922 *Crucifixion*

(*The Red Christ*) renders Christ from head to foot splattered with blood, which also streams out of the Sun, streaks across the sky, discolors a body of water, covers the terrain, and concentrates around the spear thrust into Christ's side. The Dadaist Francis Picabia's *The Holy Virgin* from 1920 (Figure 11.11) impugns the holiness of Christ by besmirching his mother, who is represented symbolically by a messy black ink stain under the painting's ironic title *La Sainte Vierge*. In *The Blessed Virgin Chastising the Christ Child in Front of Three Witnesses: André Breton, Paul Éluard and the Painter* from 1926 (Figure 11.12), the Surrealist Max Ernst depicts a naughty preteenage Jesus being whipped by his mother in front of three witnesses, while his naked buttocks and thighs have red welts from the whipping, and his halo has fallen onto the ground, a desacralized symbol indicating that "the Christ child" is stripped of divine authority.

This chapter concentrates on three prominent modernist writers who were raised in devout homes but eventually turned against various denominations of Christianity. André Gide was disturbed by Calvinist belief in predestination and infant damnation, James Joyce was troubled by Catholic emphasis on the punishments of hell, and D. H. Lawrence was discouraged by lackluster Congregationalist churchgoing. These men were also troubled by Christian metaphysics, intolerance, revealed truth, and dogma, but the main reason for their contempt for Christianity was that its morality, its suspicion of bodily desires and judgment of sex as a sin, the original sin, threatened to degrade their mature love relations, and that potential effect was the supreme outrage.[18] In spite of Christianity's claim to be a religion of love, its foundational morality made the fullest expression of that love, in the form of sexual love, nearly impossible.

Gide struggled against Calvinist sexual morality that condemned his masturbation as evil, his homosexuality as perverse, and his adultery as sinful. Joyce resolved to "make open war" on Catholicism, figuratively spit on Pope Pius X, and characterized the Catholic Church as "a big whore." He had contempt for Catholicism, which he came to view as a blight on his sexuality and a net that threatened to ensnare his creativity. Lawrence looked back on his childhood "douching" in Congregationalism as a source of mind-numbing ritual and debilitating sexual guilt and proclaimed his antipathy to conventional religion in novels and essays.

When these men were coming of age, the Christian foundation of their morality was broadly critical of sexual desire. The "miraculous" impregnation of Jesus's "immaculate" "virgin" mother came from a disembodied and "pure" Holy Spirit, and there is no record of Jesus ever having had sexual feelings or sexual relations. Jesus urged love of God and love of one's neighbor, but those loves were nonsexual. In Mark 12:25, Jesus said, "When the dead rise, they will neither marry nor be given in marriage; they will be like the angels in heaven." In Matthew 19:12 Jesus honored sexual renunciation in extreme cases of people "which have made themselves eunuchs for the kingdom of heaven's sake." In 1 Corinthians 6:9 Paul proclaimed that "fornicators," "adulterers," and "sodomites" will not inherit the kingdom of God. In 1 Corinthians 7:2 Paul degraded marriage in urging men and women to marry but only in order "to avoid fornication," and if men cannot remain celibate like himself, then, he urged grudgingly, "it is better to marry than to burn with passion." Later in that chapter he advised men who do have wives to try to practice a celibate marriage and "live as if they had none" (7:29).

Although the Bible, and especially the *Old Testament*, presents a multifaceted picture of sexual activity, beginning early in Christianity a dominant trend developed that variously minimized or maligned the sexual aspect of humanity, and over the centuries, especially in the Western Church, Christian morality became increasingly antisexual. While early theologians saw the Fall as the sin of disobedience, Western theologians following Augustine in the early fifth century reinterpreted it as the sin of concupiscence, primarily sexual lust felt by Adam and Eve during coitus, reenacted in every subsequent act of coitus, and thereby transmitted to succeeding generations who are therefore conceived and born in sin. As a result, humanity was a "mass of perdition" (*massa damnata*), and Catholic Church leaders in the West took a vow of celibacy. Catholic infants required a baptism to wash away their "sins" as they are admitted into the church. In Protestantism, according to Luther and Calvin, original sin persisted even after baptism.

Although Protestants had more positive views of sexual activity, at least in marriage, many continued to intone against the sinful nature of sex. By the late Victorian period this trend, intensified in Catholicism, offered no acceptable method of sexual gratification except in reproductive sex between married couples. Catholic disapproval of sex was so uncompromising that in 1854 Pope Pius IX proclaimed the Dogma of Immaculate Conception, requiring all Catholics to believe that Mary was uncontaminated during her conception by the sexual intercourse of her parents, St Anne and St Joachim, which without that new dogma would have meant that Jesus's mother was tainted with original sin. The pope followed in 1870 with his Decree of Papal Infallibility, which held that anything a pope announced about faith or morals ex cathedra – most notably the immaculate conception of Mary – was infallible and could not be challenged for all time. In the course of this long history, the sensuous impetus for loving relations was resoundingly stigmatized, typically along gender lines, with men viewed as sexual aggressors and women as sexual tempters. Such was the moral foundation for thinking about sexuality and its role in the genesis of love during the impressionable adolescent years of Gide, Joyce, and Lawrence, who spent the rest of their life impugning the sexual morality that was instilled in them during childhood. Their contempt for that morality tainted their conception of Christianity, as is evident in their life and work.

André Gide

The major source of Gide's early Calvinist instruction was his pious mother, Juliette Gide, who had a restrictive sexual morality and a punitive demeanor. She prevented Gide from playing Chopin because it was too sensuous, and until just before her death she blocked him from marrying his cousin Madeleine Rondeau. At age seventeen he went to Paris to prepare for his First Communion, but the boring instruction drove Gide to undertake his own religious education. After his confirmation, his piety increased. As he recalled, "I rose at dawn and plunged into a tub of icy water . . . Then, before beginning my work, I read a few verses of the Scriptures . . . Out of self-mortification I slept on boards."[19]

In his early twenties sexual guilt fired outbursts of religiosity. In 1891, with veiled reference to his habit of masturbation, he wrote: "I have followed tortuous ways and have thought to enrich myself with false goods . . . O Lord, keep me from evil."[20] *If It Die . . . An Autobiography* (1926) reports a breakthrough insight of 1893. "Up to the present I had accepted Christ's code of morals . . . It dawned on me at last that this discordant duality might possibly be resolved into harmony," but the harmony could not be achieved in accord with Calvinism.[21] That religion enforced a forbidding sexual morality, especially toward his uniquely elaborate and tortuous homosexual practices that led him to learn to masturbate and have an orgasm but without ejaculating,[22] and so emotional harmony was impossible.

Gide's writings on sexuality and love drew from efforts to reconcile his deeply meaningful but sexless marriage with his sexually intense but superficial homosexual adventures. In March of 1893 he recorded: "I lived until the age of twenty-three completely virgin and utterly depraved; crazed to such a point that eventually I came to seek everywhere some bit of flesh on which to press my lips."[23] He found an outlet for his desires during a trip to North Africa in 1893, two years before he married Madeleine. Before departing, he left behind his Bible, because, as he explained, "up till that time not a day had gone by without my going to the Holy Book for both sustenance and counsel." One afternoon during that trip, on the sand dunes of Tunisia, his guide, a boy named Ali, ripped off his clothes, threw himself naked on the sand, stretched out his arms, and laughed. "In the lovely splendor of that evening light," Gide recalled, "what radiance clothed my joy!"[24] The restraint of that confession should not obscure its pathbreaking significance, because it was the first time that a leading cultural figure had publicly disclosed his homosexuality, thirty years after Oscar Wilde had been ruined by a similar but involuntary disclosure.

The publication of *If It Die* ended a long struggle to acknowledge his homosexuality to the world, which first began with an indirect hint of his pederasty in the final lines of *The Immoralist* in 1902. The narrator protagonist Michel frequented a prostitute, who joked "about my preferring the boy [her brother] to her. She claims that he's the main thing keeping me here. Maybe she's not altogether wrong." In 1910 he began to realize how Calvinism had impaired his work, split his personality, and driven him to atheism. "I do not believe that one can imagine a form of thought more contrary to the work of art (and to my work in particular) than Calvinism. This is what made me break away from it as soon as I began to write."[25] During the war his faith wavered even more. In January of 1916 he assailed Christian emotional bifurcation. "No religion, so far as I know, has dwelt like Christianity with such earnestness on the bisection of man—on the distinction within him, vital to the very last degree, between the higher and the lower, heaven and hell." By June of 1916, he had lost contact with God: "I am no longer able either to pray or even to listen to God. If he perhaps speaks to me, I do not hear." By 1927 his anti-Christianity became militant atheism: "It strikes me as monstrous that man should need the idea of God in order to feel steady on earth; that he should be forced to accept absurdities in order to construct something solid."[26] By 1940 he was disgusted: "It would be much more likely—from noting the inadequacy of His goodness, the failure of His justice, or His helplessness (if I believed in Providence)—that my cry of despair would burst forth. It never occurs to me to regret 'not believing'; but it often occurs to me to say to myself: 'Fortunately I do *not* believe!'"[27]

Gide dramatized his anti-Christianity with tragedy, farce, and satire. *Strait Is the Gate* (1909) spins his early encounter with Calvinism into a tragic story of how fanatical love of God can destroy earthly love and lead to suicidal despair. Gide was especially vulnerable to the emotions triggered by Madeleine, the great love of his life. At age thirteen he came upon her sobbing after she discovered that her mother was having an affair, a scene that sparked his love and resolve to care for her. *Strait Is the Gate* transposes Madeleine into Alissa Bucolin and himself into Jérôme Palissier, who in narrating the story reconstructs Gide's feelings beginning as a thirteen-year-old. "My whole life was decided by that moment [Alissa's trauma] . . . Drunken with love, with pity, with an indistinguishable mixture of enthusiasm, of self-sacrifice, of virtue, I appealed to God with all my strength—I offered myself up to Him, unable to conceive that existence could have any other object than to shelter this child from fear, from evil, from life." A few days later in church, Jérôme hears the pastor's sermon: "Enter ye in at the strait gate; for wide is the gate and broad is the way that leadeth to destruction." Jérôme recalls seeing the strait gate to heaven through which he and Alissa must pass as a narrow passage that ultimately squeezes the life out of them.[28]

Jérôme desires unity with Alissa through love mediated by a shared religiosity, as he explains: "[C]ommunion in God means . . . being rapturously united in the worship of the same thing. I think it is just because I want to be united with you that I worship what I know you worship too." But she sees his desire for unity with her as a weakness and a sin. She proposes that "[w]e must each of us find God by ourselves." Early on she hopes that they may be united in death, then she puts him off because her sister Juliette loves him, and then because she thinks she is too old for him. He leaves, and each time he is scheduled to return she postpones their reunion. Once she writes that she does not want to see him at all: "Goodbye, my much-loved brother; may God keep and guide you! To Him alone can we draw near with impunity."[29] By this time, mere physical proximity had become sinful and deadly.

Alissa becomes increasingly thin and weak. After she wastes away and dies, Jérôme reads her journal – a record of suffocating emotions and twisted reasoning in pursuit of an increasingly fanatical Calvinist faith centered on self-denial before God. He learns that she missed him but kept him away. "I should like," she wrote, "to love him without knowing that I love him [and] to love him without his knowing it." This scenario runs the gamut of Calvinist self-abnegation. His love for her originally led him to God, but now his love led him away from God, so she prayed to God to teach him to stop loving her so he could relearn to love God by himself. After a short visit she asked him to leave but desperately wanted him to stay. The reason she gave herself for leaving him she no longer believed but left him anyway without knowing why. She'd like to walk toward God with him, but, she wrote, "the way Thou teachest, Lord, is a narrow way—so narrow that two cannot walk it abreast." The journal breaks for three years and resumes with Alissa's increasingly self-destructive faith and an appeal to God to let her see Jérôme one more time before she gives herself to God in death. Her last journal entry reveals that at the end of her life she was gripped by anguish over "the sudden and disenchanting illumination of [her] life," which was that she saw "for the first time the walls of [her] room in their atrocious barrenness" and thought to herself, "I am alone."[30]

Gide conceived of *The Immoralist* and *Strait Is the Gate* at the same time and viewed them as explorations of contrasting sexually caused moral dangers. In *The Immoralist* Michel immorally drives his wife to death while pursuing sexual desires that in the end he acknowledges may be homosexual, while Alissa's hypermorality drives her to death while pursuing a sexless relationship with Jérôme. These works are cautionary tales about the dangers of excessive sexual indulgence and excessive sexual restraint, respectively. While Gide was frightened about excessive indulgence, he was contemptuous of excessive restraint, especially about the punitive role of Christian asceticism as he was forced daily to see its emotional devastation on his wife. He was afraid that self-indulgence would separate him from God, decency, and morality, as well as his wife, and he was outraged over the negation of life that resulted from sexual suppression and fear of sinning.

Gide's *Lafcadio's Adventures* (1914) ridicules a Catholicism in which the devout hypocritical believers are duped, and personalities and icons are fake. The French title, *Les caves du Vatican*, refers to caves under the Vatican where the popes are buried but also the underlying tangle of subterfuge that sustains the Church. At the outset Anthime Armand-Dubois suffers from rheumatism and is an outspoken atheist. In an angry gesture he swings his crutch and breaks off the arm of a statue of Virgin Mary carved out of fake marble called Roman plaster. That night in a dream Mary appears to him with a broken arm and strikes him with her empty sleeve. He awakens miraculously cured of his rheumatism and immediately reverts to Catholicism. Meanwhile the fake cardinal Protos has invented a story about a fake pope in place of the real pope, who has supposedly been kidnapped by Freemasons, and he fraudulently solicits money to fund a plot to restore the "real pope." To recruit the naïve Amédée Fleurissoire to bring money to Rome in support of the plot, Protos appeals for Amédée's help. While returning from Rome, Amédée is murdered by a character who has no motive for his crime other than to murder without a motive. Learning of this seemingly senseless murder, Anthime becomes disillusioned and returns to his role as Church critic, suggesting to a friend that when Amédée gets to heaven he might find "that his Almighty isn't the *real* God either."[31] In this farce, Catholicism is a religion of naïve believers, fake marble, fake icons, fake miracles, fake clerics, and, as suggested by the plot of a fake conspiracy about a fake pope, perhaps also a fake god.

Gide's *The Counterfeiters* (1925) satirizes the commercialization of religion, beginning with the name of the hypocritical jurist Albéric Profitendieu (Profit-in-God). That surname captures Gide's abiding anger against the crude financial interests and banality of institutionalized Christianity. Profidendieu's bogus faith is underscored by his bogus parenting, because the first event in the novel is young Bernard's discovery that he had a different biological father than Profitendieu, and he leaves home with a farewell note cursing the "ridiculous name" he got from his father as well as his counterfeit parental love. Gide extends his titular metaphor of counterfeiting to characterize clerics and psychoanalysts, personal and family relationships, writing and churches, as well as paper money.

James Joyce

In 1888, at age six, James Joyce began his Catholic schooling and by graduation was the head of his class. In 1895 he was admitted to an elite Catholic devotional group and became the school's top student.[32] His education followed strict Jesuit standards – prayers and mass in the morning, catechism and Bible study during the day, philosophical analysis modeled after Aristotle and Aquinas, and night prayers. But during his adolescence he was deeply troubled by the Catholic bias against sex. During his fourteenth year the school's rector warned his mother of her son's "evil ways" and arranged for a three-day retreat on the punishments of hell for mortal sinners.[33] That retreat along with several others were sources for the sermons delivered by Father Arnall during a three-day retreat in Joyce's autobiographical novel *A Portrait of the Artist as a Young Man* (1916).

That retreat is a brutal attack on the adolescents' sexuality that reads as a brutal attack on Catholicism. Arnall attempts to annihilate the instinct for loving in his boys' budding sexual desires that they had been gratifying by masturbation, possibly also by visiting prostitutes. Since death may come suddenly, Arnall warns, one must be sinless at all times because at the instant of death God passes judgment indelibly and eternally, and then it will be too late to repent. While God forgives the faithful, he is unforgiving against those who reject his grace and do not lead a virtuous life or atone for their sins and receive absolution.

Arnall dwells on the torments of hell for the lustful. The walls of hell's prison for masturbators are four thousand miles thick. Sinners are crushed together "not even able to remove from the eye a worm that gnaws it." Each sense is tormented: smell, by an intolerable stench from rotting bodies; touch, by fires from brimstone in a lake that burns forever; taste, by leprous corruption and suffocating filth; hearing, by agonizing howls; and sight, by hideously ugly devils that are somehow visible in the darkness. God also punishes sinners socially. They are without family or friends and are surrounded by others who shriek blasphemy against God.[34]

The next day Arnall details spiritual suffering. In hell sinners know the pain of conscience, as when they recall "the filthy pleasures in which they delighted." They fume when they realize that they have lost the bliss of heaven for money, vain honors, or "a tingling of the nerves." To evoke the overwhelming eternity of these torments Arnall asks his boys to imagine a mountain of sand on a beach a million miles high, wide, and deep, and then imagine if a bird removed one grain of sand every million years how long it would take to clear the beach. "Yet at the end of all those billions and trillions of years eternity would have scarcely begun," and sinners would be suffering excruciating physical and mental tortures the whole time. He concludes with verbal lashes, asking his young boys in a vicious string of threats to imagine an eternity of "endless bodily and spiritual torment, without one ray of hope, without one moment of cessation, of agony limitless in extent, limitless in intensity, of torment infinitely lasting, infinitely varied, of torture that sustains eternally that which it eternally devours, of anguish that everlastingly preys upon the spirit while it racks the flesh, an eternity, every instant of which is itself an eternity" (*P*, 112–16).

Stephen staggers out of the retreat flooded with fear and guilt. The morality of his Catholic upbringing clashes with the memory of his sexual sins. "Yes, he had done them, secretly, filthily, time after time, and, hardened in sinful impenitence, he had dared to wear the mask of holiness before the tabernacle while his soul within was a living mass of corruption." Next day at confession he tries to talk around the key sexual sins by admitting to lesser ones, but finally "the last sins oozed forth, sluggish, filthy" (*P*, 119).

With this cruel retreat, Joyce leads readers to wonder accusingly what kind of mind, what kind of religion could think up such a scenario of eternal damnation for sexual activity, elaborate it with such vivid detail, and drive it into the minds of youths who are just learning to deal with their sexuality while learning to love. An unstated scandal of the retreat is the glaring omission of the one real mortal danger – from syphilis, which Stephen might have contracted from prostitutes and of which Joyce was acutely aware. As he wrote his brother Stanislaus in 1906, "I presume that there are very few mortals in Europe who are not in danger of waking some morning and finding themselves syphilitic."[35] Joyce's fictional retreat is a high point of anti-Christian sentiment in modernist fiction.

Joyce further indicts Christianity in *Ulysses* (1922) with mockery, blasphemy, parody, and heresy. In that novel Buck Mulligan refers to Jehovah as a "collector of prepuces" and to the Holy Spirit as a "gaseous vertebrate." Leopold Bloom parodies Jesus's miracles with some laughably useless miracles as he covers his left eye with his left ear, hangs from the top ledge of Nelson's Pillar by his eyelids, and eats twelve dozen oysters including the shells.[36] Thus does Joyce mock Catholicism, especially as directed against Catholic efforts to hamper sexuality in the service of its central judgment on the mortal sin of lust, which in turn stifled the expression of sensuous love that he cherished.

Joyce's contempt for Christian sexual morality became evident during his relationship with Nora Barnacle, his youthful lover and eventual mother of his children, lifelong companion, and wife. In 1904 Joyce fell in love with Nora as he became increasingly antireligious.[37] He announced that position to her in a letter of 1904: "Six years ago I left the Catholic Church, hating it most fervently. I found it impossible for me to remain in it on account of the impulses of my nature. I made secret war upon it when I was a student and declined to accept the positions it offered me. By doing this I made myself a beggar, but I retained my pride. Now I make open war upon it by what I write and say and do." He underscored the connection between his love for Nora and his contempt for Catholicism when in 1904 he wrote to an aunt that he is bonded to Nora, who trusted him enough to accompany him unmarried into exile, and he concluded forcefully, "I spit upon the image of the tenth Pius."[38]

During the "Ithaca" episode of *Ulysses* Joyce adapts the technique of catechism to explore a possible recovery of love between Bloom and his wife Molly, who have been sexually alienated for over ten years since the death of their infant son Rudy. This expansive use of catechism contrasts with its constrictive use in Catholic ritual. The episode concludes with a repeated celebration of *natural* loving instincts that portend a renewal of Bloom's love for Molly as he works through jealousy over her adulterous liaison that very afternoon to achieve a "natural state of equanimity as natural as any and every natural act of a nature expressed or understood executed in natured nature by natural creatures in accordance with his, her, and their natured natures" (*U*, 603). That emphatically natural experience rejects the paralyzing effect that an emphatically anti-natural Catholic sexual morality threatened to have on Bloom's love life.

The arc of Joyce's love life was from the early warnings about sinful sexual desire to the years of Jesuit moralizing and his transformative relationship with Nora, through which he cultivated an understanding of the unifying function of natural sexual love that he dramatized in fiction. *Ulysses* is a story of the potential loving reunification of Bloom and Molly during a day when they start off exceptionally disunited, because on that day she is unfaithful to him and during a nighttime interior monologue recalls in detail her climactic sexual pleasure with Boylan. Her account of that rupture from Bloom is, however, followed by an account of her love for him. Joyce dramatizes an array of fragmenting experiences that made up their respective lives over the years, a process that proceeds to the end of the novel in a climax that affirms the promise, if not the actuality, of unifying love with Molly, following her resounding final "Yes." *Ulysses* is an assault on the paralyzing effect of Catholic sexual morality on love as well as a unique probing of the origins, foundations, and possibilities of love in the modernist age.

D. H. Lawrence

Lawrence probed the origins and foundations of love in his life and work that took shape as an assault on the Congregationalism of his early years. The early source of that religious foundation was similar to Gide's – an emotionally demanding and puritanical mother who had a deep sense of sin that interfered with her son's budding love relationships. He claimed to have "gotten over" Christian dogma by age fifteen, but into his late teens he still considered the ministry as a profession.[39] In 1929, he looked back resentfully on his religious indoctrination:

> From earliest years into manhood, like any other nonconformist child I had the Bible poured every day into my helpless consciousness, till there came almost a saturation point. Long before one could think or even vaguely understand this Bible language, these "portions" of the Bible were *douched* over the mind and consciousness, till they became soaked in, they became an influence which affected all the processes of emotion and thought. So that today, although I have "forgotten" my Bible, I need only begin to read a chapter to realize that I "know" it with an almost nauseating fixity. And I must confess, my first reaction is one of dislike, repulsion, and even resentment.

The repulsion came from the way the Bible was "verbally trodden" into his consciousness. As an artist and lover, he also recoiled from the sanctimoniousness of religious instruction: "the pie-pie [pious] mouthing, solemn, portentous, loud way in which everybody read the Bible, whether it was parsons or teachers or ordinary persons."[40]

In later years he had increasing contempt for Christian sexual morality that he believed imperiled the emotional foundations of his ability to love and that of the larger English society. In *Sons and Lovers* (1913) the autobiographical protagonist Paul falls in love with Miriam, who "was cut off from ordinary life by her religious intensity, which made the world for her either a nunnery garden, a Paradise where sin and knowledge were not, or else an ugly, cruel thing." Their love is mired in sexual guilt and fear of sin. "The fact that [Paul] might want her as a man wants a woman had in him been suppressed into a shame. When she shrank in her convulsed, coiled torture from the thought of such a thing, he winced to the depths of his soul." When they finally have sex, the act is corrupted by guilt. "Then she would submit, religiously, to the sacrifice . . . And at the thought her whole body clenched itself involuntarily hard."[41]

The real-life basis for Miriam was Jessie Chambers, Lawrence's childhood friend, who was traumatized by her loss of virginity with him, which he described in the poem *Lilies in the Fire*.[42] It depicts Jessie on the ground like a "stack of white lilies." She sighs through the ordeal that concludes with a brutal image of "God stepping through [their] loins in one bright stride."[43] Along with belief in a savior Christian God, this poem also jettisons a strictly prudish Victorian sexual morality, which Lawrence devoted the rest of his life to extirpating from himself and, he hoped, his world. The loins that had been divinely debased in the twilight of Lawrence's orthodox belief became the source of a redemptive desire and love that he dramatized in novels and analyzed in essays. In rejecting the desexualizing of Christian churches, he cultivated a persistently sensuous aesthetic wonder that stimulated his creativity.

Lawrence believed that orthodox Christianity undermined the ability to love in a humanizing way. In spite of Jesus's central commandments to love God and love one's neighbor, Christianity had a negative view of the sexual desire that corrupted any physical expression of love into a mortal sin that condemned unconfessed transgressors to eternal hell. Humans were corrupted by the Fall, remain tainted by original sin as sexual desire and sexual activity, and require Jesus's crucifixion to save them from a horrible punishment. In gratitude, they ought to follow Jesus's model and avoid sin. For Lawrence, Christian morality was a "perpetual mean thou-shalt-not."[44] While that proscriptive instruction addressed all sorts of instincts, the most challenging thou-shalt-not was sexual.[45]

Lawrence especially disliked the idolatry of the crucifixion that symbolized everything he abhorred – sin, sacrifice, the Fall, despair, torture, pain, and death. While the crucifixion was supposed to symbolize forgiveness, salvation, and hope, it actually symbolized the enormity of human sinfulness that required such a horrific divine sacrifice to atone for it. That dark aspect was evident in carved wooden crucifixes jutting out of the snow, which Lawrence saw in the Tyrolean Alps in 1912. His descriptions of them make up a lexicon of morbid emotions. In one, the neck of Jesus was stiff "as if in resistance to the fact of the nails and the cross, which he could not escape." Another's expression evoked "an underlying meanness, secretive, cruel." Yet another crucifix had a barren face "with a dead expression of weariness and brutalized with pain."[46]

Lawrence's most powerful fictional treatment of the love-negating symbolism of the crucifixion is *The Escaped Cock* (1929), a retelling of the story of Jesus not as Christ the savior but as a man who bears emotional scars of the crucifixion. It interlaces the Egyptian myth of Isis with the crucifixion and resurrection of Jesus. Isis believes that Jesus (not mentioned by name), who wanders into her temple, is her husband Osiris, who had been killed and torn apart by enemies. She must reassemble Osiris's body to enable him to make love with and impregnate her. She has assembled all the parts except the genitals, which she has finally found in Jesus, but he is afraid of any touch, because the last time humans touched him, they nailed him to a cross. He meets Madeleine (Mary Magdalene) and, as she attempts to kiss his feet, he says "Noli me tangere" (touch me not), the words that Jesus supposedly said to Mary Magdalene when she saw him after the resurrection, meaning that once the resurrection is accomplished, the link between human beings and his person must be no longer physical but a bond of heart to heart. Later Jesus repeats that response as he resists Isis's touch. He realizes that the love he had offered to Mary Magdalene and all other women before the crucifixion had been deadly: "I asked them all to serve me with the corpse of their love. And in the end, I offered them only the corpse of my love. This is my body—take and eat—My corpse."[47]

The crucified Jesus was essentially a castrated Jesus, who, according to the Gospels, never had sex, but Lawrence resurrects his genital function. Isis revitalizes his desire with tenderness and love by anointing his "slain penis" with oil and pressing her breast against his abdominal scar. As he recovers from his wounds, he announces his sexual arousal with a blasphemous pun, "I am risen." This Jesus's resurrection is an erection, and he is ready to impregnate Isis. This story's title has multiple meanings in that the escaped cock signifies an actual rooster that escaped from his peasant owner, Osiris's missing penis, Jesus's lifetime celibacy, and the suppressed phallus of Christendom. The restoration of these missing cocks enables the peasant to get eggs from his hens, enables Jesus to come to life as a whole man capable of sexual desire and sexual intercourse, enables Isis to become pregnant with a newly potent god Osiris, and enables contemporary men and women damaged by Christian sexual morality to read about what it is like to be resurrected as a whole human being capable of fulfilling sensuous love.

Lawrence was a sounding board for revolutionary ideas on sex and love in the modernist era. In *Women in Love*, he dramatized Nietzsche's contempt for Christian asceticism with fictional characters who suffer from its excesses. He perhaps unintentionally misinterpreted Nietzsche's central conception of the will to power as a bullying force, but the larger body of his work celebrated the life-affirming, creative power that Nietzsche generally intended it to be. In *Lady Chatterley's Lover* (1928) Lawrence was emphatically concerned about men and women achieving mutual understanding and reciprocal love with passionate sex and simultaneous orgasms.

Lawrence evidently read little if any Freud but drew inspiration from popularizations of his works and got a personal infusion of his ideas through his wife Frieda. In spite of little direct knowledge, Lawrence somehow managed to write two books about Freud: *Psychoanalysis and the Unconscious* (1921) and *Fantasia of the Unconscious* (1922). He found much to dislike in psychoanalysis: Freud acquired knowledge from clinical interviews and rational analysis, while Lawrence acquired it from lived experience and intuition; Freud sought the unconscious origins of conscious experience, while Lawrence sought "blood consciousness," a metaphor for the pulsating movement of life. He also disapproved of Freud's focus on the foundations of sexuality in erotogenic zones and rather focused on foundations in a chakra at the base of the spine.

On balance, however, Lawrence was a kindred spirit to Nietzsche and Freud in championing the importance of sex and love in unifying individuals emotionally and making possible fulfilling relationships between them. These three men shared central modernist goals of probing the origins and foundations of love life and in fighting sexual prudery, ignorance, and fear, which cut off men and women from sensuous foundations of emotional wholeness. Collectively they took the traumatized, wounded body off the cross and celebrated in their respective ways the joyous Dionysian body, the creative power of libido, and the resurrection of the flesh.

The life and work of Gide, Joyce, and Lawrence align as a trio of cultural challenges to the respective Christian faiths in which they were raised. Those faiths were drummed into them during their childhood and created lasting religious foundations that they came to resent but still returned to in variously recreated forms in later years.[48] Gide had an abiding appreciation of the words of Jesus and found solace in praying to a mystical god, but he decried the asceticism and moral judgments of contemporary clerics. In *Portrait* Joyce's autobiographical protagonist announces in prophetic language that as an artist he will secularize the spirit of Catholicism as "a priest of the eternal imagination, transmuting the daily bread of experience into the radiant body of everlasting life" (*P*, 195). Lawrence continued to enjoy Christian hymns, images of resurrection, the majesty of churches, and parts of the Bible, especially Genesis, Exodus, the Gospel of John, and the Book of Revelation. Nevertheless, at the core of their views of Christianity remained deep resentment over the damage that Christian sexual morality did to their ability to experience fulfilling love.

The focus of this chapter on modernist opposition to Christian moral instruction with its emphasis on the sinful nature of sexual desire aligns with my larger argument regarding modernist ideas about the origins and foundations of human existence generally. The Christian doctrine of original sin identified the origin of human existence at conception, resulting from a sinful act that could only be absolved by church officials in accord with the unfathomable gift of God's grace. Taking ultimate responsibility for giving life moral respectability and meaning away from human beings and relocating it as the responsibility of priests and Gods made atheism a tempting way for individuals to "save" themselves from the taint of sin.

I have focused on the historical significance of modernist objections to Christian sexual morality because it offers the most passionately held basis of modernist secularist thinking about the foundations of love. Another reason for modernist anti-Christianity was also of major historical significance - the rejection of Christian ideas about the origin of existence generally, which was succinctly surveyed in the opening chapters of Genesis when, readers are told, God created all of existence out of nothing in six days. That historical explanation was quintessentially comprehensive and easy to grasp, and it answered the most basic originary questions. While Victorians did not always take that thinking as literally true, they were more inclined to accept it as a life-affirming myth than were modernists.

This chapter on religion has focused on the microhistory of modernist secularism in three major novelists. The broader historical record of modernist secularism also includes the macrohistory of Germany, Austria-Hungary, and Russia, where, during the course of World War I and its aftermath the ruling monarchs lost their authority based on the doctrine of the divine right of kings and were driven from power. That scenario was enacted in the extreme with the Russian tsar, who was brutally killed and incinerated along with his entire family and entourage to obliterate any possible return of his empire or the divine right upon which it might be based. That emphatic rejection of divine right and of the presence of God behind monarchical rule was a momentary gesture of rejection of a religion that continued to celebrate the death of Jesus in Russia and elsewhere with representations of his execution on the cross. While the cross also implied his resurrection and eternal reign and was worn as an inspiration to emulate his goodness, the underlying indelible image was still a memento of his torture and execution.

Notes

1. David Friedrich Strauss, *The Life of Jesus Critically Examined* (C. F. Osiander, 1835) and Ernst Renan, *Life of Jesus* (Michel Lévy Frère, 1863).
2. Friedrich Nietzsche, *The Gay Science*, trans. Josefine Nauckhoff (Cambridge University Press, 2001), 120.
3. Charlotte Brontë, *Jane Eyre* (Penguin, 2006), 471.
4. Ibid., 480, quoting from Revelation 21:8.
5. The American Historical Association was founded in 1884, the American Anthropological Association in 1902, and the American Sociological Association in 1905.
6. These agnostics and atheists include Antonin Artaud, A. J. Ayer, Henri Barbusse, Georges Bataille, Clive Bell, Vanessa Bell, Ruth Benedict, Gottfried Benn, Annie Besant, Franz Boas, Niels Bohr, Bertolt Brecht, André Breton, Luis Buñuel, Rudolf Carnap, Anton Chekov, Ivy Compton-Burnett, Joseph Conrad, Benedetto Croce, Clarence Darrow, John Dewey, Otto Dix, Theodore Dreiser, W. E. B. DuBois, Marcel Duchamp, Isadora Duncan, Émile Durkheim, Albert Einstein, Sergei Eisenstein, Max Ernst, Anatole France, James Frazer, E. M. Forster, Sigmund Freud, André Gide, Kurt Gödel, Emma Goldman, Maxim Gorky, Robert Graves, George Grosz, Martin Heidegger, Ernest Hemingway, Hannah Höch, Langston Hughes, Aldous Huxley, Henrik Ibsen, James Joyce, Franz Kafka, John Maynard Keynes, Alexander Kojève, Nella Larsen, D. H. Lawrence, Wyndham Lewis, H. P. Lovecraft, Desmond MacCarthy, Gustav Mahler, Bronislaw Malinowski, Katherine Mansfield, Filippo Marinetti, Roger Martin du Gard, Henri Matisse, Margaret Mead, H. L. Menken, G. E. Moore, George Moore, Otto Neurath, Friedrich Nietzsche, José Ortega y Gasset, George Orwell, Francis Picabia, Pablo Picasso, Luigi Pirandello, Sergei Prokofiev, Marcel Proust, A. R. Radcliffe-Brown, Dorothy Richardson, Rainer Maria Rilke, Bertrand Russell, Margaret Sanger, George Santayana, Jean-Paul Sartre, Moritz Schlick, Erwin Schrödinger, George Bernard Shaw, Gertrude Stein, Wallace Stevens, Lytton Strachey, August Strindberg, Italo Svevo, Algernon Swinburne, Paul Valéry, Carl Van Doren, Sylvia Townsend

Warner, Max Weber, H. G. Wells, Oscar Wilde, Leonard Woolf, Virginia Woolf, and Frank Lloyd Wright.

7. Peter Corviello and Jared Hickman, "Introduction: After the Postsecular Turn," *American Literature* 86, no. 4 (2014): 646.
8. Friedrich Nietzsche, *The Anti-Christ, Ecce Homo, Twilight of the Idols, and Other Writings* (Cambridge University Press, 2005), 45, 38, 63, 66.
9. Peter Baehr and Gordon C. Wells eds., *The Protestant Ethic and the "Spirit" of Capitalism and Other Writings* (Penguin, 2002), 3, 2, 73, 89.
10. Quoted in S. J. McGrath, *The Early Heidegger and Medieval Philosophy: Phenomenology for the Godforsaken* (Catholic University of America Press, 2006), 49.
11. Bertrand Russell, "On the Notion of Cause" [1912], in *Mysticism and Logic* (Doubleday, n.d.), 174.
12. Bertrand Russell, "Why I Am Not a Christian," in *Why I Am Not a Christian and other Essays on Religion and Related Subjects* (Allen & Unwin, 1956), 4.
13. Letter to Emma Vaughan, December 1902, *The Letters of Virginia Woolf, Volume I: 1888–1912*, ed. Nigel Nicolson (Harcourt, 1975), 64.
14. Ibid., 448.
15. Virginia Woolf, *Letters Volume Three, 1923–1928* (Harcourt, 1978), 457–58.
16. Virginia Woolf, *The Voyage Out* (Harvest, 1920), 228–29.
17. Virginia Woolf, *Mrs. Dalloway* (Harcourt, 1925), 124–25.
18. On these developments see Stephen Kern, *Modernism after the Death of God: Christianity, Fragmentation, and Unification* (Routledge, 2017).
19. André Gide, *If It Die . . . An Autobiography*, trans. *Dorothy Bussy* (Random House, 1935), 178–79.
20. André Gide, *The Journals of André Gide*, trans. Justin O'Brien (Knopf, 1947), vol. 1, 16.
21. Gide, *If It Die*, 238–39.
22. In 1921, Gide's friend, the novelist Roger Martin du Gard, also a homosexual, recorded Gide's preferred method of gratification:

> "Gide needs to empty himself out completely of sperm, and he reaches this state only after coming five, six, or even eight times in succession . . . First he comes twice . . . characterized by two distinct ejaculations, one immediately after the other . . . The third one happens soon after. He can rarely come more than three times with the same person . . . So when he leaves the place where he's come three, four, or even five times in less than an hour, he goes away unsatisfied. He is in such an unbearably tense, uncertain, unsatisfied state that he can think of only one thing: getting home and masturbating as many times as it takes to reach the point of total emptiness.
>
> "Gide then explained how his unusual sexual practices might derive from his Calvinist upbringing. He found only one possible explanation: the fact that as a child he had tried repeatedly to cure himself of masturbation, especially at a period of his life when this self-discipline was mixed-up with a religious prohibition. During that period, which went on beyond his adolescence, the idea of sin was fixed so powerfully in his

mind that he felt he was committing a lesser wrong if he did not reach a complete orgasm. Confusing the sin with the actual ejaculation, he acquired a skill in onanism, so that he could reach the exact point where he had almost reached orgasm but the ejaculation, the spasm had just not yet started. This series of shockwaves without culmination, which had become a regular habit with him, may have been the cause of the peculiarities of his disposition today, for he would repeat them often, sometimes a whole night long, without ever allowing himself to commit the 'full sin.'" Roger Martin du Gard, *Journal* (Gallimard, 1993), vol. II, 232–33. Quoted in Naomi Segal, *André Gide: Pederasty and Pedagogy* (Clarendon, 1998), 43–45. If Martin du Gard's account is correct, full sexual satisfaction could only be achieved by means of a demanding ritual that exceeded the bounds of normal human suppression and was imposed by enormous religiously inspired guilt.

23. Gide, *Journals of André Gide*, vol. I, 23.
24. *If It Die*, 241, 252.
25. Gide, *Journals of André Gide*, vol. I, 24, 33, 259.
26. Ibid., vol. II, 120, 179–80, 415.
27. Ibid., vol. IV, 15–16.
28. André Gide, *Strait Is the Gate* (Mondial, 2007), 9–11. In a study of the destructive effect of puritanism on female characters that compares Gide's Alissa and Lawrence's Miriam, Richard T. Driskill argues that "both authors understood that Christianity had somehow enlisted certain strains of romance to form a pervasive cultural mythology that encouraged the sensitive young women of their epoch to be virginal, passive, and receptive to suffering . . . Gide and Lawrence reveal the destructive presence of these cultural influences upon both their female and male characters." *Madonnas and Maidens: Sexual Confusion in Lawrence and Gide* (Peter Lang, 1999), 2.
29. Gide, *Strait Is the Gate*, 15, 64.
30. Ibid., 90, 93, 94, 96, 102.
31. André Gide, *Lafcadio's Adventures* (Vintage, 1953), 230.
32. Kevin Sullivan, *Joyce among the Jesuits* (Columbia University Press, 1958), 116.
33. Richard Ellmann assumes that Joyce was consorting with prostitutes in 1896 at age fourteen based on references at the end of Part 2 of *Portrait* where Stephen Dedalus succumbs to a woman of the streets "body and mind" and at the beginning of Part 3 where Stephen wanders among "whores" in "the squalid quarter of the brothels." *James Joyce* (Oxford, 1959), 48, 30. On differences between the sexual experiences of Joyce and Stephen see Peter Costello, *James Joyce: The Years of Growth, 1882–1915* (Pantheon, 1992), 141, 151 and *Joyce among the Jesuits*, 124–26.
34. James Joyce, *A Portrait of the Artist as a Young Man* (Norton Critical Edition, 2007), 105–8. Further text references are to this edition.
35. *Letters of James Joyce*, Richard Ellmann ed. (Faber, 1996), vol. II, 192.
36. James Joyce, *Ulysses* [1922] (Vintage, 1986), 165, 28, 175, 404. Subsequent text references are to this edition.
37. Gert Lernout concludes that "Joyce's most active anticlerical phrase was certainly between 1904 and 1914," the years following his devout mother's

death when he lived in exile in Pola, Trieste, and Rome and was writing *Portrait.*" *Help My Unbelief, James Joyce and Religion* (Continuum, 2010), 103.

38. Ibid., 49.
39. John Worthen, *D. H. Lawrence: The Early Years 1885–1912* (Cambridge University Press, 1991), 172.
40. D. H. Lawrence, *Apocalypse*, written in 1929–30 and published posthumously in 1931 in *Apocalypse and the Writings on Revelation* (Cambridge University Press, 2002), 59, 61.
41. D. H. Lawrence, *Sons and Lovers* (Cambridge University Press, 2002), Part I, 179, 216; Part 2, 328.
42. Worthen argues that Lawrence was a virgin when he first had sex with Jessie in 1910. *D. H. Lawrence: The Early Years*, 248–52.
43. D. H. Lawrence, *The Poems*, Christopher Pollnitz ed. (Cambridge University Press, 2013), vol. I, 52–54.
44. Lawrence, *Apocalypse*, 72.
45. "Throughout his career," Paul Paplowski concludes, "Lawrence had defined and refined his own religious views against the Christian framework of concepts and beliefs that he had been brought up with." *Promptings of Desire: Creativity and the Religious Impulse in the Works of D. H. Lawrence* (Praeger, 1993), 156.
46. D. H. Lawrence, "The Crucifix Across the Mountains" [1916] in Paul Egger ed., *Twilight in Italy and Other Essays* (Cambridge University Press, 1994), 91–100.
47. D. H. Lawrence, "The Escaped Cock" in *The Virgin and the Gypsy and Other Stories* (Cambridge University Press, 2005), 157.
48. For a fuller survey of their "ambivalence about Christianity" see the sections under that heading in chapters on these men along with other chapters on Nietzsche, Heidegger, Freud, and Woolf in my earlier study, *Modernism after the Death of God.*

Chapter 10

LITERATURE: Narrative Beginnings in Faulkner, Joyce, Proust, Woolf, Gide, Stein, Rhys, Richardson, and Hall

The historical distinctiveness of modernists' literary technique is most evident in their criticism of the realist technique of their Victorian predecessors. Gide criticized the psychological and social determinism of Émile Zola and Paul Bourget. Woolf rejected the "materialist" novels of H. G. Wells, Arnold Bennett, and John Galsworthy for similar reasons and concluded dismissively, "the sooner English fiction turns its back upon them . . . the better for its soul."[1] Willa Cather called for a "defurnished novel" stripped of the descriptive excess that clutters Balzac's realist novels.[2] Marcel Proust rejected "the falseness of so-called realist art" that dramatizes how people live in space rather than focusing on how they live in time.[3] The strong plots that structure Victorian realist novels were further rejected by Henry James, E. M. Forster, and Gertrude Stein.[4] In 1914 the British painter and novelist Wyndham Lewis took aim at the years of Victoria's reign when he announced in the journal *Blast* on the eve of World War I that "the years 1837–1901" are "blasted." Ezra Pound implied the obsolescence of everything that came before with his repeated injunction for artists to "make it new."[5]

Leading modernists saw themselves as an avant-garde that explored formerly neglected topics, as in Virginia Woolf, or tabooed topics, as in D. H. Lawrence, or used new formal techniques, as in Joyce, Proust, Stein, Richardson, and Faulkner. One particularly innovative originary technique was the antibeginning. Victorian writers, particularly third-person narrators, typically wrote with the confidence that their novels had a precise beginning that they knew and could relate clearly, as for example: "In the year 1815 Monseigneur Charles-François-Bienvenu Myriel was Bishop of Digne" (Victor Hugo, *Les Misérables*, 1862); or "On the 15th of September 1840, at six o'clock in the morning, the *Ville-de-Montereau* was lying alongside the Quai Saint-Bernard, ready to sail" (Gustave Flaubert, *Sentimental Education*, 1869); or "Towards the end of November, during a thaw, at nine o'clock one morning, a train on the Warsaw and Petersburg railway was approaching the latter at full speed" (Fyodor Dostoevsky, *The Idiot*, 1868–9). The first paragraph of George Eliot's *Adam Bede* (1859) begins with a clear statement of authorial confidence: "With this drop of ink at the end of my pen I will show you the roomy workshop of Mr. Jonathan Burge, carpenter and builder in the village of Hayslope, as it appeared on the eighteenth of June, in the year of our Lord 1799." Eliot is confident that she knows everything she needs to know about the origin of her story and precisely when it occurred and can communicate it to readers who expect confident narratorial authority. She is also confident that she shares with her readers the one and only collective time frame of "our Lord." While Victorians were comforted by such a notion, modernists felt oppressed by it.

Fifteen years later, in the epigraph to *Daniel Deronda*, Eliot questioned her earlier confidence. "Men can do nothing without the make-believe of a beginning . . . Even Science, the strict measurer, is obliged to start with a make-believe unit, and must fix on a point in the stars' unceasing journey when his sidereal clock shall pretend that time is at Nought." Ultimately, she concluded, "no retrospect will take us to the true beginning."

That later Eliot speculation, uncharacteristic of most Victorians, became a strong conviction in the modernist period – that lives do not have a single or clearly determinable beginning. The idea that literary beginnings are a pretense contrasts with the Victorian presumption that time zero of novels is a reliable starting moment, a narrative nucleus from which the rest of the story follows ineluctably. Modernists learned to do without the "make-believe" of a "true beginning" and reworked formal foundations of novels more generally with new narrative techniques including the continuous present, stream of consciousness, tunneling, free indirect discourse, and beginning again.[6]

Modernists also became increasingly skeptical about Victorian presuppositions of the sacred origin and divine foundation of nations, families, and individual lives. That skepticism liberated modernists to appreciate the miraculous nature of a miracle-less world. They were increasingly able to accept that they came into existence from accidental biological processes such as mutations and lived governed by random environmental and social circumstances. While some Victorian novels also began in medias res, they generally circled back to reveal a decisive beginning event, as modernists resisted doing. Modernists started their novels with innovative introductions to settings, selfhood, families, religion, and homes.

Settings

One striking shift from Victorian objective descriptions of intelligible settings to modernist subjective immersions in ambiguous settings is illustrated by a comparison of beginnings by Thomas Hardy and William Faulkner.

The beginning of Hardy's *The Return of the Native* (1878) evokes an omniscient-like knowability of the setting. "Overhead the hollow stretch of whitish cloud shutting out the sky was as a tent which had the whole heath for its floor." The chapter continues with a god's-eye survey of the heath's light, climate, terrain, and vegetation that are being displaced by encroaching civilization. A half century later, Faulkner's *The Sound and the Fury* (1929) begins with a landscape that is also being displaced by encroaching civilization but is stunningly discordant with the person observing it. While Hardy uses objective description to invite his readers to grasp the degradation of his beloved countryside, Faulkner uses subjective immersion to confuse readers about how his first-person narrator is experiencing the degradation of his beloved countryside.

> Through the fence, between the curling flower spaces, I could see them hitting. They were coming toward where the flag was and I went along the fence. Luster was hunting in the grass by the flower tree. They took the flag out, and they were hitting. Then they put the flag back and they went to the table, and he hit and the other hit. Then they went on, and I went along the fence. Luster came away from the flower tree and we went along the fence and they stopped and we stopped and I looked through the fence while Luster was hunting in the grass.
>
> "Here, caddie." He hit. They went away across the pasture. I held to the fence and watched them going away.

The beginning of this text and the origin of the story occur in the chaos of Benjy Compson's disabled mind. Confusion abounds in the first sentence regarding the "I" who is narrating, what the fence encloses, who is hitting, and what is being hit. Readers can eventually work out that the fence encloses what had been a pasture that Benjy viewed as his own. It had been sold to become a golf course to raise money to send his brother to Harvard. The hitters are golfers, and the flags on poles mark golf holes. The call "caddie" agitates Benjy because his troubled emotional life centers on love for his long-gone sister Caddy.

The first seventy pages shift over a hundred times among eight moments between 1898 and 1928, sometimes in the middle of a sentence, as these moments tumble about in Benjy's disorganized mind.

After several readings aided by commentaries, one can gradually figure out that the day on which the novel opens, 7 April 1928, is Benjy's thirty-third birthday, although his mental development is that of a three-year-old child. Faulkner crafts this jumpy chronology to dramatize not just Benjy's condition but also the unraveling of the Compson family that formed around Victorian legacies that were losing their ability to sustain the foundational values of the Old South.

Hardy's third-person narrator sees clearly across the horizon from sky to Earth and easily reflects across time, while Faulkner's narratorial voice in Benjy's mind is trapped in a confusing present moment, trying to comprehend with the limited range of his mental disability, mixing up everything he sees. A pasture that he associates with an earlier sense of being at home in the world and loved by his sister is being desecrated by strangers who are hitting something on that pasture, mysteriously taking flags on sticks out of holes and then putting them back, calling out the name of his beloved sister who has disappeared.

Hardy's omniscient third-person narrator records what he sees like an eye panning a vast scene, while Faulkner's narrator cannot coherently process a relatively small patch of space that is right in front of him. Both initial settings portend the disruption of human relations, but Hardy's setting is mappable and understandable to his narrator and his reader, while Faulkner's setting is unmappable and incomprehensible to his character as well as, initially, to his reader. Faulkner uses Benjy's confusions to defamiliarize normative ways of conceiving the world. His attention to Benjy's conceptions of space and time underscores by exaggerating how the identification of the precise beginning of any story in space or time is indeed a pretense in the service of make-believe.

That contrast illustrates a broader divide between Victorians and modernists over the understanding of space and human experience generally. Walter Houghton shows that, while leading figures in Victorian culture were engaged in serious self-questioning about the intellectual and moral basis of life, they still hoped to find a basis and a single and unchanging point of view. They tended "to divide ideas and people into tight categories of true-false, good-bad, right-wrong."[7] The conviction that a clear spatial ordering underlay experience was tersely expressed by that quintessential Victorian Samuel Smiles, who in 1875 celebrated the Victorian ideal of "A place for everything, and everything in its place."[8]

In the face of such complacency about the secure scaffoldings of life and thought, some leading modernists reacted to a breakdown of conventional forms. In 1905 Hugo von Hofmannsthal wrote that the nature of his epoch was "multiplicity and indeterminacy" and that "what other generations believed to be firm is in fact sliding."[9] Continental drift theorists suggested that even continents were sliding around on the face of the Earth. In the wake of the explosive Armory Show in New York in 1913, the poet Mina Loy quoted Mabel Dodge as saying that "nearly every thinking person nowadays is in revolt against something, because the craving of the individual is for further consciousness, and because consciousness is expanding and bursting through the molds that have held it up to now."[10] The Austrian novelist Robert Musil characterized the change as one in which "sharp borderlines everywhere became blurred, and some new indescribable capacity for entering into hitherto unheard-of relationships threw up new people and new ideas."[11]

Selfhood

Victorians viewed the beginning of a person's life as a sacred moment that initiated a purposive development to maturity like a tree growing from a seed. That conventional Victorian simile framed novels as a *Bildungsroman*, meaning about the *Bildung* or education that harmonizes the maturation of an individual within society.

In *David Copperfield* (1850), Dickens records his narrator's birth with the authoritative tone of a public records office. "To begin my life with the beginning of my life, I record that I was born (as I have been informed and believe) on a Friday, at twelve o'clock at night." This classic *Bildungsroman* about the emergence of a self begins with this record of the precise time of David's birth, but the parenthetical remark including the passive voice indicates that individuals do not directly or actively recall their birth and that knowledge of it comes from others. Modernists crafted beginnings of characters at imprecise moments out of imaginative mental processes.

One example of a radically innovative modernist beginning is Joyce's fictionalized autobiography as Stephen Dedalus in *A Portrait of the Artist as a Young Man* (1916), which begins with an interlacing of three voices – a third-person narrator, baby Stephen, and baby Stephen's father, Simon. It begins like a Victorian fairy tale, with the third-person narrator saying, "Once upon a time," and, then uses "moocow" and "nicens little boy named baby tuckoo" as if speaking to baby Stephen.

> Once upon a time and a very good time it was there was a moocow coming down along the road and this moocow that was down along the road met a nicens little boy named baby tuckoo . . .

> His father told him that story: his father looked at him through a glass: he had a hairy face.
>
> He was baby tuckoo. The moocow came down the road where Betty Byrne lived: she sold lemon platt.
>
> *O, the wild rose blossoms*
> *On the little green place.*
>
> He sang that song. That was his song.
>
> *O, the geen wothe botheth.*
>
> When you wet the bed, first it is warm then it gets cold.

The fairy-tale beginning shades into the consciousness and language of the "nicens little boy" Stephen. Baby Stephen then sings with an infantile lisp that renders "rose blossoms" as "wothe botheth." The sensations from bed wetting are baby Stephen's, but the voice describing it is the narrator's assuming the baby's perspective. These multiple voices and views capture the confused beginnings of consciousness. Joyce implies that lives emerge out of the many impressions that babies register before they have language, mature sensory receptors, or mature emotions and that continue to take shape as they struggle to identify and respond to those impressions and voices. Anthony Burgess characterized Joyce's opening as "the first big technical breakthrough of twentieth-century prose writing."[12]

Stephen's early effort to sing portends his development as an artist, the main subject of this pioneer modernist *Künstlerroman*. Characters in the traditional *Bildungsroman* learn to fit in, while characters in a *Künstlerroman* struggle to stand out. The beginning of this novel with the protagonist swamped by input from others, including language that he did not originate and cannot process, anticipates the originary existential of "worldhood" that Heidegger proposed as the ontological foundation of human existence. The typical Victorian self was shaped by qualities that were conceived and imposed from without. Musil identified such "pseudoselves" in *The Man without Qualities*. This novel depicts an unusual modern man who rejects Victorian-style, socially determined qualities (*Eigenschaften*) in order to forge his own qualities.

Another dynamic modernist beginning is in Proust, who builds the emerging consciousness of his first-person narrator out of experiences of waking up in imprecisely identified times past, stimulated by disparate fantasies, memories, and emotions along with visual, auditory, gustatory, and tactile sensations. One time he believed that he was a character in a book he had been reading before falling asleep. Another time, while his sight was emerging, he realized that he was in darkness. Once he heard a distant train and recalled a trip and the joy of going home. Another time he felt his cheek against the pillow, struck a match to look at his watch, and thought that soon servants would come to look after him. Another time he heard the creaking of the wainscot and, while falling asleep again, recalled a childhood terror when his great-uncle pulled on his curls. Sometimes during sleep a woman would emerge from a strain in the position of his thighs, and he would awaken to discover the pleasure he was on the point of consummating with his body warmth permeating hers. These experiences have a common function, the narrator reports, to "gradually piece together the original components of my ego."[13] Proust's piecing together his ego from experiences of waking portends the more expansive piecing together of his entire life as a process of search and discovery of the significance of vast stretches of time lost and regained that is his novel's main subject.

Unlike the beginning of numerous Victorian plots about the emergence of selfhood that are recalled by a first-person narrator's voluntary memory, the beginning of Proust's main plot occurs at an unspecified time after the opening when the first-person narrator has a key *in*voluntary memory. Voluntary memories are efforts of the intellect to intentionally recall the past in some specific moment that has been forgotten, while involuntary memories occur by chance when someone has an unexpected experience (a taste, vision, touch, or sound) that unleashes powerful memories that unlock multiple ingredients of pastness as, for example, the most famous involuntary memory triggered for the narrator by tasting a petite madeleine dipped in lime-flavored tea. The pleasure of that taste made the narrator wonder, "Where did it come from? What did it mean? How could I seize and apprehend it?"[14] The launch of the main narrative begins with those originary questions and continues through seven volumes about his attempt to trace all the experiences they embody. Victorians had plenty of involuntary memories, but it took Proust's acute literary sensitivity to identify and appreciate their existential and literary value.

Later in the introductory "Overture," the narrator makes a preliminary intentional effort to recapture his lost past by again tasting the tea and madeleine. After several failures, he clears his mind and allows the memory to recur; however, that voluntary effort violates the strict definition of a truly involuntary memory. He then glimpses the brief joy he seeks to recapture but realizes that he must postpone discovering how to make it endure, a postponement that extends across the entire novel. The beginning of the story starts with the involuntary memory; its ending is triggered by five other involuntary memories in quick succession, throwbacks to multiple beginnings that initiate insights into how to sustain the joy of searching for the past, namely by making it into a work of art, in his case a novel. The novel is a series of beginnings that progress toward an ending that culminates in a deeper insight into how to transform the joy of searching for the past into an enduring work of art.

Another contrast in rendering an individual self suggests other historical differences between a pre-Victorian and a modernist beginning. The first sentence that Jane Austen devotes to her protagonist in *Emma* (1816), albeit with a touch of irony, details her name, appearance, character, wealth, home situation, and age.

> Emma Woodhouse, handsome, clever, and rich, with a comfortable home and happy disposition, seemed to unite some of the best blessings of existence; and had lived nearly twenty-one years in the world with very little to distress or vex her.

In contrast, Woolf's *Mrs. Dalloway* (1925) begins with a protagonist whose age is at first ambiguous and who, readers will learn, is *not* particularly handsome, clever, young, happy, or comfortable in her world and has much to vex her.[15] The novel dramatizes a modernist antibeginning. It starts with indirect speech to an unnamed interlocutor followed by narration that conflates events at two unidentified places and times in a free indirect discourse that shifts without traditional authorial notices distinguishing narrator and protagonist.

> Mrs. Dalloway said [to whom?] she would buy the flowers herself.
>
> For Lucy [who?] had her work cut out for her [Clarissa Dalloway thought]. The doors would be taken off their hinges [why?]; Rumpelmayer's men were coming [who are they and why are they coming?]. And then, thought Clarissa Dalloway, what a morning – fresh as if issued to children on a beach.
>
> What a lark! What a plunge! [thought Clarissa] For so it had always seemed [tense shift to unidentified past] to her, when, with a little squeak of the hinges [a sound that links present with past, but reader does not yet know how], which she could hear now [when?], she had burst open the French windows and plunged at Bourton [location of her parents' summer home] into the open air. How fresh, how calm, stiller than this of course, the air was in the early morning [thirty-three years ago]; like the flap of a wave; the kiss of a wave; chill and sharp and yet (for a girl of eighteen as she then was) [she is currently fifty-one], solemn, feeling as she did, standing there at the open window, that something awful was about to happen.

Woolf's title suggests *Mrs.* Dalloway's dependency as Richard Dalloway's wife, but the fourth sentence suggests her more independent self as *Clarissa*. A few pages later Clarissa thinks explicitly about that dependency: "this being Mrs. Dalloway; not even Clarissa any more; this being Mrs. Richard Dalloway" (*MD*, 11). Something awful did happen thirty-three years earlier when Peter Walsh interrupted "the most exquisite moment of her whole life" (*MD*, 35) just after Sally Seton kissed Clarissa on the terrace at Bourton, but in these beginning lines readers do not learn that. Austen's *Emma* begins with the protagonist clearly defined and securely stationed in the single time and space of a world of heterosexual pairings. *Mrs. Dalloway* begins with a protagonist conflicted about her marital dependency and remorseful about a rudely interrupted expression of lesbian love. In Woolf's beginning we do not know where Clarissa is, to whom she is speaking, why she is buying flowers, when these actions occur, what happened thirty-three years ago, what awful thing was about to happen back then, or what might be amiss in her marriage.

While the beginning of Victorian novels was readily intelligible on first reading, Woolf's beginning is only intelligible on a second or, better, third reading. She implies that people do not understand their own beginning as well as their subsequent life as they live it first time through, and because they live it only the first time through, they never understand it very well. Woolf sought to capture the capricious atoms of experience "as they fall upon the mind, in the order in which they fall." For her, life is indeed "like that."[16] The ultimate goal of her novels was to achieve a deeper seeing into the nature of things. She makes that goal explicit in *Mrs. Dalloway*, when the narrator explains Clarissa's reason for giving parties and, more fundamentally, her reason for existing: "to go deeper, beneath what people said (and these judgments, how superficial, how fragmentary they are!)" (122). The one-day time frame of that novel enabled Woolf to dig deeper into human experience, a feature of the one-day frame of Joyce's *Ulysses* in addition to several other one-day portions of other modernist novels.[17] As Woolf explained later in *A Room of One's Own*, the ultimate function of any novel has been fulfilled if "one sees more intensely afterwards; the world seems bared of its covering and given an intenser life."[18]

A transitional text from Victorian to modernist thinking about foundations of the self in literature is the preface to the play *Miss Julie* (1888), where August Strindberg explains: "My characters are conglomerations of past and present states of civilization, bits from books and newspapers, scraps of humanity, rags and tatters of fine clothing, patched together as is the human soul." Strindberg's view of the fragmentary and capricious nature of the individual self and the patchwork process of stitching it together anticipate a similar later view of character formation in Faulkner's *Absalom, Absalom!* (1936). In recreating events Faulkner explains how his narrators Shreve and Quentin stare at one another, "the two of them creating between them, out of the rag-tag and bob-ends of old tales and talking, people who perhaps had never existed at all anywhere, who, shadows, were shadows not of flesh and blood which had lived and died but shadows in turn of what were (to one of them at least, to Shreve) . . . quiet as the visible murmur of their vaporizing breath."[19] The historical contrast is striking. Victorians saw themselves born at a divinely blessed moment and emerging out of a coherent biological ancestry with a singular heritage as well as a personal destiny, while modernists saw themselves as born at an innocuous moment out of capricious biological processes destined to an open-ended search for meaning. That view of beginnings aligns with Heidegger's notion that individuals do not begin at any special place or time but are *thrown* willy-nilly into existence at an arbitrary place and time.

Family

Victorian novels about families are typically structured around what Patricia Tobin aptly calls "the genealogical imperative" that drove family histories deriving from an ancient couple whose union was transacted with exchanges of money and property, celebrated in a ceremony sanctioned by God, embodied in family heirlooms, recorded in family Bibles, secured by wills, and locked in by formidable divorce restrictions. The Victorian family was indeed a fortress, not the "shelter . . . from all terror, doubt, and division"[20] that John Ruskin praised in "Of Queen's Gardens" (1865) but an institution that constricted women's as well as men's ideas, emotions, bodies, movements, and opportunities.

Victorian novels driven by the genealogical imperative begin with introductory sketches of parents or more distant ancestors. For example, the first chapter of Dickens's *Martin Chuzzlewit* (1842–4) is titled "Introductory, Concerning the Pedigree of the Chuzzlewit Family." Although the tone is ironic, the opening line confirms that the family's pedigree was of "extreme antiquity," and the remainder of the chapter elaborates that ancestry. The narrator of Gottfried Keller's *Green Henry* (1855) begins by explaining, "My father belonged to the peasantry of an ancient Alemannic village." In "the passing of centuries . . . a feudal lord adopted the name [of the race in that village] as his title and built a castle." In Hardy's *Tess of the d'Urbervilles* (1891), Tess's troubles start with her father's learning that he is "the lineal representative of the ancient and knightly family of the d'Urbervilles." Tobin projects her argument ahead historically in concluding that "the demise of genealogical thinking is roughly coterminous with the swerve away from linear narrative in twentieth-century novels."[21]

Two novels by André Gide dramatize that swerve. In *Lafcadio's Adventures* (1914), Gide challenges the nineteenth-century convention of reconstructing a character's ancestry back several centuries by deriding Victorian novels about families that are legitimized by a family pedigree and sustained by filial devotion. Gide mocks those conventions by tracing Lafcadio's pedigree to an obscure ancestor "who in 1514 married his second wife Filippa Viconti, a few months after the annexation of the Duchy to the papal States." Interrupting this tedious and irrelevant family history, the narrator critiques himself: "it would be . . . not very interesting to trace the family fortunes up till 1807."[22] So, he skips two centuries and resumes but with equally tedious, more recent family background that is still irrelevant to Lafcadio's character, because, Gide jokes, Lafcadio was sired by an indeterminable one of five "uncles" whom his mother kept about for amusement. Gide augments this assault on traditional lineages by having the novelist Julius express his philosophy of *inconséquence* – that individuals ought to act freely, that is inconsequentially in response to the present moment and not determined by ancestors.

In *The Counterfeiters* (1925) Gide further contests conventional family values in writing that he preferred to see his characters as orphans, "unmarried and childless," and has a stand-in for himself, the novelist Edouard, proclaim bitterly that "It is to bastards that the future belongs."[23]

The broad cultural implications of Gide's title are the focus of Jean-Joseph Goux's argument that *The Counterfeiters* questions the legitimate authority of the father in the family as well as language and money because they were all counterfeit at a time when France and England were in the process of converting from gold to paper money. The increasingly complex needs of international banking with rapid transfers of money over vast distances facilitated by new communication technologies made gold an impractical standard. The system collapsed in 1914 as the strain of war and national rivalries necessitated suspending convertibility back from paper to gold. After the war, when Gide was writing the novel, Europe and the United States tried to return to the gold standard to retrieve prewar stability, but that attempt instead destabilized economies. Gide historicizes this shift by setting the novel in the prewar years when gold was the primary currency and of "real value" but writing it after the war when Western countries attempted to replace the authority of gold by the "counterfeit" authority of paper. As Goux notes, "The 'counterfeiting' of the title reaches beyond monetary fraudulence to broach the question of the ground upon which values and meaning are based: counterfeiting becomes the central metaphor for calling into question the role of general equivalents . . . [Not] only do language and money . . . come under attack, but in addition the value of paternity, and all other values that regulate exchange."[24]

Gide's formal intention in *The Counterfeiters* is that the story should remain in a perpetual present moment, forever beginning again. To achieve that effect, he explains, "the action *would not* get under way."[25] Characters appear without any seeming connection to one another and without strong initiatory motives for their actions from their past; all acts should be rather "gratuitous."[26] Two years after the publication of *The Counterfeiters* he explained how while writing his novel he was "constantly discovering that this or that which has happened previously ought to be included. Thus, the chapters are not added one after the other; but they are continually pushing back the chapter I originally conceived as the first."[27] But of course he never gets back to any ultimate originating moment.

In *Lafcadio's Adventures*, the main character attempts to will himself into significance by committing a motiveless murder to underscore the fundamentally capricious nature of human action, but he fails because he did have a motive, namely to kill without a motive. Gide's insistence that the end of *The Counterfeiters* "must not be neatly rounded off, but rather disperse, disintegrate" implies an opposite process at its start – that the novel should slide into action and have no definitive beginning.[28]

Another modernist skeptic about knowing beginnings and their role in family life was Gertrude Stein. Her novel *The Making of Americans: Being the History of a Family's Progress* (1911) shows why we cannot know the beginning of our families or our lives. The "Making" in her title is ironic because the making of the Hersland family repeatedly returns to beginnings with her innovative narrative technique of "beginning again." Moreover, *The Making of Americans* is actually about the *un*making of Americans as well as about increasing skepticism about knowing the beginning of families or individuals.

Early in this novel Stein tries to identify the foundation of an individual's personality or, as she termed it, their "bottom nature," but after much study she concludes that knowledge of anyone's bottom nature is unattainable.[29] Failing that inquiry, she probes the origins of the Hersland family (a version of her own family), but that story repeatedly doubles back with her narrative technique of beginning again. She repeatedly theorizes about such repetitive beginnings as, for example, her account of the family patriarch David Hersland: "Beginning was all of living to him, he had it in him in his beginning to be as big as all the world around him. He had it in him to be always beginning,

beginning was living to him, and this will came out in the history of him. He had it in him to be always beginning," and on and on, again and again.[30]

Stein's repetitions force the reader to think about the meaning of *beginning* until the reader begins to do so in a fresh and productive way. She makes the same move with roses in her famous phrase "a rose is a rose is a rose."[31] Her rationale for that repetition is that over centuries the word *rose* had become dulled by countless clichéd usages so that people no longer experienced the richness and beauty of an actual rose. She attempts to resurrect that richness and make a similar move with understanding the meaning of beginnings. In always wanting to know what happens next in strongly plotted novels or in living their lives, people remain oblivious to a thoughtful examination of their beginning, but in Stein's novel the reader is forced to reflect on the difficulty of understanding any family's beginning. Modernist novelists' interrogations of the genealogy of families deepened understanding of how families do or do not begin.

Religion

Few main characters in Victorian novels are profoundly religious, and those who are religious have glaring deficiencies: Claude Frollo's cruelty in *The Hunchback of Notre Dame*, Robert Brocklehurst's sadism in *Jane Eyre*, Bradley Headstone's jealousy in *Our Mutual Friend*, and Abbé Bournisien's aloofness in *Madame Bovary*. A number of novels about religious crises appeared later in the century: Mrs. Humphrey Ward's *Robert Elsmere* (1888), Harold Frederic's *The Damnation of Theron Ware* (1896), and Samuel Butler's *The Way of All Flesh* (1903, completed in 1885), but the protagonists in these novels remain respectful of Christian faith even though Christian churches and dogmas make their life difficult. One is hard pressed to find a Victorian novel that begins with and centers on a serious religious matter and is grounded in religious faith. Anthony Trollope's *Barchester Towers* (1857) is an exception, with its beginning: "In the latter days of July in the year 185—, a most important question was for ten days hourly asked in the cathedral city of Barchester, and answered in every hour in various ways – Who was to be the new Bishop?" The novel exposes the hypocrisy of individual clergy but remains respectful of the institution of Christianity, as did most Victorian novels.

While modernists were increasingly skeptical of religious zealotry and contemptuous of religious hypocrisy, they did not go as far as Joyce, who in the opening to *Ulysses* ridiculed the most sacred foundational rite in Christendom with a mock Mass.

> Stately, plump Buck Mulligan came from the stairhead, bearing a bowl of lather on which a mirror and a razor lay crossed. A yellow dressinggown, ungirdled, was sustained gently behind him on the mild morning air. He held the bowl aloft and intoned:
>
> Introibo ad altare Dei.
>
> Halted, he peered down the dark winding stairs and called out coarsely:
>
> Come up, Kinch! Come up, you fearful jesuit!
>
> Solemnly he came forward and mounted the round gunrest. He faced about and blessed gravely thrice the tower, the surrounding land and the awaking mountains. Then, catching sight of Stephen Dedalus, he bent towards him and made rapid crosses in the air, gurgling in his throat and shaking his head ...
>
> [Mulligan continues] – For this, O dearly beloved, is the genuine christine: body and soul and blood and ouns. Slow music, please. Shut your eyes, gents. One moment. A little trouble about those white corpuscles.

Joyce begins his parody with Buck Mulligan as a mock celebrant. He is wearing a dressing gown instead of an alb and chasuble, and its yellow color is unsuited for a priest because it suggests degradation and deceit. Being ungirdled leaves Mulligan's genitals open to the air, contradicting chaste priestly dress. The gun rest of the Martello tower is the altar, suggesting war and not peace, and his shaving bowl is the chalice. The mirror represents the paten that would normally hold the Host, and Mulligan uses it prosaically to avoid cutting himself. The razor symbolizes slaughter, suggesting animal sacrifices from pagan cults and priests as butchers. During the elevation Mulligan refers to a blasphemous feminized Host "Christine" that exalts the profane female body in place of the sacred "body and soul and blood and ouns [wounds]" of Jesus. The chalice contains shaving lather instead of red wine, and the entire ceremony belittles the holiest moment in the Mass, the transubstantiation of wine into blood, as merely "a little trouble."

This beginning of Joyce's signature modernist novel undermines the fundamental structuring principle in the Western world that was anchored by an ultimate beginning in divine creation, followed by the originary moral event of the Fall, followed by the originary Incarnation of a divine Jesus that restarted human time redeemed from the punishments of sin by His sacrifice that was subsequently celebrated and reenacted in the Mass. Joyce profanes this ultimate originary sacrament with this burlesque of a Mass, which is a warm-up for his later ultimate vengeance against a vengeful God with a hallucination of a Black Mass in the "Circe" episode. This takes place on the body of a naked and pregnant woman as an altar and is performed by a celebrant with two left feet back to front and a carrot stuck in his gray hairy buttocks, who begins with a prayer welcoming the devil and elevates a Host dripping with blood (*U*, 489). Thus does Joyce magnify his opening blasphemy hurled against the possibility of an ultimate sacred beginning with a subsequent blasphemy hurled against any hope for divine salvation. This travesty of Catholicism could not justify the beginning or foundation of human existence or offer any hope for divine salvation.

Home

In addition to challenging temporal beginnings, modernists also challenged ontological foundations. Victorians tended to believe that their lives were destined to occur in some enchanted place, a secure foundation that they called "home." The quintessential Victorian interpretation of the Victorian home is John Ruskin's essay "Of Queen's Gardens" (1865), which defines "the true nature of home [as] the place of Peace, the shelter, not only from all injury, but from all terror, doubt, and division." It is "a temple of the hearth watched over by Household Gods."[32] In 1936 the historian G. M. Young looked back on the family through the eyes of Victorians as "a stable and fortified center."[33] More recently Peter Cominos identified a challenge to that notion of the Victorian family: "The revolt, creating tensions, conflicts, deviant social characters, and incongruencies, plunged the Respectable Social System into a condition of crisis that reached a climax in 1894–5."[34] That crisis resulted from the explosive atmosphere that the family imposed upon its members. The family was indeed "fortified" against the pressures of the outside world, as Young argued, but that fortification, like neurotic defense mechanisms, often stifled more than it protected or comforted and generated an atmosphere of "explosive intimacy."[35]

Modernists countered that the security of a divinely blessed home was a fantasy. Thomas Wolfe slammed the door on that fantasy in *You Can't Go Home Again* (1940): "you can't go home again . . . You can't go back home to your family, back home to your childhood . . . back home to the old forms and systems of things which once seemed everlasting, but which are changing all the time," and, I would add, forms and systems that were particularly unclear and confusing the first time anyone went through them.[36] A philosophical analysis of the capricious setting for human existence generally is Heidegger's notion announced in *Being and Time* that Dasein is fundamentally never truly at home but is always "thrown" (*geworfen*) into the world from the outset without any transcendent reason for being wherever it happens to be. Dasein does not completely belong anywhere, certainly not where an omnipotent God placed it to start a life whose ultimate purpose was to celebrate His glory.

The beginnings of modernist novels by women writers have especially profound implications for women in questioning the "protective" security of the home and the values that reigned in it. The Victorian novel conventionally introduced the heroine within a stable interior with her role securely but oppressively locked into roles as a wife and mother. Whether idealized or denigrated, the home operated as the narrative's emotional ground and narrative center. In *Jane Eyre*, the opening declaration of spatial confinement – "There was no possibility of taking a walk that day" – establishes enclosure as the condition under which female subjectivity comes into being. In *Cranford* (1853), the community is defined through female domestic management and what Elizabeth Gaskell calls "elegant economy."[37] The traditional home thus functioned as moral foundation, social system, and financial base, as well as narrative anchor. Even when the Victorian home is contested, as in Eliot's *The Mill on the Floss* or Emily Brontë's *Wuthering Heights*, it remains the origin from which conflict radiates. In the Victorian novel, life's coherence is centered in a well-defined domestic space.

By the early twentieth century, women's changing roles as wives and mothers weaken the ideological authority of the traditional home. Women's novels open in homes that do not confer identity and security but rather impermanence and uncertainty if not peril. At the beginning of Dorothy Richardson's *Pointed Roofs* (1915), the idea of home is already degraded for Miriam Henderson. She does not depart from a place of protection or emotional rootedness; she leaves rather because her family home can no longer sustain her economically or emotionally. It is a space of constraint and limited prospects rather than security. When she arrives at the German girls' school, the structure that replaces home is institutional and defined by corridors, routines, and surveillance rather than by intimacy or belonging. However, it is precisely from the failure of the traditional home that Miriam sets on a course of existential self-discovery that Richardson renders with an innovative stream-of-consciousness technique to render the workings of a distinctive female consciousness.

At the beginning of Jean Rhys's *Quartet* (1928), the degradation of home for Marya Zelli is characterized by sexual vulnerability. After her husband is imprisoned, Marya is stripped of material protection and social legitimacy. The married home that might have offered safety is replaced by a series of precarious rooms that mark her availability rather than her privacy. The apparent sanctuary offered in the home of Hugh Heidler (based on the writer Ford Madox Ford) crystallizes this replacement. His apartment simulates the outward forms of home with meals, routine, and hospitality but quickly becomes the setting for Marya's sexual exploitation. Heidler's pursuit of her operates through a scenario of sympathy and protection: he casts himself as rescuer, then gradually converts that role into entitlement. Marya's resistance is worn down by relentless emotional manipulation and the threat of renewed homelessness.

The abuse is especially disturbing because it is embedded within the language of hospitality inside a space that purports to be a place of kindness. The corruption of Heidler's home is magnified by the complicity of his wife Lois, who oscillates between protective gestures and cold withdrawal, effectively sacrificing Marya to preserve her own marital equilibrium. This triangulated structure transforms the apartment into a theater of moral abdication in which Marya's sexual exposure is normalized as the price of her continued shelter. From its opening movement, *Quartet* presents sexual abuse as the direct consequence of domestic collapse. Once the marital home is destroyed by her husband's arrest, every substitute domestic space that Marya enters subjects her body to risk. Rhys exposes how for an economically unprotected woman the loss of home entails not only homelessness but the sexualization of survival.

The opening of Radclyffe Hall's *The Well of Loneliness* (1928) establishes Stephen Gordon's childhood home at Morton Hall as a vast inherited estate governed by wealth, tradition, and paternal authority. Stephen's early life unfolds in an apparent framework of protection and social privilege, but from the moment of her birth this order is destabilized because of her father's disappointment at having a daughter rather than a son. This early contradiction of protection and rejection produces the novel's crucial degradation of the mythic home. Stephen inherits her father's intellect but cannot inherit his social position as heir. As a result, her "home" becomes a space in which Stephen learns, long before she understands her lesbian sexuality, that her deepest inclinations are disallowed. After her father's death, Stephen's mother experiences her daughter's lesbian tendencies as a social threat and personal humiliation. Her confrontation with Stephen is an extreme example of the disintegration of home in modernist fiction.

> "All your life I've felt very strangely towards you"; [her mother] was saying, "I've felt a kind of physical revulsion, a desire not to touch or to be touched by you – a terrible thing for a mother to feel. . . . I shall never be able to look at you now without thinking of the deadly insult of your face and your

> body to the memory of the father who bred you. I can only thank God that your father died before he was asked to endure this great shame. As for you, I would rather see you dead at my feet than standing before me with this thing upon you – this unspeakable outrage that you call love in that letter which you don't deny having written. In that letter you say things that may only be said between man and woman, and coming from you they are vile and filthy words of corruption – against nature, against God who created nature. My gorge rises; you have made me feel physically sick."[38]

This impassioned expulsion from the family home forces a decisive transformation in Stephen's identity. Whereas earlier she experienced her difference as a confused discrepancy between inner nature and domestic expectation, she begins to experience it as a truth that cannot survive within the home at all. Stephen's embrace of lesbian identity is born of rebellion and compelled self-recognition. If the traditional home cannot contain her, she must reconstitute herself without it.

In this way, emotional ejection from the paternal home becomes the condition of possibility for Stephen's proud self-affirmation. The very force that attempts to invalidate her – conventional domesticity – drives her toward identification with others like herself. Once she accepts that she cannot be reconciled with the heterosexual order that the home represents, her sense of self shifts from passive shame to an active ethical imperative. Loving women becomes not merely a private inclination, but a mode of moral truthfulness opposed to the falsifying demands of domestic normativity.

Stephen's later insistence on living openly as a lesbian thus emerges directly out of the home's failure to accept her. The ancestral estate, once the emblem of security, becomes the first institution to declare her existence as unacceptable. Yet this declaration enables her to reframe her identity as someone who does not require domestic authorization. By the time Stephen claims her sexuality with pride, she does so with a determination forged by exclusion. Unlike the heroines of Richardson and Rhys, whose homelessness is economic and institutional, Stephen's is ontological and symbolic. She loses not just a roof but a nameable place within the moral architecture of the home, and this emotional exile from the paternal household becomes the ground of her most resolute self-possession. Home's rejection produces not erasure but the conditions for a defiant and ethically grounded new lesbian identity. Together, these beginnings of novels by women show that, with Richardson, Rhys, and Hall, the traditional home is no longer an acceptable origin of female identity. But that new reality of homelessness forces protagonists to redefine more authentic and self-reliant personal identities as women.

The preceding comparisons of Victorian and modernist novels suggest that a major change was under way across these two periods that transformed the beginning of novels, particularly evident in the way they rendered settings, selfhood, families, religion, and home. Faulkner begins *The Sound and the Fury* in the mind of a mentally disabled character to dramatize his unstable setting and sense of himself. Woolf begins *Mrs. Dalloway* with her main character ambiguously divided between different places, times, and personal identities. Gide's two novels question the originary legitimacy of patriarchal families and the Catholic Church. Stein's technique of beginning again questions the possibility of establishing secure foundations of individual selves or families. Joyce's *Ulysses* uses an array of innovative narrative techniques to dramatize the foundational qualities of a nontraditional hero, the cuckold Leopold Bloom as he tries to get back home to his wife and reestablish their love. By beginning their novels in spaces of displacement rather than in the home, Richardson, Rhys, and Hall historically redefine female identity from one grounded in traditional domestic dependency and conformity to one grounded in increasing self-reliance and struggle.

The beginnings of Victorian novels established reassuringly clear, secure, and ultimately divinely sanctioned originary circumstances that governed human experience. In contrast, modernists' innovative beginnings destabilized those circumstances, but out of those destabilizations modernists probed new variations of settings, selfhood, and families, along with new attitudes toward religion and new ways for women to define themselves outside the "protection" of the traditional Victorian home.

Coda

While this chapter has focused on dramatic changes in the novel, other dramatic changes took place in modernist poetry, which I here indicate in brief. In the period 1890–1930, poetry rejected its domination by conventional "poetic diction" as it came to embrace colloquial or "low," even antipoetic registers of language. Poetry liberated itself from the constraints of meter and rhyme and expanded the range of permissible subject matter to include taboo or "unpoetic" content. In 1921 T. S. Eliot embraced difficulty for its own sake in arguing that "[p]oets, in our civilization as it exists at present, must be difficult."[39] Poets also demoted the discursive or argumentative as they focused on the image, while the Dadaists and Surrealists opened poetry to the brazenly irrational. While thousands of poems written about World War I greatly expanded the number of people who wrote and read poetry, that development tended to reinforce traditional diction and formal techniques, although in the aftermath of the war innovators such as William Butler Yeats (*The Second Coming*) and T. S. Eliot (*The Waste Land*) revolutionized poetic forms as well as subject matter.

Notes

1. Virginia Woolf, "Modern Fiction" [1919] in *The Common Reader* (Harvest, 1953), 151.
2. Willa Cather, "The Novel Démeublé," *New Republic* 30 (12 April 1922): 5–6.
3. Marcel Proust, *In Search of Lost Time*, trans. Andreas Mayor and Terrence Kilmartin, revised by D. J. Enright (Modern Library, 2003), vol. VI, 277.
4. Henry James, "Preface" to *The Portrait of a Lady*, written for his collected works in 1908; Wyndham Lewis, *Tarr* [1918] (Penguin, 1982), 311; E. M. Forster, *Aspects of the Novel* [1927] (Harcourt, 1954), 26; Gertrude Stein, "Portraits and Repetition" [1934], in Patricia Meyerowitz ed., *Look at Me Now and Here I Am: Writings and Lectures 1901–1945* (Penguin, 1967), 110.
5. Ezra Pound, "The Classicist" [1913], *The New York Times*, 1934.
6. On these formal techniques see Stephen Kern, *The Modernist Novel: A Critical Introduction*, 126–39.
7. Walter Houghton, *The Victorian Frame of Mind 1830–1870* (Yale University Press, 1957), 162.
8. Samuel Smiles, *Thrift* (John Murray, 1875), 70.
9. Hugo von Hofmannsthal, "Der Dichter und diese Zeit" in *Erzählungen und Aufsätze* [1905] (Fischer, 1957), 445.
10. Mina Loy, "Aphorisms on Futurism," *Camera Work* 45 (January 1914): np.
11. Robert Musil, *The Man without Qualities* [1930] (Perigree, 1965), I, 62.
12. Anthony Burgess, *Here Comes Everybody* (Faber and Faber, 1965), 50.
13. Proust, *In Search of Lost Time*, vol. I, 5.
14. Ibid., 60.
15. Virginia Woolf, *Mrs. Dalloway* (Harvest, 1981), 3. Subsequent text references are to this edition.
16. The source of these two quotes is Woolf's literary manifesto, "Modern Fiction," where she states her intention to "record the atoms [of experience] as they fall upon the mind," and asks, challengingly, about the lives described by her "materialist" predecessors H. G. Wells, Arnold Bennett, and John Galsworthy, "Is life like this? Must novels be like this?" Her implied negative answers resound throughout her novels. *The Common Reader: First Series* [1925] (Harvest, 1953), 154, 155.
17. Malcolm Lowry's *Under the Volcano* (1947) is set on the Day of the Dead in 1939. While Proust's novel covers many decades, it nevertheless moves at a historically slow pace. Gérard Genette compared Balzac's *Eugénie Grandet* (1833) and Proust's *In Search of Lost Time* and calculated that

Balzac covers around ninety days per page in comparison with Proust's 5.5 days. By going one sixteenth as fast as Balzac, Proust was able to delve deeper into experience. Gérard Genette, *Narrative Discourse: An Essay in Method* (Cornell University Press, 1980), 95.

18. Virginia Woolf, *A Room of One's Own* [1929] (Harcourt, 1981), 110.
19. William Faulkner, *Absalom, Absalom!* [1936] (Vintage, 1986), 243.
20. "A woman's power is in her ability to make a home a fortress of peace and safety, a refuge from the world, a place where a man may find solace and purpose." John Ruskin, *Sesame and Lilies* [1865] (John Wiley &Sons, 1886), 123.
21. Patricia Tobin, *Time and the Novel: The Genealogical Imperative* (Princeton University Press, 1978), 5–8, 20.
22. André Gide, *Lafcadio's Adventures* [*Les caves du Vatican*], trans. Dorothy Bussy (Vintage, 1953), 14.
23. André Gide, *The Counterfeiters*; with *Journal of "The Counterfeiters"* [1926] trans. Dorothy Bussy (Vintage, 1973), 114.
24. Jean-Joseph Goux, *The Coiners of Language* [1984], trans. Jennifer Curtiss Gage (University of Oklahoma Press, 1994), 3, 4.
25. *Journal of "The Counterfeiters,"* 414.
26. Gide explained that his references to the gratuitous act, "*l'acte gratuit*," do not mean entirely without motive, just that they cannot be accounted for by "ordinary psychological explanations." *Faits-divers," La nouvelle revue française* 30 (1 June 1928): 841.
27. *Journal of "The Counterfeiters,"* 431.
28. Ibid., 449.
29. Gertrude Stein, *The Making of Americans: Being a History of a Family's Progress* [1925] (Dalkey Press, 1995), 311.
30. Ibid., 120.
31. Gertrude Stein, "Sacred Emily," in *Geography and Plays* (Four Seas, 1922), 187.
32. Ruskin, *Sesame and Lilies*, 100. For an elaboration of Victorians' notion of "Home, Sweet Home," see Walter E. Houghton, *The Victorian Frame of Mind 1830–1870* (Yale University Press, 1985), 341–48.
33. G. M. Young, *Portrait of an Age* (Oxford University Press, 1953), 134.
34. Peter T. Cominos, "Late Victorian Sexual Respectability and the Social System," *International Review of Social History* VIII (1963): 18–48, 216–50.
35. For an elaboration of this argument see Stephen Kern, "Explosive Intimacy: Psychodynamics of the Victorian Family," *History of Childhood Quarterly* 1, no. 3 (Winter 1974): 437–61.
36. Thomas Wolfe, *You Can't Go Home Again* (Harper & Row, 1940), 706.
37. Elizabeth Gaskell, *Cranford* (Chapman & Hall, 1853), 1.
38. Radclyffe Hall, *The Well of Loneliness* [1928] (Avon, 1981), 200–201.
39. T. S. Eliot, "The Metaphysical Poets," *Times Literary Supplement*, 20 October 1921.

Chapter 11

ART: Formal Foundations in Fauvism, Cubism, Futurism, Collage, Abstraction, Expressionism, Dadaism, and Surrealism

In art history the Victorian period is typically known as the Age of Realism, a time when artists depicted subjects from their current "real" world instead of the long ago, the far away, or the imaginary and did so as "realistically" as possible. In 1848 the novelist George Henry Lewes argued confidently that "Realism is . . . the basis of all Arts, and its antithesis is not Idealism, but Falsism."[1] The formal foundations of Realism included naturalistic colors, accurate representation of recognizable forms, clear spatial relations among those forms depicted in clear perspectives, high finish and attention to detail, carefully staged compositions, and paint applied on a flat surface, while the subject of their paintings included ordinary domestic interiors, peasant and rural labor, urban working-class life, contemporary social issues, deference to religious values, and narrative subjects that illustrate the slogan "every picture tells a story" along with an overall suggestion that what is depicted is "real."[2] Modernists revolutionized the formal foundations of art with the striking innovations of Fauvism, Cubism, Futurism, Collage, Abstractionism, Expressionism, Dadaism, and Surrealism.

I use "Realist" interchangeably with "Victorian" to sharpen the historical role of modernist artists who challenged conventional "Realist" foundations across the board, although the Victorian period included subdivisions of Pre-Raphaelitism, Impressionism, Symbolism, and Postimpressionism. In the modernist period, Fauvists took liberties with naturalistic colors, Cubists distorted accurate forms and rejected art's narrative role, Futurists scorned historical subjects and art for permanent display in museums, Collagists ignored the convention of applying thin layers of artistic materials to a flat picture surface, Abstractionists replaced identifiable subjects and narratives with unidentifiable colored forms, Expressionists worked to capture inner emotions and even underlying bones and veins of their subjects, Dadaists assailed the rationality of a world gone mad during World War I, and Surrealists subverted the "reality" of Victorian images generally. In challenging conventions about what is real and the suitability of trying to depict it as it "really" was, Modernist probed ever deeper into the emotions and bodies of their subjects and themselves, explored new possibilities of artistic techniques, and created a new sense of reality altogether.

Fauvist Color

In 1900 the critic Maurice Denis opened the new century by announcing a revolution in art: "a picture—before being a war horse, a nude woman, or an anecdote—is essentially a flat surface covered with colors assembled in a certain order."[3] Five years later in the French coastal city of Collioure, Henri Matisse, André Derain, and Maurice de Vlaminck pioneered the "wild" color technique of Fauvism.[4] Colors compose their paintings not as surfaces of objects in their world but as expressive tools that emerge from creative interactions with their world and their paint, and so, in *Portrait of Matisse* (1905), Derain painted the right side of his subject's face turquoise, and in *The Green Line* (1905) Matisse painted a green stripe down the bridge of his wife's nose that separates a right cheek in greenish yellow from a left cheek in warm pink.

Derain's *The Turning Road, L'Estaque* (Figure 11.1) transforms a "real" place into a creative pattern of colors and marks a turning away

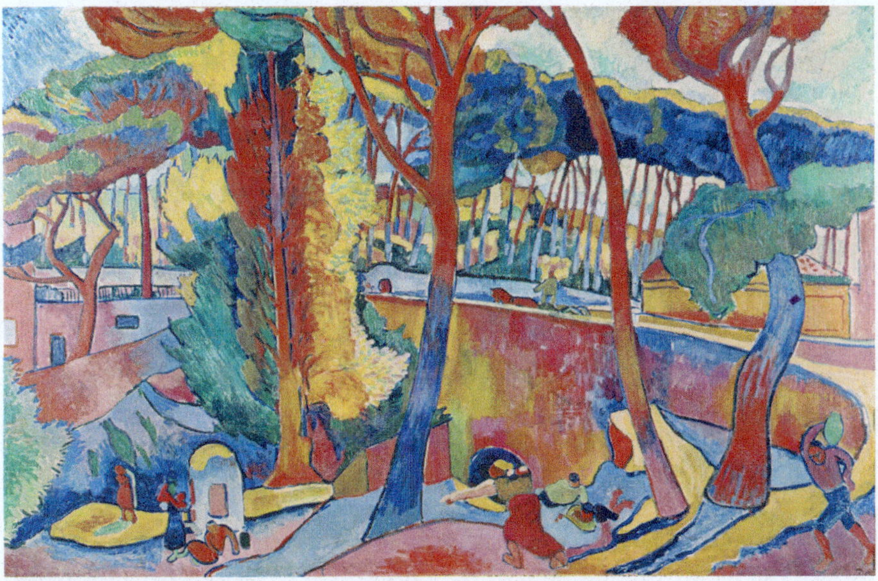

Figure 11.1 André Derain, *The Turning Road, L'Estaque*, 1905.

from naturalistic color toward the use of color as a purely compositional element. The tree trunks are not naturalistic brown but shades of cobalt blue, powder blue, vermilion, and purple to complement the colors of the terrain behind them. The leaves are undulating patterns of saturated greens and blues laced with streaks of caramel in deference to the overriding purpose of the painting as an autonomous expressive display. The tree on the left has a brick-red trunk and a leafy mix of marigold, azo green, and cobalt blue. The structure on the left shades from pink to light blue with the same colors of the surface in front of it. The diminutive people in the foreground are colored in accord with the landscape and are subordinate to the massive colored forms that tower over them. The man at the lower right is colored in shades of red and blue that match the colors of the tree trunks next to him that project bright blue shadows. He holds a celadon green jug like a colored punctuation mark in the rhythmic distribution of blues and greens throughout the composition. Patches of caramel distributed rhythmically throughout the painting from the ground and leaves to the treetops in the background evoke the dazzling light and heat of an exuberant terrain under a Mediterranean sun that lights up the tree in the center that is accented with Naples yellow as if it were on fire. In contrast to conventional Victorian painting that used color to make objects identifiable as they "really" appeared to the eye, Derain used color to express aesthetic sensibilities. The painting obliges viewers to reexamine many possibilities of color beyond slavishly rendering objects the way they appear naturally. As an artist and not a copyist, Derain offers a declaration of independence on behalf of artistic color over the dictates of natural color.

In *The Joy of Life* (Figure 11.2) Matisse uses color to revolutionize the foundations of art by accenting coloring along with drawing. The painting depicts a mythic place where the joys of life and love as well as color emerge in brilliant sunlight. Myths surface in human experience as if conceived by gods communicating eternal truths. According to Hesiod, the Golden Age was the first in a series, followed by the Silver, Bronze, Heroic, and Iron Ages, each increasingly corrupted by historical developments. The uncorrupted Golden Age was a time when life was an uninterrupted enjoyment of harmony and happiness, and Matisse's naked figures are dancing, making music, and making love, while he celebrates making art. With this bold use of colors, Matisse seems to be dancing among them.

The primordial "Golden Age" offers an apt setting to express the foundational role of color. The pinks, oranges, mauves, yellows, and greens do not depict any actual landscape but express Matisse's joy in a world flooded by sunlight. Colors appear to emerge from the artistic process

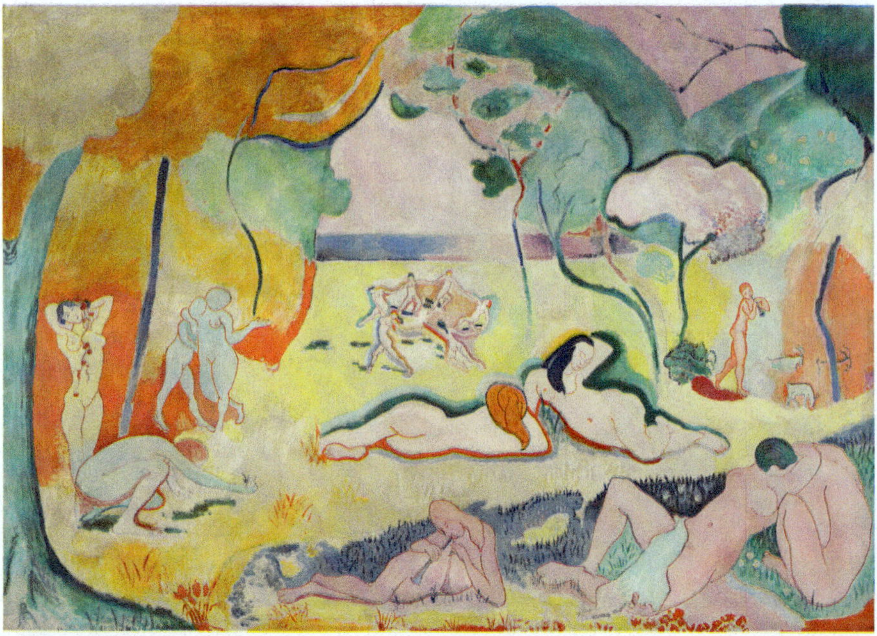

Figure 11.2 Henri Matisse, *The Joy of Life*, 1906.

itself, analogous to the way figures and trees appear to emerge out of the land. The reclining pair of women in the center have orange and blue hair, and their bodies are outlined by those same colors as are the diminutive six dancing figures in the background. Matisse ignores the dictates of accurate scale and does not record what he knows about the "actual" colors of these figures but rather projects his impressions as if he had just discovered colors and was in thrall to their beauty as they oozed out of tubes. His bold artistic "voilà!" is for viewers to share his joy over the magic of color as if to say, "Don't be distracted by trying to figure out the actual colors of these objects but enjoy this composition of the beauty of the colors themselves." In 1908, influenced by Henri Bergson's theory of intuition, Matisse explained that the main function of color should be to serve artists' intuitions, not because artists know what color suited a season of the year, or what color their subjects' faces "really" were, or because of any scientific theory of color, but because of their "felt experiences" of color.[5] Gertrude Stein predicted that in this painting Matisse "created a new formula for color that would leave its mark on every painter of the period."[6]

Matisse's *The Red Studio* (Figure 11.3) is a pictorial manifesto of Fauvism. It proclaims the ultimate supremacy of color by rendering a studio's floor, walls, table, and stools plus a wine glass, clock, chair, and

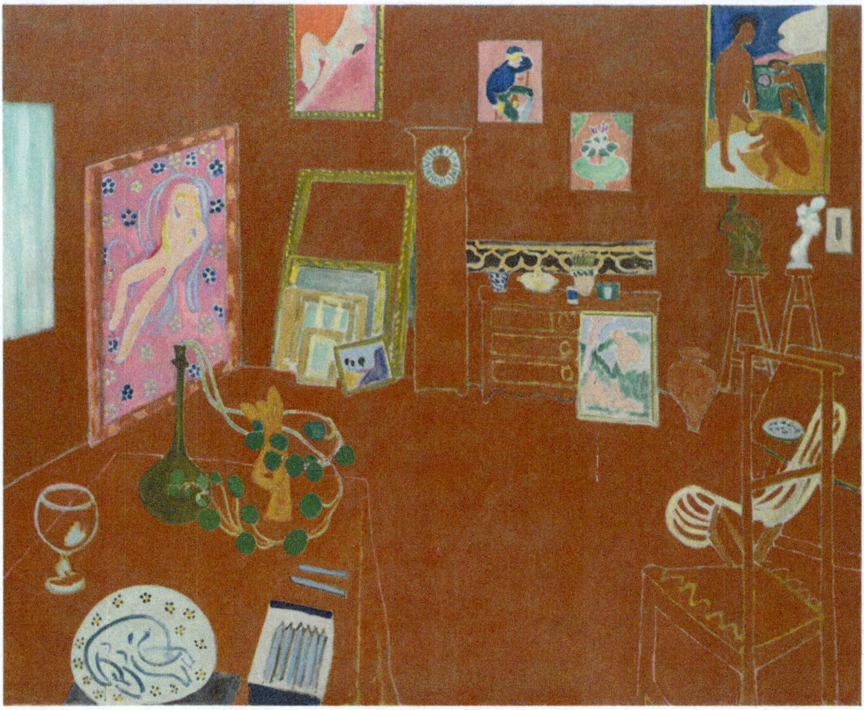

Figure 11.3 Matisse, *The Red Studio*, 1911.

chest of drawers in the same shade of Venetian red. Matisse scraped the outline of those objects out of red paint that constitutes those images along with the floor and walls. Red is the substance of space as well as time as the red disk at the center of a clockface. In 1945 Matisse looked back on his contribution to the use of color in noting that since the Renaissance drawing came first with color added, but beginning with Gauguin, Van Gogh, and Cézanne there had been a "rehabilitation of the role of color and a restitution of its emotive power." "Each age brings with it its own light," and in his age color was more fully realized when it expressed the artist's pure aesthetic sensibility.[7] Fauvist revisions of conventional uses of color opened new expressive capabilities.

Cubist Innovations

The Cubists further transformed the foundations of modernist art as they reconstituted realistic forms, reduced pictorial depth, depicted multiple perspectives and light sources, and rejected narrative art. Some of these key re-formations were initially worked out by Georges Braque.

In *Landscape at La Ciotat* (1907) he simplified a landscape into a pyramidal composition that spilled down the picture plane toward the viewer. In *Houses at L'Estaque* (1908) he geometricized the houses and trees and flattened the picture surface, innovations that also appear that year in works by Picasso.

Throughout the "Realist" period, the value of an artist's ability to draw a likeness was increasingly diminished by photography that could achieve a likeness with a shutter click, but photography, in liberating artists from the obligation to portray visual appearances, also inspired them to alter natural forms, including most notably the human form that Picasso radically reconstructed in *Les Demoiselles d'Avignon* (Figure 11.4). This painting of 1907 that presaged Cubism presents stunning deformations of the human face and body that are reconstructed as rhythmic compositions of simplified forms. The standing nude on the far left was originally a clothed male looking at the scene in front of him. In the final composition he is re-formed into a female in profile with long hair and breasts, facing the other figures but with a lidless right eye rotated toward the viewer. In 1935 Picasso evaluated his own revolutionary historical role: "A picture used to be a sum of additions. In my case a picture is a sum of destructions."[8] *Demoiselles* is more accurately a deconstruction because it does not just take bodies apart and deform them, let alone destroy them, but reconstructs them into a new sort of artistic whole. The founder of Surrealism André Breton celebrated this painting as "the capital event in the onset of the twentieth century," adding that "with the *Demoiselles d'Avignon* it's good-by to all the art of the past."[9]

John Golding summarized Picasso's formal deconstructions. "In the *Demoiselles* Picasso began to shatter the human figure and the pictorial conventions for rendering it. He spent the rest of his artistic life dissecting, reassembling, and reinventing it."[10] The deconstruction of objects further implied a deconstruction of social structures and political institutions. Years later Picasso made that connection explicit, no doubt with Cézanne's apples in mind: "It is not necessary to paint a man with a gun. An apple can be just as revolutionary."[11] In the next several years Picasso and other Cubists increasingly re-formed the foundations of art as they re-formed figures in it as well as landscapes and still lifes.

The seriousness with which Picasso approached *Demoiselles* and evidence of his intention to make it a major artistic statement are the over 400 studies associated with its creation over nine months from 1906 to 1907, recorded in sixteen sketchbooks that he kept in his possession until his death.[12] That quantity of preparatory work, noted William Rubin, "was not only unique in Picasso's career, but without parallel for a single picture in the entire history of art."[13]

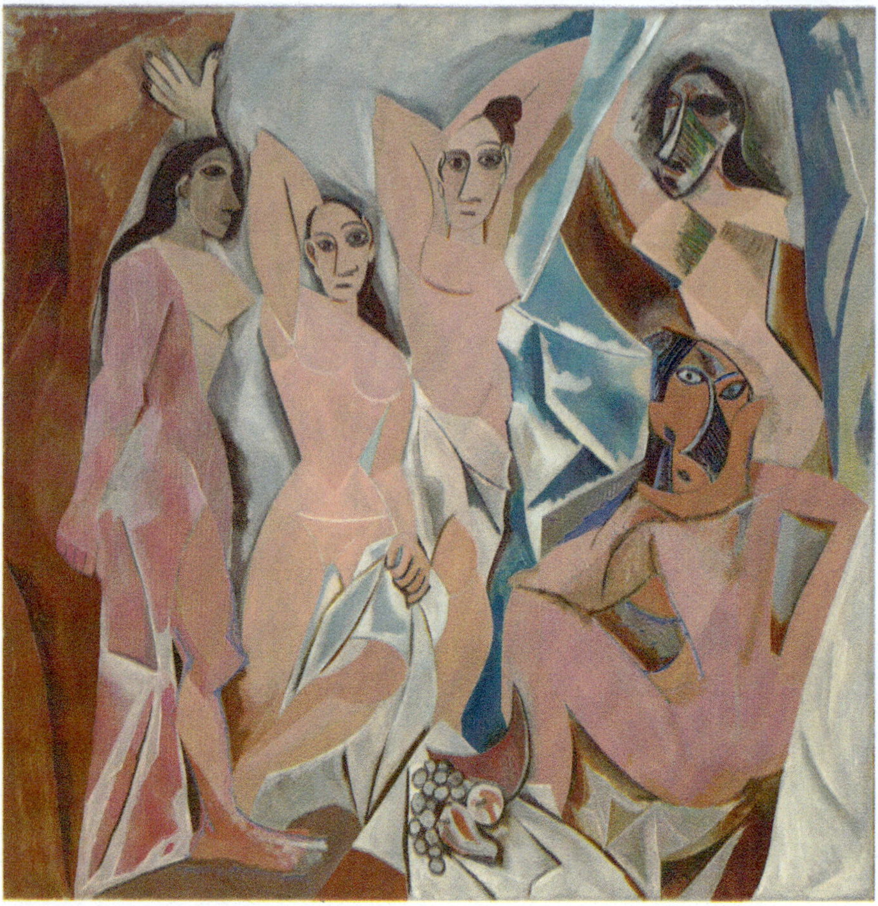

Figure 11.4 Pablo Picasso, *Les Demoiselles d'Avignon*, 1907.

Cubist Rejection of Narrative

Among many motives for modernists' revolutionary shift, a significant one was the desire to reestablish painting as an art form independent of storytelling to which it had been traditionally subservient. In the long development of *Demoiselles*, Picasso worked to undo its original powerful narrative and, by implication, the status of narrative art generally. Nineteenth-century images of prostitutes implied a variety of clear narratives: the glamorized implied narratives of harem scenes in Delacroix's *Women of Algiers* (1834) and Ingres's *Turkish Bath* (1863), the disturbing implied narratives of degraded prostitutes in Holman Hunt's *The Awakening Conscience* (1853) and John Stanhope's *Thoughts of the*

Past (1859), and later narratives about the unglamorized prostitutes in Toulouse-Lautrec and Degas. Picasso's early sketches for *Demoiselles* show that he originally intended to depict a strong narrative with five prostitutes in a brothel displaying their sexual charms to a seated sailor, with the entire scene observed by a second man. Picasso repeatedly referred to this painting as "my brothel" and in 1933 explained that "in the beginning it was called *The Brothel of Avignon*."[14]

In earlier compositions of that painting the faces of the prostitutes, looking to stimulate business, are directed at a seated sailor or at the standing male figure, who in some ensemble sketches holds open a curtain for viewers to observe the scene. In detail sketches that figure resembles Picasso, who was indeed presenting this scene to viewers of his painting as a pictorial narrative. In one sketch that figure is holding a medical book, in another a skull, suggesting together with the seated sailor and the prostitutes a strong narrative of sexual desire, venereal disease, and death. Biographical sources for that narrative include that Picasso frequented brothels, had friends who contracted syphilis, may have contracted it himself or at least was perpetually afraid of it, and in 1901 was traumatized by the suicide of his close friend Carles Casagemas over an unrequited love, an incident that Picasso painted that year with a bullet hole in Casagemas's temple.

During the long creation of *Demoiselles*, Picasso gradually eliminated traces of narrative, and in the final composition nothing refers to a brothel, sex, or death aside from a hint in the final title's reference to Avignon Street, the red-light district of Barcelona. The seated sailor is gone. The other clothed man has been transformed into a naked woman and is not holding a medical book or a skull, and none of the five women is identifiable as a prostitute. The painting does not tell a story but creates what Leo Steinberg calls an "anti-narrative counter-principle: neighboring figures share neither a common space nor a common action, do not communicate or interact, but relate singly, directly, to the spectator."[15] The painting is an invitation to look again with care at this pathbreaking transformation of traditional renderings of form and narrative.

The painting evinces Picasso's prolonged meditation on the foundations of art while he worked on this painting. As Rubin suggested, the painting allowed Picasso "to forge new tools adequate to excavate the deeper strata of his mind." Instead of conventional nineteenth-century narrative art, Picasso made an artistic record of the long creative process in his notebooks. They were "a kind of visual diary" illustrated by the fact that he wrote on the cover of one of them "I am the NOTEBOOK." In an obvious reference to Freud's self-analysis, Rubin added that the metamorphoses of the painting evince "Picasso's journey into the recesses of his psyche." "This self-analysis, peeling away layers of consciousness, was associated with . . . a search into the origins of man's way of picturing himself, an attempt very selectively to retrace the history of image-making."[16] In place of an implied pictorial narrative starting with sexual desire, Picasso painted a vivid product of his own self-analysis as he reworked the artistic process. While his concerns were primarily formal, he was also aware of his work's epistemological significance, as, looking back on his career in 1944, he explained how he "wanted, through drawing and through color, since those were my weapons, to penetrate always more deeply into knowledge of the world and of men."[17]

Futurist Rejection of Historical Subjects and Celebration of the Future

Since the sixteenth century, artistic genres had been ranked with the noblest subject of history painting at the top followed by portraits, landscapes, still lifes, and scenes of everyday life progressively down toward the bottom. Modernists rejected that hierarchy, particularly the primacy of historical painting and its strong narrative foundation. Between 1909 and 1915 Futurists produced their most notable achievements before they lost members and authority after Italy entered the war in 1915 and suffered enormous losses. They directed their antihistorical ire back to classical antiquity, which they found to be disturbingly revered in contemporary Italian culture. Living where the glorious ancient Roman Empire had been the seat of a cultural triumph that present-day Italians could not match, Futurists were disgusted by the idolatry of the past. They idolized present-day technologies such as racing cars and big cities, and instead of fighting for immortality on museum wall-space fought to create an art of the present pitched toward the future.

The Victorian period was the golden age of museum building in response to the rise of nationalism, imperialism, and bourgeois public culture. The South Kensington Museum first opened in 1852 and was renamed the Victoria and Albert Museum in 1899. The Neues Museum in Berlin opened in 1855, followed by the Museo Arqueológico Nacional of Madrid in 1867, the Metropolitan Museum of Art in New York in 1872, the Art Institute of Chicago in 1879, the Rijksmuseum in Amsterdam in 1885, and the Kunsthistorisches Museum of Vienna in 1891. The Galleria Nazionale d'Arte Moderna e Contemporanea, founded in Rome in 1883, was particularly irksome to the Futurists because it was located in their own national capital. In their founding manifesto, Marinetti announced their goal to free Italy from "numberless museums that cover her like so many graveyards." Museums were "absurd abattoirs of painters and sculptors ferociously slaughtering each other with color-blows and line-blows, the length of the fought-over walls!"[18] In 1914 Carlo Carrà further challenged the goal of being hung on those walls in a museum's permanent collection: "While our predecessors, without exception (including Cézanne and Renoir) had as their dream and ultimate destination the MUSEUM, we Futurist painters are nauseated by the presence before us of embalmed cadavers to contemplate."[19]

Along with the rejection of historical subjects and their narrative foundation, the Futurists sought to capture energy and speed and move toward abstraction – all realized in Giacomo Balla's series of speeding cars from 1912 to 1913. Marinetti had famously announced that "the world's magnificence has been enriched by a new beauty, the beauty of speed. A racing car whose hood is adorned with great pipes . . . is more beautiful than the *Victory of Samothrace*."[20] The title of Balla's *Speeding Car. Abstract Speed* (1913) captures a car speeding along a road, generating abstract lines of force through the surrounding air. The full series of images captures how speeding cars impact light and sound along with the road underneath, which in several of the paintings is churned into spirals of dust as "force lines" streaming around them, symbols of a dynamic world.

In a diary entry of 1907 Umberto Boccioni recorded that he wanted "to cancel all the values" he had known in order to "reconstruct on new foundations."[21] On those foundations he crafted *The City Rises*, "a great synthesis of labor, light, and movement" (Figure 11.5).[22] That painting replaces Realist description with a new visual language for the modern world. Rather than presenting horses from a fixed vantage point, as the Realist Rosa Bonheur did in *The Horse Fair* (1852–3), Boccioni organizes the canvas as a field of explosive forces that include wildly moving horses.[23] Perspective does not govern space; instead, space functions like an energizing continuum in which foreground and background transition one into the other. Realism's commitment to accurate objective representation gives way to Futurists' capturing motion and speed along with industrial transformation. The overall effect is of an emotionally charged atmospheric environment built from color and gesture rather than clearly identifiable representational imagery.

The rendering of distinct forms marks another decisive break. Realist art encases figures in anatomically coherent silhouettes, but *The City Rises* dissolves human bodies and horses into blurred shapes and stretched limbs. Forms are crafted not to represent objects at rest but to visualize velocity. Boccioni's four horses are not in realistic motion; they

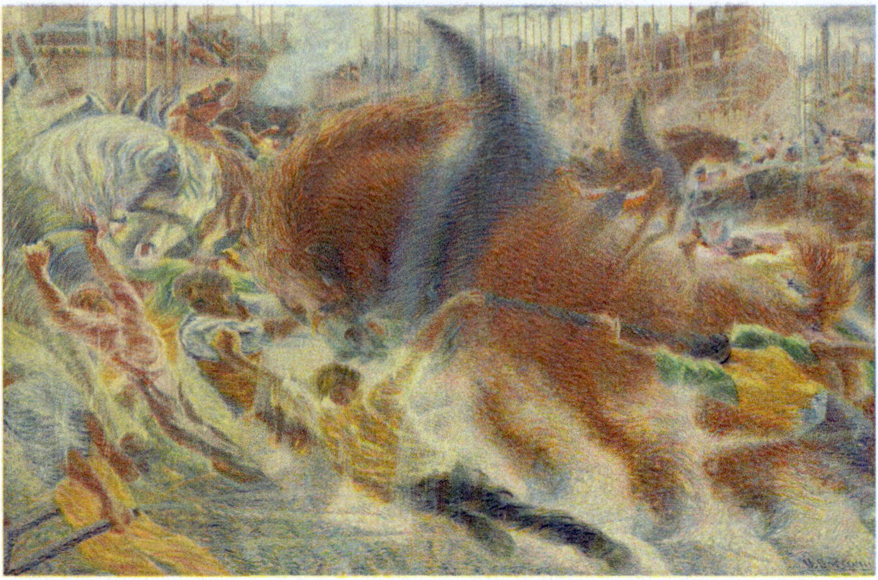

Figure 11.5 Umberto Boccioni's *The City Rises,* 1910–11.

rather erupt out of the picture surface, energized by driving forces. This image also transforms color and brushwork. While Realist painters used naturalistic colors, Boccioni's palette is intensified to produce emotional impact. Varieties of red, blue, and yellow undulate across the surface, generating rhythms rather than representing objects with local colors. Brushwork also varies from the Realist ideal of invisible technique to Boccioni's conspicuous dashes that accentuate the painting's dynamism as visualized speed. "The ground itself," writes Christine Poggi, "swells and roils in brilliant red and yellow waves that simultaneously suggest the sea, fire, or the flux of an energized field."[24]

The relation between figures and environment further exemplifies Futurism's break with Realism. In earlier Realist works, horses, humans, buildings, and landscape remain distinct elements of a visually coherent world. In *The City Rises*, workers blend into horses, construction scaffolding dissolves into atmospheric turbulence, and industrial expansion merges with bodily exertion. This integration reflects a core Futurist belief that modern life is defined by flux, interpenetration, and the annihilation of traditional boundaries. As Boccioni put it vividly, "[t]he motor bus rushes into the houses which it passes, and in their turn the houses throw themselves upon the motor bus and are blended with it."[25] By embodying these principles at every technical level – composition, form, color, brushwork, and spatial logic – Boccioni transformed painting from a window onto reality into an engine of modern sensation. *The City Rises* is emblematic of the Futurists' radical departure from Realist precedent. Theirs was an art influenced by a half dozen Futurist manifestos, several of which Boccioni wrote, along with Freud's libido theory, Bergson's theory of *élan vital*, and Nietzsche's theory of the will to power.

Collage and the Picture Surface

Five years after completing *Demoiselles*, Picasso inaugurated another formal artistic revolution with collage, a technique that extended the location of art above the picture surface and expanded the materials that could be applied to it.[26] Collages subverted the singular identity of things because in them a newspaper cutting could be a source of information, a wine glass, or a surface on which an artist could draw, or all three of those components. Collage was also a medium for Picasso to express ideas about current social and political issues. The newspaper articles glued onto his *Bottle of Suze* (1912) were cut out to preserve their legibility so that viewers could look at the whole collage as well as read the newspaper clippings in it. From 1912 to 1914, when he made the bulk of his collages, roughly half of them depicted the Balkan Wars and European political and economic activity, while another quarter referenced suicide and murder along with anarchist concerns about war profiteering and government corruption.[27] Patricia Leighton explores that pairing between collagist form and anarchist content in her aptly titled *Re-ordering the Universe: Picasso and Anarchism, 1897–1914*.[28]

Such a reordering of the aesthetic foundations of the artistic universe is evident in this collage in which the label of a Suze aperitif bottle is glued to a white paper cutout of a bottle along with cutouts of a newspaper clipping in the shape of a glass and a turquoise-colored paper tabletop, the setting in Parisian cafés for discussions of politics and art. The newspaper cuttings reveal Picasso's concerns over the First Balkan War of 1912. One clear link between form and content is a news column about a cholera epidemic that is uniquely upside down, suggesting a world flipped upside down. In later years Picasso explained the intent of these collages.

> If a piece of newspaper can become a bottle, that gives us something to think about in connection with both newspapers and bottles. This displaced object has entered a universe for which it was not made and where it retains, in a measure, its strangeness. And this strangeness was what we wanted to make people think about because we were quite aware that our world was becoming very strange and not exactly reassuring.[29]

Picasso led viewers to defamiliarize newspapers and bottles similar to the way Stein led readers to defamiliarize words and grammar. In defamiliarizing objects Picasso also engendered a profound reflection on their nature.[30] Throughout his career formal innovations accompanied revolutionary content. In his Barcelona years from 1897 to 1904, he was immersed in anarchist politics. *Demoiselles* opened a deeper analysis of colonialism, racism, and the discriminatory treatment of prostitutes. In 1945 he explained his commitment to revolutionary politics.

> What do you think an artist is? An imbecile who has only his eyes if he's a painter . . . On the contrary, he's at the same time a political being, constantly alive to heart-rending fiery or happy events. No, painting is not done to decorate apartments. It is an instrument of war for attack and defense against the enemy.[31]

The enemy included everything from ignorance and philistinism to warmongering and chauvinism, and Picasso repurposed picture surfaces to fight those enemies in "reordering the universe."

Abstraction

In 1861 the leading Realist painter Gustav Courbet explicitly rejected abstract art: "Painting is essentially a *concrete* art and consists only in the representation of *real* and *existing* things. . . An *abstract* object, not visible, nonexistent, is not of the domain of painting."[32] That view prevailed until modernists insisted that representational art constricts expression and that abstraction ought to be an acceptable element of painting's domain, if not its key element.

While abstraction emerged from aesthetic pressures within the artworld, it was also influenced by scientific discoveries about vision, matter, space, and energy.

In abstract art, light passes through, lines cross over, and colors bleed into and out of "solid" material objects. Those effects reflect the thinking of researchers who questioned the solidity of matter and found new constitutive energies in the void of space. In 1896, Bergson argued that the division of matter into independent solid bodies or atoms is wrong. An atom is not a plenum of matter but rather a crossing point for energy radiating across the continuity of space; it is "a vortex ring, ever whirling in this continuity."[33] In 1897 J. J. Thomson identified the electron, which he characterized as a vortex of matter and energy, a component of the atom that was largely empty space. In 1898 "empty" space took on a positive constitutive role for the American philosopher

Hiram Stanley, who theorized that space was not an empty container of matter but a forum for dynamic energy exchanges. He argued, with language that anticipated Heidegger's view of human existence as fundamentally "worlded," that space was "not full of things, but things are spaceful."[34] In the late 1880s, Heinrich Hertz produced electromagnetic waves, previously predicted by James Clerk Maxwell's equations for electromagnetism, which could transmit information across space. By the mid-1890s Marconi's wireless telegraph was transmitting messages all over the world through a supposed luminiferous ether that no experiment was able to detect. This science influenced artists who embraced abstraction as a liberation from the requirement for accurate representation and a way to enhance artistic expressivity and spiritual revival.

Wassily Kandinsky was troubled by wide-spread spiritual decline. In 1911 he produced his first abstract painting and published the major abstract artists' manifesto *Concerning the Spiritual in Art*, which begins with a summary of the spiritual barrenness of a contemporary culture that is grounded in material things and lacks a clear sense of purpose.[35] Museums display spiritless images – naked women, silver dishes, and crucifixions by artists who do not believe in Christ. People call themselves Jews, Catholics, or Protestants but are unable to believe in God. For them God *is* dead, and heaven *is* empty. In politics they are essentially anarchists, and in science they are positivists who value only what can be measured or weighed. That same year in "Content and Form," Kandinsky developed his abstract style to enable his inner spirit to "vibrate" with "the soul of another person, the receiver," uninterrupted by the distractions of recognizable figures or narratives.[36]

Kandinsky continued to note how music and painting offer direct sources of "the spiritual." He was inspired by the way music uniquely approaches a purely spiritual beauty in Wagner's *leitmotifs*, Debussy's "spiritual impressions," and Schoenberg's atonal compositions. In painting Cézanne's vases and apples suggest "the inner life of things." Matisse probed the new possibilities of color, while Picasso explored new possibilities of form. Their paintings are "seeds of the struggle toward the nonmaterialistic, the abstract, toward inner nature" (*CSA*, 153). Abstract art conforms to a "principle of inner necessity" that the more abstract the form, the purer its message. Analogous to Freud's self-analysis that tore away the veil over his Oedipus complex and Picasso's self-analysis during the painting of *Demoiselles* that tore away the foundations of traditional artistic forms, Kandinsky's artistic journey into abstraction tore down artistic commitment to the depiction of recognizable figures.

His paintings *Composition I* to *Composition IV* include images that can be identified without scholarly commentary. In these works he started with conventional Christian imagery that would initiate powerful responsiveness in himself and his viewers but then abstracted that imagery like a jazz musician riffing a melody. The result was intended to be pure resonance between his spirit and that of his viewers. "By hiding apocalyptic motifs based on folk sources," explained Rose-Carol Washton Long, "Kandinsky hoped to avoid the materialism of representational art and still involve the spectator through familiar key motifs. He also hoped to make the spectator take part in the creation of the work by unraveling its mysterious, ambiguous images."[37]

Mann's *Death in Venice* opens with a warning about 1911, the year when Germany sent a gunboat to Agadir to support German interests in Morocco, creating a diplomatic crisis. Kandinsky's barely representational paintings from that year are crafted with guns and boats.[38] At the upper left of *Composition IV* (1911), two horsed riders battle with sabers, and below them two lines of guns clash. The theme of battle is also indicated by the fortress on top of the blue mountain in the middle. In contrast, on the right side a reclining couple symbolizes love, two standing saints symbolize hope, and at the lower left two boats with oars row to salvation, suggested by a rainbow. But while the painting's content references mythic and historic events, the most historically significant achievement is its abstract form. As Kandinsky explained, the color is balanced between the cool blue in the center and the warm reds and yellows on the right side. The entangled lines of the battling horses in the upper left contrast with the parallel lines of the harmonious figures on the right. Summarizing how these colors and lines suggest a battle he explained: "The juxtaposition of this bright-sweet-cold tone with angular movement (battle) is the principal contrast in the picture."[39] Colors and lines clash but also resolve aesthetically as he hoped modern culture would do politically, with himself as St. George doing battle with the dragon of materialism, by which he meant representational art as well as a chaotic and unspiritual modern world. *Composition IV* is a farewell to representational art with horses, riders, boats, and reclining lovers before passing into a new kind of art composed of pure artistic forms.

Later in 1911, with *Composition V* (Figure 11.6), Kandinsky became fully abstract in that the painting's imagery cannot be identified without comparison with other works and at first glance looks like just oddly shaped colors and black lines. Its title gives away nothing, but the underlying subject becomes clearer after considering Kandinsky's comments on this composition and viewing preliminary works with religious

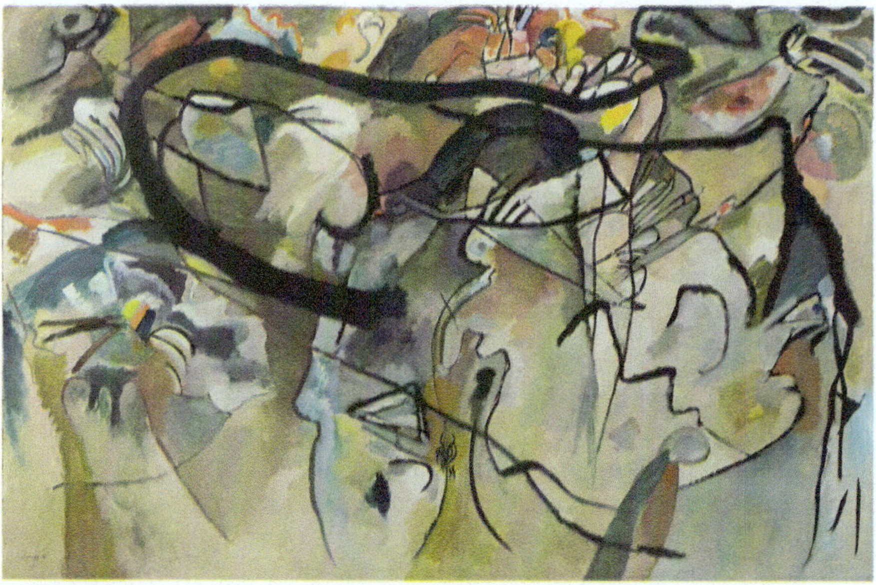

Figure 11.6 Wassily Kandinsky, *Composition V*, 1911. Private collection; formerly Solomon R. Guggenheim Museum.

themes such as *Apocalypse*, *Sound of Trumpets*, and *Resurrection*. *Composition V* depicts the apocalyptic Day of Judgment when seven trumpets ring out God's messages. The first six trumpets alert sinners to plagues of hail, fire, poisoned water, darkness, and death that will punish them for their sins, which are symbolized at the top by a city with broken towers. Two angels suggested by streaming hair and rudimentary wings blow on trumpets at the upper left (with a blue horn) and upper right (with a red horn). Center left is a trumpet that is colored red and white. Below the black mountain in the center are two crisscrossing trumpets. A sixth trumpet in the lower center is depicted by a triangle on its side. The seventh trumpet is the conspicuous back shape that snakes from the upper right under the falling towers, then drops down and doubles back as its thickness increases like a trumpet's bell. It announces the glory of God and His imminent judgment that will punish sinners and reward the virtuous by resurrecting the dead, who are surging up from the bottom of the composition, one of whom reaches up to retrieve his severed head. The theme of resurrection is further underscored left of center by three men colored yellow, red, and black, rowing to salvation and at the bottom right where a brown and green couple are rising from their grave.

This painting and *Composition VI* (1913), which originated from the motif of "The Deluge," were the last of Kandinsky's Composition paintings that had biblical narrative origins. As he explained in 1914, "the pictures painted since then have neither any theme as their point of departure nor any forms of corporeal origin." In later years, "forms that have arisen of their own accord right from the beginning have gained an ever-increasing foothold, and I immersed myself more and more in the manifold value of abstract elements."[40] Repeatedly Kandinsky emphasized the long struggle to enable pure abstract compositions to emerge in his work.

While abstraction at first glance seems to lessen communicative power, the loss of recognizable content made for new ways to communicate artists' inner spirit. Kandinsky's abstractions approached the free expressive capabilities of music, as in *Composition V*, where unbound patches of color swirl in spatial rhythms over the surface inside as well as outside the outlines of material objects. *Considering the Spiritual in Art* codified the capabilities of abstract colors and lines to express warm and cold, near and far, open and closed, harmony and conflict. In accord with Walter Pater's frequently cited observation that "all art constantly aspires towards the condition of music,"[41] Kandinsky's abstractions created harmonies of color and rhythms of line, and *Composition V* was fundamentally a pictorial expression of the beauty of music as God's message of hope for salvation expressed by trumpets.[42]

The synesthesia between music and art is evinced in an exchange of creations by Arnold Schoenberg and Kandinsky. Schoenberg created a musical version of abstraction in a musical composition, while Kandinsky expressed the visual equivalent of atonality in a painting, and their cross-genre influences were documented in a lively exchange of letters between them from 1911 to 1914, a prime example of the cross-cultural patterns I survey in my conclusion.[43]

Expressionism

One goal of Impressionism was to capture how sunlight and atmosphere affect the outer appearance of subjects. In reaction, Expressionism emerged to capture the inner feelings and character of subjects beneath their surface appearance as well as the feelings of the artist rendering them. The term "Expressionism" was used earlier but officially coined in 1910 in contrast to Impressionism by the Czech art historian Antonin Matějček. As he explained, "[a]n expressionist wishes, above all, to express himself . . . [He rejects] immediate perception and builds on more complex psychic structures, expressing more fully his feelings about nature."[44] An early spokesperson of that school, Hermann Obrist, proposed that the goal of art was to create "a deepened expression and intensification of the essence, instead of a hasty impression."[45] More recently the literary critic Daniel Albright argued that, "[i]nstead of providing news of the look of the outer world, Expressionist pictures provide glimpses of the inner workings of the mind," and, I would add, inner workings of the body.[46] The most intense artistic probes of that interiority were by the Austrian Expressionists Oscar Kokoschka and Egon Schiele, who were influenced by physiological probing from autopsies and X-rays, psychological probing from psychoanalyses, and pressure for greater interiority and expressivity from the artistic community.

Beginning in 1834, Carl von Rokitansky laid the foundation for modern pathological anatomy at the Vienna School of Medicine by performing or overseeing throughout his career around 70,000 autopsies to confirm diagnoses. That practice became integral to the cultural community in which intellectuals and artists trained. Rokitansky appointed Freud's teacher Theodor Meynert, while Freud's colleague Breuer worked directly under Rokitansky. The Viennese physician Arthur Schnitzler, who dramatized unconscious mental processes in literary characters, was a student in the medical school in the last years of Rokitansky's leadership. These pioneers of the interior view who absorbed "the Rokitanskian spirit of the age"[47] were fascinated by dissecting bodies and observing subcutaneous structures. Kokoschka and Schiele were influenced by Gustav Klimt, who personally witnessed autopsies at the University of Vienna. In 1925 Albert Ehrenstein characterized Kokoschka's art as "a kind of gruesome psychotomy [anatomy of the psyche] reminiscent of vivisection," while, more recently, another art critic suggested that Kokoschka's portraits were "a tearing away of the surface of an individual to expose the raw psyche beneath."[48]

Expressionist interiority was also augmented by X-rays. Röntgen's discovery of them in 1895 penetrated the opacity of solid objects and had a huge impact on popular views of vision, as in 1896, notes Linda Dalrymple Henderson, "more than fifty books and pamphlets and well over a thousand papers were published on the subject of x-rays."[49] In František Kupka's painting *The Dream* (1906–9) overlapping translucent forms float as in a dreaming mind, while in his *Planes by Colors* (1909) a reclining figure reveals bones underlying multicolored strips of flesh. In *Portrait of Chess Players* (1911) Marcel Duchamp depicted the inner thoughts of his brothers while playing chess.[50] In 1910, Boccioni wondered "Who can still believe in the opacity of bodies, since our sharpened and multiplied sensitiveness has already penetrated the obscure manifestations of the medium? Why should we forget in our creations the doubled power of our sight, capable of giving results analogous to those of the x-rays?"[51] Futurists extended that fascination with underlying foundations to the scaffolding of buildings, which Marinetti celebrated in manifestos and Boccioni depicted in art. Marinetti declared that "nothing is more beautiful than the scaffolding of a house under construction" and, he elaborated, "scaffolding symbolizes our burning passion for the coming-into-being of things."[52] Boccioni included conspicuous scaffoldings of a factory under construction in the upper section of *The City Rises* (Figure 11.5).

Between 1909 and 1911, Kokoschka painted over fifty portraits that attempted to capture his subjects' underlying feelings and bodily processes. Several critics noted the likely influence of X-ray photography on his art. Paul Westheim characterized his work as a "laying bare of the spiritual physiognomy of his models. It is as plain as an x-ray through which the inner being is visible." Kokoschka himself recalled that his mentor Adolf Loos "was convinced I had x-ray eyes."[53] As Claude Cernuschi observed, "Just as a decorative façade can be removed from the underlying framework of an architectural structure (revealing the 'reality' of its materials underneath), Kokoschka associates the 'truth' of a person with the internal physical layers revealed underneath the skin."[54]

Kokoschka's portraits feature his subjects' hands and face in a way that suggests underlying muscles, bones, tendons, and veins. They also render textures of skin, which, in his elderly subjects, is thin and nearly translucent. For example, his portrait of August Forel (1909) reveals swollen arthritic knuckles and dislocated joints, and his right thumb is locked in his jacket's left sleeve, suggesting that he may have had a stroke, an interpretation that Kokoschka confirmed. Other subjects' fingers are variously clawlike, broken, or twisted. The elongated fingers of *Hans Tietze and Erica Tietze-Conray* (1909) expand as if reaching out to express their long intimate attachment. His subjects' faces are even more expressive. The skeletal face of *Ludwig Ritter von Janikowski* (1909) is bony and emaciated, with vacant eyes, prominent red ears, and blotchy skin suggesting malnutrition or disease, possibly mental illness. In the portrait of Rudolph Blümner (1910), the subject's eyes are unfocused, and his mottled face suggests his complex personality, inner turmoil, and the effects of advanced age. Kokoschka personally noted the cultural historical significance of his approach to interiority. Expressionism, he wrote, is a "reaction against Jugendstil or Art Nouveau, which set out only to beautify the surface and made no appeal to the inner life. Expression was a contemporary and rival of Freud's development of psychoanalysis and Max Planck's discovery of quantum theory. It was a sign of the times, not an artistic fashion."[55]

Kokoschka's *The Bride of the Wind* (Figure 11.7) portrays the artist's anguished feelings for a pregnant Alma Mahler sleeping peacefully on his chest during a turbulent period a few months before they eventually broke up over her decision to abort their fetus. The pair are framed in spheres of tension from a swirling storm and heaving sea that accent his agitated feelings. Her warm flesh tones with touches of green suggest peaceful sleep, while her left hand rests calmly on his chest. His inner tension is evident in his wide-open and deeply set dark eyes, tightly pursed lips, a fixed jaw, and a dash of red at his right temple, evidence of a throbbing troubled mind. Short strokes of color streak down his right arm, indicating tense muscles leading to agonizingly intertwined fingers of both hands that reveal the turmoil in his gut, which continues down his body through his genital area to the splayed toes of his right foot. He is a quintessential bundle of nerves.

Images of sexual excitation are particularly well suited to capturing feelings. In turn-of-the-century Vienna, sexual feelings, especially in women, were collectively oppressed by public censorship, personally suppressed by self-restraint, and intrapsychically repressed by psychosomatic or, as they were called in that time, hysterical neuroses. Freud

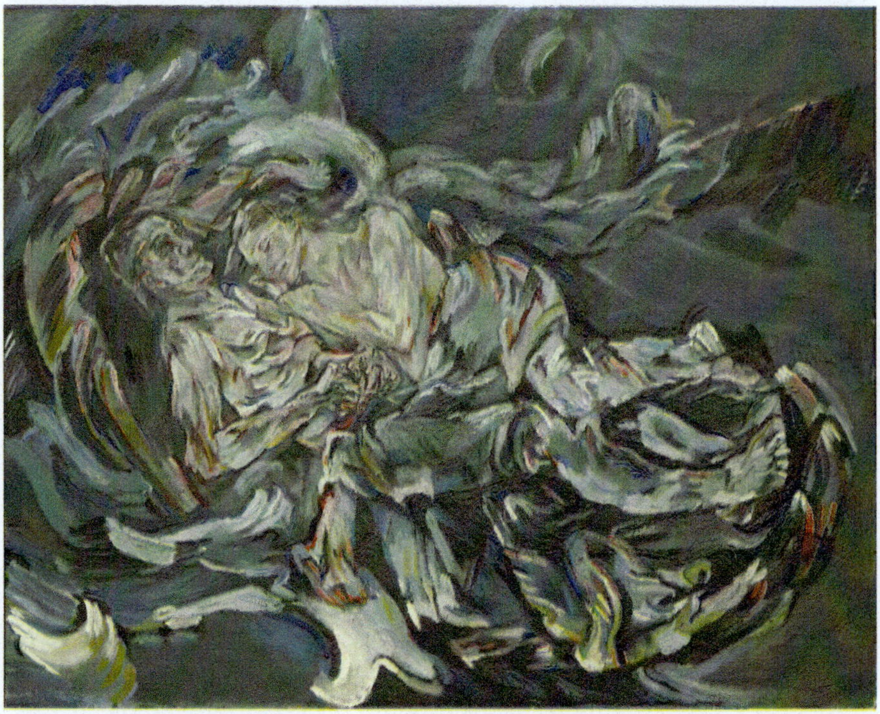

Figure 11.7 Oskar Kokoschka, *The Bride of the Wind*, 1914.

reported on the frequency of sexual problems in his practice, recalling how, in consulting over a neurotic woman, his teacher Charcot generalized memorably that "in these cases it is always the genital thing," and he bounced up and down on his toes as he reiterated, "always . . . always . . . always."[56]

The suppression of female sexual excitation as well as its increasing expression is evident in Vienna Secessionist painters. Klimt depicted women in sexually suggestive compositions but muted explicit references to actual sexual arousal by selecting them from Greek and Roman mythology, the Bible, and history. Kokoschka depicted women from classical mythology enjoying sexual feelings as they are observed from without. While he limited his artistic depiction of the unclothed body to the face and hands, Schiele depicted the entire body, both on and beneath its surface. He made dozens of paintings and drawings of women displaying their naked body in many positions. Some of his portraits and self-portraits look like fleshed out X-rays. As Margaret MacMillan noted, "One of the goals of the Viennese Secession was *not* to show the world as it actually was but to probe beneath the surface into the life of instinct and emotion."[57]

Schiele's quintessentially Expressionist *Portrait of Eduard Kosmack* (Figure 11.8 and cover image detail) depicts a publisher whose captivating gaze projects the deep thought and determination of its subject and hints at the state of mind of the artist portraying it. Kosmack's staring eyes, highlighted by white below the irises and intriguing black accents on the lower lids, are those of a weary, overworked man and appear to also radiate toward the viewer as if they projected X-ray beams. His left eye has a red lid indicating sleep deprivation and possibly underlying disease. Schiele ingeniously crafted its iris slightly off center and paler than the iris of the right eye, giving his gaze an eerie mix of the faraway focus and distraction of that left eye with the forward-looking and more proximate concentration of the right eye. His gaze suggests that he is looking at you but at the same time over your shoulder, at something or someone possibly more engaging than yourself. Kosmack's intense inner emotions are further indicated by the bony, arthritic hands with super-elongated fingers that are pressed together between his knees, suggesting anguish along with self-control. That intensity is magnified by the repeated symmetry of his ill-fitting suit that pulses up from his knees and hands to his elbows and shoulders, culminating with crushed ears. That bilateral symmetry suggests that he is keeping it together with enormous self-restraint right to left and top to bottom. The striking "V" of Kosmack's white shirt seems to prop up his neckless head like a vase. His exaggerated high forehead betokens intelligence, and a conspicuous

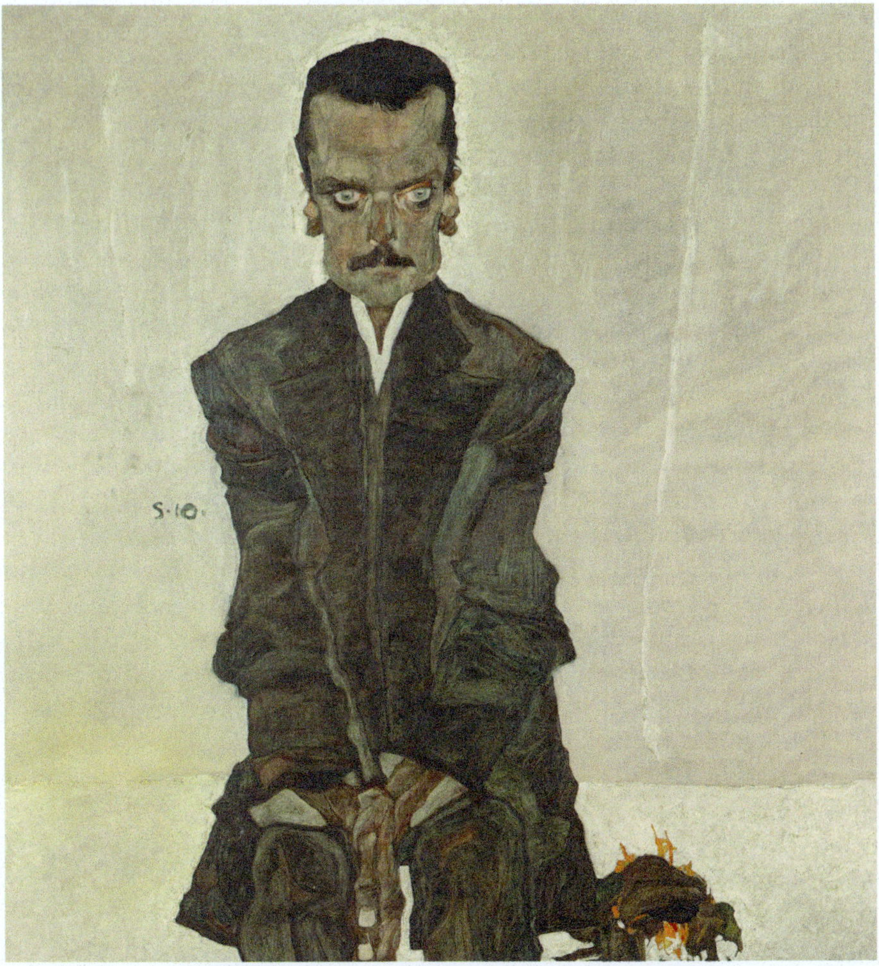

Figure 11.8 Egon Schiele, *Portrait of Eduard Kosmack*, 1910.

blue vein on it is another indication of deep emotional tension and possibly high blood pressure. The prominent bones in his narrow face evident under the skin along with the hawk-like stare suggest a predatory bird. He is clearly someone to admire and respect but also to fear. His intent look indicates that he is self-occupied as well as engaged with the artist for whom he is posing. This painting of a subject also suggests self-portraiture, a duality that Schiele made explicit in several double self-portraits of himself as artist and subject titled "Self-Seers," which he painted along with *Kosmack* in the years 1910–11.[58]

The surrounding empty space accents Kosmack's isolation from the surrounding world and others, mirroring Schiele's own isolation.

The dried-up sunflower at Kosmack's left is a reminder of the ephemeral nature of life, of the processes of decay and disintegration that are also at work in this relatively young man. That homage to Van Gogh's sunflowers that symbolize vibrant beauty and radiant life here symbolizes Kosmack's fragility and vulnerability.

The painting also suggests Schiele's sense of himself, evident in other self-portraits done around that time, most notably *Nude Self-Portrait, Grimacing* (Figure 11.9), which shows him, like the sunflower, desiccated and twisted.

Earlier artists sometimes indicated veins and muscles underneath the skin, but in those works such as Michaelangelo's *David* those features are elements of a perfectly formed muscular body in a classically flattering pose. Schiele's *Nude Self-Portrait, Grimacing* is a plunge into the possibilities of radically frank and unflattering self-portraiture. In contrast to the subject of Manet's Impressionist *Self-Portrait with Palette* (1878–9), who is calm, confident, and fully clothed, and whose palette and brushes give him a protective identity as a respectable artist, Schiele's self-portrait is unprotected, without any clearly visible professional identity and exposed to potential public mockery. His grimacing face has squinting eyes highlighted by raised eyebrows and a toothy open mouth. The awkward posture of his daring self-revelation with one arm crossed behind his body suggests a crucifixion, symbolizing the way he was treated by the authorities for scandalous portraits of himself and others.

The painting is a modernist "ecce homo" of a man exposing himself to the crowd, tortured and vulnerable, sacrificing himself for his beliefs and determination to pursue uninhibited the expressive possibilities of art while also proclaiming of himself "behold the man" – as much as it is possible to show. Visible underneath his skin are his clavicles, ribs, and sternum as well as his shoulder, elbow, and pelvic bones. Also visible beneath the skin are muscles, veins, and nerves leading into his bony fingers, his instruments of creativity that are darkened by a rich blood supply from overwork or possibly inflammation. They generally hold his palette and brushes but are here empty and spread open to accent the tension that energizes his art and runs throughout his body. The bristling hairs on his thighs and even his scrotum further indicate nervous sensitivity as well as vulnerability. The painting is an emphatically candid self-exposure of a man who holds nothing back. It is a forthright challenge to the long history of flattering self-portraits, as if to say, "you want to depict yourself, then depict all of yourself—full body and without clothing, even underneath your skin in a moment of emotional intensity and with a highly expressive pose." This grimacing

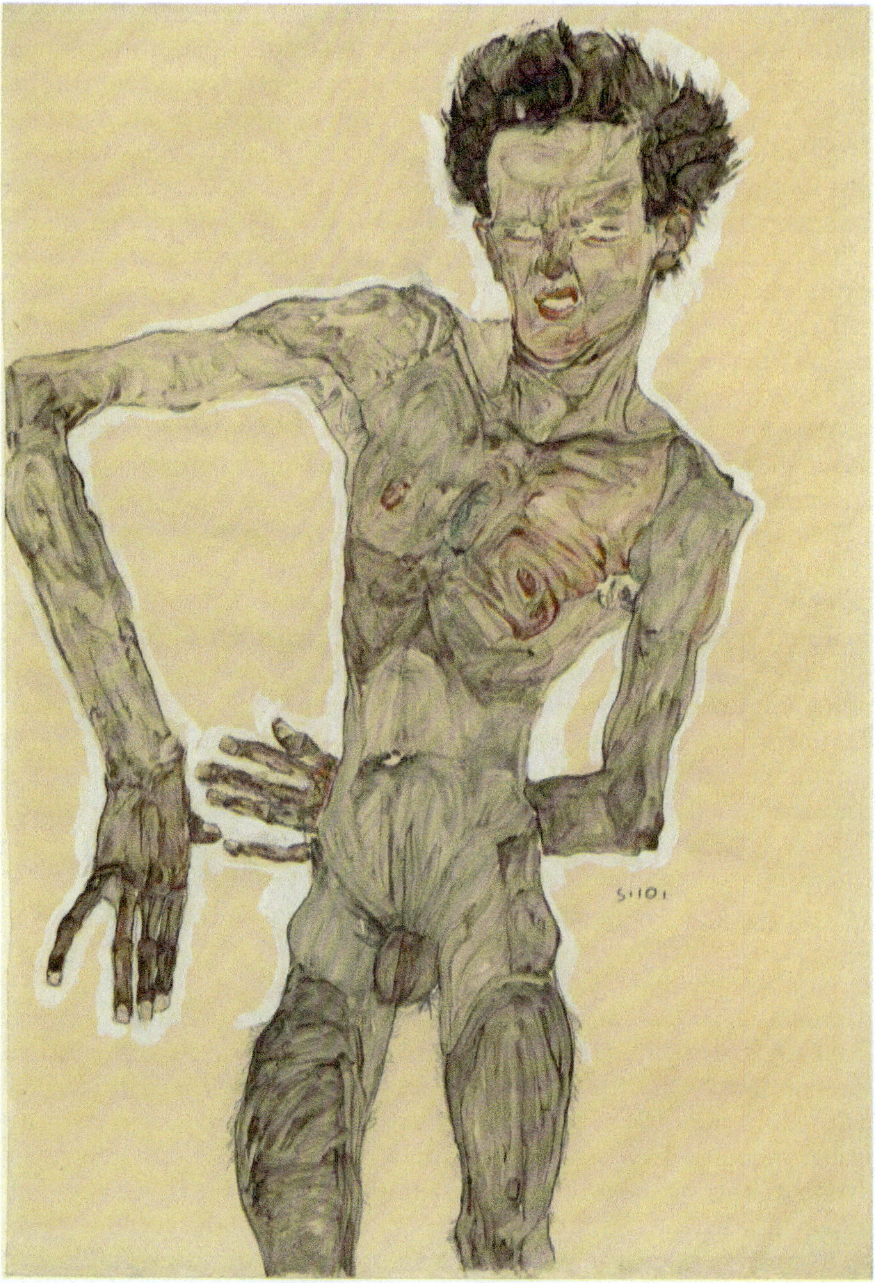

Figure 11.9 Egon Schiele, *Nude Self-Portrait, Grimacing*, 1910.

image announces the struggle he had wrestling his art from the innermost recesses of his mind and body. He does not depict his heart, but his left breast has a reddish tint, and the painting screams that his art indeed comes from the heart. If ever there was a deep dive into the inner foundations of artistic creation, this is it. In 1910, the year he completed this self-portrait, he identified his place in history: "The new artist must be unconditionally himself; he must be a creator; without needing all the relics of tradition and the past, he must have in himself, immediate and entire, the foundation on which he builds."[59]

The exceptionally productive rush of Schiele's images is evinced by the fact that in his short life of twenty-eight years (he died of influenza in 1918) he produced 300 paintings and several thousand drawings and watercolors, many highly expressive renderings of himself or his female models, frequently naked and sometimes masturbating or as couples engaging in coitus. While Kokoschka raised critical eyebrows with a portrait of two fully clothed and comparatively innocent intertwined children (*Children Playing*, 1909), Schiele depicted completely naked children, starting with his younger sister Gerti and then other pubescent and even prepubescent teens as models in dozens of paintings and drawings, sometimes engaged in self-stimulation, but his fundamental motivation for these works was not to depict forbidden sexual activities or titillate observers but to explore myriad possibilities of human embodiment, the energizing force of sexual desire, and his own passion for uninhibited artistic expression.

Art's Status in Dadaism

In an atmosphere of radical cynicism about the values of Western civilization, three years into an insane war when some artists used their skills to create war recruitment posters or, like the Futurists, glamorized weapons and war, the Dadaist Marcel Duchamp upended traditional aesthetic standards with a notoriously irreverent artwork – a men's urinal mounted on a pedestal and titled *Fountain* (1917). In traditional art, fountains symbolize purity and life. This fountain suggests impurity and men's detritus. Some viewed it as vulgar, while others thought it to be possibly the most important work of the century, which redirected thinking about the nature of art or at least proposed fundamental questions about art's fundamental nature. It raised serious questions as to whether art requires specific skills learned in art academies and long years of dedication or can simply be an imaginative personal choice of an insightful artist. Others thought that the fountain's shape suggested the

Madonna or Buddha.[60] The irony of this Dadaist "readymade" is that, while Duchamp intended it to challenge conventionally "refined" taste in the arts, it became an iconic statement on the nature of art in the modernist period precisely because Duchamp had highly developed artistic skills and was acutely aware of the conventions that he set out to reconsider.

While this "found object" pugnaciously transposed a "low" object into a "high" work of art suitable for exhibition in a gallery, Duchamp also reversed that rehierarchization by lowering the status of "high" art, Leonardo da Vinci's *Mona Lisa*, by drawing on a cheap postcard reproduction of it the subject's face sporting a moustache and goatee. Its title, *L. H. O. O. Q.*, when read aloud in French, sounds like "she has a hot ass." That is, the painting's subject Lisa del Giocondo, an exquisitely refined young woman from the Italian Renaissance, is horny. In expressing the irreverence, absurdity, and anti-art message of the Dadaist movement, these two works interrogated the foundations of art by undermining the conventional standards of high and low art and questioning whether "great art" was at all possible after such a horrific war.

Hannah Höch's Dadaist photomontage with a conspicuously long and vivid title, *Cut with the Kitchen Knife through the Last Weimar Beer Belly Epoch of Germany* (Figure 11.10) cuts down a patriarchal and sexist Weimar Germany by artistically beheading and otherwise degrading several prominent men. Cutout photographs of oversized gears, wheels, and roller bearings signify how transformative industries overwhelm the diminutive cut-up human figures in their midst. A locomotive engine rolls over Einstein's head, while an oversized rubber tire crushes a truck. Höch puts heads of prominent men onto the headless bodies of female dancers and athletes. She also celebrates women artists, writers, and politicians.

The upper right section shows the former Kaiser Wilhelm II, who played a major role in the outbreak and conduct of World War I and abdicated in November of 1918 but still appears in full-dress military uniform with a high collar, sword-belt, sash, and gold chain. Photo cutouts literally deface his former stature as Kaiser. On his head is not a crown or a helmet adorned with an eagle (as he had appeared in photos wearing the absurdly pompous Golden Eagle Pickelhaube) but a prosaic businessman's top hat. Across his forehead is the word "anti" from the phrase "*Die Anti Dadaistische Bewegung*" (The Anti-Dadaist Movement). A machine gun mounted on the top wing of a biplane points at his forehead as if he were about to have his head shot off.[61] His famous Kaiser moustache is reconstituted by the legs of two upturned wrestlers, and the left foot of one of them appears to hold up a tiny

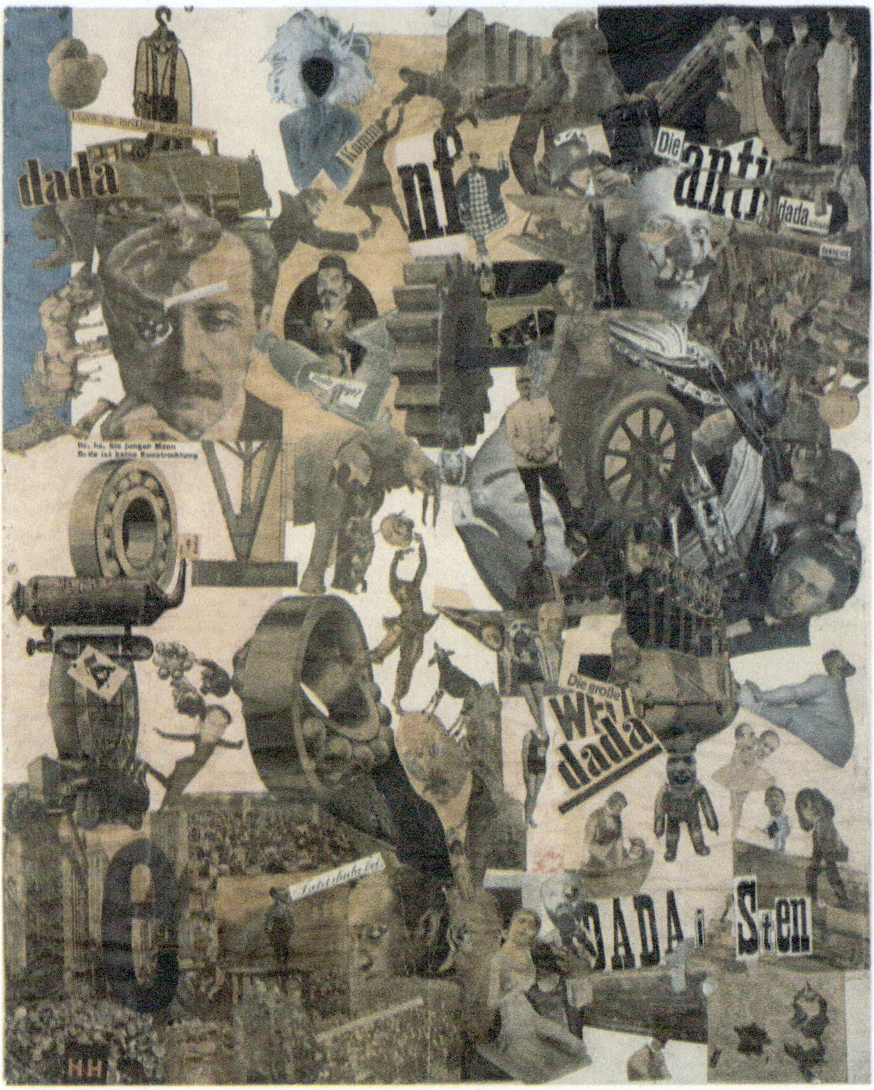

Figure 11.10 Hannah Höch, *Cut with the Kitchen Knife Dada through the Last Weimar Beer Belly Epoch of Germany*, 1919–1920.

tub containing an infant, blocking Wilhelm's right eye like a ludicrous monocle. The head of the Jewish Expressionist poet Else Lasker-Schüler is pasted like a medal on the rabidly anti-Semitic Wilhelm's chest.[62] The buttock of another woman covers his right elbow. Dominating the entire right half of the picture and trailing down from Wilhelm's right shoulder is a parody of the chain of command that led Germany into a ruinous war. At the top of that chain is the head of Field Marshal Hindenburg pasted onto the headless body of a Swedish exotic dancer, whose outstretched left hand appears to tickle Wilhelm's chin, although in the original photo she was holding a tambourine.[63] Her right hand rests on the shoulder of General von Pflazer-Baltin (an unsuccessful general in the Austro-Hungarian army), who appears to stand on the head of Reichsminister Gustav Noske (a supporter of the right-wing paramilitary Freikorps), just above the head of Lenin, who together with the Dadaist Johannes Baader and the Communist Party leader Karl Radek are positioned on the bodies of the two headless women in bathing suits standing one on top of the other.[64] This photomontage indeed cuts through the bodies and cuts off the heads of men drunk with power in the army, government, and industry in the turbulent early years of the Weimar Republic and makes a special effort to deface Wilhelm's head and torso with this Dadaist equivalent of throwing rotten tomatoes. Höch cuts up the male-dominated Weimar Republic, wielding a pair of scissors like a "kitchen knife," using the technique of photomontage.

Modernist writers made repeated assaults on the spiritual and historical foundations of Christianity, as I have noted in Nietzsche's philosophy, Freud's psychology, Weber's sociology and the novels of Gide, Joyce, and Lawrence. Modernist artists also assailed the foundations of Christianity with irreverent Expressionist crucifixions as in Max Beckmann's *Descent from the Cross* of 1917, which depicts a crucified Christ as a lifeless, emaciated corpse on which rigor mortis seems to have already set in, being manhandled by four people with vacant expressions, and in Lovis Corinth's *The Red Christ* of 1922, which shows blood from the Crucifixion splattered all over Christ's body as well as streaming out of the sun and staining the water and land, like the aftermath of a bloody firefight during the previous war.

Dadaists and Surrealists made stunning formal artistic innovations as they also assailed the foundations of Christianity with images that debased its most sacred persons. The most irreverent among these works is by the Dadaist Francis Picabia, which blasphemes the "Holy Virgin" (Figure 11.11) by representing her as a messy black ink drop, looking like a pool of muddy water on the ground.[65] This image, produced two years after the end of World War I, captures Dadaist disdain for the religion that sanctioned the mass killings with blessings to the troops and prayers for their success. The image also suggests the Dadaist affirmation of chance in human affairs, although the three lines streaming to the left of the central globule and another line streaming to the upper right corner were clearly made by carefully repositioning the paper in two directions to allow the ink to run and produce those effects. Still, the entire image captures a single irreverent moment as one can almost

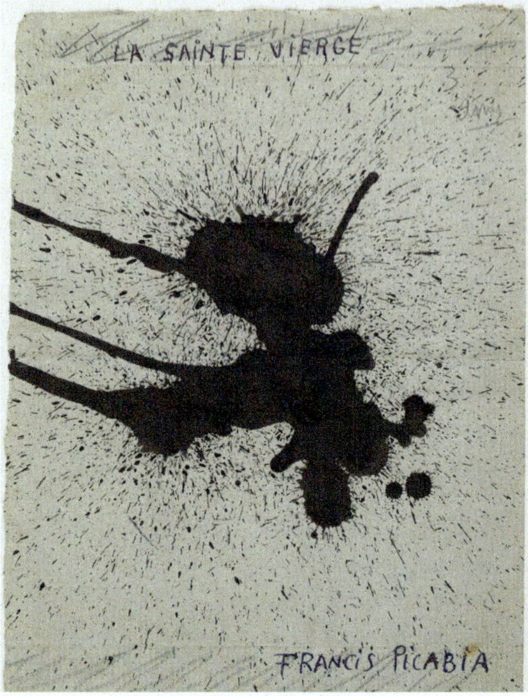

Figure 11.11 Francis Picabia, *The Holy Virgin*, 1920.

hear the sound that the central black mass made as it went "splat" and threw off ink splatter as if it was blood from one of the bullets that struck a human body during a firefight. Critics have proposed other implications – that the stain suggests blood from childbirth to ridicule the Virgin birth or a shot of black sperm to ridicule the miracle of the Incarnation. Others suggest that the act of staining a virginal white sheet of paper with a black ink puddle symbolizes female sexual defilement. To underscore the directness of his insult, Picabia wrote the subject's name, "La Sainte Vierge," at the top of the image.

This insulting effacement of the Holy Mother recalls Nietzsche's tirades against Christianity in his last book *The Anti-Christ: A Curse on Christianity* (1888), where he claimed that, "[i]n fact, there have never been any Christians."[66] Picabia's painting proclaims that there was no Holy Virgin and therefore no savior in Christ and no Christian faith, with the entirety of Christianity blotted out by a drippy ink stain. But, for all their negations of conventional artistic foundations, Dadaists' larger purpose was to celebrate a renewal of life as well as art, a sentiment captured by Hans Arp, a founding member of Dadaism who proposed hopefully that "Dada is the primal ground of all art."[67]

Beyond Realism in Surrealism

Modernists ridiculed the presumption of Realists that they knew what was real. That interrogation became particularly sharp with the Surrealists, who put scare quotes around the word "reality" itself. In an essay of 1934, Max Ernst did just that when he wrote, "In yielding quite naturally to pushing back appearances and disturbing the relationships of 'realities,'" Surrealists accelerated "the general crisis of conscience which must come to a head in our time."[68]

In *Manifesto of Surrealism* (1924), Breton defined a new way of thinking that he called Surrealism, a mix of dreaming and waking life that he intended to capture in art.[69] Breton's second Surrealist Manifesto of 1930 elaborated his intention "to lay waste to the ideas of *family, country, religion*."[70] By that time Ernst had already taken on family and religion in *The Blessed Virgin Chastising the Christ Child in Front of Three Witnesses: André Breton, Paul Éluard and the Painter* (Figure 11.12), which achieves several Surrealist objectives: mixing reality and fantasy, expressing forbidden erotic impulses, and religious sacrilege.

The painting depicts Christ's mother spanking her naughty boy whose halo is on the ground, propped against a wall and producing a realistic shadow. In the "reality" of Christian mythology, halos are immaterial objects that do not have shadows. Ernst sur-realizes the event in several ways. The mother's red dress and blue shawl match the colors of clothes in sixteenth-century portraits of the Mother and Child by Raphael, but, while Raphael's delicately featured Mary tenderly and lovingly holds the infant Jesus, Ernst's Mary has powerful arms and a right hand swung back to deliver a buggy whip of a smack to boy Jesus, who already has realistic red welts on his naked buttocks and thighs from the beating, while the fingers of his left hand splay out from realistic human pain.

The painting also mocks Ernst's own family. The painting was denounced by Ernst's devoutly Catholic father, also a painter, and the Archbishop of Cologne closed the exhibition in which the painting was shown and excommunicated Ernst from the Catholic Church in front of a crowd in the cathedral that included Ernst's parents.[71] The marginalized husband Joseph of the original Holy Family is here represented by three marginalized artist-fathers witnessing the scene: Éluard on the left has closed eyes, Breton on the right looks away, and Ernst sneaks a peak between their profiled faces.[72] The image depicts not tender motherly love but perverted sexual pleasure: the mother's sadistic pleasure in inflicting pain, the child's masochistic pleasure from being beaten, and the men's voyeuristic pleasure from witnessing the beating. A possible

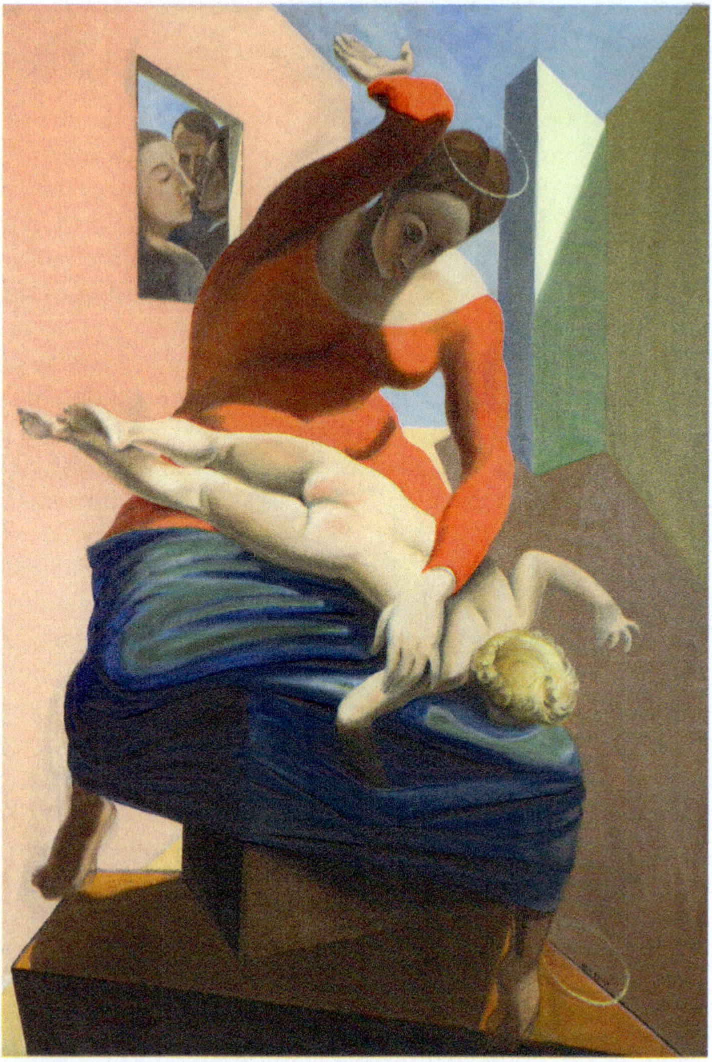

Figure 11.12 Max Ernst, *The Blessed Virgin Chastising the Christ Child in Front of Three Witnesses: André Breton, Paul Éluard and the Painter*, 1926.

explanation for this motif is the Surrealists' fascination with Freud, who defined voyeurism, sadism, and masochism as sexual perversions in *Three Essays on Sexual Theory* (1905). In a 1919 article, "'A Child Is Being Beaten': A Contribution to the Study of the Origin of Sexual Perversions," Freud analyzed patients who masturbated while fantasizing that "a child is being beaten."[73]

A further sacrilegious detail is that, while the Holy Mother's halo remains miraculously hovering above her head, indicating that she is still blessed, Jesus's halo has been knocked off by the whipping, indicating that his blessedness is on the ground. It is tempting to also interpret his grounded halo as a symbol of the Russian, German, and Austro-Hungarian royal crowns that metaphorically hit the dirt after World War I. Despite her halo, the Holy Mother does not appear to be free of sin, and Jesus does not appear to be the future all-powerful forgiver of sins but rather guilty of the sin for which he is being punished, and he is far from omnipotent or on high but nearly upside down – powerless, vulnerable, and degraded. Another sacrilege is of the imagery of the *Pieta*, in which Mary sadly contemplates the adult Jesus whom she lovingly holds in her lap after he was removed from the cross. With this image Ernst has travestied the reality and truth of Christian iconography with an image of "real" Christian characters acting in unchristian ways but with realistic painterly skills, while the Surrealist painter himself watches approvingly.

Artists in the modernist period were not a single community and at times challenged one another's conceptions as is evident in their manifestos with different statements of purpose. Nevertheless, from a broad cultural historical perspective, their work shared the function of transforming how artists experienced their world and recreated it. They differed from Realists with new skills for rendering the artistic foundations of color, light, form, space, time, history, the picture surface, artistic materials, identifiable objects, the hierarchy of motifs, and their contemporary world. Their most significant collective achievement was a reinterpretation of "reality," specifically their questioning the presumption of Courbet's earlier dictum that painters should depict "real and existing things." The modernists asked in a radical way what exactly real and existing things were and how any representation of them was possible without transforming the foundations of their art. For these modernists, Picasso's women of Avignon with deformed eyes were just as real as Courbet's peasants witnessing a burial at Ornans. Modernist artists came to see that the Realists' reality was based on historically contingent aesthetic conventions, not on an ultimate reality. Modernists pushed back against those conventions. They sought to strip away traditional conventions and techniques to probe deeper into fundamental stylistic bedrock with nonnatural color, nonrepresentational imagery, nonnarrative content, and, among Dadaists and Surrealists, a stunningly unconventional view of the world.

Notes

1. George Henry Lewes, *Realism in Art* (1859), quoted by Linda Nochlin, *Realism* (Penguin, 1971), 35.
2. On the strong narrative of Victorian courtship imagery see John Hadfield, *Every Picture Tells a Story: Images of Victorian Life* (Herbert Press, 1985).
3. Quoted in Judith Wechsler ed., *Cézanne in Perspective* (Prentice-Hall, 1975), 7.
4. The art critic Louis Vauxcelles disparagingly referred to these painters as "*Fauves*" (wild beasts), but the name caught on and became a source of pride.
5. Henri Matisse, "Notes of a Painter" (1908) in Jack D. Flam ed., *Matisse on Art* (University of California Press, 1994), 38.
6. Quoted in Claudine Grammont, "*Le Bonheur de vivre (The Joy of Life)*," in Yve-Alain Bois ed., *Matisse in the Barnes Foundation* (Thames & Hudson / Barnes Foundation, 2015), vol. 2, 46–60.
7. Henri Matisse, "The Role and Modalities of Color" [1945] in ibid., 98–100.
8. Pablo Picasso, interview by Christian Zervos in 1935, quoted in Patricia Leighton, *Re-ordering the Universe: Picasso and Anarchism, 1897–1914* (Princeton University Press, 1989), 96.
9. Quoted in William Rubin, "The Genesis of *Les Demoiselles d'Avignon*" in William Rubin et al. eds., *Studies in Modern Art 3, Les Demoiselles d'Avignon* (Abrams, 1994), 24.
10. John Golding, "*Les Demoiselles d'Avignon* and the Exhibition of 1988," in Christopher Green ed., *Picasso's* Les Demoiselles d'Avignon (Cambridge University Press, 2001), 29.
11. Quoted in Leighton, *Re-ordering the Universe*, 97.
12. Golding, "*Les Demoiselles d'Avignon* and the Exhibition of 1988," 19.
13. Rubin, "Genesis of *Les Demoiselles d'Avignon*," 14, n6.
14. From an interview of 2 December 1933 with Daniel-Henry Kahnweiler, quoted in Herschel Browning Chipp, *Theories of Modern Art: A Source Book by Artists and Critics* (University of California Press, 1968), 266.
15. Leo Steinberg, "The Philosophical Brothel," *October* 44 (Spring 1988): 13.
16. Ibid., 15.
17. Quoted in Leighton, *Re-ordering the Universe*, 97. In *Jane Eyre* (1847), Charlotte Brontë introduced her protagonist reading a book on the history of birds in which "Each picture told a story." The phrase was used frequently in the nineteenth century to describe highly popular art depicting historical subjects with evident moral themes.
18. Filipo Marinetti, "The Founding Manifesto of Futurism" [1909] in R. W. Flint ed., *Marinetti: Selected Writings* (Farrar, Strauss, 1972), 22.
19. Quoted in Christine Poggi, "Destroying the Cult of the Museum: The Futurist Collage Aesthetic," in *Defiance of Painting: Cubism, Futurism, and the Invention of Collage* (Yale University Press, 1992), 165.
20. Marinetti, "Founding Manifesto of Futurism," 41.
21. Diary entry of 14 March 1907, quoted in Christine Poggi, *Inventing Futurism: The Art and Politics of Artificial Optimism* (Princeton University Press, 2009), 73.

22. Quoted in ibid., 83.
23. Ironically Boccioni was killed in a horse-riding accident while serving in the Italian army in 1916 during World War I.
24. Poggi, *Inventing Futurism*, 91.
25. Umberto Boccioni et al., "Futurist Painting: Technical Manifesto," [1910] in Umbro Apollonio ed., *Futurist Manifestos* (Viking, 1070), 28.
26. Technically Georges Braque preceded Picasso in collage by a couple of months when in September of 1912 he pasted pieces of imitation-woodgrain wallpaper to a drawing on paper.
27. Leighton, *Re-ordering the Universe*, 121.
28. Ibid., 59.
29. Pablo Picasso, quoted in Christopher Butler, *Early Modernism: Literature, Music, and Painting in Europe, 1900–1916* (Clarendon Press, 1994), 167.
30. Russian formalists argued that literature exists to defamiliarize the world – to make habitual perceptions strange so that readers genuinely *see* rather than merely recognize objects or experiences. In "Art as Technique" (1917) Viktor Shklovsky claimed that everyday life dulls perception through automatization, and art counteracts that reaction by disrupting ordinary language and forms. Techniques such as unusual syntax, distorted chronology, or unexpected metaphors slow perception, restoring the vividness and intensity of experience that routine normally erases. Viktor Shklovsky, "Art as Technique" in *Russian Formalist Criticism: Four Essays*, ed. Lee T. Lemon and Marion J. Reis (University of Nebraska Press, 1965), 3–24.
31. Quoted in *Re-ordering the Universe*, 97.
32. Gustave Courbet letter of 25 December 1861 in Joshua C. Taylor ed., *Nineteenth-Century Theories of Art* (University of California Press, 1987), 347.
33. Henri Bergson, *Matter and Memory* [1896] (George Allen & Unwin, 1911), 196–97.
34. Hiram M. Stanley, "Space and Science," *The Philosophical Review* (November 1898): 161.
35. Wassily Kandinsky, *Concerning the Spiritual in Art* [1911], in Kenneth C. Lindsay and Peter Virgo eds., *Kandinsky: Complete Writings on Art* (Capo, 1994). Further text references are to this edition.
36. Wassily Kandinsky, "Content and Form" (1910–1911) in Lindsay and Virgo, *Kandinsky: Complete Writings on Art*, 87.
37. Rose-Carol Washton Long, "Kandinsky's Abstract Style: The Veiling of Apocalyptic Folk Imagery," *Art Journal* 34, no. 3 (Spring 1975): 227.
38. Yule F. Heibel argues that Kandinsky's paintings "ought to be seen in light of the renewed validity of anarchism as a form of political and cultural revivification in the immediate prewar period." Heibel also notes the link between the Morocco Crisis and Kandinsky. "'They Danced on Volcanoes': Kandinsky's Breakthrough to Abstraction, the German Avant-Garde and the Eve of the First World War," *Art History* 12, no. 3 (September 1989): 342–61.
39. Lindsay and Virgo, *Kandinsky: Complete Writings on Art*, 384–85.
40. Wassily Kandinsky, "Cologne Lecture," in Lindsay and Virgo, *Kandinsky: Complete Writings on Art*, 339.
41. Walter Pater, "The School of Giorgione," *The Renaissance*, 1873.

42. On the broader expansion of abstraction across the arts see Jeff Wallace, *Abstraction in Modernism and Modernity: Human and Inhuman* (Edinburgh University Press, 2023), and Nell Andrew, *Moving Modernism: The Urge to Abstraction in Painting, Dance, Cinema* (Oxford University Press, 2020).
43. Jelena Hahl-Koch, *Arnold Schoenberg Wassily Kandinsky: Letters Pictures, and Documents* (Faber and Faber, 1984).
44. Quoted in Donald E. Gordon, *Expressionism Art and Idea* (Yale University Press, 1987), 175.
45. Quoted in Daniel Albright, *Putting Modernism Together: Literature, Music, and Painting, 1872–1927* (Johns Hopkins University Press, 2015), 74.
46. Ibid., 75.
47. Eric R. Kandel, *The Age of Insight: The Quest to Understand the Unconscious in Art, Mind, and Brain, from Vienna 1900 to the Present* (Random House, 2012), 16.
48. For these sources see Claude Cernuschi, "Anatomical Dissection and Religious Identification: A Wittgensteinian Response to Kokoschka's Alternative Paradigms for Truth in His Self-Portraiture Prior to World War I," in Tobias G. Natter ed., *Oskar Kokoschka: Early Portraits from Vienna and Berlin 1909–1914* (Neue Galerie, n.d.), 44.
49. Linda Dalrymple Henderson, "X Rays and the Quest for Invisible Reality in the Art of Kupka, Duchamp, and the Cubists," *Art Journal* (Winter 1989): 324.
50. Ibid., 323–40.
51. "Futurist Painting: Technical Manifesto," 28. Linda Dalrymple Henderson surveys the impact of these developments on a wide range of modernist artists in "Cubism, Futurism, and Ether Physics in the Early Twentieth Century," *Science in Context* 17, no. 4 (2004): 445–58.
52. From Marinetti, "Great and Traditional and Futurist Milan," quoted in Poggi, *Inventing Futurism*, 69–70.
53. Oskar Kokoschka, *My Life*, trans. David Britt (Macmillan, 1974), 51.
54. Claude Cernuschi, *Re/Casting Kokoschka: Ethics and Aesthetics, Epistemology and Politics in Fin-de-Siècle Vienna* (Farleigh Dickinson University Press, 2002), 35.
55. Kokoschka, *My Life*, 66.
56. Sigmund Freud, "On the History of the Psychoanalytic Movement" in *Standard Edition*, vol. XIV, 14.
57. Margaret MacMillan, *The Road to 1914: The War that Ended the Peace* (Random House, 2013), 247.
58. Alessandra Comini, "1910–1914: Isolation and Thematic Absorption in Self," in *Egon Schiele's Portraits* (University of California Press, 1974), 79–112.
59. Quoted in Wolf-Dieter Dube, *Expressionism* (Praeger, 1972), 193.
60. On these references see William A. Camfield, *Marcel Duchamp, Fountain* (Houston Fine Art Press, 1989), 34–36.
61. That image is from the cover photo on the *Berliner Ilustrierte Zeitung* from 3 March 1918, showing a pilot firing a machine gun mounted on his biplane, reproduced in Gertrude Jula Dech, *Schnitt mit dem Küchenmesser DADA durch die letzte weimarer Bierbauchkulturepoche Deutschlands:*

Unterzuchen zur Fotomontage bei Hannah Höch (LIT Verlag, 1981), Bildteil, XXIV.

62. Ibid., 100.
63. Ibid., Bildteil, XXVIII.
64. Maud Lavin, *Cut with the Kitchen Knife: The Weimar Photomontage of Hannah Höch* (Yale University Press, 1993), 19–23, 30–32, 34–35.
65. Simon Marginson suggests that the painting may also depict Joan of Arc. "The Blood of France: Joan of Arc and Francis Picabia's *La Sainte-Vierge*," *Dada/Surrealism* 22 (2018).
66. Friedrich Nietzsche, *The Anti-Christ, Ecce Homo, Twilight of the Idols, and Other Writings*, trans. Judith Norman (Cambridge University Press, 2005), 35.
67. Hans Arp, "Dada-Sprüche," quoted in Richard Sheppard, *Modernism—Dada—Postmodernism* (Northwestern University Press, 2000), 197.
68. Max Ernst, "Comment on force l'inspiration," in *Le surréalisme au service de la revolution*, Nr. 6 (15 May 1933): 45.
69. André Breton, "Manifesto of Surrealism" [1924] in André Breton, *Manifestos of Surrealism* (University of Michigan Press, 1972), 14.
70. Ibid., 128.
71. Caled Froelich, "The Scandal of God Incarnate," *Transpositions* (10 October 2016).
72. On the Oedipal dynamics in this painting and others by Ernst see Elizabeth M. Legge, *Max Ernst: The Psychoanalytic Sources* (UMI Press, 1989), 64–71 and passim.
73. Sigmund Freud, "A Child Is Being Beaten," *Standard Edition*, vol. XVII, 179–81.

CONCLUSION: Cross-Field Patterns

In 1924 Virginia Woolf interpreted a major shift in modernist culture with an outrageously bold generalization "that on or about December, 1910, human character changed." She gave scant evidence for that change of fourteen years earlier – a first sign in Samuel Butler's *The Way of All Flesh*, the plays of George Bernard Shaw, and an illustration of seemingly little significance – "the character of one's cook" who might emerge from the basement to ask advice about a hat. With no other evidence she still insisted sweepingly that "all human relations have shifted – those between masters and servants, husbands and wives, parents and children" along with "a change in religion, conduct, politics, and literature."[1] She did not indicate what those changes signified collectively, although she implied that the seemingly trivial example of a servant asking advice about a hat might have wider implications about a transformation of far-reaching social or cultural hierarchies.[2]

This conclusion offers an example of a general history based on a cluster of generalizations about the modernist period derived from cross-field cultural patterns of similar originary ideas from many researchers and artists, some trivial and some of epic significance, surveyed in the preceding chapters. Those generalizations support a general argument that modernist culture was a radical break from Victorian culture in three respects: (1) that leading modernist physical scientists probed the physical world more deeply and accurately than did the Victorians, (2) that leading modernist social scientists probed human experience more reasonably and humanely than did the Victorians, and (3) that leading modernist novelists and painters probed new depths of experience with innovative techniques. The overall image of modernist culture as it pertains to origins and foundations is not a unified picture but rather a group of cross-field patterns that this conclusion surveys. This conception of culture aligns with the anthropologist Clyde Kluckhohn, who argued that culture refers to explicit and implicit as well as rational and irrational designs for living evinced in "ideal patterns" that include "clusters that persistently turn up."[3]

Relativity, Probability, Simultaneity

Three key originary ideas from physics that probed the fundamental nature of the physical world – relativity, probability, and simultaneity – reverberated across the cultural record.

In contrast to the dominant Victorian notion that time and space are unchanging frames for tracking events, Einstein argued that time and space are relative to the motion of the person measuring those events. Relative times and spaces are not perspectival distortions of "real time" and "real space" that in themselves are fundamentally stable and absolute everywhere in the universe; they are fundamentally relativistic conditions of knowledge of events everywhere and for all time, although they are only relative for extremely high relative velocities. He did not mean, as he was often misinterpreted to mean, that "everything is relative," and he was especially critical of applications of relativity to ethics.

The concept of relativity aligned thematically with several modernist ideas that challenged the more singularly focused Victorian theory of knowledge. In 1887, Nietzsche characterized his version of relativity as a philosophy of "perspectivism." He explained: "There is *only* a perspectival seeing . . . and the *more* affects we are able to put into words about a thing, the *more* eyes, various eyes, we are able to use for the same thing, the more complete will be our 'concept' of this thing."[4] In *A Pluralistic Universe* (1909) William James vented his contempt for Victorian "monistic dogmatism," a singular and absolute world view of a deterministic "block universe." His philosophy is radically pluralistic and sees individual parts "engaged in all the other relations simultaneously." In a properly conceived pluralistic universe "all the parts are essentially and eternally co-implicated." Even God is only "a part when the system is conceived pluralistically."[5] An analogous new way of thinking about personal life was dramatized in Musil's *The Man without Qualities* with the lifestyle of "essayism," which involved exploring life's possibilities from many points of view to counter the limits of philosophical monism and cultural chauvinism.

The anthropologist Franz Boas championed "cultural relativism," which celebrated widely diverse cultures and challenged the arrogant superiority of Victorian cultural monism. One crucial difference between Einstein and Boas is that, while Einstein's relativity theory applied only to specific events involving extremely high relative velocities, Boas's cultural relativism called for value free analyses of every culture. The modernist historian Herbert Butterfield criticized a chauvinistic "Whig interpretation of history" that viewed the entire historical record as progressing up to and culminating in the singular achievement of modern liberal democracy and constitutional government.[6] Just as Boas celebrated different "primitive" cultures, Butterfield called for an evaluation of all societies in terms of their own relative cultural context.

Nineteenth-century sciences were increasingly probabilistic. Darwin's theory of evolution, Mendel's theory of hereditary transmission, Boltzman's theory of thermodynamics, and sociological studies of suicide and crime by means of "moral statistics" all involved probabilistic calculations. Nevertheless, the modernist "probabilistic revolution"[7] refined statistical techniques for dealing with such probabilistic explanations. It centered on developments among statisticians, who, in the generation around 1900, developed new techniques for calculating probabilities. In 1893 Karl Pearson introduced the modern concept of standard deviation, which enabled researchers to quantify the amount of variability in a set of data, and in 1900 he invented the concept of the chi-square, which laid the foundation for statistical hypothesis testing.[8] These techniques made it possible to assign magnitudes to causal factors impinging on a single event and thereby determine the statistical probability and magnitude of their specific causal role. Whereas for many Victorians chance was thought to be determined by destiny or God, for modernists destiny and divine providence lost their explanatory function.

Victorian physicists believed that it was possible to conceive of the entire universe simultaneously at any given moment. In contrast, Einstein argued in his special theory of relativity that no exact determination of simultaneity is possible for an observer in motion with respect to any observable event and therefore that one cannot attach any absolute status to the concept of simultaneity throughout the universe. However, Einstein's special theory applied only to cosmic events involving enormously high relative velocities and did not affect the everyday experience of anyone. So, for most people, as one historian generalized, on the eve of World War I "succession gave way to simultaneity."[9]

That shift was made possible technologically by the millions of miles of new telephone wires that multiplied experiences of everyday simultaneity. Those wires crisscrossed territories, carrying personal telephone calls and signals for the first public broadcasts; revolutionized newspaper reporting, business transactions, crime detection, and courting; made possible the extension of suburbs and the upward thrust of skyscrapers; impacted the conduct of politics and accelerated the pace of diplomacy; forced military leaders to leave their lofty promontories and follow increasingly expanded battles from telephone headquarters; and eroded the exclusivity of social and religious hierarchies as they penetrated homes and churches.

The wide reach of this new sense of simultaneity was dramatized in *Ulysses*, which sought to capture myriad simultaneous events all over Dublin, particularly in the Wandering Rocks episode. There Joyce used cinematic quick-cut and literary stream-of-consciousness techniques to capture a vast network of simultaneous activities. He used sudden interpolations to capture simultaneous events that occurred in different places but had contrasting meanings such as Leopold Bloom buying soft porn for his wife Molly while elsewhere at the same time her lover Blazes Boylan buys her fruit. Many people widely scattered around Dublin hear the same sounds simultaneously such as an ambulance siren and the last-lap bell of a bicycle race. Simultaneity is also signified by the river Liffey, which winds through Dublin sequentially but has an overall simultaneous presence in it. Simultaneity is further suggested by journeys through Dublin that begin and end the episode. At the beginning Father Conmee's journey by foot and tram enacts the role of the Church throughout the city and with the many citizens who interact with him supportively, while the episode ends with a cavalcade that occasions a sequence of interactions throughout Dublin between the British viceroy and many Irish inhabitants who express contempt for him. These two journeys interlace the city with their exchanges regarding the simultaneous presence of the ubiquitous institutions of church and state that they symbolize. The entire novel strains to capture simultaneous happenings of Dublin on 16 June 1904 and as such offers a glimpse of the city's essential cultural meaning compressed into a single day, including even its architectural foundation as suggested by Joyce, who proposed that, "if Dublin one day were destroyed by an earthquake, it could be reconstructed out of my book."[10]

A Centerless Universe

While Victorian cosmologists never located the center of the universe, they did not question the possibility of doing so. Modernist cosmologists affirmed that the universe has no center because the space it occupies is itself expanding in every direction from no discernable center. That existentially challenging cosmic notion aligned with several literary dramatizations of the idea that human beings lack a clear existential center or home.

Strindberg's *A Dream Play* (1901) is about characters with diverse specialties who want to know what is behind a conspicuous closed door visible on center stage. In the end they discover that nothing is behind it, that the play has no central meaning. The Dean of Theology interprets the door's significance: "Nothing. That is the key to the riddle of the world." The Dean of Philosophy observes: "Out of nothing comes nothing." The Dean of Medicine makes an analogous medical diagnosis: "Bosh! Nothing. Period." Dos Passos's *Manhattan Transfer* (1925) begins with Bud Korpening coming into Manhattan and repeatedly asking how to get to the center of town, but he never finds it because Manhattan has no center. In *Ulysses*, Bloom spends the entire day of the novel striving to get back home even though that home, indeed his marital bed, has been "usurped" by his wife's lover, Blazes Boylan. The loose springs of Bloom's violated bed go "jig-a-jig," in contrast to the solid bed of the classical hero Odysseus, who built his bed, symbol of his secure marriage and home, out of the trunk of a living olive tree.

In Victorian novels mansions and castles are centers from which sources of authority emanate clearly. In Charlotte Brontë's *Jane Eyre* (1847), Thornfield Hall radiates Mr. Rochester's financial and social authority, while in Elizabeth Gaskell's *Cranford* (1853) the home is a moral and narrative center, providing a stable domestic interior in which social relations are regulated. In Gaskell's *North and South* (1855), Lovelace Castle represents authoritative power vested in class distinctions. In all three novels there is no question as to where these buildings are located or the centering hierarchies they symbolize. By contrast, in Kafka's *The Castle* (1926), the ironically named castle symbolizes the lack of any solid, stable, or centering hierarchy as it is situated in a space that cannot be easily accessed or even seen. The castle's new employee, the land surveyor K, has trouble gaining access to the castle and difficulty establishing his credentials for entering it. The castle's appearance shifts from moment to moment, but, unlike Monet's Rouen Cathedral, which changed its appearance perceptually, Kafka's castle changes its fundamental nature ontologically. It is nothing specific, rather a projection of the villagers' need for domination by an authority that has no legitimate ground. The castle is an illusory "center" of a new kind of Kafkaesque space that nevertheless generates the forms, distances, and directions that structure village life. While Victorian mansions and castles were mysterious, they were nothing like Kafka's castle. In questioning the traditional idea of an ultimate center to the universe as well as everyday life, modernists were also obliged to consider more broadly that no god created their world out of nothing at the beginning and center of creation and that they were responsible for fashioning their own existential center.

A striking evisceration of the protective function of the Victorian home is depicted in Rainer Maria Rilke's fictional journal, *The Notebooks of Malte Laurids Brigge* (1910), which describes the exposed walls of a disintegrating building that still bear traces of those who in years past attempted with questionable success to be at home in it. "You could see bedroom walls with wallpaper still sticking to them, [and] along the entire length of the outer wall, a dirty-white space through which, in unspeakably nauseating, worm-soft, digestive movements, the open, rust-spotted channel of the toilet pipe crawled, [and you could sense] the exhalations, and the smoke of many years, and the sweat that trickles down from armpits and makes clothing heavy, and the stale breath of mouths, and the oily smell of sweltering feet."[11]

Leading female novelists dramatized alternative ways that men and especially women from the modernist period felt increasingly estranged from the secure Victorian home. Richardson, Rhys, and Hall crafted originary settings in homes that were not secure sources of love and friendliness but rather unstable, temporary, or hostile regions, such as the one Hall crafted most powerfully as a "well of loneliness." Across their work, homes shift from locales of comforting attachments to sites of estrangement where men and women search for alternative forms of attachment in cafes and bars, boarding houses and cheap rented rooms, nocturnal city streets and expatriate enclaves.

Ideas about a centerless world radiated to philosophy, psychology, art, and music, as modernists came to accept that there is indeed no place like home. Thomas Wolfe dramatized how "you can't go home again," while Heidegger argued that human beings never had a true home in the first place. Rather he insisted on the ultimate "thrownness" (*Geworfenheit*) of human existence in an uncanny (*Unheimlich*) and centerless world. Although Jung did believe that the psyche had a center that he sought to reveal with meditation on mandalas, Freud did not, because there is no center to the unconscious. As Richard Rorty noted, "Freud, by helping us to see ourselves as centerless, as random assemblages of contingent and idiosyncratic needs . . . opened up new possibilities for the aesthetic life."[12]

Cubists rejected the conventional spatial center of Victorian art that was framed by a single vanishing point perspective that grounded realism, orderliness, and rationality. Cubists dismantled conventional spatial coherence with multiple perspectives that aligned with modernist pluralism and uncentered their artistic subjects. Picasso's *Demoiselles d'Avignon* has no clear perspectival center: the lighting of the five women comes from multiple sources, and the flattening of its pictorial space has replaced any traditional single vanishing point perspective.

Modernist music rejected conventional notions of a tonal center or home. In 1909, Arnold Schoenberg pioneered atonality with *Five Pieces for Orchestra*. In tonal music, notes and chords have hierarchical roles such as dominant and subdominant that direct the music toward the dominant single pitch. That dynamic aims toward resolution and home. In atonal music that hierarchy is missing. Another feature of tonal music is that dissonance resolves to consonance, generating another sense of direction toward a tonal home. As Carl Schorske notes, "[b]y establishing a democracy of tones, [the emancipation of dissonance] has vastly enlarged *all* the expressive possibilities, thematic and rhythmic as well as coloristic and tonal. The composer can create any community or cluster of tones he wishes, now crowding them all into confined musical space, now associating them across distances that are positively astral."[13] In 1923, Schoenberg devised a new technique of twelve-tone rows (using all twelve pitches of the chromatic scale before repeating any pitch), which ensured tonal neutrality and further subverted any clear sense of a tonal home. Without the familiar sense of a single dominant pitch and the resolution of dissonance toward consonance, listeners experienced disorientation and frustration. They wanted to "go home" but could not.

The purpose of such "decentralized music" was not just to resist aesthetic closure and frustrate listeners but to make possible an "emancipation of dissonance"[14] and create new expressive possibilities. In the centuries leading up to Schoenberg, European composition had been governed by stabilizing structural elements including a tonal center or key, a resolution of the piece into its definitive key, and a fixed scale of tones on which the melody was constructed. Schoenberg's atonal music dispensed with those features and freed itself from any other content beyond its own formal relations and pure sound. Its purpose was rather to probe the essentials of music and capture "the inexplicable emotions of turmoil, agitation, and unease." As Thomas Harrison concluded, in Schoenberg's music "a new compositional method is at work in the dark disorder of this free atonal music."[15]

Modernists also experienced the loss of a centering home in a material way during and after World War I. Gertrude Stein characterized the war's decentered combat with a decentering metaphor:

> Really, the composition of this war, 1914–1918, was not the composition of all previous wars, the composition was not a composition in which there was one man in the center surrounded by a lot of other men but a composition that had neither a beginning nor an end, a composition of which one corner was as important as another corner, in fact the composition of cubism.[16]

That is, the battles were not directed by a commander at the center giving orders based on visual surveys of a battlefield but by commanders at telephone headquarters deciphering reports from multiple points scattered across multiple fronts over many miles. That lack of a center for commanding officers was multiplied a millionfold by the men who fought in the war and the women who suffered from it. Nearly 10 million men were absent from home as they fought on the battle fronts. Millions of women were absent from traditional homes as they worked in newly available "men's jobs." War widows struggled to feed themselves and their children under food shortages that plagued the population of the Central Powers subject to the Allied naval blockade, especially after 1916. Young men neglected school and home to work in factories to make up for the loss of adult male incomes. In Germany the loss of a sense of homeland (*Heimat*) was particularly intense following the Treaty of Versailles, which stripped Germany of 10 per cent of its prewar German population and 13 per cent of its prewar European territory.

Further radical decentering came from discoveries in cosmology. The Earth was not a divinely conceived center for specially created human beings but a miraculous concentration of flukey accidental occurrences, located in a place that no one could precisely identify, moving at compounded velocities by rotating on its axis and orbiting through the solar system, which was orbiting through the Milky Way Galaxy, which was itself speeding through the universe.[17] During the modernist period it became increasingly clear that in any ultimate sense there was indeed no place like home.

Sexual Reproduction and Hereditary Transmission

The impact of Darwin's *Origin* reverberated throughout the Victorian scientific world and everyday life as it was translated into twenty-four languages by the end of the nineteenth century, transformed biological and social sciences, and widely impacted religion and popular culture. It announced its subject as the origin of species, presumably all of them, but its earth-shaking news pertained especially to human origins. That feature is surprising because Darwin mentions "the origin of man and his history" only once in that book and because his theory of variation by means of pangenesis, which was the first step in natural selection, was wrong.

In 1900 Mendel's theory of sexual reproduction came into wider notice in three countries when rediscovered by the independently working Dutch botanist Hugo De Vries, the German geneticist Carl Correns, and the Austrian agronomist Erich von Tschermak, suggesting a broad shift in thinking about sexual reproduction and hereditary transmission. An accurate interpretation of Mendel's findings undercuts erroneous Victorian notions of male or female primacy, the inheritance of complex behaviors in toto, and the conflation of evolution with growth.

Modernist reproductive biologists also achieved a clearer understanding of sexual reproduction and the biological foundations of human character. Among many advances over Victorian reproductive biology, modernists made several with particularly significant consequences. They rejected Lamarckism and the notion that children inherit biologically the acquired ideas and values of their parents. They liberated mothers from believing that they carried in their bloodstream the characteristics of any man who impregnated them or, as many experts believed, even any man with whom they had had sexual relations. The Victorian obsession with virginity was based partly on misunderstanding the nature of sexual reproduction. Victorians were sexually terrified in ways that modernists were less so because of advances in the detection and treatment of syphilis. The modernist biologist August Wasserman devised a reliable test for syphilis in 1906, and Paul Erlich developed a quasi-effective, arsenic-based treatment for it with Salvarsan in 1909, which was superseded by penicillin, which Alexander Fleming discovered in 1928, although it was not widely used for treatment until the 1940s. Modernist biologists also produced the modern synthesis that integrated Darwin's theory of natural selection with Mendelian genetics in unifying the theoretical foundations of biology, paleontology, embryology, physiology, and medicine.

Not-So-Trivial Trivia

Malinowski based his analysis of the social and economic foundations of Trobriand Islanders on their exchanges of Kula bracelets and necklaces. His analysis lines up thematically with other theories and cultural products that focused on seeming trivia that rose to new significance in the modernist period. Freud analyzed seemingly trivial slips of the tongue and insisted that during psychoanalysis patients describe everything that comes to mind, no matter how trivial it might seem, as it might be a key to their disorder. Joyce valorized "trivial" epiphanies reported in *Portrait* such as the wading girl he saw on a beach who inspired his vocation to become an artist. In *Ulysses* Joyce invested seemingly unimportant objects with significance such as Bloom's pocketed potato that symbolizes his Irish identity and his lost house key that symbolizes estrangement from his wife. The chapter that introduces Bloom concludes with him sitting on an outhouse toilet relieved that his constipation is gone and hoping that "it's not too big [which might] bring on piles again." Then Joyce shares a further trivial detail – "No, just right," the sort of detail that would have been inconceivable in a Victorian novel.

Woolf appreciated "little daily miracles" as she explained: "The mind receives a myriad impressions – trivial, fantastic, evanescent. . . . Let us not take it for granted that life exists more fully in what is commonly thought big than in what is commonly thought small."[18] Her entire corpus was designed to reverse that conventional ranking of the big and the small as her novel *Jacob's Room* (1922) spells out: "it's not catastrophes, murders, deaths, diseases, that age and kill us; it's the way people look and laugh, and run up the steps of omnibuses."[19] Her short story "Kew Gardens" tracks four groups of people strolling through the gardens pondering weighty worldly matters but focuses on a not-so-trivial snail in the garden struggling to "decide" whether to go under a leaf or across it. "A Mark on the Wall" (1917) speculates about what a tiny mark on a wall might be, but finally explodes into outrage over the current war and England's patriarchal government. Another of Woolf's little daily miracles with a big outcome is in *To the Lighthouse* (1927), where the artist Lily Briscoe puts a salt cellar on a flower pattern in a tablecloth to remind her to move a tree further to the middle of her painting to avoid an awkward empty space, while meditating on Mr. Ramsay's contention that women can't paint. A few days later, when she sees the salt cellar, she says to herself consolingly that she need not marry because she can achieve fulfillment with her art. Ten years later, she recalls that earlier reflection and resumes her painting, which becomes a symbol of the meaning of life. But, she reflects, "[t]he great revelation had never come. The great revelation perhaps never did come. Instead, there were little daily miracles, illuminations, matches struck unexpectedly in the dark." This little miracle was that her earlier moving a salt cellar set off a chain of thoughts and artistic efforts that ultimately demonstrated that ordinary objects could have enormous significance such as that she need not marry, and that Mr. Ramsay was wrong because she could be an artist after all, even if her painting wound up hung in an attic.[20]

In *Absalom, Absalom!* Faulkner's textual narrators Quentin and Shreve propose that their stories are fashioned out of "rag-tag and bob-ends of old tales and talking."[21] A famous piece of nontrivial trivia is in Proust's novel about the enormous significance of an involuntary memory of the taste of a petite madeleine dipped in lime-flavored tea, a novel that also includes an account of Legrandin's extraordinary nontrivial wink.[22] In Faulkner's *The Sound and the Fury*, Quentin Compson's broken watch symbolizes the impossibility of stopping time, the decay of Southern honor, and the inexorable approach of death. The Compson brothers' seemingly trivial glimpse of their sister Caddy's muddy drawers triggers sexual arousal and remains fixed in their minds as it gains traumatic force, while Faulkner himself reported that that image was "perhaps the only thing in literature which would ever move [him] very much."[23]

Other examples include William James's philosophy of "little words" such as "and" and "if," Heidegger's phenomenology of moods such as boredom, Duchamp's Dadaist urinal fountain, Sherlock Holmes's "trivial" clues such as the dog that did not bark, Gide's character who pushes a stranger out of a moving railroad car for no other reason than to kill without a reason, Maurice Maeterlinck's "old man, seated in his armchair, waiting patiently with his lamp beside him,"[24] Ezra Pound's poetic image of "Petals on a wet, black bough," Hemingway's "trout keeping themselves steady in the current with wavering fins," and William Carlos Williams's poem about "a red wheel barrow glazed with rain water beside the white chickens," upon which, Williams proposes, "so much depends."[25] A red wheelbarrow? Really? What important matters could possibly depend on that? Such an image rejects traditional poetic symbols or abstractions. The wheelbarrow does not symbolize labor, the rain does not symbolize the Holy Spirit, and the chickens are not allegorical innocents. The image is an implied manifesto against narrative poetry and Victorian ornament. It celebrates poetry as pure vision and clears away the clutter from metaphorical embellishments. What at first glance appears to be a trivial, superficial perception suggests without elaborating the myriad associations that radiate from any single image and in that way reveal a broader as well as a deeper meaning to what is immediately and directly perceived. In attending to such "trivia" upon which so much indeed depended, modernist anthropologists, psychoanalysts, novelists, philosophers, artists, and poets reevaluated what is and is not significant in understanding the origins and foundations of human existence.

Nineteenth-century writers described simple moments such as Wordsworth's delight over a "crowd of golden daffodils" as having an enormous spiritual impact. Another Wordsworth poem made the point more explicitly:

> One impulse from a vernal wood
> May teach you more of man,
> Of moral evil and of good,
> Than all the sages can.

Victorian writers typically regarded such inspirational moments as divinely inspired. Emerson believed that "all history is sacred; that the universe is represented in an atom, in a moment in time."[26] His *Journal* records that every such moment, however "dull" or "despised," is "an Epiphany of God."[27]

"With few notable exceptions," writes Morris Beja, "the disillusion with religion has continued the shift—already evident by the end of the nineteenth century—from divine revelation, purely religious experiences, to epiphanies, for the most part regarded as secular."[28] Modernists tended to think of such seemingly trivial moments as psychologically motivated or physiologically triggered, not as religiously inspired. Beja suggests the historical significance of modernists' increasing attention to trivia: "For as men have found themselves putting less and less trust in the truths and absolutes of the past, they have more and more come to stress the *trivia* of existence. They have sought meaning in what they could see, all around them, in the apparently inconsequential objects and events of everyday life."[29] Daniel Albright makes a similar point: "Modernist artists looked for value in all sorts of under- or ill-valued places, including evil (Baudelaire), dung heaps (Chekhov), noise (Russolo), obscenity (Lawrence), triviality (Satie)."[30] The first Christian epiphany was a vision of Jesus, while modernist secular epiphanies had a wider range and deeper penetration into the meaning of ordinary, everyday human experiences.

Exploring the Depths of the Mind

While the name for the psychological theory that is most distinctive to the modernist period is *psychoanalysis*, a more accurate name is the one given to it by Freud's colleague Joseph Breuer and subsequently popularized by the Swiss psychiatrist Eugen Bleuler – *depth psychology* – which referred to the analysis of unconscious mental processes in Freud, Jung, Pierre Janet, and others.[31] Psychoanalysis denotes the analytical taking apart of neurotic defense mechanisms, although Freud ultimately sought to heal traumatic wounds and reintegrate the psyche.[32] Bleuler's term more suitably captures the method of a deep penetration into the temporal origins as well as the unconscious foundations of the psyche. Other modernist researchers pursued other depth psychologies increasingly further back in time: Theodor Reik's analysis of birth trauma and, earlier yet, Jung's exploration of archetypes of the collective unconscious.[33]

Freud's clinical work started with his patients' current symptoms and then pushed down into their unconscious origins and foundations in repressed childhood sexual traumas. Meeting with resistance, he resorted to interpretating dreams. That procedure sought to make conscious and reverse the disguises of unconscious dream work back to underlying latent sources that in turn led to the origins of deeply rooted neurotic symptoms. Along with Jung he introduced metaphors from archaeology to describe the psychoanalytic process as an uncovering of accreted layers of the mind that built up over time back to their earliest origins and basic psychodynamic foundations. His psychology thematically paralleled the writings of the Viennese physician Arthur Schnitzler who adapted techniques of direct interior monologue and stream of conscious narration in dramatizing dreams, sexuality, and unconscious mental processes. In 1922 Freud wrote to Schnitzler about their common interest in depth psychology.

> Whenever I get deeply interested in your beautiful creations, I always seem to find behind their poetic sheen the same presuppositions, interests, and conclusions as those familiar to me as my own. . . . Your deep grasp of the truths of the unconscious and of the biological nature of man, the way you take to pieces the social conventions of our society, and the extent to which your thoughts are preoccupied with the polarity of love and death; all that moves me with an uncanny feeling of familiarity. So the impression has been borne in on me that you know through intuition – really from a delicate self-observation – everything that I have discovered in other people by laborious work. Indeed, I believe that you are an explorer of the depths.[34]

Proust drew from earlier writers, most notably Chateaubriand, for subject matter and tone as well as an appreciation of the powerful influence of memory in probing the past and the depths of the mind. But, whereas Chateaubriand explored the rich possibilities of revisiting memorable places in space, Proust conceived of a new sort of *in*voluntary memory that could only be evoked by a chance happening upon simple objects or experiences such as the taste of tea with a petite madeleine that triggered deeply buried and powerful involuntary memories in time. Proust's literary depth psychology was also suggested by a distinctly modernist technology – the X-ray – as a metaphor for a new kind of vision into the depths of things, which also inspired Expressionist painters. In the novel, the narrator's housekeeper Françoise remarks about his mother: "Madame knows everything; Madame is worse than the x-rays."[35] Elsewhere the narrator reports that he was "radiographing" characters, again using the X-ray as a metaphor for his own literary version of a depth psychology.

People without Qualities

In Victorian novels characters retain their singular identity even as they respond to dramatically changing circumstances. For example, Captain Ahab changed dramatically after the whale bit off his leg, but his altered personality remained singular, as Melville accents by repeatedly referring to him as a "monomaniac" obsessed with revenge.[36] Albright contrasts Victorian depictions of more fixed personal identities with Woolf's probing of more fluid personalities as "leaky vessels." "This shapelessness of personality is astonishing, when we consider it in relation to the practice of nineteenth-century novelists: Instead of the incised gingerbread men of Dickens or Eliot, overt, with such predictability of gesture that they are almost robots, we have characters who are sponges, bladders, exocrine glands, oozing emotion on all sides. They have skin and clothing, but their insides are more potent, more visible than their outsides."[37] For Victorian forensic psychiatrists, criminals were thoroughly criminal, wall to wall criminal, even if they suffered from a split personality, in which case both personalities were unified in each mode as they switched from one to the other.

Modernists novelists drilled deep in dramatizing how no one is fundamentally a single person, but rather multiple personas oriented toward different aspects of the world, not one after the other but interlaced at the same time. The narrator in Ford Madox Ford's *The Good Soldier* (1915) wonders "who in this world can give anyone a character?"[38] Woolf offers a similar idea with the reflection of her protagonist in *Mrs. Dalloway*: "She would not say of any one in the world now that they were this or were that" and, at the end of that paragraph, more specifically, "she would not say of Peter, she would not say of herself, I am this, I am that" (*MD*, 8–9). Woolf strives to dramatize the multiplicity of personas with the technique of direct interior monologue where the voice or thoughts of a narrator and a character or multiple characters lead one into the other without quotation marks or any other indication of who is speaking or thinking, because, she realized, language circulates between two people or among several people and does not go discretely from a sender to a receiver like an arrow to a target. In Woolf's *The Waves* (1931), character traits flow in and out of six characters in wave-like undulations. One character says about another, "It is Percival . . . who makes us aware that these attempts to say, 'I am this, I am that,' which we make, coming together, like separate parts of one body and soul, are false."[39] Joyce accents the impossibility of clear self-definition when Bloom, after a seaside encounter with Gerty MacDowell, attempts to write a message about himself for her in the sand but can only come up with "I," and then a moment later, "AM. A." and finally rubs out the incomplete sentence with his boot. With that gesture Bloom sweeps away the confident conclusion of Descartes' famous argument "I think, therefore I exist." In the Lestrygonians episode, Bloom registers the blunt existential conclusion, "No-one is anything" (*U*, 135). In Faulkner's *As I Lay Dying* (1930), Darl Bundren concludes: "I don't know if I am or not. Jewel [Darl's brother] knows he is, because he does not know that he does not know whether he is or not."[40]

Concern about the way public roles drain any chance of a unified self energizes *The Man without Qualities*, a sustained challenge to the coherence and stability of a character. In contemporary Austria, the narrator explains, one's character is split into professional, national, civic, class, geographic, sexual, conscious, unconscious, and private roles, not one after the other but simultaneously. After a losing battle in youth, people accept the public persona that enfolds their lives. Some may come up with "a slogan in which they think they can recognize themselves," but such gestures fail to counteract the "heavy world, weighing on tongue, hands, and eyes . . . and inside nothing but an unstable, shifting mist." Ulrich is *the* man without qualities who attempts to shape his character as his own in a world that is constituted by the wide-ranging circuitry of culture. To do so he must divest himself of "all those prefabricated compartments and forms of life, semblances of reality, the molds set by earlier generations."[41] Those coercive molds are analogous to Heidegger's coercive impersonal force of "the 'they.'" For modernists generally, no one has a singular and stable self.

Spiraling Down into Human Existence

The main investigatory thrust of Freud and the other depth psychologists was ever deeper temporally into the past of their patients, back to earliest infancy and birth, back to antiquity or even earlier than that. In contrast, the main investigatory thrust of Heidegger is not back in time but deeper into the ontological foundations of human existence. He begins his phenomenological inquiry with a human being immersed in the world of others, and then he spirals down to explore the originary existential significance for human beings of care, time, and death.

Heidegger's philosophy of care draws from other concepts circulating in his era such as the mind's essential directedness by means of *intentionality*, a mental process analyzed by Franz Brentano in 1874 and elaborated by Heidegger's teacher Husserl in the early twentieth century, who also believed, like Brantano, that consciousness by itself is a fiction. Rather, one is always conscious *of* something. Consciousness is not a passive receptor of data but rather an active generator of meaningful experience. Intentionality also aligns with Freud's concept of an individual's investment of psychic energy by means of cathexes. Care, intentionality, and cathexes all refer to ways that individuals attach significance to everything in their world, primordially and fundamentally, before any conscious intention or serious evaluation comes into play.

Heidegger's philosophy of human existence in *Being and Time* ultimately boils down to a philosophy of temporality, an interest that was addressed by many other thinkers across modernist culture. In the 1880s Wilhelm Dilthey underscored the temporal foundation of all knowledge that is based on what individuals know best – their own life story. "Thus to impenetrable depths within myself," he wrote, "I am a historical being."[42] In 1889 Henri Bergson explored private time that in contrast to a divisible and measurable public time is a flowing stream that is intuited as an inner experience of duration that cannot be properly divided or measured.[43] In *The Phenomenology of Internal Time-Consciousness* (1905) Husserl probed the fundamental aspects of human temporality that are experienced not one after the other but as an integration of *retentions* of recent past experiences, *recollections* of distant past experiences, and *protentions* of future experiences. These composite components ground the complex experience of every momentary temporal experience including his example of experiencing a melody of seven notes. Without the capacity to experience time as an intuited flowing of these aspects of time-consciousness, one might hear the individual notes of a melody one after the other but not experience them as a melody flowing through time. That view of time was central to Heidegger's philosophy of the integration of the three "ecstasies" of time that are structured by Dasein as an integration of past, present, and future orientations simultaneously. That view of time draws on the philosophy of time in Bergson and Husserl and was succinctly captured in T. S. Eliot's *Burnt Norton* (1935): "Time present and time past / Are both perhaps present in time future / And time future contained in time past."

Heidegger's exploration of the foundational nature of lived time led, with bitter irony, to a crucial consideration of lived time's running out in death. Victorians made elaborate efforts to blunt death's absurd nature. They idealized dying as a graceful sliding from normal sleep into a fatal sleep, surrounded by loving family and friends, and absolved by clergy, followed by a traditional funeral, the culmination of a meaningful life and a moment of passage to an eternal hereafter when and where the deceased will see formerly deceased loved ones in the company of

Jesus and God. Heidegger distinguished between an inauthentic death constructed with such ideas about its divinely secured meaning in contrast to an authentic death (*Sein zum Tode*) constructed with a heightened commitment to live as fully as possible in the here and now while resolutely anticipating and accepting life's inevitable, lonely, and absurd end and accepting that there is no life after death.

Typical Victorian funerals were ritualized ceremonies with special attire, prayers, blessings, grave markers, and stationery. In sharp contrast is the Hades episode in *Ulysses*, when Bloom struggles to fathom the mystery of death while at a funeral. Joyce mixes authorial narration with Bloom's cynical and sometimes amused streaming consciousness: "Pallbearers, gold reins, requiem mass, firing a volley. Pomp of death. . . . Makes them feel more important to be prayed over in Latin. Requiem mass. Crepe weepers. Blackedged notepaper." Bloom scans the desperate denials and evasive thoughts of the dying and those around them in anticipation of death. "Can't believe it at first. Mistake must be: someone else. Try the house opposite. Wait, I wanted to. I haven't yet. Then darkened deathchamber. Light they want. Whispering around you. Would you like to see a priest? Then rambling and wandering. Delirium all you hid all your life. The death struggle." While walking past graves, Bloom observes the remnants of sculpted efforts to make death meaningful: "saddened angels, crosses, broken pillars, family vaults, stone hopes praying with upcast eyes" (*U*, 85, 91, 93). Bloom has these thoughts while periodically thinking about his father's suicide and his dead infant son Rudy, which recur to him throughout the day without the comforting mythology surrounding death that pervaded Victorian culture. For Bloom those deaths were sudden and senseless and left a persistent, aching grief. *Ulysses* was translated into German in 1927, the same year that Heidegger published *Being and Time*, so it is unlikely that he read Joyce by that time, but their similar views of what Heidegger considered inauthentic attitudes toward death suggest that they were both responding to ideas that were current in their increasingly secular world.[44]

Vanquishing the Shadow of God

Among the many terms used to reference the unifying structure of any culture, *Zeitgeist* best suggests the myriad penetrations of divine action as if by means of some culture-wide smoke pervading everything and everyone. Even for atheists, religion remained a powerful foundational force with its ubiquitous denials and persistently present absences.

Thinking in the Victorian age included many instances of divine influence such as William Paley's natural theology (1802), Richard Owen's theory of archetypes in the divine mind (1849), and Lord Kelvin's thermodynamic arguments for divine design (1862). By the modernist period physical scientists did without divine influence altogether. Social scientists analyzed social foundations driving religious influences, as in Max Weber's *The Protestant Ethic and the Spirit of Capitalism* (1904) and Émile Durkheim's *The Elementary Forms of Religious Life* (1912). Freud made several analyses of the psychological foundations of religion, which he tagged degradingly as a regressive narcissistic obsession and a caricature of a paranoid system.

The first significant human event in Judeo-Christian mythology was Adam and Eve's original sin, which was originally disobedience but by the fifth century became sexual lust, stigmatizing the primordial pair and all their descendants with the messiness and passions of sexual reproduction. The driving force behind the personal outlooks of Gide, Joyce, and Lawrence that I analyzed in Chapter 9 was contempt for the damage that Christian sexual morality did, or threatened to do, to their love life. An even stronger case could be made for gay and lesbian modernists who ran afoul of Christian sexual morality: Woolf, Stein, Proust, and Colette along with Oscar Wilde, Jean Cocteau, E. M Forster, Roger Martin du Gard, Lyton Strachey, Djuna Barnes, Radclyffe Hall, and Vita Sackville-West. The most historically significant developments in modernist thinking about religious explanations of the origins and foundations of human existence were among the more than 100 leading scholars and artists, who rejected their childhood religious upbringing and became nonbelievers.[45]

A further feature of modernist culture is residual ambivalence about Christianity even among atheists such as Nietzsche, who appreciated the discipline of priests and admired the spiritual integrity of Jesus the man even as he excoriated the significance of Christ "the savior" and the church founded in his name. In 1882 Nietzsche offered an explanation from another religion for the abiding circulation of a religious spirit in a secular age: "After Budda was dead, his shadow was still shown for centuries in a cave—a tremendous, gruesome shadow. God is dead; but given the way of men, there may still be caves for thousands of years in which his shadow will be shown—and we—we still have to vanquish his shadow, too."[46] The shadow of Catholicism influenced Joyce's mature work as an enduring ambivalence about Christianity. In *Portrait*, Stephen, speaking on behalf of Joyce, announces that as an artist he will secularize the spirit of Catholicism as "a priest of the eternal imagination, transmuting the daily bread of experience into the radiant body of everlasting life" (*P*, 195).

The decline of religious faith and authority plays a crucial role in the outbreak of World War I. The three monarchs most responsible for the failure of diplomacy and hence the war's outbreak along with the ill-conceived military strategies that followed were the Russian tsar, the Austro-Hungarian emperor, and the German kaiser. In 1914 they were at the end of a long-running erosion of the basis for their authority according to the doctrine of the divine right of kings. Their declining powers along with their egomania and incompetence led to dangerous delays and confusion during the diplomatic crisis of July 1914, while their meddling in military policy, over which they had a substantial measure of authority, contributed to battlefield catastrophes. The historical impact of the declining authority of the doctrine of the divine right of kings was of the highest order of magnitude, a centerpiece of modernist secularization.

"Exagmining" Possibilities of Language

Victorian writers were acutely aware of the difficulty of finding the right words, as Flaubert famously acknowledged: "Human speech is like a cracked kettle on which we tap crude rhythms for bears to dance to, while we long to make music that will melt the stars."[47] But even he did not question the traditional mechanics of letters, words, and sentences to the extent that modernists did. In *Ulysses* Joyce obliged readers to think about the arbitrary arrangement of letters in words by tracking five sandwich-board men, each wearing a white hat bearing a single letter of a word, who disorder the letters in that word by wandering away from one another (*U*, 127). Woolf suggested something similar in *Mrs. Dalloway* when the letters in the sky created by a skywriting plane are blown away by the wind.

Modernists also questioned the originary function of words as they coined new ones. Dos Passos exposed wartime propaganda with brutal coinages for the collection of body parts for the Tomb of the Unknown Soldier from the "puky dirtstench of the yearold dead."[48] Joyce created hundreds of new words such as enigmatic adverbs *biscuitfully* and *underdarkneath*, colorful adjectives *snotgreen* and *scrotumtightening*, and a playful description of sexuality as a "flood of warm jimjam likitup secretness" (*U*, 226). In 1929 the critic Eugene Jolas called for a solution to "worn-out verbal patterns" with new words to capture the unconscious mind analyzed by Freud, the bizarre experiences explored by the

Surrealists, and the clashing of languages among big-city immigrants. His account of Joyce's revolutionary challenge to worn-out verbal patterns in *Finnegans Wake* is evident in the playful title of his article on the subject: *Our Exagmination Round His Factification for Incamination of Work in Progress.*[49] His essay suggests a re-examination of the logic of spelling as well as what words mean.

Victorian Realists viewed language as a reasonably adequate representation of experience, while modernists became increasingly aware of how experience itself was generated by language. That shift, which came to be known as "the linguistic turn," was led in the early twentieth century by Bertrand Russell and Ludwig Wittgenstein, who saw philosophy not as a direct study of thought and reality but as an analysis of how thought and reality are constituted by language. A similar point was made by Niels Bohr, who acknowledged that we are all "suspended in language" that is inadequately equipped to describe quantum paradoxes. The enormous historical significance of this development was noted by the critic George Steiner, who concluded that, until the late nineteenth century, the most biting skeptics about language remained confident in the ability of language to at least put their case with intelligible propositions. Then a transformation took place: the modernist "break of the covenant between word and world [that constituted] one of the very few genuine revolutions of spirit in Western history."[50]

Modernist novelists increasingly interrogated the foundational role of language in communication. In *The Counterfeiters*, Gide suggested that language is essentially a counterfeit system that, like money, has no absolute backing and offers no reliable frame of reference. Even the monetary gold standard was a grand illusion, an arbitrary system of value. In his novel counterfeiting is a metaphor for the unreliable representational function of all signifiers (or coinages) including gold, paternity, nationality, religion, and, most notably, literature. His counterfeit characters who pretend to be worth more than they are include an unfatherly father, an unjust judge, an unfaithful wife, and a novelist who subverts belief in language as a reliable tool. In "Journal of the Counterfeiters," Gide listed those subversions: how in his novel "there would be unnecessary characters, ineffectual acts, pointless remarks, and the action *would not* get under way."[51] The reader must decode all the counterfeiting and construct meanings as the novel proceeds.

Josef K.'s problems with language in Kafka's *The Trial* appear first as the pornographic text that he discovers in law books, then as legal gibberish in scraps of law relayed to him by his lawyer, and finally as a legal text that the man in the "Parable of the Law" cannot access during a lifetime of futile waiting to read it. Legal language also works against

Musil's psychotic criminal Moosbrugger. Unaccustomed to metaphors, he breaks metaphorical language into literal elements and puts them together improperly. The night he stabbed the prostitute, her solicitous words attached to his fractured ego, and he felt compelled to cut them away in a desperate act of self-definition, mutilating her in the process. Musil's constructions of how Moosbrugger uses language and how legal jargon "bends his mind" dramatize how justice is derailed by syntactical conventions.[52] Dos Passos chastises the power of advertising to exploit customers and secure the hegemony of a political elite. His alter ego in *Manhattan Transfer*, the journalist Jimmy Herf, has lost faith in language exploited for commercial use, as he complains: "I know that every sentence, every word, every picayune punctuation that appears in the public press is perused and revised and deleted in the interests of advertisers and bondholders."[53]

Modernist writers also subvert conventional syntax. In *Tender Buttons* (1914) Stein explodes basic notions of causality. Some causal actions cannot possibly cause their effects as when "A curving example makes righteous finger-nails."[54] Some *because* phrases do not explain: "there is a shadow in the kitchen because every little thing is bigger" (*TB*, 480). *Why* questions are unanswerable ("Why is there so much resignation in a package . . .") or make no sense ("Why is there no necessary dull stable . . .") (*TB*, 502, 503). With "real is only, only excreate, only excreate a no since," she suggests that causal explanations are nonsense or excrement (*TB*, 496). Reading her text carefully is exhausting because of the many frustrations of normal expectations about what language is supposed to do but does not; the text is also powerfully and unforgettably illuminating in the way it exposes the arbitrary nature of language. In a study of Woolf, Eliot, and Joyce, which also includes Ford, Forster, Hurston, and Lawrence as well as Stein, Randy Malamud argues that modernists "announce the inadequacy of the old language. Their linguistic experiments reflect a conviction that language, in the modern age, had to stretch—had to say more than it had been able to say before; had to reveal more, embody more."[55]

Abstraction, the Distinctive Style of the Modernist Epoch

Abstraction was the most revolutionary artistic development of the modernist period. It explicitly rejected realism's goal of depicting identifiable objects in the "real" world and raised fundamental questions about what is real or, more profoundly, if anything is really real. Abstraction was most fully developed in the visual arts of dance and cinema and especially painting. A few poets briefly attempted it, Stein and Wyndham Lewis flirted with it, while Gide and Woolf addressed it through characters who were intrigued by its possibilities. Abstraction reflected wider cultural developments, as Hermann Broch noted in *The Sleepwalkers* (1930–32): "the style of an epoch, it is certain, affects not merely the artist; it penetrates all contemporary activities, and crystallizes itself not only in works of art but in all the values which make up the culture of an age."[56] The penetrating style of the age to which he referred was abstraction. Abstract art effaced trees and houses from landscapes as a portent of the wartime terrain of no-man's-land.

From ancient cave drawings to Cubism, one major goal of pictorial art was to capture recognizable objects. Then, in an extraordinarily compressed period from 1909 to 1914, artists went abstract, influenced by a variety of challenges to conventional ways of depicting the world. Kandinsky's abstractionist manifesto, *On the Spiritual in Art* (1911), identified this artistic innovation as part of a broad cultural rejection of materialistic representations as he bemoaned the "soulless content of modern life" with its atheism, positivism, and naturalist art.[57] He proposed an art that would enrich by communicating directly from the inner spirit of the artist to the inner spirit of the viewer, unmediated and thereby undistracted by familiar narratives or even recognizable objects. To reach a public accustomed to familiar narratives, he began with easily recognizable imagery such as angels and serpents from the Book of Revelation. His titles also conjured up familiar biblical imagery such as *Deluge*, *Resurrection*, and *Apocalypse*, but soon the narratives disappeared, images became abstract, and the paintings had nonspecific titles such as *Impression*, *Improvisation*, and *Composition*. Kandinsky's abstraction also accorded with the anarchist objectives of liberating the long-suppressed human spirit from coercion by chauvinistic familial, national, and religious institutions.[58]

Kandinsky's abstraction was the centerpiece of a cultural shift across many nations. In 1909 the Italian Futurist Boccioni dissolved objects in depictions of modern technology that by 1913 became images of speeding cars and finally "abstract speed" with no object identifiable in the composition or referenced in the title. Also in 1909, the Dutch artist Piet Mondrian began depicting trees that by 1912 became abstracted geometricized forms. His rhetoric of purity through abstraction was echoed by the Russian artist Kasimir Malevich, who in 1913 developed the abstract style of Suprematism. In a manifesto of 1915, Malevich announced that Suprematists rejected art that depicted recognizable things and figures. They were inspired by Italian Futurist renderings of machines and movement but eventually rejected even that art because, even when it represented "abstract speed," its inspiration came from and remained attached to the speeding car that inspired it. Purely non-objective Suprematism begins with nothing but the inner processes of the unconscious.[59] It clears from art any accidental objects or narratives of a corrupt world of moribund nations, commercialized religions, and manufactured products.

Malevich characterized his painting of a black square within a white border, titled *The Black Square*, as "the icon of our times."[60] In 1915 he hung it at the upper corner of an exhibition room, facing down from a position like the spot in Russian homes where icons of Christ were traditionally placed. By representing Christianity's holiest icon as a black square, Malevich proclaimed that art need not reference any real object or story but only constitute itself as a pattern of shapes. In 1927 he explained the historical significance of his Suprematist goals: "Art no longer cares to serve the state and religion. . . . It wants to have nothing further to do with the object, as such, and believes that it can exist in and for itself, without 'things.'"[61] Across the Western world, other artists also went abstract: František Kupka in Czechoslovakia, Wyndham Lewis in the UK, and Sonia Delaunay in France. In this same period modern dancers freed the body from the melodramatic narratives of classical ballet with its dramatic plot twists, tragic loves, and supernatural elements and instead created movement dictated by artfully disciplined bodily impulses and expressive purposes.

In ten "Disintegration of Values" essays interspersed throughout *The Sleepwalkers*, Broch related abstract style to the secular and anarchist tendencies of his age. Each age is defined by its style, and the dominant style of the modernist age is abstraction. Although abstract art began before the war, Broch links it to the destruction of that war, which is the historical context of the last part of the novel set in 1918. As he explains, "[a]n epoch which is completely under the dominion of death and hell must live in a style that can no longer give birth to ornament" (*S*, 398). The defining ornament of the medieval church, the cross, was replaced by an abstract art that signified the ornament-less emptiness of the modern world, spinning out of control in a plurality of debased values. The current slogans "war is war" and "business is business" embody the secularist and anarchist "values and non-values of our age," which are graphically represented in abstract art (*S*, 446). Whether or not abstract art engendered the war or the war gave new meaning to abstraction, Broch sees the two phenomena linked as components of the larger cultural historical record.

The confluence of such a wide range of artists from different countries, all of whom went abstract around the same time, is strong evidence of the broad international connections of modernist culture. Their effacing of identifiable subject matter and narratives in the years before the war suggests patterns of change that include the effacing of national borders, monarchies, and empires along with traditional aesthetic practices. It is conceivable that Abstractionist artists were responding to destructive developments in the prewar period (gunboats and guns, bombs and torpedoes, military rearmaments and military alliances) that were "in the air" and in some ways portended and may have ultimately contributed to the outbreak of the war or in some way made it possible for so many soldiers to continue killing so many enemy soldiers all dressed alike as abstracted and in that respect dehumanized human beings. However, Abstractionists' most significant contribution to the arts and to culture generally was not destructive but creative, as they worked to liberate spiritual impulses that they believed had been trapped in the requirement that paintings depict some recognizable moment in a narrative or at least some recognizable object so that viewers could easily recognize and for that very reason seriously misinterpret "what it is about."

Writers also tried to write abstracted, if not entirely abstract, texts. Stein's *Tender Buttons* uses identifiable words grouped in phrases that jar conventional syntax but are not completely abstract. In later years she explained her dilemma while working on *Tender Buttons*: "I took individual words and thought about them until I got their weight and volume complete and put them next to another word, and at this same time I found out very soon that there is no such thing as putting them together without sense. . . . I made innumerable efforts to make words write without sense and found it impossible."[62] While she despaired of producing a fully abstract text without sense, critics saw her work as moving in that direction. Richard Kostelanetz viewed many of her postwar writings as "nonrepresentational, lacking even a suggestion of anything outside themselves." He also related her "acoherent" style to "atonal" music of the time.[63] Both innovations in art and music challenged the responsiveness of a tradition bound public and reconstituted the foundations of their arts.

The overpowering narrative of the war made literary abstraction seem especially insensitive, but one group of wartime writers used abstraction to indict that war by being as irreverent as possible. Their poetry was part of a movement whose name itself was a pair of abstract syllables – Dada. In a Zurich cabaret on 14 July 1916, the Dadaist poet Hugo Ball read his first poem of abstract nonsense. He violated rational semantics and syntax to rescue language from what he believed was its subservience to the "rational" insanity that led to a senseless war. As he explained, "[t]he human figure is progressively disappearing from pictorial art, and no object is present except in fragmentary form. This is one more proof that the human countenance has become ugly and outworn and that the things which surround us have become objects of revulsion. The next step is for poetry to discard language as painting has discarded objects and for similar reasons."[64] Ball eventually gave up nonsense poetry, and Dadaists went on to produce unorthodox but not abstract texts. Poetry could not entirely dispense with words that always evoke some meaning, because abstract sounds could not sustain audience interest, not even for Dada enthusiasts.

Neither Gide nor Woolf wrote an abstract novel, but they were influenced by abstract art and created characters who contemplated abstraction's purified literary potential. Abstract art radically effaces the clichés associated with conventional themes and familiar objects, but literature can never entirely dispense with words without becoming unintelligible gibberish. In Gide's *The Counterfeiters*, when offered a chance to found a new literary journal, the misanthropic student Strouvilhou proposes to get rid of "those promissory notes which go by the name of words." Literature, he explains, has failed to keep up with the abstractionist impulse. "In painting today, just see how the 'motif,' as it used to be called, has fallen into discredit. A fine subject! It makes one laugh. Painters don't even dare venture on a portrait unless they can be sure of avoiding every trace of resemblance. . . . I don't ask for more than two years before a future poet will think himself dishonored if anyone could understand a word of what he says" (*C*, 332–33). Gide conveys the radical abstractionist impulse of the prewar years that tempted writers but never could become a viable literary style, because of their abiding commitment to meaning.

Woolf puzzled over abstraction in *To the Lighthouse*. Her diary records how the second section, "Time Passes," was "the most difficult abstract piece of writing" she undertook because she had to render "an empty house, no people's characters, the passage of time, all eyeless & featureless with nothing to cling to [and wonders] Is it nonsense, is it brilliance?"[65] That section is not entirely abstract, but its effort to capture passing time and featureless characters is abstractionist. Woolf also explored abstraction with the artwork of her character Lily Briscoe, who is inspired by abstractionist impulses such as capturing the instinct that attracts bees to sweetness or becoming one with one's art like water poured into a jar. An early composition draws the attention of Lily's friend Mr. Bankes, who asks what a triangular purple shape signifies. She explains that it is "Mrs. Ramsay reading to [her son] James," but, the narrator adds, "she had made no attempt at likeness." She introduced the shape only because she felt the need of a dark area to offset